T3-BNY-063

The
Armored
Dinosaurs

LIFE OF THE PAST
James O. Farlow, Editor

The
Armored
Dinosaurs

Edited by Kenneth Carpenter

Indiana University Press
Bloomington & Indianapolis

ST. PHILIP'S COLLEGE LIBRARY

QE
862
.O65
A76
2001

This book is a publication of

Indiana University Press

601 North Morton Street

Bloomington, IN 47404-3797 USA

http://iupress.indiana.edu

Telephone orders 800-842-6796

Fax orders 812-855-7931

Orders by e-mail iuporder@indiana.edu

© 2001 by Indiana University Press

All rights reserved

No part of this book may be reproduced
or utilized in any form or by any means,
electronic or mechanical, including
photocopying and recording, or by any
information storage and retrieval system,
without permission in writing from the
publisher. The Association of American
University Presses' Resolution on
Permissions constitutes the only exception
to this prohibition.

The paper used in this publication meets
the minimum requirements of American
National Standard for Information
Sciences—Permanence of Paper for
Printed Library Materials, ANSI
Z39.48-1984.

Manufactured in the United States of
America

**Library of Congress Cataloging-in-
Publication Data**

The Armored dinosaurs / edited by
   Kenneth Carpenter.
      p. cm. — (Life of the past)
   Includes index.
   ISBN 0-253-33964-2 (cloth)
   1. Ornithischia. I. Carpenter,
   Kenneth, date II. Series.
   QE862.O65 A76 2001
   567.915—dc21

                        2001000533

1 2 3 4 5   06 05 04 03 02 01

Dedicated to
Walter P. Coombs, Jr.

*In recognition of his work
on Ankylosaurs—he led
the way*

CONTENTS

Paul M. Barrett, Department of Zoology and University Museum of Natural History, University of Oxford, South Parks Road, Oxford, OX1 3PS, UK

John Bird, Prehistoric Museum, College of Eastern Utah, Price, UT 84501, USA

William T. Blows, Division of Applied Biological Sciences, St. Bartholomew School of Nursing and Midwifery, City University, London EC1A 7QN, UK

Don Burge, Prehistoric Museum, College of Eastern Utah, Price, UT 84501, USA

Kenneth Carpenter, Department of Earth and Space Science, Denver Museum of Natural History, 2001 Colorado Boulevard, Denver, CO 80205, USA

H. Trevor Clifford, Queensland Museum, P.O. Box 3300, South Brisbane, Queensland 1401, Australia

Karen Cloward, Western Paleontology Lab, Lehi, UT 84043, USA

Margery C. Coombs, Department of Biology, University of Massachusetts, Amherst, MA 01003, USA

Rodolfo A. Coria, Dirección Provincial de Cultura—Museo Carmen Funes, Av. Córdoba 55 (8318) Plaza Huincul, Neuquén, Argentina

P. J. Currie, Royal Tyrrell Museum of Palaeontology, P.O. Box 7500, Drumheller, AB, T0J 0Y0, Canada

Tracy L. Ford, P.O. Box 1171, Poway, CA 92074, USA

Peter M. Galton, College of Naturopathic Medicine, University of Bridgeport, Bridgeport, CT 06601, USA

Robert W. Gaston, Gaston Design, 1943 K Road, Fruita, CO 81421, USA

James I. Kirkland, Utah Geological Survey, P.O. Box 146100, Salt Lake City, UT 84114, USA

Martin G. Lockley, Geology Department, University of Colorado at Denver, P.O. Box 173364, Denver, CO 80217, USA

Richard T. McCrea, Department of Geological Sciences, University of Saskatchewan, 114 Science Place, Saskatoon, SK, S7N 5E2, Canada

Lorrie McWhinney, Department of Earth and Space Science, Denver Museum of Natural History, 2001 Colorado Boulevard, Denver, CO 80205, USA

Christian A. Meyer, Geologisch-Palaeontologisches Institut Basel, Bernoullistrasse 32, CH-4056 Basel, Switzerland

Clifford Miles, Western Paleontology Lab, Lehi, UT 84043, USA

Ralph E. Molnar, Museum of Northern Arizona, 3101 North Fort Valley Road, Flagstaff, AZ 86001, USA

David B. Norman, Sedgwick Museum, Department of Earth Sciences, University of Cambridge, Downing Street, Cambridge, CB2 3EQ, UK

Paul Penkalski, The Geology Museum, 1215 West Dayton Street, University of Wisconsin, Madison, WI 53706, USA

Xabier Pereda Suberbiola, Universidad del País Vasco/Euskal Herriko Unibertsitatea, Facultad de Ciencias, Departamento de Estratigrafía y Paleontología, Apartado 644, 48080 Bilbao, Spain; and Museum National d'Histoire Naturelle, Laboratoire de Paléontologie, 8 rue Buffon, 75005 Paris, France

Bruce M. Rothschild, Arthritis Center of Northeast Ohio, Youngstown, OH 44512, USA

A. P. Russell, Department of Biological Sciences, University of Calgary, Calgary, AB, T2N IN4, Canada

N. Rybczynski, Department of Biological Anthropology and Anatomy, Duke University, Durham, NC 27710, USA

Leonardo Salgado, Museo de la Universidad Nacional del Comahue, Buenos Aires 1400 (8300) Neuquén, Argentina

Jennifer Schellenbach, Gaston Design, 1943 K Road, Fruita, CO 81421, USA

M. K. Vickaryous, Department of Biological Sciences, University of Calgary, Calgary, AB, T2N 1N4, Canada

Foreword

Walter P. Coombs, Jr.:
An Affectionate Perspective

MARGERY CHALIFOUX COOMBS

Born in March 1944, Walter Coombs grew up in Baldwin, New York, a Long Island suburban community outside New York City. His youthful hobbies included nature exploration, photography, classical violin, and dinosaurs. As a dinosaur enthusiast, he visited the American Museum of Natural History whenever he could and was inspired by reading the books of American Museum of Natural History curator Edwin H. Colbert. He dreamed of becoming a dinosaur paleontologist, and as a high school senior, he applied to the only university in the United States that had a major in paleontology, the University of California, Berkeley.

Walter's years at Berkeley corresponded with the tumultuous period of the mid-1960s. Always one to question authority, he identified with the free speech and antiwar movements, though he was never politically active. As a paleontology student, he was especially influenced by Joseph T. Gregory and R. A. Stirton. During the summer before his senior year, he took Comparative Vertebrate Anatomy, which was taught by a charismatic young visiting Ph.D., David Klingener.

After graduating from Berkeley in January 1966, Walter spent the next six months before beginning graduate school as an employee working in the preparation laboratories of the Department of Vertebrate Paleontology at the American Museum of Natural History. One of his responsibilities was to help in the move of large mammoth and mastodon skeletal mounts from the previous Quaternary Hall to the center of the Tertiary Hall. That fall, he enrolled in the Ph.D. program in Biological Sciences at Columbia University, New York, where he would ultimately take courses from a wide range of influential biologists and geologists, among them Drs. Edwin Colbert, Bobb Schaeffer, Malcolm McKenna, Walter Bock, Marshall Kay, and David Ehrenfeld. He also joined a lively group of graduate students, including Eugene Gaffney, Robert Hunt, Pat and Tom Rich, Eric Delson, and Robert Emry. In the fall of 1967, he met a new graduate student, Margery Chalifoux, over a fossil fish (according to family legend). Although the fish received more attention at that time, the two married in February 1969.

In the 1960s, it was customary for Columbia paleontology students to complete a written monograph before proceeding to their oral preliminary exams. Walter's substantial topic, the Ornithischia, would surely give a graduate student pause today, but the beneficial result was a broad knowledge of this group that would serve as a basis for his Ph.D. dissertation on the Ankylosauria. At that time, the ankylosaurs were a group ripe for work; both Barnum Brown and Ed Colbert had intended to do descriptions and comparisons of American Museum of Natural History specimens, but other projects had intervened. Other important comparative specimens of ankylosaurs lay relatively nearby at Yale University, the Royal Ontario Museum, the National Museum of Canada, and the Smithsonian Institution. It was a perfect Ph.D. topic. Walter completed and defended his dissertation in 1971, winning the John S. Newberry Prize from Columbia University for achievement in vertebrate zoology. A major breakthrough of his dissertation was the division of the Ankylosauria into two major groups, Nodosauridae and Ankylosauridae, on the basis of important differences recognizable in virtually every part of the skeleton.

Walter moved directly from graduate school to a position teaching comparative vertebrate anatomy as an assistant professor at Brooklyn College. This location would have allowed continued regular access to dinosaur collections at the American Museum of Natural History; but Walter, having grown up in the Long Island suburbs, was not sure that he wanted to spend his entire life in the same place. Therefore, when Margery completed her Ph.D. degree in 1973 and was offered a position as assistant professor of zoology at the University of Massachusetts, Walter left Brooklyn College and moved with her and their two-year-old son Matthew to western Massachusetts. The evolution in roles was not easy for either one, but neither has regretted the move to this beautifully semirural, historic, and culturally active part of New England. Walter switched gears, working on exhibit and collection development at the Pratt Museum at Amherst College; completing some of his most important dinosaur papers (Coombs 1975, 1978b, 1978c, 1978d, 1979); pursuing several on Tertiary mammals with Margery; doing occasional adjunct college teaching; and playing an important role in the life of his young son. In 1978 he accepted a full-time position as assistant professor of biology at Western New England College, a small, primarily undergraduate institution in Springfield, Massachusetts. In 1980 he and Margery welcomed a second child, a daughter, Alexandra.

The main mission of faculty at Western New England College is teaching, and the typical faculty load in science is three courses per semester, plus laboratories. At various times, Walter has taught courses as diverse as Introductory Biology, Physical Geology, Historical Geology, Ecology, Evolution, Comparative Vertebrate Anatomy, Animal Behavior, Dinosaurs, and Geological Hazards. During the semester, research is virtually impossible. Nonetheless, by working hard during summers and other vacation breaks, Walter kept his research program alive and moved into new areas: dinosaur footprints (Coombs 1980b, 1996), juvenile ornithischians (Coombs 1980a, 1982, 1986), and dino-

saur nesting and parental behavior based on modern avian and crocodilian analogues (Coombs 1989, 1990a). He also became involved in
fieldwork with the Milwaukee Public Museum to census dinosaurs near
the Cretaceous–Tertiary boundary in Montana and western North
Dakota. This "Dig-a-Dinosaur" project involved participation by in-
terested amateurs, and Walter's expertise and enthusiasm were important ingredients in identifying fossils and working with the volunteers.

In the early 1990s, Walter returned to several ankylosaur projects:
general systematics (Coombs and Maryańska 1990), the utility of teeth
(Coombs 1990c), tail club structure and function (Coombs 1995a), and
descriptions of new materials (Coombs 1995b; Coombs and Demére
1996). Nonetheless, he was beginning to feel the cumulative effects of
trying to combine a research career with a busy teaching schedule.
Keeping up with the literature proved more and more difficult, and he
felt less and less able to write the kinds of synthetic papers that met his
own high standards. He also felt that he had no time to follow other
pursuits that he enjoyed. These factors drove him in the late 1990s to
detach himself from dinosaur research. Partway was not enough; he
vowed to go cold turkey.

Walter's contributions to dinosaur paleontology rest on more than
a simple list of publication dates and topics. The breadth and depth of
his knowledge and his ability to use this knowledge in creative ways
have enriched both his publications and his interactions with col-
leagues. Always a critical thinker, he is quick to see flaws in evidence or
logic and can be incredibly tenacious in discussion. This characteristic,
though not always endearing, has often led to improvements in papers
he has reviewed and theories which have been aired in his presence.
Walter has also been generous with his time as a dinosaur spokesman
and is well known in the local region for his playful and informative
lectures to school groups and nature organizations. He has identified
many puzzling fossils (debunking some pseudofossils in the process)
and has helped a number of University of Massachusetts graduate
students refine theses or learn techniques of fossil photography. As a
college professor, he is intense, humorous, and rigorous, well loved by
students capable of seeing past his more intimidating aspects.

There are also many nonpaleontological sides to Walter. The most
prominent today is the nature photographer, who can spend hours
patiently taking a series of close-up pictures as a monarch butterfly
gradually emerges from its pupa. He takes photographic rambles on
calm early mornings and occasionally travels to Florida, California,
Arizona, Wales, and recently the Galápagos Islands. He develops nature slide shows for local audiences and serves actively as president of
the Ware Camera Club. Walter is also energetic in home projects. In the
1980s, he designed and completed the inside finishing of a substantial
home addition, and he continues to undertake repairs and improvements indoors and out. He has also been an active partner in the
everyday activities of cooking, cleaning, and raising a family and thus
gives constant affirmation to the ideal of gender equality. He constantly
surprises people with his knowledge of topics as diverse as medieval
history, astronomy, idiosyncrasies of plant and animal natural history,

and camera technology. Multitalented and complicated, he continues to defy easy analysis.

Chronological Bibliography of Works by Walter P. Coombs, Jr.

1972. The bony eyelid of *Euoplocephalus* (Reptilia, Omithischia). *Journal of Paleontology* 46: 637–650.

1975. Sauropod habits and habitats. *Palaeogeography, Palaeoclimatology, Palaeoecology* 17: 1–33.

With M. C. Coombs. 1977. Dentition of *Gobiohyus* and a reevaluation of the Helohyidae (Artiodactyla). *Journal of Mammalogy* 58: 291–308.

With M. C. Coombs. 1977. The origin of anthracotheres. *Neues Jahrbuch für Geologie und Paläontologie, Monatshefte* 1977: 584–599.

1978a. An endocranial cast of *Euoplocephalus* (Reptilia, Omithischia). *Palaeontographica* A 161: 176–182.

1978b. The families of the ornithischian dinosaur order Ankylosauria. *Palaeontology* 21: 143–170.

1978c. Forelimb muscles of the Ankylosauria. *Journal of Paleontology* 52: 642–657.

1978d. Theoretical aspects of cursorial adaptations in dinosaurs. *Quarterly Review Biology* 53: 393–418.

1979. Osteology and myology of the hindlimb in the Ankylosauria (Reptilia, Ornithischia). *Journal of Paleontology* 53: 666–684.

With M. C. Coombs. 1979. *Pilgrimella*, a primitive Asiatic perissodactyl. *Zoological Journal of the Linnean Society* 65: 185–192.

1980a. Juvenile ceratopsians from Mongolia—The smallest known dinosaur specimens. *Nature* 283: 380–381.

1980b. Swimming ability of carnivorous dinosaurs. *Science* 207: 1198–1200.

With R. E. Molnar. 1981. Sauropoda (Reptilia, Saurischia) from the Cretaceous of Queensland. *Memoirs of the Queensland Museum* 20: 351–373.

With P. M. Galton. 1981. *Paranthodon africanus* (Broom), a stegosaurian dinosaur from the Lower Cretaceous of South Africa. *Geobios* 14: 299–309.

1982. Juvenile specimens of the ornithischian dinosaur *Psittacosaurus*. *Palaeontology* 25: 89–107.

With M. C. Coombs. 1982. Anatomy of the ear region of four Eocene artiodactyls: *Gobiohyus*, ?*Helghyus*, *Diacodexis*, and *Homacodon*. *Journal of Vertebrate Paleontology* 2: 219–236.

1986. A juvenile ankylosaur referable to the genus *Euoplocephalus* (Reptilia, Ornithischia). *Journal of Vertebrate Paleontology* 6: 162–173.

1987. Asiatic Ceratopsidae might be Ankylosauridae. *Occasional Paper of the Tyrrell Museum of Paleontology* 3: 48–51.

1988a. Review of *Dinosaurs Past and Present*, S. J. Czerkas and E. C. Olson, eds. *Journal of Vertebrate Paleontology* 8: 238–239.

1988b. The status of the dinosaurian genus *Diclonius* and the taxonomic utility of hadrosaur teeth. *Journal of Paleontology* 62: 812–817.

With P. M. Galton. 1988. *Dysganus*, an indeterminate ceratopsian dinosaur. *Journal of Paleontology* 62: 818–821.

1989. Modern analogs for dinosaur nesting and parenting behavior. In J. O. Farlow (ed.), *Paleobiology of the Dinosaurs. Geological Society of America Special Paper* 238: 21–53.

1990a. Behavior patterns of dinosaurs. In D. B. Weishampel, P. Dodson, and H. Osmólska (eds.), *The Dinosauria*, pp. 32–42. Berkeley: University of California Press.

1990b. Review of *Dinosaur Tracks and Trails,* D. D. Gillette and M. G. Lockley (eds.). *Journal of Vertebrate Paleontology* 10: 131–132.

1990c. Teeth and taxonomy in ankylosaurs. In K. Carpenter and P. J. Currie (eds.), *Dinosaur Systematics: Approaches and Perspectives,* pp. 269–279. Cambridge: Cambridge University Press.

With T. Maryańska. 1990. Ankylosauria. In D. B. Weishampel, P. Dodson, and H. Osmólska (eds.), *The Dinosauria,* pp. 456–483. Berkeley: University of California Press.

With D. B. Weishampel and L. M. Witmer. 1990. Basal Thyreophora. In D. B. Weishampel, P. Dodson, and H. Osmólska (eds.), *The Dinosauria,* pp. 427–434. Berkeley: University of California Press.

With J. S. MacIntosh and D. A. Russell. 1992. A new diplodocid sauropod (Dinosauria) from Wyoming, U.S.A. *Journal of Vertebrate Paleontology* 12: 158–167.

1995a. Ankylosaurian tail clubs of middle Campanian to early Maastrichtian age from western North America, with description of a tiny club from Alberta and discussion of tail orientation and tail club function. *Canadian Journal of Earth Science* 32: 902–912.

1995b. A nodosaurid dinosaur (Dinosauria: Ornithischia) from the Lower Cretaceous of Texas. *Journal of Vertebrate Paleontology* 15: 298–312.

1996. Redescription of the ichnospecies *Antipus flexiloquus* Hitchcock, from the Lower Jurassic of the Connecticut Valley. *Journal of Paleontology* 70: 327–331.

With T. A. Deméré. 1996. A late Cretaceous nodosaurid ankylosaur (Dinosauria: Ornithischia) from marine sediments of coastal California. *Journal of Paleontology* 70: 311–326.

With M. C. Coombs. 1997. Analysis of the geology, fauna, and taphonomy of Morava Ranch Quarry, early Miocene of northwestern Nebraska. *Palaios* 12: 165–187.

Part I
Thyreophorans

1. *Scelidosaurus,* the Earliest Complete Dinosaur

DAVID B. NORMAN

Abstract

Richard Owen's work on the early armored ornithischian dinosaur *Scelidosaurus* is reviewed. Dinosaur paleontologists have ignored the contribution of *Scelidosaurus* to the history of dinosaur discovery and interpretation since its discovery in Dorset, England, in 1858. This simple fact has distorted the common narrative of dinosaur studies that can be traced through all general books on the topic (1968–2000). Exploring the question of why the *Scelidosaurus* has been overlooked provides insights into the intellectual work and conflicting pressures (academic, political, and social) that affected leading scientists, in this instance Professor Owen, in mid-Victorian times.

Introduction

Scelidosaurus has merited (but not received) considerable attention since its discovery in the Liassic cliffs of Charmouth in 1858. It was the very first essentially complete dinosaur ever discovered and subjected to proper scientific scrutiny; and it is of particular interest to the study of dinosaurian evolution because it is very early, geologically speaking—occurring in rocks formed during Early Jurassic (Sinemurian) times. The complete specimen was presented, for study and description, to one of the premier comparative anatomists of the time: Professor Richard Owen, the intellectual father of the Dinosauria. On these grounds alone, *Scelidosaurus harrisonii* Owen, 1861, should occupy an important place in dinosaur studies. But in truth, the year

1858 resonates in the history of dinosaurs because it marks the announcement of the discovery, by Professor Joseph Leidy (Warren 1998), of the American dinosaur *Hadrosaurus foulkii* Leidy, 1859. As Weishampel and Young (1996) have remarked, this discovery and the subsequent report by Leidy marked the beginning of serious dinosaur research in America and led inexorably to the "bone wars" that were fueled by the rivalry between Leidy's protégé Edward Drinker Cope and Yale University–based Othniel Charles Marsh. The report that was published in 1859 allowed Leidy to question Richard Owen's original vision of dinosaurs as huge scaly quadrupeds, which was made famous by the Crystal Palace models that he helped to construct (Owen 1854). In their place, Leidy created a vision, eventually supported by full-size skeletal models (1868), of the American dinosaur as a far more lithe-looking, upright biped. This history was elaborated by Professor Edwin Colbert in a book that provided a wonderfully vivid benchmark account of the history of dinosaur studies (Colbert 1968).

It is a curious fact that by 1863 Owen had published two monographs on *Scelidosaurus,* and it is quite clear that *Scelidosaurus* was a large, scaly quadruped, utterly unlike Leidy's *Hadrosaurus* and extraordinarily reminiscent of Owen's original vision.

Institutional Abbreviations. BMNH: Natural History Museum (formerly British Museum [Natural History]), London. GSM: The Geological Sciences Museum (now incorporated into the collections of the British Geological Survey, Keyworth).

Historical Context

Richard Owen (Fig. 1.1), though undoubtedly eminent and influential for many decades of the 19th century, has received scant attention in the 20th century (Rupke 1994). Owen's name only generally appears as an anachronistic footnote in accounts of paleontological works because he created the name Dinosauria in 1842 (Torrens 1992) and brought them to worldwide attention through the remarkable models that he helped to create in 1853–1854 for the landscaped park surrounding Sir Joseph Paxton's great glass and steel Crystal Palace exhibition pavilion. The eclipse of Owen and his achievements (largely the work of Owen's archenemy Thomas Henry Huxley and his acolytes) is unwarranted, as Rupke has begun to make us realize. Richard Owen remained influential and continued to publish a stream of papers on dinosaurs (and a wealth of other taxa) until the end of his long life; and he was almost single-handedly responsible for the creation of the magnificent Alfred Waterhouse–designed British Museum (Natural History) in South Kensington, London, of which he became the first director (1881–1883).

Between 1842 and the mid-1850s, the prevailing view (with the notable exception of that of Gideon Mantell 1841, 1851) concerning the anatomy and general appearance of dinosaurs had been established by Owen in what was undoubtedly a landmark report to the British Association for the Advancement of Science (Owen 1842). This work was a triumph of intellect and of anatomical brilliance, conjuring from

Figure 1.1. Professor Sir Richard Owen, KCB, FRS. A mezzotint of H. W. Pickersgill's portrait of a youthful Owen in 1845. The majority of accounts, almost without exception, show Owen as a particularly haggard-looking man of advanced years.

a few dislocated fragments a group of fossil reptiles that had been entirely unrecognized until then. Though unsupported by illustrations, his narrative created a remarkable account of these animals, their probable appearance, and even aspects of their physiology; in view of the debate concerning dinosaur physiology over the past three decades, his work was remarkably prescient. Just over a decade later, Owen's report was given further substance by his supervision of the construction of life-size models of prehistoric animals by Benjamin Waterhouse Hawkins; the models formed part of the geological section (McCarthy and Gilbert 1994) of the landscaped area in front of Joseph Paxton's Great Exhibition Hall when it was re-erected at Sydenham on what were then the southern outskirts of London in 1853–1854 (Anonymous 1853, 1854; Desmond 1975). This extraordinarily popular public exhibit brought dinosaurs, and their appearance, to the attention of the world, or at least those parts of the world influenced by the British Empire. The terminology used passed into the popular language of the

ST. PHILIP'S COLLEGE LIBRARY

day through its adoption by such notable authors as Charles Dickens (Desmond 1975).

Just four years later, Joseph Leidy announced the discovery of a new dinosaur, *Hadrosaurus foulkii,* on the basis of a partial postcranial skeleton collected from a site of old marl diggings at Haddonfield, New Jersey. As recounted more recently (Desmond 1975; Weishampel and Young 1996), bones had been discovered considerably earlier at this site but had been removed as curiosities and were not studied systematically. On the rumor of these bones, William Parker Foulke, a member of the Academy of Natural Sciences of Philadelphia, relocated and reexcavated this pit and discovered some teeth and more bones, including vertebrae and, very importantly, bones of the limbs. All of these remains were passed to Foulke's close friend (and fellow member of the Academy of Natural Sciences) Joseph Leidy for study and description. Within two months of their arrival in Philadelphia, a verbal account of this discovery, during which the new name for the dinosaur was created by Leidy, was given at the December meeting of the Philadelphia Academy of Natural Sciences (December 14, 1858).

Leidy was able to observe and report in his article (published in 1859) that *Hadrosaurus* possessed teeth that bore considerable resemblance to those described for *Iguanodon* (one of three fossil reptiles that constituted Owen's original Dinosauria) and seemed for that reason to be a close relative. However, he went on to point out that

> the great disproportion of size between the fore and back parts of the skeleton of *Hadrosaurus,* leads me to suspect that this great extinct herbivorous lizard may have been in the habit of browsing, sustaining itself, kangaroo-like, in an erect position on its back extremities and tail. (Leidy 1859: 218)

This particular part of the article resulted in the reconstructions (again by Waterhouse Hawkins) of New World dinosaurs such as those shown to the public in 1868 at the Philadelphia Academy where Leidy worked (Fig. 1.2), and those planned and built (but sadly not exhibited) for New York's Central Park (Colbert 1968; Desmond 1975). These reconstructions were immortalized by Waterhouse Hawkins in paintings at Princeton University (Desmond 1975; Weishampel and Young 1996). However, Leidy's opinions were not nearly as unequivocal as those quoted above, because in the next section of the article, he goes on to say:

> As we, however, frequently observe a great disproportion between the corresponding parts of the body of recent, well known extinct saurians, without any tendency to assume such a position as that mentioned, it is not improbable that *Hadrosaurus* retained the ordinary prostrate condition. . . .
>
> *Hadrosaurus* was most probably amphibious; and though its remains were obtained from a marine deposit, the rarity of them in the latter leads us to suppose that those in our possession have been carried down the current of a river, upon whose banks the animal lived.

These citations from Leidy exemplify the *style* of argument he used,

Figure 1.2. Hadrosaurus foulkii *Leidy, 1859. Skeletal restoration at the Academy of Natural Sciences, Philadelphia, as it appeared in 1868, in a kangaroolike pose. This represents the first dinosaur skeleton ever mounted for exhibition and was based on the work of Professor Joseph Leidy and on one particular part of his conclusions in the paper he read in December 1859.*

which was equally as important as the strictly scientific content. To paraphrase: Leidy first drew a strong comparison between *Hadrosaurus* and Owen's *Iguanodon*, then used limb proportions to propose a radical interpretation concerning the dinosaur's posture (adding a proviso in case he was wrong!); he then introduced an explanation of this dinosaur's appearance in marine deposits and its likely amphibious mode of life. This was not only clever and insightful (Ostrom 1964), it was also artful and reminiscent of Richard Owen's equally circuitous style of observation and induction evident in Owen's own writings (probably best described as sophistry). This style makes it very difficult to decide precisely what Owen's views were on any given subject, because the observations and his logical inferences were littered with provisos. We get a glimpse of his style in the passages cited below. Many of the themes that Leidy mentioned in his 1859 article are picked up in Owen's writings on *Scelidosaurus*, even though no mention whatever is made of Leidy's work.

With the discovery of Cope's *Laelaps* a few years later (Cope 1866), which showed an even more extreme disparity between forelimbs and hindlimbs, the interpretation of New World dinosaurs veered away from the "prostrate condition" toward Leidy's "kangaroo-like" vision. This is, in general, one of the dominant views that was passed on to us at the close of the twentieth century. Thus, the views of the preeminent comparative anatomist Richard Owen were being challenged and largely overturned by new materials and new ideas. In a late postscript to this time that emphasizes the fact that Owen's views did not endure, Professor O. C. Marsh reflected rather ungraciously in 1895 (Schuchert and LeVene 1940) that

> so far as I can judge, there is nothing like unto them [the Crystal Palace dinosaurs] in the heavens, or on the earth, or in the waters under the earth. We now know from good evidence that both *Megalosaurus* and *Iguanodon* were bipedal, and to represent them as creeping, except in their extreme youth, would be almost as incongruous as to do this by the genus *Homo*.

And yet Marsh was also offering an edited view of Owen's dinosaurian models. He did not mention the third of Owen's founding dinosaurs displayed at Crystal Palace: *Hylaeosaurus* (Fig. 1.3), an armored species recovered from the Wealden of Tilgate Forest and named by Gideon Mantell (Mantell 1833). As will be seen later, this dinosaur had an equally important role to play in the search for a proper understanding of dinosaur anatomy, posture, and diversity, particularly in light of the fortuitous and timely discovery of yet another dinosaur, *Scelidosaurus*.

In 1858 James Harrison of Charmouth, Dorset (Lang 1947), collected or was given a selection of vertebrate fossils exposed during quarrying in the soft, crumbling cliffs of limestone, marl, and shale along the coast just below the village of Charmouth (probably from a geological unit known as the Black Ven Marl). These remains were passed to Richard Owen for study and description (Owen 1861a) and included an almost complete skull (lacking only the tip of the snout and some shards from the upper part of the side of the face), which was

illustrated in detail (see Figs. 1.6, 1.7). During the following year, the limestone band from which the skull was removed was further excavated, and a fully articulated skeleton was gradually recovered (Owen 1863; see Fig. 1.11).

Palaeontographical Society Monograph for 1859 (Owen 1861a)

In the volume of the Palaeontographical Society's Monograph series for 1859 (published in 1861), Owen announced the discovery of the new dinosaur *Scelidosaurus harrisonii* (Greek: limb lizard of Mr. Harrison). In truth, the generic name *Scelidosaurus* had appeared in print earlier (Owen 1859, 1860); but the publications provided neither a specific name nor type material, and the name could be regarded as a *nomen nudum* until the date of the more detailed publication in 1861. The status of this taxon was further complicated by the fact that the material collected and illustrated was composite. It included not only genuine scelidosaur remains but also those of a theropod (whose remains formed the basis for Owen's choice of generic name; Fig 1.4); Owen's first plate illustration is of an incomplete theropod left femur (GSM 109560; Fig. 1.4A). Plate II of Owen illustrates a partial knee joint and ungual phalanx, which appear to be theropod (Fig. 1.4B). Plate III illustrates hindlimb bones from a very small ?juvenile scelidosaur (Fig. 1.5), and plates IV–VI provide excellent lithographs of an almost complete scelidosaur skull (Figs. 1.6, 1.7).

Newman (1968) reviewed this material and demonstrated the

Figure 1.3. Hylaeosaurus armatus Mantell, 1833. *This animal is one of the three specimens that first described Richard Owen's Dinosauria. This restoration was created for the Great Exhibition by Benjamin Waterhouse Hawkins under the supervision of Professor Owen in 1853 and can be seen today in Crystal Palace Park, South London.*

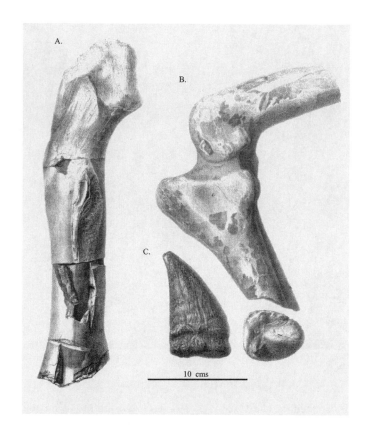

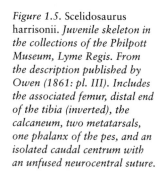

Figure 1.4. "Scelidosaurus" *from the illustrations on plates I and II of Owen (1861); these are the incomplete remains of an undescribed Liassic theropod. (A) GSM 109660, an imperfect left femoral shaft. (B) BMNH 39496, the articulated femur and tibia of a theropod dinosaur designated the type of* Scelidosaurus harrisonii *by Lydekker (1889). (C) GSM 10956. Ungual phalanx of a ?theropod.*

Figure 1.5. Scelidosaurus harrisonii. *Juvenile skeleton in the collections of the Philpott Museum, Lyme Regis. From the description published by Owen (1861: pl. III). Includes the associated femur, distal end of the tibia (inverted), the calcaneum, two metatarsals, one phalanx of the pes, and an isolated caudal centrum with an unfused neurocentral suture.*

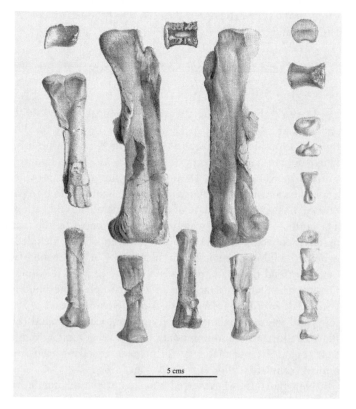

10 cms

Figure 1.6. Scelidosaurus harrisonii. *BMNH R.1111. The articulated skull, lacking the rostral tip, as illustrated by Owen (1861: pl. IV). This is the first articulated dinosaur skull ever described.*

10 cms

Figure 1.7. Scelidosaurus harrisonii. *BMNH R.1111. The skull from an illustration in Owen (1861: pl. V), showing details of the teeth, and the cross section of the broken end of the rostrum.*

Figure 1.8. Scelidosaurus harrisonii. *BMNH R.1111. As illustrated by Owen (1863: pl. III), showing the articulated shoulder, humerus, and proximal ends of the radius and ulna, as well as some of the scattered skin armor (dermal ossicles).*

50 cms

Figure 1.9. Scelidosaurus harrisonii. *BMNH R.1111. Articulated lower limb bones and foot in articulation and reconstruction drawing (Owen 1863: pl. XI).*

Figure 1.10. (opposite page) Scelidosaurus harrisonii. *BMNH R.1111. Parts of the articulated series of tail vertebrae illustrated by Owen (1863: pl. IX), including articulated dermal ossicles.*

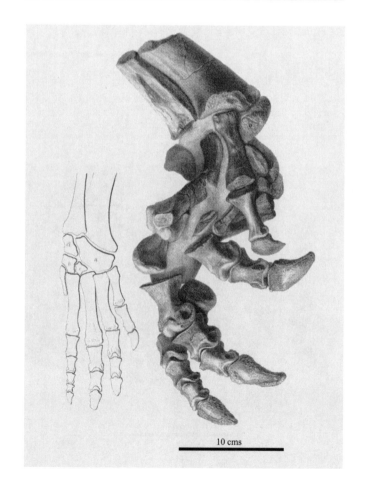

10 cms

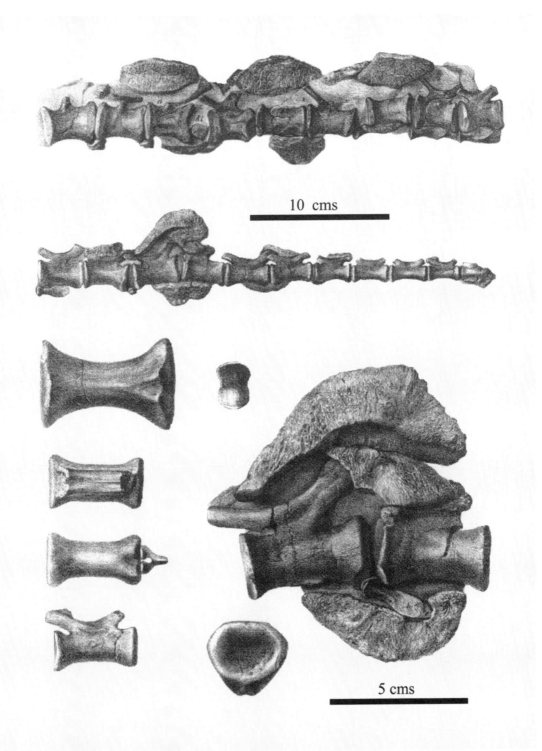

10 cms

5 cms

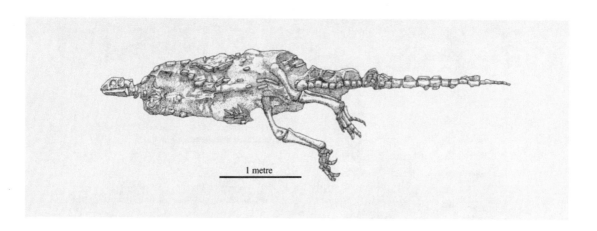

Figure 1.11. Scelidosaurus harrisonii. *BMNH R.1111. A sketch of the full skeleton of this dinosaur is based on a poor-quality photograph of an old display at the British Museum (Natural History), London (from Newman 1968).*

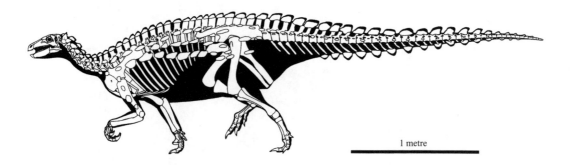

Figure 1.12. A skeletal restoration of BMNH R.1111 by the artist Gregory S. Paul. Copyright G. S. Paul.

theropod affinities of the femur, knee joint, and phalanx. Subsequently, Charig and Newman (1992) stabilized the nomenclature by designating a lectotype for *Scelidosaurus harrisonii* from the syntypic series illustrated and described by Owen; they nominated the skull (Figs. 1.6, 1.7) and its associated skeleton (BMNH R1111; Figs. 1.8–1.11).

The illustrated material was fully described by Owen (1861a). The remains of the "young Scelidosaur" were illustrated and described in some detail, and the fossil's status as a dinosaur was confirmed by the structure of the femur. The crushing and distortion to which the small specimens had been subjected accounted for any differences between this specimen and the earlier illustrated material.

The bones referred to as the "young Scelidosaur" (see Fig. 1.4) were, for the most part, identified correctly (though the tibia was inverted vertically, and one of the metatarsals was incorrectly identified as a fibula). Their relatively small size led Owen to suspect that these bones belonged to a newly born individual, and he even speculated that dinosaurs were not only ovoviviparous but also that this specimen might have represented a fetus carried by a "gravid Scelidosaur" (Owen 1861a: 7). In this respect, he was demonstrating some of the imaginative interpretation relating to the biology of dinosaurs of which he had shown himself capable in the 1842 article, although this was later qualified by an admission that he suspected that newly born scelidosaurs (by comparison with oviparous crocodiles) would be even smaller than the remains he had described.

The most interesting paragraph is Owen's discussion of the mode of locomotion of this dinosaur (Owen 1861a: 6). Owen pointed out that even though the limb bones attested to a terrestrial existence, nothing precluded their use for swimming. In what can only be seen as a direct discussion of Leidy (1859) without any direct reference, Owen considered limb proportions and body posture in dinosaurs:

> the disproportionate shortness of the fore limbs, even in the Iguanodon, leads to the suspicion that they might be short in reference to diminishing the obstacles to propelling the body through water by actions of the strong and vertically extended tail; and that, as in the living land lizard of the Gallopagos [sic] Islands, called *Amblyrhynchus,* the fore limbs might be applied close to the trunk in the Iguanodon, when it occasionally sought the water of the neighbouring estuary or sea.

That these remains were recovered from marine deposits (a very similar circumstance to Leidy's *Hadrosaurus*), further reinforced the interpretation that this dinosaur was capable of amphibious activity.

In 1861 the crowning achievement of this monograph was the opportunity to describe the first ever complete skull of any dinosaur. The entire period between 1824 and 1858 had been marked by the persistently fragmentary nature of dinosaur remains (but see Norman 1993), to the enormous frustration of Mantell and Owen in particular. The skull was described and well illustrated (Figs. 1.6, 1.7) and the structure of the teeth considered at some length, allowing Owen to establish for the first time the anatomical configuration of a dinosaur skull and comparing it with most other known fossil reptiles. The breadth of comparison would be considered unacceptably large by modern standards. However, this simply emphasizes the fact that the relationships between dinosaurs and other fossil forms were neither well understood nor well established. The same was true of Owen's original dinosaurs: *Megalosaurus, Iguanodon,* and *Hylaeosaurus,* and newer forms such as *Echinodon* and *Regnosaurus* were included in discussion as well as modern reptiles (a variety of modern lizards and the crocodile) and a range of far less closely related fossil forms (the synapsid *Oudenodon,* and the ichthyopterygian *Ichthyosaurus*). The teeth of *Scelidosaurus* (Fig. 1.7) were well preserved and allowed close comparison with living forms, notably iguanid lizards, which suggested that this animal was herbivorous. However, even this interpretation was considered provisional, allowing for the possibility that if sharp and elongate teeth were found on the missing front part of the jaws, it might turn out to be an aquatic predator. *Echinodon* (Owen 1861b) from the English Purbeck Limestone (Berriasian, basal Cretaceous), which Owen believed to be a lizard, had teeth similar in shape to those of *Scelidosaurus* and showed some evidence of longer, pointed, anterior teeth in its jaws that might have been used for catching slippery fish (Norman and Barrett, in press).

Palaeontographical Society Monograph for 1860 (Owen 1863)

During 1858–1859, the remainder of the skeleton belonging to the skull noted in Owen's first monograph on *Scelidosaurus* was discovered in a series of consecutively positioned limestone blocks, all of which were passed on to Owen for description. Thus, by some time in 1859 Owen had in his possession the first essentially complete dinosaur skeleton. The vertebral column was described in detail and comparisons were drawn wherever possible with other dinosaurs and crocodiles (including *Streptospondylus* and *Cetiosaurus,* which Owen still believed—incorrectly, as Mantell had previously shown—were fossil crocodilian species). Owen (1863: 12) concluded this section with the a telling observation:

> Upon the whole, I find the closest agreement to be between *Scelido-* and *Hylaeo-saurus* in the characters of the vertebral column; and I infer for both, but especially for *Scelidosaurus,* a greater aptitude for swimming than in the larger Dinosauria.

The next sections of the monograph were devoted to the girdles and limbs. Though not complete, the forelimb and shoulder (Fig. 1.8) were described as being comparatively similar to those seen in marine crocodilians, and relatively reduced and not as important in locomotion as in the forelimbs of either *Iguanodon* or *Megalosaurus.* The reasoning involved in drawing such conclusions is even more involved than usual and thereby appears to succeed in obscuring Owen's meaning; comparisons are drawn very widely and include reference to dicynodonts and the armored Triassic archosauromorph *Stagonolepis*.

The hindlimb is well preserved and is described in intricate detail. It demonstrated not only the structure of the ankle but that the pes contained four functional toes with a slightly reduced first toe and considerably reduced fifth digit (Fig. 1.9). In this respect, *Scelidosaurus* appeared to be a structural predecessor of the three-toed *Iguanodon* and prompted Owen to describe a transformation in pedal morphology that has an almost evolutionary cadence.

> From the abortion of the fifth digit, and the disproportionate shortness of the first, we have in *Scelidosaurus* the example of a reptile manifesting a tendency to the tridactyle type of the hind foot, and this is effected in its remote successor of the Wealden period,—the *Iguanodon*. (Owen 1863: 19)

Finally, the description turned to the dermal skeleton, which was also described in very considerable detail (Fig. 1.10). This allowed Owen to build a picture of the dermal armor pattern, which was composed of several paired bilateral rows of bones extending from the neck along the flanks as far as the hips. The tail was composed of a median row dorsally and ventrally and a much smaller row down the sides of the tail.

In the concluding section, Owen (1863: 26) summarized his understanding of this animal as follows:

> The general condition of this almost entire skeleton of a reptile, organised, as seems by the structure of its hind foot, for terrestrial rather than aquatic life, or at least for amphibious habits on the margins of a river rather than for pursuit of food in the open sea, I infer that the carcass of the dead animal has been drifted down a river, disemboguing in the Liassic ocean, on the muddy bottom of which it would settle down when the skin had been so far decomposed as to permit the escape of the gases engendered by putrefaction.

Owen used this death scenario to allow himself a few comments on the nature of preservation of the skeleton and the circumstances that had probably led to its survival in the fossil record. So ended Owen's description of the first essentially complete dinosaur skeleton ever discovered.

This account of *Scelidosaurus* is striking for its attention to anatomical detail and its simultaneous brevity and lack of larger-scale interpretation of this animal in relation to the understanding of the anatomy and biology of dinosaurs. This was the same author who was notable, earlier in his career, for making breathtaking leaps of imagination—not the least of which had been his recognition of dinosaurs as a distinct group and their probable ecology and physiology (Owen 1842). It is curious that Owen made no attempt to reconstruct this animal even though, as his monograph on *Dimorphodon* (Owen 1870) and many other forms had shown, Owen was more than capable of attempting a reasonably accurate restoration on the basis of a good skeleton (Fig. 1.11).

Richard Owen—The Years 1856–1863

Nicolaas Rupke (1994) has provided the first reasonably comprehensive reassessment of Owen's life and work, and this is ably supported by reference to earlier work by Jacob Gruber and John Thackray (1992) and Roy MacLeod (1965). In addition, the longer accounts by Adrian Desmond (1982, 1989), in which Owen's personality and contributions figure significantly in the narrative, add further supportive information. The reader should refer to these latter accounts to gain a more balanced view of the man and his writings, as well as the social context within which he worked.

On the basis of information in this scholarship, as well as the bibliography verified by C. D. Sherborn in Owen's biography (Owen 1894), it is possible to get an overview of the extent of his activities in the years immediately preceding the presentation of *Scelidosaurus* to Owen. In this way, it is possible to fit this particular facet of his work into the context of his wider responsibilities and ambitions. A brief chronology of Owen's activities reads as follows:

• 1856. Owen resigns from his curatorship at the Hunterian Museum of the Royal College of Surgeons after continued and increasing disagreements with the Council. In the same year he is appointed first Superintendent of the Natural History Collections at the British Museum (Bloomsbury).

Publishes 15 papers, articles, monographs, and books.

• 1857. Owen, in the annual report on the collections for 1856, emphasizes the need for space for the Natural History Collections at the British Museum. Commences publications on the structure of the brain as a means of classifying animals followed by additional papers on the topic in 1858–1860. His work leads to the "Hippocampus debate," which came to dominate the meetings of the British Association for the Advancement of Science at Oxford in 1860, Manchester in 1861, and Cambridge in 1862, and represented a strikingly heated and public controversy between Owen and Huxley. The debates culminate in victory for Huxley, which is represented by the publication of his book *Man's Place in Nature* in 1863; Lyell's *The Antiquity of Man,* also 1863; and a little later Darwin's *Descent of Man* (1871).

Publishes 13 papers, articles, monographs, and books.

• 1857–1861. Gives annual courses of lectures on paleontology at the Royal School of Mines.

• 1858. Rising controversy over the Natural History Collections. Scientific lobby formed to prevent removal of collections from Bloomsbury. Owen elected president of the British Association for the Advancement of Science; begins to campaign to separate the Natural History Collections and to build a new museum to house them. Another lobby group, supported primarily by "Darwinians," advocates disbanding and disbursing the Natural History Collections around the country.

Publishes 21 papers, articles, monographs, and books.

• 1859. Submits formal plan for a Museum of Natural History. Delivers the Rede Lectures. Darwin's *Origin of Species* is published.

Publishes 20 papers, articles, monographs, and books.

• 1859–1861. Appointed Fullerian Professor at the Royal Institution and commences a course of public lectures.

• 1860. Gives a course of lectures to the royal household at Buckingham Palace. The Gregory Committee appointed to assess the British Museum. Writes hostile anonymous review of *Origin of Species* for the *Edinburgh Review*. Publishes his book *Palaeontology* as a compilation of his lectures at the Royal School of Mines.

Publishes 8 papers, articles, monographs, and books.

• 1860–1863. Parliamentary debates on the British Museum.

• 1861. Delivers Royal Institution lecture, "On a New Natural History Museum." British Museum trustees vote in favor of removal of the Natural History Collections from Bloomsbury. Publishes three articles "On a National Museum of Natural History" in the *Athenaeum*.

Publishes 13 papers, articles, monographs, and books.

• 1862. Publishes a book on the extent and aims of a national museum of natural history. Arranges purchase of the first skeleton of a feathered reptile (*Archaeopteryx*) from the private collector Karl Haberlein in Germany.

Publishes 7 papers, articles, monographs, and books.

• 1863. Publishes a monograph on the fossil *Archaeopteryx*.

Publishes 6 papers, articles, monographs, and books.

From this simple synopsis, it should be abundantly clear that Owen was exceedingly busy in a number of areas during this critical period: not only was he involved in a heavy and extraordinarily diverse teaching load through his various appointments, he was also heavily involved in governmental lobbying for a new natural history museum in the face of a vociferous opposition headed by Huxley and his supporters; he was fighting a rearguard action against Darwin's theory of evolution; and he was embroiled in a public and heated controversy over the structure of the brain and the distinctiveness of the human being.

As Rupke and others have pointed out, Owen was not specifically and simplemindedly against evolution per se, or "transmutation," as it was termed at this time. He was simply unconvinced that natural selection was either a correct or appropriate mechanism by which organic change could be achieved; his was a view supported by many scientists and philosophers at this time. One of the primary reasons for skepticism was the absence of an appropriate hereditary mechanism and Darwin's reliance on the tenets of Lamarckism. Owen continued for many years to search for alternative "secondary causes" that might be responsible for transmutation, but without success.

In addition to these more diverse threads, Owen was publishing books, scientific monographs, and shorter descriptive articles on anatomy and paleontology at a rapid rate; they included such notable contributions as the detailed descriptions of the giant ground sloth *Megatherium;* a sequence of newly discovered mammal-like reptiles from

South Africa; numerous British dinosaur monographs in the Palaeontographical Society series; the anatomy of the chimpanzee, orangutan, gorilla, and Madagascan primates; a full description of the giant flightless bird, the moa (*Dinornis*), from New Zealand; descriptions of new fossil mammals from Australia; and finally, a continuous series of notes on new British fossil mammals. Given this range of activities, the discovery of another fossil animal (*Scelidosaurus*) would perhaps seem to be of lower priority, from Owen's perspective, than from the narrower view of dinosaur specialists of the present day.

Discussion

The Dinosauria: Anatomy, Biology, and Diversity

In the period starting in 1824 with the description by Buckland of the Stonesfield reptile *Megalosaurus* (Buckland 1824) and ending in 1863, a great deal of progress was made in the general understanding of large fossil reptiles. Central to this progress was the review of British Fossil Reptiles that was published in two parts: 1840 (Owen 1840) and 1842 (Owen 1842), which did much to establish a scientific understanding of the range and variety of fossil vertebrates. Two of Owen's most notable contributions were to summarize what was then known of the giant marine reptiles that had been named the Enaliosauria many years before by Conybeare (primarily plesiosaurs and ichthyosaurs) and to pair them up with their terrestrial equivalents, Owen's newly invented Dinosauria. Owen's 1842 paper not only established the anatomical criteria by which dinosaurs could be recognized, which were based on a remarkably limited selection of anatomical remains (Torrens 1997), but it also provided an opportunity for him to speculate on the anatomy and physiology of these animals in a way that presaged much of the contemporary debate on these topics. Whether such work is part of the political debate concerning the rising progressionism of the time (see Desmond 1979) is largely immaterial to this discussion; what is striking is the logic and vision implicit in Owen's work. This purely descriptive vision (unaccompanied by illustrations in its 1842 published form) of the dinosaur was given substance by the Crystal Palace dinosaur models when they were devised and built, with Owen's guidance, in 1853–1854.

Within four years, this Owenian view of dinosaurs was to be challenged by new discoveries in America (Leidy 1859). Leidy's interpretation has been portrayed as radical and revolutionary in the literature describing the history of dinosaur studies (Colbert 1968; Desmond 1975); however, it is clear that Leidy's vision was not as clearly expressed as it has been portrayed. The kangaroolike image of *Hadrosaurus* was only one of a range of postures that Leidy proposed for this animal. What emerges from Leidy's work is a belief in the variable posture of this animal (from kangaroolike to "prostrate") and their amphibious nature. These interpretations were built in part at least on its anatomical features and the marine depositional environment in which the remains were embedded. It was only considerably later (1868) that the kangaroo model became preferred—again as a result of model-making for a popular exhibition (Weishampel and Young 1996).

With impeccable timing, the 1858 discovery of a complete articulated dinosaur made possible Owen's papers of 1861 and 1863 on *Scelidosaurus*, which followed in the wake of Leidy's paper on *Hadrosaurus*. The descriptive anatomy is appropriately thorough and comparative in the tradition of the time, given the nature and quality of the material. The interpretations, however, focus almost exclusively on its locomotor potential (terrestrial, aquatic, or amphibious) and the circumstances (taphonomy) that would lead to the burial in marine deposits of animals that were supposedly terrestrial (Owen 1842). What is striking is that Owen appears not to have thought it necessary to demonstrate, by use of the vehicle provided by *Scelidosaurus,* the apparent wisdom of his 1842–1854 vision of dinosaurian anatomy and form by reference specifically to *Hylaeosaurus*—the dinosaur and the model so clearly supported by the discovery of *Scelidosaurus* (compare Figs. 1.2 and 1.12).

Conclusions

In short, the discovery in 1858 of *Scelidosaurus* offered a remarkably timely opportunity for Richard Owen, the founder of the Dinosauria, to describe the first complete skeleton of any dinosaur and to reaffirm his original vision on the basis of pioneering work on this group of fossil reptiles (Owen 1842). Owen's published reports on this dinosaur describe the anatomy in detail, but his interpretation confines itself to rather restricted matters relating to the probable mode of life of the animal (locomotion and diet) and discussion of its normal habitat. Owen's approach, in the light of Leidy's work (of which he must have been aware), would seem to have been quite remarkable in two particular respects: it appears to lack vision when compared with his earlier work on dinosaurs, and it lacks any obvious and pointed criticism of interpretations that run counter to his own opinions—the result of a certain arrogance for which Owen is justifiably renowned.

From a modern perspective, the discovery and examination of *Scelidosaurus* would seem to have provided Owen with a multiplicity of opportunities. For example, he had the potential to describe and provide the first unequivocal reconstruction of any dinosaur on the basis of a virtually complete skeleton; and to show that his remarkable vision of dinosaurs with a heavy-limbed quadrupedal posture (very similar to that rendered in the form of the Crystal Palace dinosaurs—specifically *Hylaeosaurus*) was confirmed by the anatomy of *Scelidosaurus*.

It might also have been an opportunity for Owen to begin to elaborate on the diversity of dinosaurian form, which only began to emerge once Harry Seeley had established the split between the Ornithischia and Saurischia in 1887 (Seeley 1887); to discuss in detail the implications of the postural interpretations of dinosaurs raised by the work of Joseph Leidy (1859); and to discuss the origin of dinosaurs since *Scelidosaurus* was arguably the earliest dinosaur known with any confidence at this time.

Though the latter points are clearly the result of hindsight, I mention them here because such issues would seem to have lain within Owen's intellectual ambit, given the remarkably insightful nature of his earlier work on dinosaurs. That Owen "failed" to use the discovery of *Scelidosaurus* to promote the intellectual rigor of his earlier vision of the dinosaurs, or to have any significant impact on dinosaur studies at this time (before the major dinosaur discoveries made in North America after the late 1870s) seems paradoxical given the self-evident importance of *Scelidosaurus*. Opportunities were undoubtedly missed by Owen in this respect; but if we simply highlight such shortcomings, we fail to grasp the fact that Professor (later Sir) Richard Owen was, from the mid-1850s onward, concerned with a wide variety of issues in addition to his paleontological work. Such matters were commensurate with his high standing among the Victorian intellectual elite and related to institutional power-brokering, the public understanding of science, government lobbying for a new natural history museum for London, and the rise of Darwinism.

In short, I suspect that although Owen never lost his interest in dinosaurs—he published papers about them until well into his retirement—his other duties and ambitions simply left him little time to think about dinosaurs in depth, and certainly not with the freedom that he had had in the 1830s. In mid-Victorian Britain, at the height of the British Empire, Owen was undoubtedly among the most successful and important of scientists. The monument to his labors, the Natural History Museum in South Kensington, is a magnificent and permanent testament to Owen's eminence and his driving ambition. That Owen failed to fully capitalize on the significance of the discovery of *Scelidosaurus* is, to me, a matter of some regret because it was clearly a most important and timely discovery.

The history of the discovery of dinosaurs cannot be rewritten on the basis of these revelations; but the realization of the importance of the discovery of *Scelidosaurus* adds to our appreciation of the richness and complexity of dinosaurian research during the middle decades of the last century.

Acknowledgments. This work would not have been possible without the generosity of the late Dr. Alan Charig, who allowed me to study the *Scelidosaurus* that he had originally intended to study after supervising its acid preparation in the laboratory of the Natural History Museum, London. I also thank Dr. Angela Milner (Head of Fossil Vertebrates), Sandra Chapman (Curator of Reptiles), and Carol Gokce (Natural History Librarian) of the Natural History Museum, London, for access to materials in the collections for which they are responsible. Carole Holden (British Library) kindly checked and verified the date of acquisition of the *Proceedings of the Academy of Natural Sciences,* Philadelphia, by the British Museum. This manuscript has benefited from comments and advice from Professor Hugh Torrens (Keele University), as well as commentary from two reviewers: Professor Peter

Dodson (University of Pennsylvania) and Dr. Kenneth Carpenter (Denver Museum of Natural History).

References Cited

Anonymous. 1853. The Crystal Palace at Sydenham. *The Illustrated London News. London* 23: 599–600.

———. 1854. The Crystal Palace at Sydenham. *The Illustrated London News. London* 24: 22.

Buckland, W. 1824. Notice on the *Megalosaurus* or great fossil lizard of Stonesfield. *Transactions of the Geological Society of London,* n.s. 1: 390–396.

Charig, A. J., and B. H. Newman. 1992. *Scelidosaurus harrisonii* Owen, 1861 (Reptilia, Ornithischia): Proposed replacement of inappropriate lectotype. *Bulletin of Zoological Nomenclature* 49(4): 280–283.

Colbert, E. H. 1968. *Men and Dinosaurs: The Search in Field and Laboratory.* London: Evans Brothers.

Cope, E. D. 1866. Remains of a gigantic extinct dinosaur. *Proceedings of the Philadelphia Academy of Natural Sciences.* 1866: 275–279.

Darwin, C. R. 1871. *The Descent of Man, and Selection in Relation to Sex.* London: John Murray.

Desmond, A. J. 1975. *The Hot-Blooded Dinosaurs. A Revolution in Palaeontology.* London: Blond and Briggs.

———. 1979. Designing the dinosaur: Richard Owen's response to Robert Edmond Grant. *Isis* 70: 224–234.

———. 1982. *Archetypes and Ancestors: Palaeontology in Victorian London 1850–1875.* London: Blond and Briggs.

———. 1989. *The Politics of Evolution: Morphology, Medicine and Reform in Radical London.* Chicago: University of Chicago Press.

Gruber, J. W., and J. C. Thackray. 1992. *Richard Owen Commemoration.* London: Natural History Museum.

Huxley, T. H. 1863. *Evidence as to Man's Place in Nature.* London: John Murray.

Lang, W. D. 1947. James Harrison of Charmouth, geologist (1819–1864). *Proceedings of the Dorset Natural History and Archaeological Society* 68: 103–118.

Leidy, J. 1859. *Hadrosaurus* and its discovery. *Proceedings of the Academy of Natural Sciences, Philadelphia,* 10 (for 1858): 213–218.

Lydekker, R. 1888. *Catalogue of the Fossil Reptilia and Amphibia in the British Museum (Natural History), London (Part 1).* London: Trustees of the British Museum (Natural History).

Lyell, C. 1863. *Geological Evidences of the Antiquity of Man with Remarks on Theories of the Origin of Species by Variation.* London: John Murray.

MacLeod, R. 1965. Evolutionism and Richard Owen. *Isis* 56: 259–280.

Mantell, G. A. 1833. *The Geology of the South East of England.* London: Longman, Rees, Orme, Brown, Green and Longman.

———. 1841. Memoir on a portion of the lower jaw of the *Iguanodon,* and on the remains of the *Hylaeosaurus* and other saurians, discovered in the strata of Tilgate Forest, in Sussex. *Philosophical Transactions of the Royal Society of London* 131: 131–151.

———. 1851. *Petrifactions and Their Teachings; or, a Hand-Book to the Gallery of Organic Remains of the British Museum.* London: Henry G. Bohn.

McCarthy, S., and M. Gilbert. 1994. *The Crystal Palace Dinosaurs: The Story of the World's First Prehistoric Sculptures.* London: Crystal Palace Foundation.

Newman, B. H. 1968. The Jurassic dinosaur *Scelidosaurus harrisonii,* Owen. *Palaeontology* 11(1): 40–43.

Norman, D. B. 1993. Gideon Mantell's "mantel-piece": The earliest well-preserved ornithischian dinosaur. *Modern Geology* 18: 225–245.

Norman, D. B., and P. M. Barrett. In press. Ornithischian dinosaurs from the Early Cretaceous (Berriasian) of England. In A. R. Milner and D. G. Batten (eds.), *Life and Environments in Purbeck Times.* Special Papers in Palaeontology.

Ostrom, J. H. 1964. A reconsideration of the paleoecology of hadrosaurian dinosaurs. *American Journal of Science* 262: 975–997.

Owen, R. R. 1840. Report on British fossil reptiles. Part 1. *Report of the British Association for the Advancement of Science* 9: 43–126.

———. 1842. Report on British fossil reptiles. Part 2. *Report of the British Association for the Advancement of Science* (Plymouth) 11: 60–204.

———. 1854. *Geology and the Inhabitants of the Ancient World.* London: Crystal Palace Library.

———. 1859. Palaeontology. *Encyclopaedia Britannica,* Vol. 17, pp. 91–176. Edinburgh: Adam and Charles Black.

———. 1860. *Palaeontology or a Systematic Summary of Extinct Animals and Their Geological Relations.* Edinburgh: Adam and Charles Black.

———. 1861a. Monograph of the fossil Reptilia of the Liassic Formations. Part 1. A monograph of a fossil dinosaur (*Scelidosaurus harrisonii* Owen) of the Lower Lias. *Palaeontographical Society Monographs* Part 1: 1–14.

———. 1861b. Monograph of the fossil Reptilia of the Wealden and Purbeck Formations (Lacertilia). *Palaeontographical Society Monographs* Part 5: 31–39.

———. 1863. A monograph of the fossil Reptilia of the Liassic Formations. Part 2. A monograph of a fossil dinosaur (*Scelidosaurus harrisonii* Owen) of the Lower Lias. *Palaeontographical Society Monographs* Part 2: 1–26.

———. 1870. Monograph of the fossil Reptilia of the Liassic formations. Part 3. *Dimorphodon. Palaeontographical Society Monographs* Part 3: 41–81.

———. 1894. *The Life of Richard Owen.* London: John Murray.

Rupke, N. A. 1994. *Richard Owen: Victorian Naturalist.* New Haven, Conn.: Yale University Press.

Schuchert, C., and C. M. LeVene. 1940. *O. C. Marsh: Pioneer in Paleontology.* New Haven: Yale University Press.

Seeley, H. G. 1887. On the classification of the fossil animals commonly named Dinosauria. *Proceedings of the Royal Society of London* 43: 165–171.

Torrens, H. S. 1992. When did the dinosaur get its name? *New Scientist* 134(4 April 1992): 40–44.

———. 1997. Politics and palaeontology: Richard Owen and the invention of dinosaurs. In J. O. Farlow and M. K. Brett-Surman (eds.), *The Complete Dinosaur,* pp. 175–190. Bloomington: Indiana University Press.

Warren, L. 1998. *Joseph Leidy: The Last Man Who Knew Everything.* New Haven, Conn.: Yale University Press.

Weishampel, D. B., and L. Young. 1996. *Dinosaurs of the East Coast.* Baltimore, Md.: Johns Hopkins University Press.

2. Tooth Wear and Possible Jaw Action of *Scelidosaurus harrisonii* Owen and a Review of Feeding Mechanisms in Other Thyreophoran Dinosaurs

PAUL M. BARRETT

Abstract

Feeding mechanisms in thyreophoran dinosaurs are poorly understood. Tooth wear and possible jaw action of the Early Jurassic thyreophoran *Scelidosaurus harrisonii* are examined and a preliminary review of the feeding mechanisms in other thyreophorans is provided. The limited evidence available suggests that some thyreophorans (*Scelidosaurus* and various ankylosaurs) had a puncture-crushing jaw mechanism that involved tooth–tooth contact during jaw closure. There is also some evidence to support the proposition that feeding mechanisms within members of the Thyreophora were more diverse than suggested previously.

Introduction

Herbivorous dinosaurs were the most abundant and diverse large vertebrates in late Mesozoic terrestrial ecosystems (Wing and Sues 1992). Deducing the feeding mechanisms and behaviors of these animals is therefore fundamental to our understanding of Mesozoic paleo-ecology, dinosaur paleobiology, and dinosaur evolution. Functional morphological studies on ornithopods (Crompton and Attridge 1986; Norman 1984a; Norman and Weishampel 1985, 1991; Ostrom 1961; Weishampel 1984), ceratopians (Dodson 1993; Ostrom 1964, 1966), and sauropods (Barrett and Upchurch 1994, 1995b; Calvo 1994; Upchurch and Barrett 2001) have greatly improved our knowledge of the feeding mechanisms, and hence the paleobiology and paleoecology, of these animals. In contrast, the feeding mechanisms of thyreophoran dinosaurs have been neglected. This may partly be the result of the difficulties posed by working with thyreophoran cranial material. Although craniodental material of thyreophorans is not rare per se, specimens that yield useful functional data are extremely uncommon; for example, many otherwise complete skulls often lack in situ dentition, hampering discussion of jaw action and diet. The current consensus view holds that the majority of thyreophorans used a pulping or slicing jaw mechanism with limited tooth–tooth contact, a simple orthal jaw action, and little oral food processing (Galton 1986; Weishampel and Norman 1989).

Tooth wear and jaw action can now be described for *Scelidosaurus harrisonii*, a primitive thyreophoran from the Early Jurassic (Sinemurian) of southern England (Norman 2001; Owen 1861, 1863). *Scelidosaurus* has been regarded as either a primitive stegosaur (e.g., Steel 1969), a primitive ankylosaur (e.g., Carpenter 2001; Norman 1984b), or an ornithopod (Thulborn 1977); but recently it has been described as a basal thyreophoran and the sister taxon to the Eurypoda (Stegosauria + Ankylosauria) (Fig. 2.1; Coombs et al. 1992; Sereno 1986, 1997, 1999). No comprehensive description of this important taxon has been produced since Owen's original monographs were published, though the type and referred material of *Scelidosaurus* are currently being re-studied (Norman 2001). A more detailed understanding of feeding in *Scelidosaurus* should provide a number of insights into the evolution of feeding mechanisms within the Thyreophora.

I present a preliminary review of the tooth wear and jaw action in other thyreophorans to assess the diversity of feeding mechanisms within this poorly understood clade of Mesozoic herbivores.

Institutional Abbreviations. AMNH: American Museum of Natural History, New York. BMNH: Natural History Museum (formerly British Museum [Natural History]), London. BRSMG: Bristol City Museum and Art Gallery, Bristol, UK. CMNH: Carnegie Museum of Natural History, Pittsburgh, Pennsylvania. DMNH: Denver Museum of Natural History, Denver, Colorado. HMN: Humboldt Museum fur Naturkunde, Berlin. IVPP: Institute of Vertebrate Paleontology and Paleoanthropology, Beijing. NSM: National Science Museum, Tokyo. ROM: Royal Ontario Museum, Toronto. SMC: Sedgwick Museum of Earth Sciences, University of Cambridge, Cambridge. USNM: National Museum of Natural History (formerly United States National Museum), Washington, D.C. YPM: Peabody Museum of Natural History, Yale University, New Haven, Connecticut. ZDM: Zigong Dinosaur Museum, Zigong, China.

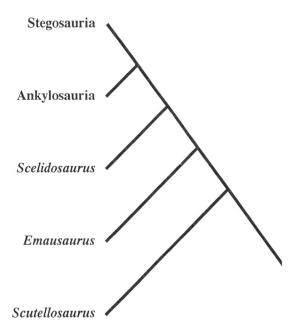

Stegosauria

Ankylosauria

Scelidosaurus

Emausaurus

Scutellosaurus

Figure 2.1. Thyreophoran phylogeny (after Sereno 1986, 1997, 1999).

Tooth Wear and Jaw Action in *Scelidosaurus*

Material

Cranial material of *Scelidosaurus* is known from two individuals. BMNH R1111 (the lectotype) includes an almost complete, three-dimensionally preserved skull. Before preparation, the skull and mandibles were fully articulated (Owen 1861: pls. IV–VI). Acid preparation of the skull has resulted in the partial disarticulation of the skull and mandibles, and the complete removal of the surrounding matrix has allowed detailed anatomical study (Fig. 2.2). The snout of BMNH R1111 was sheared off before collection; the premaxillae, anterior parts of the dentaries, and the predentary of this individual are therefore unknown. BRSMG Ce12785 is a juvenile specimen that possesses a partial skull. The skull is crushed mediolaterally and is missing most of the left-hand side. The median skull roof elements, the braincase, and the left mandible are also missing. The preserved portion includes the following: right and left premaxillae, right and left maxillae, right lachrymal, right jugal, right quadratojugal, right postorbital, right quadrate, various crushed palatal elements, partial right dentary, and right surangular. Most of the following account is based on BMNH R1111, supplemented with additional information from BRSMG Ce12785.

General Comments

The skull is relatively long and shallow in lateral view, and the snout slopes anteroventrally from a point just dorsal to the orbits. In dorsal view, the skull narrows anteriorly and the snout is narrow and pointed (see Owen 1861, pls. IV–VI). The premaxillae of BRSMG Ce12785 meet in the midline to form a small premaxillary secondary

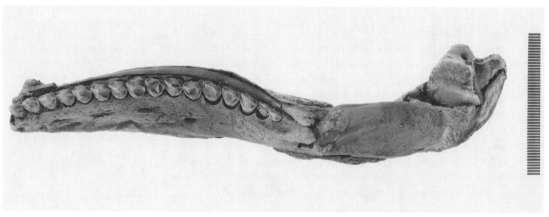

Figure 2.2. (top) Partial left facial skeleton and left mandible of Scelidosaurus harrisonii *(BMNH R1111) in lateral view. Scale bar = 50 mm.*

Figure 2.3. (bottom) Left mandible of Scelidosaurus harrisonii *(BMNH R1111) in dorsal view. Scale bar = 50 mm.*

palate. The mandible is extremely robust, lacks an external mandibular fenestra, and bears a coronoid eminence of moderate height (Fig. 2.2). Maxillary and dentary tooth rows are bowed medially, as in ankylosaurs (Coombs 1978; Fig. 2.3). Both the maxillary and dentary tooth rows are deeply inset, and there is a large buccal emargination (Figs. 2.2, 2.3). The jaw joint is offset slightly ventrally relative to the tooth row.

Cranial Kinesis

The skull is essentially akinetic. Streptostyly was prevented by extensive overlapping contacts between the quadrate, squamosal, quadratojugal, and pterygoid. The head of the quadrate is subtriangular in transverse section and fits into a deep socket on the ventral surface of the squamosal. This socket is bounded by two elongate processes that originate from the ventral surface of the squamosal. One process is anterior to the socket and has an overlapping contact with the anterior surface of the dorsal part of the quadrate; the other process is posterior to the socket and contacts the posterodorsal surface of the quadrate. The jugal wing of the quadrate is overlapped anteriorly by the dorsal process of the quadratojugal and articulates with a shallow recess on the posterior surface of this process. The dorsalmost part of the dorsal process of the quadratojugal is overlapped anteriorly by the ventralmost part of the anterior ventral process of the squamosal. The anterior surface of the quadrate shaft ventral to the quadrate and quadratojugal foramen has a broad contact with the posterior part of the quadratojugal. This series of contacts results in a continuous bar of bone that extends along the entire length of the anterior surface of the quadrate; this structure would have prevented anteriorly directed rotation of the quadrate around its junction with the squamosal. An extensive overlapping contact between the pterygoid wing of the quadrate and the quadrate wing of the pterygoid would have further stabilized the quadrate. Posterior rotation of the quadrate would have been prevented by contact with the posterior ventral process of the squamosal. Connective tissues may also have had a role in stabilizing the posterolateral part of the skull.

The elements of the median skull roof (nasals, frontals, and parietals) are either firmly sutured to each other or have extensive overlapping contacts that would have prevented any flexion in this region of the skull.

Tooth Morphology

In the preserved tooth rows of BMNH R1111, there are 16 or 17 maxillary teeth and 16 dentary teeth. The right and left upper tooth rows of BRSMG Ce12785 each possess 6 premaxillary teeth and 17 maxillary teeth. The anterior premaxillary teeth are conical in cross section and have bulbous crown bases. The crown apex is pointed and recurved. Denticles are generally absent, but small denticles are present on the mesial margin of the second tooth in the left premaxilla. The sixth premaxillary teeth are coarsely denticulate, labiolingually compressed, and mesiodistally expanded: they are very similar to the maxillary teeth.

The maxillary and dentary teeth of *Scelidosaurus* are similar to each other (Fig. 2.4). Unworn tooth crowns are mesiodistally expanded, tall, and approximately diamond-shaped in labial view. The crown apices are oriented slightly distally, making the teeth asymmetrical in labial view. The mesial and distal edges of the crown are coarsely denticulate. No well-defined primary or secondary ridges are present on either the labial or lingual surfaces of the crown. A broad central eminence is present on the labial surface of the maxillary teeth and on the lingual surface of the dentary teeth. This eminence extends from the base of the tooth crown to the tip of the crown apex; the mesiodistal width of the eminence diminishes toward the apex. Cingula are present on both maxillary and dentary teeth and are continuous with the basalmost denticles on the mesial and distal edges of the crown.

There is one major difference between dentary and maxillary teeth. In apical view, the maxillary teeth are simple, labiolingually compressed cone-shaped structures. In contrast, the dentary teeth have a distinct "shelf" on the labial surface (Fig. 2.3), caused by marked labial expan-

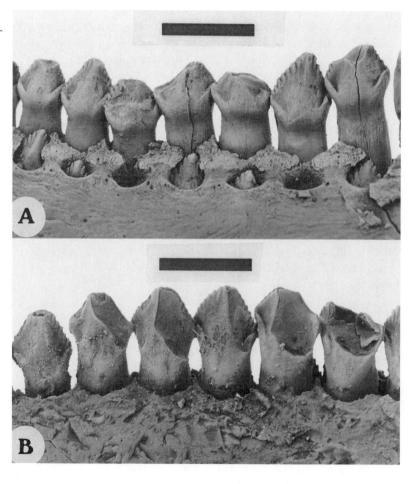

Figure 2.4. Details of the tooth-bearing bones of Scelidosaurus harrisonii *(BMNH R1111). (A) Left maxilla in lingual view, tooth positions 3–9. (B) Left dentary in labial view, tooth positions 4–9. Anterior is toward the left. Scale bars = 10 mm.*

sion of the crown adjacent to the crown–root junction. The functional implications of this tooth morphology are addressed below.

The pattern of tooth replacement is currently under study by Dr. D. B. Norman (University of Cambridge). Although the pattern of tooth eruption influences the distribution of wear along the tooth row (i.e., newly erupted teeth display less wear than older, functional teeth), it does not affect the mechanism by which wear is produced. This topic will not be dealt with here.

Tooth Macrowear

In BMNH R1111, wear is present on many, but not all, of the maxillary and dentary teeth. Each of the tooth-bearing bones is dealt with separately because several different types of wear are present along the tooth row. In general, the heaviest tooth wear is found on the teeth in the center of the tooth row; teeth in the anteriormost and posteriormost portions of the jaws usually show limited wear. Tooth positions are counted from the anterior end of the preserved tooth row.

Left Maxilla (Fig. 2.4A). Wear facets are present on the lingual surfaces of the apices of many teeth, though only a few of the teeth display heavy wear. Small, low-angled planar wear facets are present on the very tip of the crown in teeth 1 and 4. On other crowns, these apical facets are more steeply inclined and extend further down the lingual crown surface, so that the wear facet may cover up to 20% of this surface (tooth positions 3, 5, 7, 13, and 14). These larger apical wear facets are often best developed on the distal half of the lingual surface of the tooth crown; wear mesial to the apical denticle is much less pronounced. One tooth (tooth position 6) possesses a high-angled mesial wear facet, similar to those seen in *Lesothosaurus* (Thulborn 1971). A large apical wear facet and a high-angled mesial wear facet are found in combination on tooth 9. Several teeth show no wear (e.g., tooth positions 8 and 12).

Right Maxilla. Large wear facets are absent from most of the teeth in the right maxilla, and many of the teeth show no sign of wear. However, three teeth possess steeply inclined apical wear facets on their lingual surfaces (tooth positions 6, 7, and 12). These wear facets vary in extent; that on tooth 6 covers approximately 50% of the lingual crown surface, whereas the facet on tooth 12 is limited to a small area adjacent to the apical denticle. Several other teeth (e.g., tooth positions 4, 8, and 11) display very small low-angled wear facets that are confined to the lingual surface of the apical denticle.

Left Dentary (Fig. 2.4B). A number of teeth in the left dentary display labial wear facets. Six of the teeth (tooth positions 5, 6, 7, 8, 11, and 12) possess an extremely large high-angled planar wear facet. These facets extend down toward the dorsal border of the cingulum and cover between 50% and 100% of the labial crown surface. On teeth with wear facets that cover less than 90% of the labial crown surface, the facets are most strongly developed on the distal part of this surface; the mesial edge of some of these heavily worn teeth may be totally unworn. The extent of these wear facets, in combination with the labially expanded crown base, creates a shallow bowllike depression near the root–crown junction (Figs. 2.3, 2.4B). Several other teeth (e.g., tooth positions 2 and 4) possess very small, low-angled apical wear facets that are limited to the area immediately adjacent to the apical denticle, and tooth 10 displays a small high-angled wear facet that extends from the apex along the distal edge of the crown. Several teeth (tooth positions 1 and 3) show no wear.

Right Dentary. The teeth in the right dentary are poorly preserved, and it is not possible to examine the distribution of wear along the tooth row in detail. Some teeth possess small, low-angled, apical wear facets (e.g., tooth positions 7 and 8), and small wear facets are present on the distal edges of teeth 9 and 12. Two teeth (tooth positions 5 and 6) possess large wear facets that cover the labial surface of the crown. These wear facets are almost identical to those in the left dentary in tooth positions 5–8, 11, and 12.

Several premaxillary teeth in BRSMG Ce12785 show small amounts of wear. The distal margin of tooth 2 in the left premaxilla possesses a small high-angled wear facet that may have been formed by occlusion against a predentary beak. Tooth 3 in the left premaxilla and teeth 2, 3, and 6 in the right premaxilla display small wear facets that are limited to the apical region of the crown. In BRSMG Ce12785, the lingual surfaces of the left maxillary teeth are obscured by their apposition to the rest of the skull; the lingual surfaces of right maxillary teeth 9–17 are still embedded in matrix and cannot be studied. Small, low-angled apical wear facets, similar to those described from the left and right maxillary teeth of BMNH R1111 (e.g., right maxillary teeth 4, 8, and 11) are present on teeth 1, 5–8, and 9 in the left maxilla of BRSMG Ce12785. The lack of larger apical wear facets that extend further down the length of the crown may be due to the immaturity of the specimen.

None of the teeth in either BMNH R1111 or BRSMG Ce12785 display double high-angled, steep-sided wear facets (*contra* Galton 1986; *contra* Weishampel and Norman 1989).

Tooth Microwear

Examination of the teeth of BMNH R1111 with a stereoscopic microscope revealed the presence of fine wear striae within the wear facets of several dentary teeth. These striae are present on teeth 6–8 in the left dentary and on tooth 8 in the right dentary. In all cases, the striae are oriented parallel or subparallel to the long axis of the tooth crown. No wear striae were observed on any of the maxillary teeth.

Because of the fragility of the material, a detailed study of microwear on the teeth of *Scelidosaurus* was not possible. A study was made of several representative left maxillary teeth (teeth 7, 13, and 14); they were cleaned and molded in dental rubber (by L. Cornish of the Natural History Museum, London) and examined with a scanning electron microscope. Only a few surface details were discernible, and the possibility that some of these features are artefacts of preparation cannot be ruled out. The accompanying electron micrograph (Fig. 2.5) is of a mold; as a result, the microwear appears in negative relief. Wear striae are present within the wear facets of all three teeth (on both the dentine and enamel), but are rare and only extend for short distances (up to 400 μm) along the crown. All of the wear striae observed are oriented parallel or subparallel to the long axis (apex–cingulum) of the crown. Small pits, ranging in size from 20 to 40 μm in diameter, are fairly common and are found on the wear facets and on adjacent areas of the crown surface. They are randomly distributed (Fig. 2.5). The pits have a range of outlines, from subcircular to completely irregular, and they appear to have been approximately as deep as they were wide on the basis of the strength of their negative relief.

Morphology of the Jaw Joint

The distal end of the quadrate is divided into lateral and medial condyles (BMNH R1111 and BRSMG Ce12785). The division between the two condyles is not strongly developed, however, and consists of a shallow, anteroposteriorly directed groove. In posterior view, the medial condyle is wider mediolaterally than the lateral condyle and projects further ventrally. In medial view, the medial condyle is expanded anteroposteriorly relative to the lateral condyle.

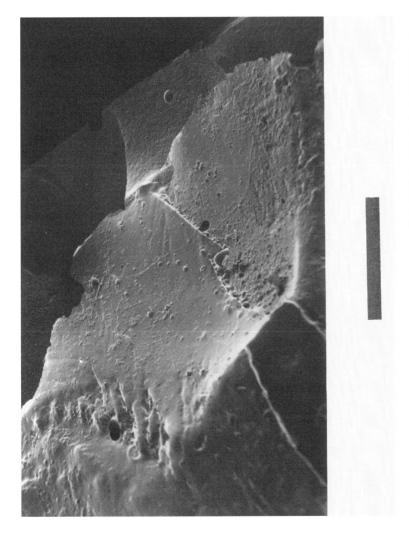

Figure 2.5. Scanning electron micrograph of the wear facet of the 14th left maxillary tooth. Because the electron micrograph is of a mold, the tooth appears reversed and the microwear appears in negative relief. The apex of the tooth is toward the top of the picture (the wear facet is confined to the distal part of the crown). Note the presence of pits and striae on the surface of the wear facet and on adjacent areas of the crown. Scale bar = 1 mm.

The mandibular glenoid fossa is composed of the articular posteriorly and medially and of the surangular anteriorly and laterally (BMNH R1111; Fig. 2.3). The articular makes the greatest contribution to the surface area of the fossa. The glenoid fossa is bounded anterolaterally by the coronoid eminence; the surangular projects upward abruptly approximately 90° to the long axis of the mandible, forming a vertical sheet of bone anterior to the glenoid. A ridge formed by the surangular (laterally) and the articular (medially and posteriorly) surrounds the rest of the glenoid fossa. This ridge is particularly prominent along the posteromedial edge of the articular and clearly defines the posterior margin of the glenoid fossa. In dorsal view, the fossa encompasses two sulci that are separated from each other by a low, rounded ridge. This ridge extends anteromedially from the posterolateral corner of the articular and traverses the dorsal surface of the articular, until it terminates at the anteromedial margin of that element. The anterolateral sulcus is the smaller of the two and faces dorsally and slightly medially. The larger posteromedial sulcus faces dorsally and slightly laterally and is situated slightly more ventrally than the anterolateral sulcus.

This configuration mirrors the morphology of the distal end of the quadrate. The smaller lateral condyle articulates with the small anterolateral sulcus on the glenoid, whereas the larger medial condyle articulates with the ventrally offset posteromedial sulcus. The glenoid fossa is not anteroposteriorly expanded relative to the anteroposterior length of the quadrate condyles; when the jaws are articulated, the quadrate fits snugly into the glenoid fossa and is restricted anteriorly by the posterior margin of the coronoid eminence and posteriorly by the ridge extending along the posterior margin of the articular.

Interpretation of Jaw Action and Feeding Mechanism

Comparison between Owen's figures (1861, 1863), casts of BMNH R1111 made before preparation, photographs of the skull at various stages of preparation, and the prepared skull elements allows the identification of opposing teeth in the upper and lower tooth rows. This demonstrates that there is some correspondence between tooth wear on the upper teeth and wear on the opposing lower teeth; but in almost all of these instances, the upper and lower wear facets have different morphologies (small apical wear facets on the maxillary teeth versus large bowllike facets on dentary teeth). Moreover, one tooth of an opposing pair may be completely unworn, although the corresponding tooth in the other tooth row may be heavily worn (e.g., between the eighth left maxillary and dentary teeth). The type and extent of wear are highly suggestive of occlusal contact. But the distribution of wear along the tooth row, the lack of correspondence between some wear facets, and the variability in the shape of the wear facets—both along the tooth row and between opposing pairs of teeth—are difficult to reconcile with each other. An additional difficulty is the presence of significantly heavier wear on the dentary teeth than on the maxillary teeth. What mechanism or mechanisms could have produced this pattern of wear?

The absence of regularly distributed, high-angled mesial and distal wear facets suggests that the tooth rows did not interdigitate during jaw closure, as has been proposed for *Lesothosaurus* (Thulborn 1971) and various sauropods (Barrett and Upchurch 1995b; Calvo 1994; Upchurch and Barrett 2001). Instead, the tooth crowns of *Scelidosaurus* seem to be partially realigned so that they meet in almost direct opposition to each other, with the lingual surfaces of the maxillary teeth occluding against the labial surfaces of the dentary teeth. This arrangement partially explains the distribution of wear along the tooth rows; small apical wear facets on the maxillary teeth often correspond with large labial wear facets on opposing dentary teeth, and it is possible that some of this wear may have been formed by tooth–tooth contact during occlusion, a suggestion supported by the apparent presence of microwear.

However, this occlusal model does not account for the difference in the extent of wear between the upper and lower tooth rows. Part of the answer may lie in the contrasting morphologies of maxillary and dentary teeth. As mentioned above, the bases of dentary teeth are labially expanded. As a result, the wear facets on the dentary teeth are often bowl-shaped as they extend over the labial surface of the tooth and onto the dorsal surface of the labially expanded basal region of the crown (Figs. 2.3, 2.4B). During occlusion, the maxillary teeth would, in effect, fit into a basin walled lingually and ventrally by the labial surface of a dentary tooth crown. This arrangement would provide the animal with an effective puncturing and crushing mechanism, with the dentary teeth acting as a row of mortars and the maxillary teeth representing a series of pestles. This mechanism may account for the differences in wear between the tooth rows: the lower teeth are continually undergoing the stresses associated with crushing food, whereas the upper teeth simply push food down into the basins. The presence of pits on the surfaces of the tooth crowns (see above and Fig. 2.5) is consistent with this kind of puncture crushing. It appears, therefore, that dental macrowear in *Scelidosaurus* was produced by a combination of tooth–tooth and tooth–food wear, providing an efficient puncture-crushing system rather different from the *Lesothosaurus*-like orthal-slicing mechanism proposed by Galton (1986), or the orthal-pulping mechanism suggested by Weishampel and Norman (1989). The wear on the premaxillary teeth of BRSMG Ce12785 is probably the result of abrasional tooth–food wear (cf. Popowics and Fortelius 1997).

The configuration of the jaw joint and the presence of vertically oriented microstriae indicate that the jaw action was strictly orthal and had no transverse or translational component.

Review of Feeding Mechanisms in Other Thyreophoran Dinosaurs

Scutellosaurus

The skull of *Scutellosaurus* is very poorly known and is represented by a few maxillary and dentary fragments (Colbert 1981). The maxillary and dentary tooth rows are straight, in contrast to the medially bowed tooth rows of *Scelidosaurus* and ankylosaurs (Coombs 1978). Colbert (1981) notes that a buccal emargination was absent but that the posterior maxillary and dentary teeth are inset (Galton 1986; Sereno 1991). Colbert (1981) has suggested that there may have been a low coronoid eminence.

The dentition is similar to that of the basal ornithischian *Lesothosaurus* (Sereno 1991; Thulborn 1970). Six premaxillary teeth are present; the tooth crowns are labially convex, sharply pointed, and recurved. The maxillary and dentary teeth are almost identical to those of *Lesothosaurus*, differing only in the number of denticles on the mesial and distal surfaces of the crown (Colbert 1981). The mesial and distal edges of several teeth are slightly worn, possibly as a result of tooth–food contact (Galton 1986). Large wear facets are absent.

The absence of information on the form of the jaw joint prevents deduction of the jaw action of *Scutellosaurus,* but the lack of attritional tooth wear (cf. Popowics and Fortelius 1997) suggests that a precise occlusion was absent.

Emausaurus

Emausaurus is represented by an almost complete skull with some associated postcranial material (Haubold 1990). The skull and dentition are superficially similar to those of *Scelidosaurus,* but these taxa differ in several respects. For example, in *Emausaurus,* there is no accessory dermal ossification on the lower jaw (with the corresponding presence of an open external mandibular fenestra), the posterior part of the skull is much broader in dorsal view than in *Scelidosaurus* (although it narrows considerably anteriorly), and the tooth rows are not bowed medially to the same extent as those in *Scelidosaurus* (see Haubold 1990).

The jaw joint of *Emausaurus* is anteroposteriorly short; Haubold (1990) suggested that there was a simple orthal jaw action, though he provided no information on tooth wear to support this view. There is a low coronoid eminence and the jaw joint is slightly ventrally offset. A precise occlusion appears to have been absent (Haubold 1990).

Stegosaurs

Complete skulls are known for *Stegosaurus* (USNM 4934 [Gilmore 1914], DMNH 2818, and NSM PV 20380), *Huayangosaurus* (ZDM T7001 and IVPP V6728; Sereno and Dong 1992; Zhou 1984), and *Tuojiangosaurus* (Dong et al. 1983); partial skulls of these and other stegosaur taxa are also available. The stegosaur skull has an extremely elongate snout and a low temporal region; in dorsal view, the skull tapers to a narrow beak (Fig. 2.6). The skull is akinetic (Galton 1997; Weishampel and Norman 1989). The skull roof bones are firmly sutured to each other, and streptostyly is prevented (in adults) by fusion of the quadrate head to the squamosal (Galton 1992, 1997). A well-developed maxillary buccal emargination is present in *Huayangosaurus* (ZDM T7001) and *Paranthodon* (BMNH 47338), and a reduced emargination is present in *Stegosaurus* (USNM 4934). Complete stegosaur

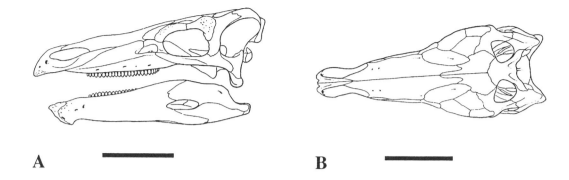

A   B

Figure 2.6. Skull of Stegosaurus
stenops *in (A) lateral and
(B) dorsal views (from Sereno
and Dong 1992).*
Scale bars = 120 mm.

tooth rows are extremely rare, and even those specimens with complete tooth rows are usually preserved in such a way that it is not possible to examine the teeth in detail. For example, USNM 4934 has complete maxillary and dentary tooth rows; but as the mandibles are preserved in tight articulation with the skull, the maxillary teeth obscure most of the dentary teeth and the maxillary teeth are only visible from one side. As a result, it is not possible to assess the distribution of wear facets along the tooth row or to see if opposing maxillary and dentary teeth interlock.

The mandibles of both *Huayangosaurus* (ZDM T7001) and *Stegosaurus* (USNM 4934 and CMNH 41681) possess an external mandibular fenestra (Berman and McIntosh 1986; Galton 1992; Sereno and Dong 1992). The mandible is long and slender in lateral view and bears a well-developed coronoid eminence (Fig. 2.6). The predentary is similar to that of *Lesothosaurus* (Thulborn 1970) and bears a long ventral process and short, triangular lateral processes (Gilmore 1914; Sereno and Dong 1992). Berman and McIntosh (1986) suggest that the dentary tooth row of *Stegosaurus* was directed medially, on the basis of their study of CMNH 41681. However, comparison of this mandible with the mandibles of USNM 4934 suggests that the medial inclination of the tooth row in CMNH 41681 is the result of postmortem distortion. The mandibles of *Huayangosaurus* (ZDM T7001) and "*Regnosaurus*" (BMNH 2422; Barrett and Upchurch 1995a) possess a large buccal emargination. The buccal emargination in *Stegosaurus* (CMNH 41681 and USNM 4934) and *Kentrosaurus* (HMN WJ60; Galton 1988) is reduced. The reduction of the buccal emargination in *Stegosaurus* and *Kentrosaurus* is correlated with the presence of the "dorsal lamina" (Sereno and Dong 1992). This lamina is produced by a dorsal expansion of the dentary, which arises lateral to the tooth row and which merges with the coronoid eminence posteriorly. The lamina obscures the posterior part of the dentary tooth row in lateral view. When the animal's mouth was closed, the lamina may have acted as a cheek, helping to retain food or guide food into the mouth. The jaw articulation of all stegosaurs is set slightly ventral to the tooth row. The glenoid fossa is a gently concave, cup-shaped structure. The anterior margin of the glenoid fossa is marked by a low ridge at the anterior margin of the articular (CMNH 41681). When the jaws are articulated, the shaft of the quadrate lies in close proximity to the posterior margin of the coronoid eminence. The posterolateral and posteromedial margins of the fossa are formed by a low ridge that originates on the surangular and extends to the articular. The anteroposterior length of the glenoid fossa is approximately equal to that of the quadrate cotylus. This morphology would have prevented lateral and translational movements of the mandibles, and the jaw action was strictly orthal.

Huayangosaurus is the only stegosaur known to retain premaxillary teeth (Sereno and Dong 1992). Seven premaxillary teeth are present. The anterior teeth are small, subconical, and slightly recurved. Posterior premaxillary teeth are larger, more symmetrical in labial view, and more labiolingually compressed. Wear is absent. Denticles are present on all of the premaxillary teeth (Sereno and Dong 1992).

The maxillary and dentary teeth of stegosaurs are similar to those of other primitive ornithischians. They are subtriangular in labial view and usually bear cingula (Fig. 2.7). A prominent primary ridge is present is some genera (e.g., *Paranthodon*: BMNH 47338 and BMNH R4992), but absent in others (e.g., *Stegosaurus*: USNM 4934 and YPM 1263). The total number of marginal denticles present is highly variable, from between 7 (*Kentrosaurus*) to 15 or more (*Stegosaurus*). The denticles are usually rounded at their tips, in contrast to those of other ornithischians (such as ankylosaurs), in which the denticles end in a sharp point. In comparison with the size of the skull, the teeth of stegosaurs are extremely small. Maximum mesiodistal width of the tooth crown rarely exceeds 8 mm (personal observation).

Tooth wear is usually absent or limited in extent (Galton 1992). Some *Stegosaurus* teeth (DMNH 2818, USNM 4934, USNM 419656, YPM 1263, and YPM 1498), one of the dentary teeth of the holotype of *Huayangosaurus* (IVPP V6728; Sereno and Dong 1992), and several *Paranthodon* maxillary teeth (BMNH 47338) display wear facets. YPM 1263 and YPM 1498 are isolated teeth that possess small planar wear facets situated on the apical tip of the tooth crown. These facets are very low angled with respect to the tooth crown and are blunt and almost horizontally inclined. It seems likely that these facets are the result of tooth–food wear (cf. Popowics and Fortelius 1997). Similar wear facets are present on some of the teeth of *Stegosaurus* (DNMH 2818). Several of the anterior maxillary teeth of USNM 4934 display slightly larger wear facets with a sharper leading edge and a larger exposure of dentine (Gilmore 1914), as do several *Paranthodon* maxillary teeth (BMNH 47338). The sharp leading edge of these facets suggests that they may have been formed by tooth–tooth contact (cf. Popowics and Fortelius 1997), although the small size of the facets may also indicate that they formed as a result of an extended period of tooth–food wear. The worn *Huayangosaurus* tooth (IVPP V6728) has a large, almost vertically inclined wear facet on the lingual surface of the tooth crown that appears to be the product of tooth–tooth wear. Wear is absent from most of the teeth of *Paranthodon* and *Huayangosaurus* and from the limited

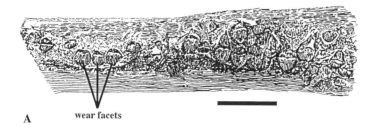

A   wear facets

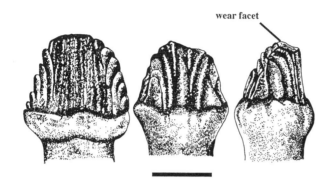

wear facet

B

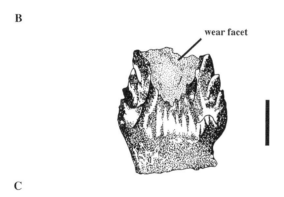

wear facet

C

Figure 2.7. Stegosaur teeth.
(A) Right maxillary tooth row of
Stegosaurus stenops *(USNM*
4934) in lingual view (redrawn
from Gilmore 1914). Anterior is
toward the left.
Scale bar = 20 mm.
(B) Isolated cheek teeth of
Stegosaurus sp. *in three views*
(redrawn from Galton 1992).
Scale bar = 4 mm.
(C) Dentary tooth of
Huayangosaurus taibaii *(IVPP*
V6728; redrawn from Barrett
1998b) in labial view.
Scale bar = 2 mm.

material attributed to *Kentrosaurus* (Galton 1988). No studies on stegosaur dental microwear have been attempted to date.

The mechanism or mechanisms that produced this tooth wear cannot be determined with confidence from the stegosaur dental material
that is currently available, though examination of this material allows
some tentative conclusions to be drawn. The random distribution of
wear along the maxillary tooth row of *Paranthodon* (BMNH 47338)
and the small size of the wear facets (where present) indicate that a
systematic occlusion was absent. The observed wear either formed as a
result of tooth–food contact or as a consequence of individual variation
in the direction of tooth eruption (leading to rare tooth–tooth wear).
Similar mechanisms may have produced the observed tooth wear in

Stegosaurus, but more information is needed on the distribution of wear along the tooth rows. The general absence of wear in *Huayangosaurus* (IVPP V6728 and ZDM T7001) suggests the absence of a precise occlusion; the presence of one tooth with a single large wear facet may be due to a unique variation in the direction of eruption of this tooth. Discovery of better preserved cranial material and more intensive study and preparation of existing material are necessary in order to gain a fuller understanding of stegosaur feeding mechanisms.

Ankylosaurs

The ankylosaur skull is dorsoventrally compressed in lateral view and extremely wide in dorsal view; the snout is wide and bluntly rounded (Fig. 2.8; Coombs 1978). Fusion of plates of dermal bone to the exterior surface of the skull closes the supratemporal, infratemporal, and antorbital fenestrae and renders the skull roof akinetic. In most ankylosaurids, the head of the quadrate fuses to the underside of the skull roof, though fusion does not occur in *Talarurus* or in juvenile

Figure 2.8. Ankylosaur cranial material. Skull of Saichania chulsanensis in (A) dorsal, (B) palatal, and (C) right lateral views (from Maryańska 1977). Scale bars = 200 mm. (D) Right mandible of Euoplocephalus tutus in lateral view (from Coombs and Maryańska 1992). Scale bar = 100 mm.

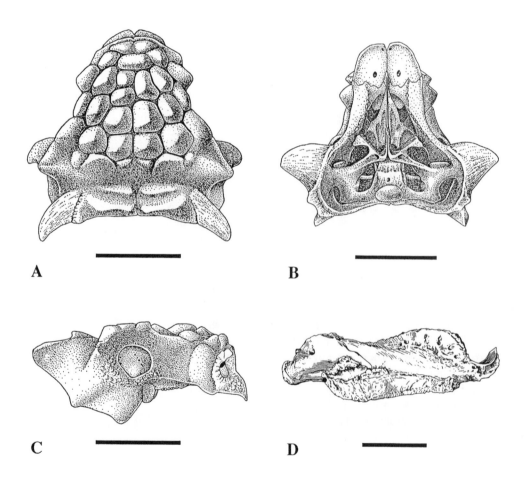

A B

C D

Pinacosaurus (Tumanova 1987); in nodosaurids and the ankylosaurids *Shamosaurus* and *Saichania*, the head of the quadrate fuses to the ventral tip of the paroccipital process (Tumanova 1987). The quadrate ramus of the pterygoid is fused to the pterygoid ramus of the quadrate (except in juvenile *Pinacosaurus*), and in nodosaurids and *Shamosaurus*, the pterygoids are fused to the braincase (Tumanova 1987). All of these features would have prevented streptostylic movements of the quadrate. There is a moderately developed buccal emargination lateral to the maxillary tooth row, and the maxillary tooth row is bowed medially in palatal view (Coombs 1978). The adductor chambers of ankylosaurids are small (Coombs 1971), and they were probably capable of only a weak bite.

In lateral view, the ankylosaur mandible is deep and possesses a low coronoid eminence (Fig. 2.8). The dentary tooth row forms an S-shaped curve in lateral view and is bowed medially in dorsal view. There is a large buccal emargination, and the jaw joint is slightly offset ventrally. In some ankylosaurs (e.g., *Sauropelta*: YPM 5340, YPM 5391, and YPM 5502; *Sarcolestes*: BMNH R2682; and *Panoplosaurus*: BMNH R16010 [cast of ROM 1215]), the articular fossa is rather short anteroposteriorly, shallow, cup-shaped, and bounded anteriorly by the coronoid eminence and posteriorly by a ridge extending across the articular. The distal end of the quadrate almost completely occupies the glenoid fossa when the skull and mandibles are articulated (Barrett 1998b; Tumanova 1987). The mandibular glenoid fossa in *Euoplocephalus* (AMNH 5403) is not delimited by marked anterior and posterior ridges; the articular surface is gently curved and shows slight anteroposterior expansion. The predentary is wide in dorsal view and bears a short posteroventrally directed median process (Coombs 1971).

Premaxillary teeth are known in the nodosaurids *Silvisaurus* (Eaton 1960), *Pawpawsaurus* (Lee 1996), and *Sauropelta* (Coombs and Maryańska 1992). The premaxillary teeth of *Silvisaurus* are subconical and terminate in a slightly recurved, pointed apex. They possess three or four denticles on the distal edge. Eaton (1960) suggested that the premaxillary teeth might have been useful in tearing leaves and stems from plants. With the exception of *Gargoyleosaurus* (Carpenter et al. 1998) and a new taxon from the Early Cretaceous of Utah (Carpenter 2001), premaxillary teeth are absent in all known ankylosaurids.

The maxillary and dentary teeth of ankylosaurs are similar to those of stegosaurs and other primitive ornithischians (Figs. 2.9, 2.10). Ankylosaur teeth can usually be distinguished from stegosaur teeth by the form of the denticles. Although complete ankylosaur skulls are reasonably common, few skulls possess in situ teeth. Skulls with a complete, erupted in situ tooth row are particularly rare, confounding deduction of the distribution of wear along the tooth row.

Wear facets are present on a number of isolated cheek teeth (e.g., AMNH 21763, AMNH 22599, BMNH R2940, BMNH R3682, BMNH R4285, BMNH R5257, USNM 5944, and USNM 8437). The wear facets are variable in form. Some teeth possess high-angled wear facets on the mesial and distal margins of the crown (e.g., AMNH 21763), whereas others display large high-angled planar wear facets that al-

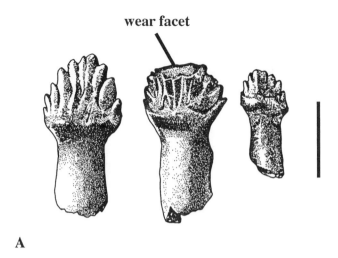

wear facet

A

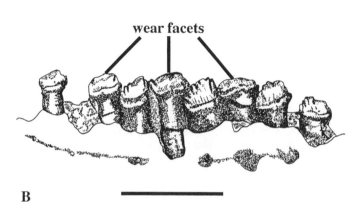

wear facets

B

Figure 2.9. (right) Ankylosaur teeth. (A) Teeth of Edmontonia rugosidens *(redrawn from Coombs and Maryańska 1992). Scale bar = 10 mm. (B) Left maxillary teeth of* Panoplosaurus mirus *(ROM 1215) in lingual view (redrawn from Carpenter 1990). Scale bar = 20 mm.*

Figure 2.10. (below) Stereopair of left maxillary tooth of Priodontognathus phillipsi *(SMC B53408) in lingual view. Scale bar = 8 mm.*

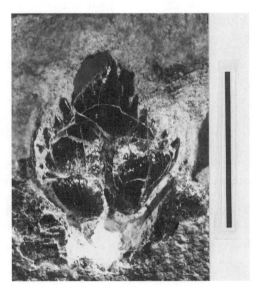

most cover the labial (or lingual) crown surface (e.g., AMNH 5381 and USNM 358113) and are rather similar to the large wear facets seen on some *Scelidosaurus* dentary teeth. The sharp edges of these wear facets suggest that they were formed by tooth–tooth (attritional) contact (cf. Popowics and Fortelius 1997). Some teeth possess small, steeply inclined planar wear facets covering the apical region of the labial (or lingual) crown surface (e.g., BMNH R2940, BMNH R3682, and BMNH R4285). These wear facets are probably the result of tooth–food (abrasional) contact (cf. Popowics and Fortelius 1997).

In situ tooth rows with wear facets are known from *Panoplosaurus* (Fig. 2.9B; ROM 1215; Carpenter 1990; Coombs 1990; Russell 1940) and *Euoplocephalus* (AMNH 5405; Rybczynski and Vickaryous 1998, 2001). In *Panoplosaurus* (ROM 1215), a number of the maxillary teeth display large, obliquely inclined planar wear facets (Carpenter 1990; Coombs 1990; Russell 1940); the apicalmost denticles have been completely worn away, giving the tips of the worn teeth a beveled appearance (Fig. 2.9B). The degree of wear reflects the age of the tooth; fully erupted teeth are the most heavily worn, with the beveled top of the tooth extending down toward the cingulum. More recently erupted crowns show a lesser degree of wear that is confined to the area adjacent to the apex. The extent of this wear is strongly suggestive of tooth–tooth contact. The lower dentition of this specimen is missing (Russell 1940), and consequently the distribution and extent of tooth wear in the upper and lower dentitions cannot be compared at present. Rybczynski and Vickaryous (1998, 2001) have reported wear facets on the maxillary dentition of *Euoplocephalus* (AMNH 5405) that appear to be continuous between adjacent teeth. In situ maxillary teeth of *Tarchia* also display small wear facets (Tumanova 1978, 1987), but the extent and distribution of the wear has not yet been described in detail. Several other Mongolian ankylosaur specimens have well-preserved tooth rows (*Pinacosaurus* and *Saichania*; see Maryańska 1977; Tumanova 1987), but wear has not been described in these taxa; reexamination of these specimens (which was not possible during the course of this study) should offer important additional information on feeding in ankylosaurs.

Only two studies have attempted to identify microwear on ankylosaur teeth. Coombs (1971) briefly described vertical wear striae on the teeth of various ankylosaurs, but in *Euoplocephalus* (AMNH 5405), these striae have been reinterpreted as preparation artefacts (Rybczynski and Vickaryous 2001). Vertically oriented wear striae are present on BMNH R2940 (personal observation). In their study of *Euoplocephalus,* Rybczynski and Vickaryous (1998, 2001) reported the presence of mesiodistally oriented wear striae.

In most ankylosaurs, the jaw action appears to have been orthal; translational and transverse movements of the mandible were prevented by the morphology of the glenoid fossa and the unusual curvature of the tooth rows. The presence of vertically oriented wear striae on some ankylosaur teeth supports this view (Coombs 1971; personal observation). Current information on tooth macrowear indicates the presence of tooth–tooth contact in some ankylosaurs but cannot be used to determine the direction of jaw movement in most cases because no specimen is known to include sufficiently well-preserved upper and lower tooth rows. Analysis of available data on tooth macrowear does suggest that an interlocking occlusion (as in *Lesothosaurus*; Thulborn 1971) was absent in most ankylosaurs because high-angled mesial and distal wear facets are extremely rare. The disposition of wear on the labial or lingual crown surfaces suggests a partial realignment of the tooth rows so that the teeth met in almost direct opposition (as suggested for *Scelidosaurus*). This hypothesis cannot be confirmed, however, until the discovery of better preserved material.

Tooth macrowear, the presence of mesiodistally oriented microstriae, and the anteroposterior expansion of the glenoid fossa relative to the length of quadrate condyle, suggest that *Euoplocephalus* may have used translational (propalinal) movements of the mandibles during the power stroke (Rybczynski and Vickaryous 1998, 2001). To date, this represents the only evidence for complex jaw mechanisms within the Thyreophora. This jaw action may be classified as a propalinal-grinding jaw mechanism (cf. Weishampel and Norman 1989).

The teeth of the ?Late Jurassic–?Early Cretaceous nodosaurid *Priodontognathus* (SMC B53408, Fig. 2.10; Galton 1980) are rather unusual in comparison with those of other ankylosaurs. The maxillary tooth crowns possess two subcrescentic cingula on their lingual surfaces. These cingula bear a large number of fine denticles. The cingula and the lingual surface of the tooth crown form the lingual and labial boundaries of a shallow basin; could this basin be analogous (or even homologous?) to the bowllike dentary wear facets of *Scelidosaurus*? If the teeth of most ankylosaurs (including *Priodontognathus*) met in direct opposition, the apices of the dentary teeth would fit into the basin formed by the cingula. Furthermore, if the dentary teeth of *Priodontognathus* possessed the reverse arrangement (with the cingula on the labial surfaces of the crowns), as might be expected, the maxillary teeth would also fit into a basin formed by the cingula. This would produce an efficient puncture-crushing mechanism that would result in the production of tooth–tooth wear on the labial and lingual surfaces of the opposing tooth crowns. This suggestion is based on limited material, however, and should be regarded as speculative until the discovery of more complete material allows the rejection or confirmation of this hypothesis.

Discussion

Feeding mechanisms within the Thyreophora are usually considered to be uniform (Weishampel and Norman 1989). However, consideration of tooth macrowear and microwear, the morphology of the jaw joint, and the shape of the snout, demonstrates that thyreophorans utilized a variety of feeding mechanisms.

Extensive high-angled planar wear facets on the teeth of *Scelidosaurus* and various ankylosaurs (*Euoplocephalus* and *Panoplosaurus*) indicate the possession of a precise and systematic occlusion in these animals. Isolated ankylosaur teeth display a variety of tooth macrowear that may reflect taxonomic differences in jaw mechanics or diet. *Scutellosaurus* and at least some stegosaurs (*Huayangosaurus* and *Paranthodon*) appear to have lacked precise occlusion, with tooth wear resulting from contact with food alone.

The possession of an anteroposteriorly elongate glenoid fossa and the presence of horizontally oriented microstriae on the teeth suggest that *Euoplocephalus* utilized a complex jaw action that included translational movements of the mandible (Rybczynski and Vickaryous 1998, 2001). Other ankylosaurs, stegosaurs, and *Scelidosaurus* had anteroposteriorly short glenoid fossae that would have prevented translation of the lower jaw. Jaw closure in these animals was strictly orthal.

In extant ungulates, the width of the snout (or muzzle) is correlated with dietary preferences (Janis and Ehrhardt 1988; Jarman 1974). Narrow-snouted forms (e.g., dik-diks) tend to be highly selective and only eat plant organs with high nutritive value (e.g., flowers, fruits, and new shoots). This suggests that stegosaurs and *Scelidosaurus* may have been selective feeders. However, a narrow snout is found in most archosaurs, many of which are exclusively carnivorous. Also, narrow-snouted antelope are usually small animals (1 or 2 m in length) with high mass-specific metabolic rates, which require high rates of food acquisition (Jarman 1974). Thyreophorans are, in general, much larger animals and would have had lower mass-specific food requirements. Therefore, their diet would not need to be as selective as that of small antelope. The diets of *Scelidosaurus* and stegosaurs may have been more selective than those of similarly sized ankylosaurs, however. Ankylosaurs, with their broad, rounded snouts, analogous to those of large grazing ungulates (such as wildebeest; Janis and Ehrhardt 1988), are likely to have been unselective feeders (Carpenter 1982).

Several soft tissue structures are often referred to in discussions of ornithischian feeding mechanisms (e.g., Crompton and Attridge 1986; Galton 1973, 1986; Norman and Weishampel 1991). All ornithischian dinosaurs (including all thyreophorans) are thought to have possessed a horny beak (rhamphotheca) that covered the tips of the upper and lower jaws, and many ornithischians have been reconstructed with muscular cheeks. Inferring the presence of soft tissue structures in extinct vertebrates is fraught with difficulties, however, and is possible with reasonable certainty only when "osteological correlates" of these soft tissue structures can be identified (Witmer 1995).

The presence of a rhamphotheca in thyreophorans, and in other ornithischian dinosaurs, is uncontroversial (e.g., Crompton and Attridge 1986; Galton 1973, 1986; Norman and Weishampel 1991; Sereno 1991). The anterior margins of both the upper and lower jaws of ornithischians are vascularized and have a roughened surface texture; these features are found on the jaws of extant vertebrates that are known to possess a rhamphotheca (turtles and birds). Fossil evidence provides direct support for the presence of a rhamphotheca in at least some ornithischians; beaks have been found preserved on several hadrosaur skulls (Morris 1970).

Galton (1973) suggested that the buccal emargination of ornithischians (including thyreophorans) was delimited laterally by muscular cheeks. This interpretation has since been followed by a large number of researchers (see Czerkas 1999 for a brief review). Galton's hypothesis was based largely on a functional interpretation of chewing in ornithischian dinosaurs; he reasoned that cheeks would be necessary in these animals in order to prevent food falling out of the mouth during mastication (Galton 1973). Czerkas (1999) has posited an alternative hypothesis for stegosaurs, according to which the margins of the buccal emargination did not support cheeks but provided attachment sites for a posteriorly elongated rhamphotheca. The latter hypothesis is based on similarities between the lower jaws of *Stegosaurus* and those of turtles. The turtle mandible possesses sharp ridges of bone that extend along the dorsomedial and dorsolateral edges of the dentary; these ridges support the rhamphotheca. Czerkas (1999) has suggested that a small, sharp ridge that extends anterior to the dentary tooth row of *Stegosaurus* and that the dorsal lamina are functionally analogous to the ridges on the dorsal surface of the turtle mandible. Czerkas (1999) therefore concluded that *Stegosaurus* might have possessed a turtlelike rhamphotheca.

The dorsomedial and dorsolateral ridges on the turtle mandible might be termed "osteological correlates" (cf. Witmer 1995) of a posteriorly elongated rhamphotheca, and the presence of similar structures in *Stegosaurus* are suggestive, though not necessarily indicative, of soft tissue relations in *Stegosaurus* that are similar to those observed in extant turtles. As stegosaurs do not appear to rely on extensive oral processing of food, it may be that cheeks were unnecessary in these animals. The size reduction of the maxillary and dentary teeth might also suggest that the teeth were relatively unimportant in food collection and processing—a turtlelike rhamphotheca may have been the main structure that served these functions. The presence of a posteriorly elongate rhamphotheca in other thyreophorans (and other ornithischians) is doubtful, however, because the mandibles of these taxa do not possess the osteological correlates of this structure; dorsal laminae are unique to derived stegosaurs (Galton 1992; Sereno and Dong 1992).

In the nodosaurids *Panoplosaurus* and *Edmontonia,* elliptical dermal ossifications ("cheek" plates; Lambe 1919) lie within the buccal emargination (Carpenter 1990; Lambe 1919). These ossifications can be regarded as osteological correlates of cheeks; they must have formed in a sheet of dermal tissue that extended from the maxillary ridge delimiting the dorsal boundary of the emargination to the ridge on the dentary that forms the ventral border of the emargination. This demonstrates that at least some ornithischians possessed sheets of dermal tissue in the region of the buccal emargination and that the reconstruction of ornithischian cheeks cannot be entirely dismissed. Other ankylosaurs, *Scelidosaurus,* primitive stegosaurs (such as *Huayangosaurus*), and other ornithischians might also have possessed muscular cheeks (Galton 1973), an interpretation that accords well with the puncture-crushing, transverse-grinding, propalinal-grinding and shearing jaw mechanisms found in these dinosaurs (Barrett 1998a, 1998b, this study; Crompton and Attridge 1986; Dodson 1993; Norman 1984a; Norman and Weishampel 1985, 1991; Ostrom 1961, 1964, 1966; Rybczynski and Vickaryous 2001; Weishampel 1984; Weishampel and Norman 1989). However, because the jaws of most ornithischian dinosaurs lack incontrovertible osteological correlates of cheeks (by comparison with the jaws of extant mammals; Papp and Witmer 1998), this interpretation will always be subject to some doubt.

The phylogenetic distribution of feeding mechanisms within the Thyreophora is difficult to assess on the basis of current evidence. Basal thyreophorans (*Scutellosaurus* and *Emausaurus*) appear to lack a precise occlusion, suggesting that the absence of tooth–tooth contact is primitive for the clade. If *Scelidosaurus* is the sister taxon of Stegosauria + Ankylosauria, as is currently held (Sereno 1986, 1991, 1997), then the puncture-crushing mechanism of *Scelidosaurus* and ankylosaurs must either have evolved independently or been secondarily lost in stegosaurs, a scenario that requires two evolutionary steps. If *Scelidosaurus* is the sister taxon of ankylosaurids (Carpenter et al. 1998), however, then the puncture-crushing jaw mechanism of *Scelidosaurus* might represent the primitive condition of the ankylosaur feeding system, a scenario that would only invoke one evolutionary step. These hypotheses will remain tentative, however, until more is known of the jaw mechanisms in stegosaurs and primitive ankylosaurs and until the phylogenetic position of *Scelidosaurus* has been resolved.

Conclusion

The feeding mechanisms of many thyreophorans remain enigmatic, largely due to the absence of specimens with well-preserved tooth rows. Analysis of the limited evidence that is available, however, indicates that at least some thyreophorans had more sophisticated puncture-crushing and propalinal-grinding jaw mechanisms than previously supposed. Occlusal contact was present in at least some thyreophorans (*Scelidosaurus* and various ankylosaurs), with the tooth rows meeting in almost direct opposition. In *Scutellosaurus* and stegosaurs, a precise occlusion appears to have been absent. Feeding mechanisms within Thyreophora were more diverse than is usually supposed (Galton 1986; Weishampel and Norman 1989).

Considerable potential remains for studies in the following areas: thyreophoran feeding mechanisms, including careful reexamination of existing material; Mongolian ankylosaurids; detailed comparison of feeding in ankylosaurid and nodosaurid ankylosaurs; and continuing study of dental microwear. These lines of inquiry, and the discovery of better preserved craniodental material (particularly of stegosaurs), should permit significant advances in our understanding of the feeding in these animals.

Acknowledgments. I thank Dr. K. Carpenter for inviting me to contribute to the thyreophoran symposium held at the Snowbird Society of Vertebrate Paleontology annual meeting. Dr. D. B. Norman provided useful personal communications and discussions. Useful reviews were received from Dr. K. Carpenter, N. Rybczynski, and an anonymous referee. Thanks to Professor P. M. Galton, Dr. K. Carpenter, Dr. A. C. Milner, M. Vickaryous, and N. Rybczynski for various personal communications. Dr. R. Harrison translated sections from Haubold (1990). For access to material in their care, I thank Dr. A. C. Milner and S. Chapman (Natural History Museum); R. Long and M. Dorling (Sedgwick Museum of Earth Sciences); Professor Dong Zhiming (Institute of Vertebrate Paleontology and Paleoanthropology); Dr. Ouyang Hui (Zigong Dinosaur Museum); Dr. M. Brett-Surman (National Museum of Natural History); Dr. M. Norrell (American Museum of Natural History); C. Chandler (Peabody Museum of Natural History); N. Wuerthele and A. Henrici (Carnegie Museum of Natural History); Dr. W.-D. Heinrich (Humboldt Museum fur Naturkunde); Dr. M. Manabe (National Science Museum); Roger Clark (Bristol City Museum and Art Gallery); and Professor P. M. Galton for access to National Museum of Natural History and Carnegie Museum of Natural History material on loan to him. Dr. K. Carpenter provided casts and peels of stegosaur cranial material. L. Cornish (Natural History Museum) provided casts of *Scelidosaurus* teeth for scanning electron microscopy work; P. Ratcliffe and I. Marshall (Cambridge) assisted with the scanning electron microscopy. Many thanks to Phil Crabb (Natural History Museum) for his excellent photographs of BMNH R1111. Dudley Simons (Cambridge) prepared the stereophotographs of *Priodontognathus*. Sharon Capon (Cambridge) redrafted many of the figures and significantly improved the artwork in this article. This work was supported by a Natural Environment Research Council studentship (GT4/93/128/G), The Dinosaur Society, Cambridge Philosophical Society, and Trinity College, Cambridge.

References Cited

Barrett, P. M. 1998a. Feeding mechanisms in thyreophoran dinosaurs [abstract]. *Journal of Vertebrate Paleontology* 18(3, Suppl.): 31A.
———. 1998b. Herbivory in the non-avian Dinosauria. Ph.D. diss. University of Cambridge.
Barrett, P. M., and P. Upchurch. 1994. Feeding mechanisms of *Diplodocus*. *Gaia* 10: 195–204.
———. 1995a. *Regnosaurus northamptoni*, a stegosaurian dinosaur from the Lower Cretaceous of southern England. *Geological Magazine* 132: 213–222.
———. 1995b. Sauropod feeding mechanisms: Their bearing on palaeoecology. In A.-L. Sun and Y.-Q. Wang (eds.), *Sixth Symposium on Mesozoic Terrestrial Ecosystems and Biota, Short Papers,* pp. 107–110. Beijing: China Ocean Press.
Berman, D. S, and J. S. McIntosh. 1986. Description of the lower jaw of *Stegosaurus* (Reptilia, Ornithischia). *Annals of the Carnegie Museum* 55: 29–40.
Calvo, J. O. 1994. Jaw mechanics in sauropod dinosaurs. *Gaia* 10: 183–194.
Carpenter, K. 1982. Skeletal and dermal armor reconstructions of *Euoplocephalus tutus* (Ornithischia: Ankylosauridae) from the Late Cretaceous Oldman Formation of Alberta. *Canadian Journal of Earth Sciences* 19: 689–697.
———. 1990. Ankylosaur systematics: Example using *Panoplosaurus* and

Edmontonia (Ankylosauria: Nodosauridae). In K. Carpenter and P. J. Currie (eds.), *Dinosaur Systematics: Approaches and Perspectives*, pp. 281–298. Cambridge: Cambridge University Press.

———. 2001. Phylogenetic analysis of the Ankylosauria. In K. Carpenter (ed.), *The Armored Dinosaurs*. Bloomington: Indiana University Press [this volume, Chapter 21].

Carpenter, K., C. Miles, and K. Cloward. 1998. Skull of a Jurassic ankylosaur (Dinosauria). *Nature* 393: 782–783.

Colbert, E. H. 1981. A primitive ornithischian dinosaur from the Kayenta Formation of Arizona. *Bulletin of the Museum of Northern Arizona* 53: 1–61.

Coombs, W. P., Jr. 1971. The Ankylosauria. Ph.D. diss. Columbia University, New York.

———. 1978. The families of the ornithischian dinosaur order Ankylosauria. *Palaeontology* 21: 143–170.

———. 1990. Teeth and taxonomy in ankylosaurs. In K. Carpenter and P. J. Currie (eds.), *Dinosaur Systematics: Approaches and Perspectives*, pp. 269–279. Cambridge: Cambridge University Press.

Coombs, W. P., Jr., and T. Maryańska. 1992. Ankylosauria. In D. B. Weishampel, P. Dodson, and H. Osmólska (eds.), *The Dinosauria*, pp. 456–483. Berkeley: University of California Press.

Coombs, W. P., Jr., D. B. Weishampel, and L. M. Witmer. 1992. Basal Thyreophora. In D. B. Weishampel, P. Dodson, and H. Osmólska (eds.), *The Dinosauria*, pp. 427–434. Berkeley: University of California Press.

Crompton, A. W., and J. Attridge. 1986. Masticatory apparatus of the larger herbivores during Late Triassic and Early Jurassic time. In K. Padian (ed.), *The Beginning of the Age of the Dinosaurs*, pp. 223–236. Cambridge: Cambridge University Press.

Czerkas, S. A. 1999. The beaked jaws of stegosaurs and their implications for other ornithischians. In D. D. Gillette (ed.), *Vertebrate Paleontology in Utah*, pp. 143–150. Miscellaneous Publication of the Utah Geological Survey 99-1.

Dodson, P. 1993. Comparative craniology of the Ceratopsia. *American Journal of Science* 293A: 200–234.

Dong, Z.-M., S. Zhou, and Y. Zhang. 1983. Dinosaur remains of Sichuan Basin [in Chinese with English abstract]. *Palaeontologica Sinica Series C* 23: 1–145.

Eaton, T. H., Jr. 1960. A new armored dinosaur from the Cretaceous of Kansas. *Paleontological Contributions from the University of Kansas* 25: 1–21.

Galton, P. M. 1973. The cheeks of ornithischian dinosaurs. *Lethaia* 6: 67–89.

———. 1980. *Priodontognathus phillipsii* (Seeley), an ankylosaurian dinosaur from the Upper Jurassic (or possibly Lower Cretaceous) of England. *Neues Jahrbuch für Geologie und Paläontologie, Monatschefte* 1980: 477–489.

———. 1986. Herbivorous adaptations of Late Triassic and Early Jurassic dinosaurs. In K. Padian (ed.), *The Beginning of the Age of the Dinosaurs*, pp. 203–221. Cambridge: Cambridge University Press.

———. 1988. Skull bones and endocranial casts of stegosaurian dinosaur *Kentrosaurus* Hennig, 1915, from Upper Jurassic of Tanzania, East Africa. *Geologica et Palaeontologica* 22: 123–143.

———. 1992. Stegosauria. In D. B. Weishampel, P. Dodson, and H. Os-mólska (eds.), *The Dinosauria,* pp. 435–455. Berkeley: University of California Press.

———. 1997. Stegosaurs. In J. O. Farlow and M. K. Brett-Surman (eds.), *The Complete Dinosaur,* pp. 291–306. Bloomington: Indiana University Press.

Gilmore, C. W. 1914. Osteology of the armored Dinosauria in the United States National Museum, with special reference to the genus *Stegosaurus. Bulletin of the Unites States National Museum* 89: 1–136.

Haubold, H. 1990. Ein neuer dinosaurier (Ornithischia, Thyreophora) aus dem Unteren Jura des nördlichen Mitteleuropa. *Revue de Paléobiologie* 9: 149–177.

Janis, C. M., and Ehrhardt, D. 1988. Correlation of relative muzzle width and relative incisor width and dietary preference in ungulates. *Zoological Journal of the Linnean Society of London* 92: 267–284.

Jarman, P. J. 1974. The social organisation of antelope in relation to their ecology. *Behaviour* 58: 215–267.

Lambe, L. M. 1919. Description of a new genus and species (*Panoplosaurus mirus*) of an armored dinosaur from the Belly River Beds of Alberta. *Transactions of the Royal Society of Canada,* Series 3, 13: 39–50.

Lee, Y.-N. 1996. A new nodosaurid ankylosaur (Dinosauria: Ornithischia) from the Paw Paw Formation (late Albian) of Texas. *Journal of Vertebrate Paleontology* 16: 232–245.

Maryańska, T. 1977. Ankylosauridae (Dinosauria) from Mongolia. *Palaeontologia Polonica* 37: 85–151.

Morris, W. J. 1970. Hadrosaurian dinosaur bills—Morphology and function. *Los Angeles County Museum Contributions in Science* 193: 1–14.

Norman, D. B. 1984a. On the cranial morphology and evolution of ornithopod dinosaurs. *Symposia of the Zoological Society of London* 52: 521–547.

———. 1984b. A systematic reappraisal of the reptile order Ornithischia. In W.-E. Reif and F. Westphal (eds.), *Third Symposium on Mesozoic Terrestrial Ecosystems, Short Papers,* pp. 157–162. Tübingen: Attempto Verlag.

———. 2001. *Scelidosaurus,* the earliest complete dinosaur. In K. Carpenter (ed.), *The Armored Dinosaurs.* Bloomington: Indiana University Press [this volume, Chapter 1].

Norman, D. B., and D. B. Weishampel. 1985. Ornithopod feeding mechanisms: Their bearing on the evolution of herbivory. *American Naturalist* 126: 151–164.

———. 1991. Feeding mechanisms in some small herbivorous dinosaurs: Processes and patterns. In J. M. V. Rayner and R. J. Wootton (eds.), *Biomechanics in Evolution,* pp. 161–181. Cambridge: Cambridge University Press.

Ostrom, J. H. 1961. Cranial morphology of the hadrosaurian dinosaurs of North America. *Bulletin of the American Museum of Natural History* 122: 39–186.

———. 1964. A functional analysis of jaw mechanics in the dinosaur *Triceratops. Postilla* 88: 1–35.

———. 1966. Functional morphology and evolution of the ceratopsian dinosaurs. *Evolution* 20: 290–308.

Owen, R. 1861. A monograph of the fossil Reptilia of the Liassic formations. I. *Scelidosaurus harrisonii. Monographs of the Palaeontographical Society* 13: 1–14.

———. 1863. A monograph of the fossil Reptilia of the Liassic formations. II. *Scelidosaurus harrisonii* (continued). *Monographs of the Palaeontographical Society* 14: 1–26.

Papp, M. J., and Witmer, L. M. 1998. Cheeks, beaks or freaks: A critical appraisal of buccal soft-tissue anatomy in ornithischian dinosaurs [abstract]. *Journal of Vertebrate Paleontology* 18(3, Suppl.): 69A.

Popowics, T. E., and M. Fortelius. 1997. On the cutting edge: Tooth blade sharpness in herbivorous and faunivorous mammals. *Annales Zoologici Fennici* 34: 73–88.

Russell, L. S. 1940. *Edmontonia rugosidens* (Gilmore), an armored dinosaur from the Belly River Series of Alberta. *University of Toronto Studies. Geological Series* 43: 3–28.

Rybczynski, N., and M. K. Vickaryous. 1998. Evidence of complex jaw movement in a Late Cretaceous ankylosaurid (Dinosauria, Thyreophora) [abstract]. *Journal of Vertebrate Paleontology* 18(3, Suppl.): 73A.

———. 2001. Evidence of complex jaw movement in the Late Cretaceous ankylosaurid *Euoplocephalus tutus* (Dinosauria: Thyreophora). In K. Carpenter (ed.), *The Armored Dinosaurs*. Bloomington: Indiana University Press [this volume, Chapter 14].

Sereno, P. C. 1986. Phylogeny of the bird-hipped dinosaurs (order Ornithischia). *National Geographic Research* 2: 234–256.

———. 1991. *Lesothosaurus,* "fabrosaurids," and the early evolution of Ornithischia. *Journal of Vertebrate Paleontology* 11: 168–197.

———. 1997. The origin and evolution of dinosaurs. *Annual Review of Earth and Planetary Science* 25: 435–489.

———. 1999. The evolution of dinosaurs. *Science* 284: 2137–2147.

Sereno, P. C., and Z.-M. Dong. 1992. The skull of the basal stegosaur *Huayangosaurus taibaii* and a cladistic diagnosis of Stegosauria. *Journal of Vertebrate Paleontology* 12: 318–343.

Steel, R. 1969. Ornithischia. *Handbuch der Paläoherpetologie* 15: 1–83.

Thulborn, R. A. 1970. The skull of *Fabrosaurus australis*, a Triassic ornithischian dinosaur. *Palaeontology* 13: 414–432.

———. 1971. Tooth wear and jaw action in the Triassic ornithischian dinosaur *Fabrosaurus. Journal of Zoology* 164: 165–179.

———. 1977. Relationships of the Lower Jurassic dinosaur *Scelidosaurus harrisonii. Journal of Paleontology* 51: 725–739.

Tumanova, T. A. 1978. New data on the ankylosaur *Tarchia gigantea. Paleontological Journal* 11: 480–486.

———. 1987. The armored dinosaurs of Mongolia [in Russian]. *Transactions of the Joint Soviet–Mongolian Palaeontological Expedition* 32: 1–77.

Upchurch, P., and P. M. Barrett. 2001. The evolution of sauropod feeding. In H.-D. Sues (ed.), *The Evolution of Terrestrial Herbivory*, pp. 79–122. Cambridge: Cambridge University Press.

Weishampel, D. B. 1984. Evolution of jaw mechanisms in ornithopod dinosaurs. *Advances in Anatomy, Embryology and Cell Biology* 87: 1–110.

Weishampel, D. B., and D. B. Norman. 1989. Vertebrate herbivory in the Mesozoic; jaws, plants and evolutionary metrics. *Special Papers of the Geological Society of America* 238: 87–100.

Wing, S. L., and H.-D. Sues. 1992. Mesozoic and Early Cenozoic terrestrial ecosystems. In A. K. Behrensmeyer, J. D. Damuth, W. A. DiMichele, R. Potts, H.-D. Sues, and S. L. Wing (eds.), *Terrestrial Ecosystems through Time,* pp. 327–416. Chicago: University of Chicago Press.

Witmer, L. M. 1995. The extant phylogenetic bracket and the importance of reconstructing soft tissues in fossils. In J. J. Thomason (ed.), *Functional Morphology in Vertebrate Paleontology,* pp. 19–33. Cambridge: Cambridge University Press.

Zhou, S. 1984. *The Middle Jurassic Dinosaurian Fauna from Dashanpu, Zigong, Sichuan.* Volume 2, *Stegosaurs* [in Chinese with English summary]. Beijing: Sichuan Scientific and Technological Publishing House.

Part II
Stegosauria

3. New Primitive Stegosaur from the Morrison Formation, Wyoming

KENNETH CARPENTER,
CLIFFORD A. MILES, AND KAREN CLOWARD

Abstract

A new genus of stegosaurid, from the Upper Jurassic Morrison Formation, is more primitive than *Stegosaurus stenops* but not as primitive as *Huayangosaurus*. The specimen consists of a nearly complete skull and much of the skeleton, minus the forearms and hind legs. It is characterized by a short, broad skull; domed frontoparietal; low neural arches; and low, oval plates. A preliminary phyletic analysis of the Stegosauria places this new genus closest to *Dacentrurus* from the Upper Jurassic of Europe. The occurrence of this new stegosaurus 5 m above the base of the Morrison Formation makes it one of the oldest known stegosaurs in North America.

Introduction

The armor-plated stegosaurs were first named by O. C. Marsh in 1877 for elements collected from the Morrison Formation near Morrison, Colorado. Since then, several genera of stegosaurs have been found in the Jurassic and Cretaceous of North America, Europe, Africa, and Asia (reviewed by Galton 1990b). One genus, *Stegosaurus*, and three species, *S. armatus* (*sensu stricto*), *S. ungulatus,* and *S. stenops,* are recognized by Carpenter and Galton (2001) from the Upper Jurassic of North America. The obscure *Hypsirophus discurus,* named by Cope (1878), differs enough from *Stegosaurus* to warrant recognition

(Carpenter 1998). Yet another specimen of stegosaurid has recently been found that differs from all known genera (Carpenter and Miles 1997), indicating that the diversity of stegosaurids was greater in North America than previously realized.

The new specimen was semiarticulated when found (Fig. 3.1), and the limbs were missing. Considering how complete the specimen is, the limbs were probably present at the time of burial and were lost by recent erosion. Remarkably, the skull was disarticulated, allowing a description of the individual bones. Only a preliminary description of the skull and skeleton is presented below, which is made on the basis of a representative sample of elements; we are preparing a more detailed account.

The specimen was found in a fine-grained sandstone 5 m above the contact between the Morrison Formation and the underlying Windy Hill Member of the Sundance Formation (Weege, personal communication). The Morrison Formation is about 53 m in the area, just south of Buffalo, Wyoming. This occurrence of a stegosaur is the oldest in North America. Not surprisingly, it is the most primitive stegosaur in North America as well.

Institutional Abbreviations. DMNH: Denver Museum of Natural History, Denver, Colorado. HMNH: Hayashibara Museum of Natural History, Okayama, Japan.

Systematic Paleontology
Order Stegosauria
Family Stegosauridae
Hesperosaurus n. g.

Holotype. HMNH 001, consisting of a nearly complete, disarticulated skull, 1 hyoid, 13 cervicals, 13 dorsals, 3 sacrals, 44 caudals, most of the cervical and dorsal ribs, most of the chevrons, 10 dermal plates, both ilia, both ischia and both pubes, partial left scapula, and ossified tendons. A cast of the holotype is also housed at the Denver Museum of Natural History (DMNH 29431) and at the North American Museum of Ancient Life (Orem, Utah).

Etymology. Hesper, Greek for "western," in reference to its discovery in the western United States, and *saurus,* reptile.

Holotype Locality. Five meters above base (Weege, personal communication), Morrison Formation, S. B. Smith Ranch, Johnson County, Wyoming (exact locality information on file at Hayashibara Museum of Natural History, Western Paleontological Laboratories, and the Denver Museum of Natural History). Discovered by Patrick McSherry, June 1985.

Diagnosis. As for the species.

Hesperosaurus mjosi n. sp.

Etymology. For Ronald G. Mjos ("mūs"), who was responsible for collecting and preparing the specimen and for mounting a cast of the holotype skeleton.

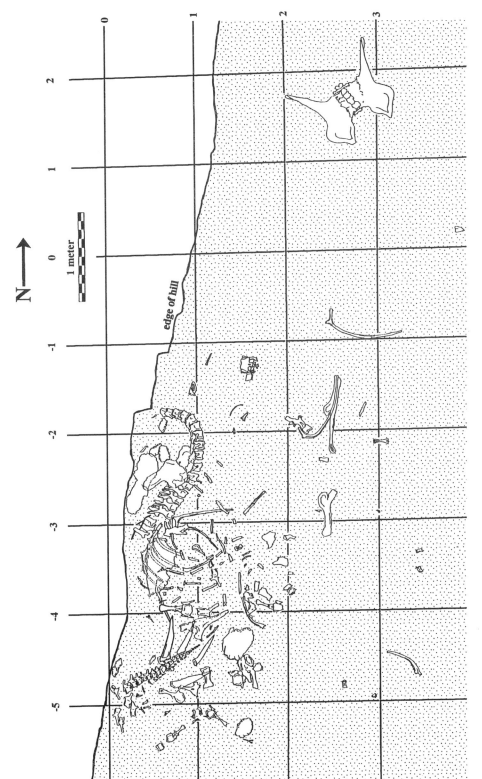

N→

1 meter

edge of hill

Figure 3.1. Quarry map showing the distribution of Hesperosaurus mjosi.

57

Diagnosis. Cranial—Skull intermediate between *Stegosaurus sten-ops* and *Huayangosaurus*; medium-size antorbital fenestra, intermediate in size between *H. taibaii* and *S. stenops;* lateral maxillary shelf broad, narrow in *S. stenops* and *H. taibaii;* frontoparietal domed, flat in all other stegosaurs; basisphenoid short, long in *S. stenops;* mandible below coronoid process proportionally deeper to length than in *S. stenops* and *H. taibaii;* alveolar border of mandible not visible in lateral profile as in *S. stenops,* visible in *H. taibaii;* foramen magnum wider than occipital condyle, narrower in *S. ungulatus, S. stenops,* and *H. taibaii;* teeth proportionally larger relative to skull size than *S. stenops.* Postcranial—Vertebral formula c-13, d-13, s-3, cd-45(?), versus *S. stenops,* c-10, d-17(?), s-3, cd-47; *H. taibaii:* c-8, d-17–18, s-3, cd-42; *Dacentrurus:* c-12, d-?, s-4, cd-?; *Kentrosaurus:* c-8(?), d-17(?), s-4, cd-43; axis with tall, posteriorly inclined neural spine as in *S. stenops,* low in *H. taibaii;* posterior cervical neural canal taller than wide, neural arch of middorsals low as in *H. taibaii,* not tall as in *S. ungulatus* and *S. stenops;* cervical ribs expanded blades distally, not expanded in *S. stenops* and *H. taibaii;* caudal neural spines not bifurcated as in *S. stenops;* distal end of ribs not expanded as in *H. taibaii;* ilia strongly divergent anteriorly as in *H. taibaii,* straight in *S. stenops.* Distal end of pubis expanded as in *H. taibaii,* nonexpanding in *S. stenops.* Cervical plates low, oval, versus tall, triangular in *S. ungulatus, S. stenops,* and *H. taibaii;* caudal spikes similar to *S. stenops* but with more rugose base.

Description

Characters found in both *Stegosaurus stenops* and *S. ungulatus* (if known) are simply presented as occurring in *Stegosaurus.* *S. armatus* is fragmentary, and a diagnosis is not yet possible (Carpenter and Galton 2001). Data for *Huayangosaurus* is from Sereno and Dong (1992) and Zhou (1984); *Dacentrurus* from Galton (1985, 1991) and Carpenter (unpublished notes); *Kentrosaurus* from Galton (1982, 1988); *Lexovisaurus* from Galton (1985, 1990a), Galton et al. (1980), and Carpenter (unpublished notes); *Wuerhosaurus* from Dong (1973, 1990); *Chialingosaurus* from Dong et al. (1983) and Dong (1990); *Chunkingosaurus* from Dong et al. (1983) and Dong (1990); and *Tuojiangosaurus* from Dong et al. (1983), Dong (1990), and Carpenter (unpublished notes). Measurements of selected bones are given in Table 3.1.

Cranial. The skull is mostly disarticulated (Figs. 3.2, 3.3), which is surprising considering how well ossified the postcrania are (a detailed description of the individual skull bones is in preparation). The skull has been reassembled from a cast of the bones. The result shows that the skull is intermediate between *Stegosaurus stenops* and *Huayangosaurus* (Fig. 3.4). Not surprisingly, the antorbital fenestra is also intermediate in size between that of *S. stenops* and *Huayangosaurus.*

The maxilla is more emarginated above the tooth row than in *S. stenops* and *Huayangosaurus* but considerably less than in ankylosaurs. There are 20 alveoli in the maxilla, versus 28 in *Huayangosaurus* and 24 in *Stegosaurus.* The lachrymal process above the antorbital fenestra

TABLE 3.1.
Measurements (maximum) of select elements (most complete or figured)
in millimeters of *Hesperosaurus mjosi* (HMNH 001).

	Length	Height	Width
Left mandible	223	81 (at coronoid)	—
Left maxilla	170	72	—
Left nasal	176	—	41
Right lachrymal	85	—	—
Left prefrontal	85	53	—
Supraorbital	50	—	27
Left medial supraorbital	57	—	28
Left posterior supraorbital	46	—	—
Right anterior supraorbital	55	27	—
Left postorbital	74	70	—
Right jugal	66	60	—
Left quadrate	85	112	—
Cranium (middle)	125	52	100
		(Above foramen magnum)	(paroccipital– paroccipital)
Atlas centrum	—	32	56
Atlas neural arch	58	54	—
Cervical 6 (centrum)	96	60	52
Cervical 12	80	66	78
Dorsal 3	83	77	75
Dorsal 6	78c	88e	84e
Dorsal 11	82	~91	92
Sacrum	31.5	—	—
Caudal 2	48	91	127
Caudal 22	63	80	57
Left pubis			
Prepubic process	466	—	—
Postpubic process	515	—	—
Right ilium	872	—	—
Ilium–ilium anterior	—	—	990
Ilium–ilium posterior	—	—	510
Plate (largest)	777	—	—
Anterior spike	445	—	—
Posterior spike	400	—	—

Note. e = estimate.

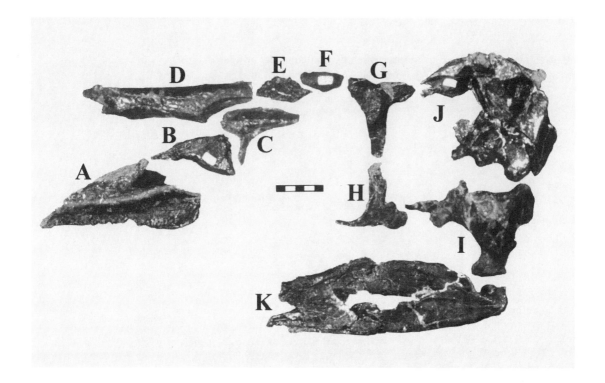

Figure 3.2. Skull bones of the left side of Hesperosaurus mjosi *(HMNH 001) in lateral view. (A) Maxilla. (B) Lachrymal. (C) Prefrontal. (D) Nasal. (E) Anterior supraorbital. (F) Middle supraorbital. (G) Postorbital. (H) Jugal. (I) Quadrate. (J) Frontal, parietal, and braincase. (K) Dentary, angular, surangular, and articular. Scale in centimeters.*

is longer than in *S. stenops* or *Huayangosaurus*; it extends back to lock into a groove on the lachrymal. The antorbital fenestra is partially walled medially by a sheet of the maxilla. The lachrymal is a triangular wedge with a long suture along its dorsoposterior edge with the prefrontal. A small portion along the anteroventral edge forms part of the posterior border of the antorbital fenestra, as it does in all stegosaurs. The prefrontal has a nearly flat, elongate, horizontal process as in *S. stenops* that forms the dorsal wall of the orbit and a thin, descending lachrymal process. A rugosity where the horizontal and lachrymal processes meet is the suture for the anterior supraorbital, and a rugosity on the posterior part of the horizontal process is for the middle supraorbital. The nasal is long, slender, and slightly curved along its long axis. It has a relatively long suture for the maxilla and an arched, underlapping suture for the prefrontal. The postfrontal has three processes: the descending jugal process, posteriorly projecting squamosal process, and a medially projecting frontal process. These processes are thicker than in *S. stenops*. There is a rugose area on the lateral surface for the posterior supraorbital and a smaller area for the medial supraorbital. The anterior or maxillary process of the jugal is incomplete on both sides, so it is not known how delicate it was. On the basis of the gap between the posteriormost part of the maxilla and the posteroventral margin of the lachrymal, the maxillary process of the jugal must have been robust, more like that of *Huayangosaurus* than slender as in *S. stenops*. The quadrate is proportionally short, with a short, deep quadratojugal coossified to it. The quadratojugal is angled anterodorsally rather than anteriorly as in *S. stenops*.

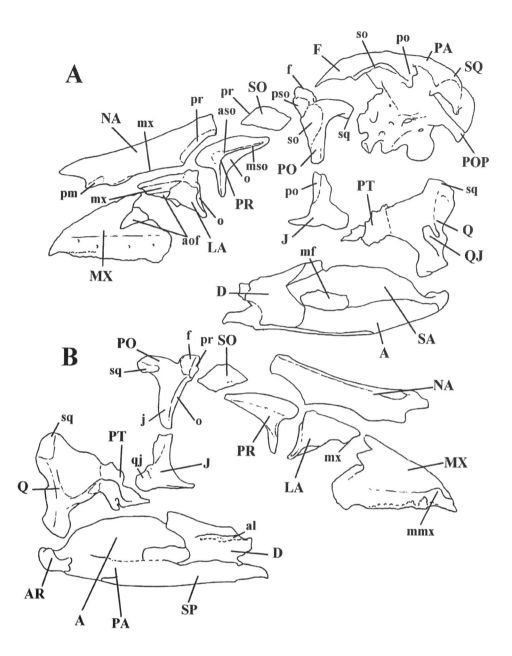

Figure 3.3. Outline sketches of skull bones from the left side of Hesperosaurus mjosi *(HMNH 001) in lateral (A) and medial (B) views. Abbreviations for bones: A = angular; AR = articular; D = dentary; F = frontal; J = jugal; LA = lachrymal; MX = maxilla; NA = nasal; PA = parietal; PO = postorbital; POP = paroccipital process; PR = prefrontal; PT = pterygoid; Q = quadrate; QJ = quadratojugal; SA = surangular; SO = supraorbital. Abbreviations for sutures and contacts: al = alveoli; aof = antorbital fenestra; aso = anterior supraorbital; f = frontal; j = jugal; mf = maxillary fenestra; mmx = median maxillary process; mx = maxillary; mso = medial supraorbital; o = orbit; pm = premaxillary; po = postorbital; pr = preorbital; pso = posterior supraoccipital; qj = quadratojugal; so = supraorbital; sq = squamosal.*

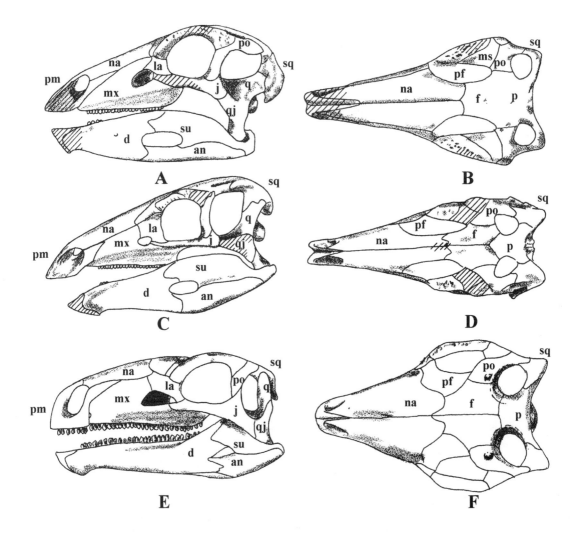

Figure 3.4. Reconstructed skull
of Hesperosaurus mjosi in left
lateral (A) and dorsal (B) views
compared with that of Stegosaurus stenops (DMNH 33365,
cast) in lateral (C) and dorsal
(D) views and Huayangosaurus
taibaii in lateral (E) and dorsal
(F) views. Missing areas are cross
hatched. D was modified in part
from Sereno and Dong (1992).

The posterior part of the cranium includes the braincase, frontals,
parietals, and squamosals coossified together. The frontoparietal is
sloped in lateral view, rather than flat as in *Stegosaurus* and *Kentrosaurus*; it is also slightly sloped in *Huayangosaurus*. The basisphenoid
is short as in *S. ungulatus* (Galton 2001), rather than long as in *S.
stenops*. The foramen magnum is wider than the occipital condyle,
whereas it is narrower in *Stegosaurus* and *Huayangosaurus*.

The posterior half of the mandibles is preserved, with partial coossification of the elements. The mandible is proportionally deeper and
the external mandibular fenestra larger than in *S. stenops* and *Huayangosaurus*. The alveolar border of the mandible is not visible in lateral
profile as it is in *S. stenops* (in part) and in *Huayangosaurus*, where it is
completely visible.

A few teeth were recovered (Fig. 3.5), and these are proportionally
larger relative to skull size than *S. stenops*. Otherwise, the teeth are
similar to *Stegosaurus*.

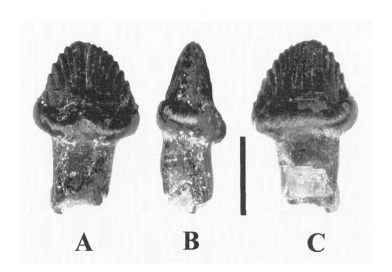

Figure 3.5. (left) *Tooth of* Hesperosaurus mjosi *(HMNH 001) in (A) ?buccal, (B) ?anterior, and (C) ?lingual views. Scale bar = 2 mm.*

Figure 3.6. (below) *Vertebrae of* Hesperosaurus mjosi *(HMNH 001) in left lateral views except where noted. (A) Atlas with rib articulated with the axis with rib. (B) Cervical 6. (C) Cervical 6 in ventral view showing the broad ventral ridge. (D) Cervical 12. (E) Dorsal 3. (F) Dorsal 6. (G) Dorsal 6 in anterior view. (H) Dorsal 11. (I) Dorsal 11 in anterior view. (J) Caudal 2. (K) Caudal 2 in anterior view. (L) Caudal 22.*

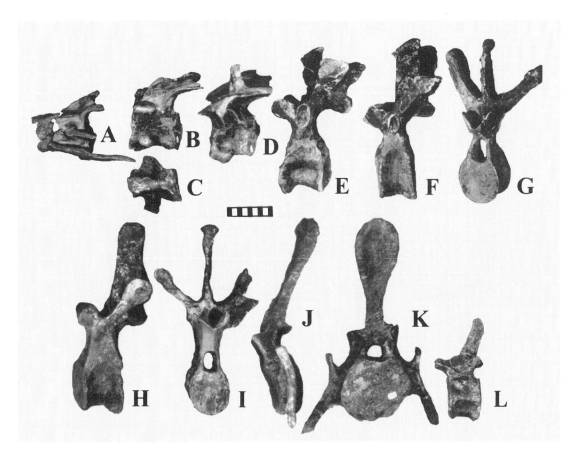

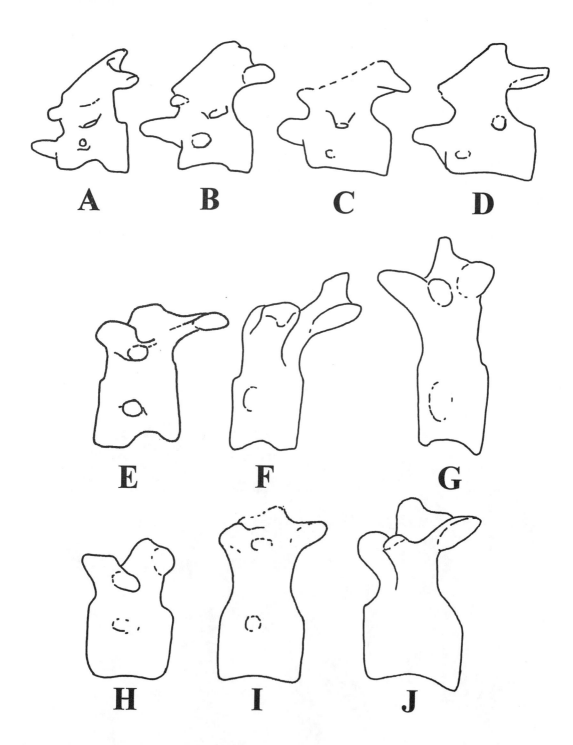

Figure 3.7. Comparison of the axes. (A) Hesperosaurus. *(B)* Stegosaurus. *(C)* Lexovisaurus. *(D)* Kentrosaurus. *Comparison of the midcervicals of (E)* Hesperosaurus mjosi, *(F)* Stegosaurus, *(G)* Huayangosaurus, *(H)* Dacentrurus, *(I)* Lexovisaurus, *and (J)* Tuojangosaurus *(to scale).*

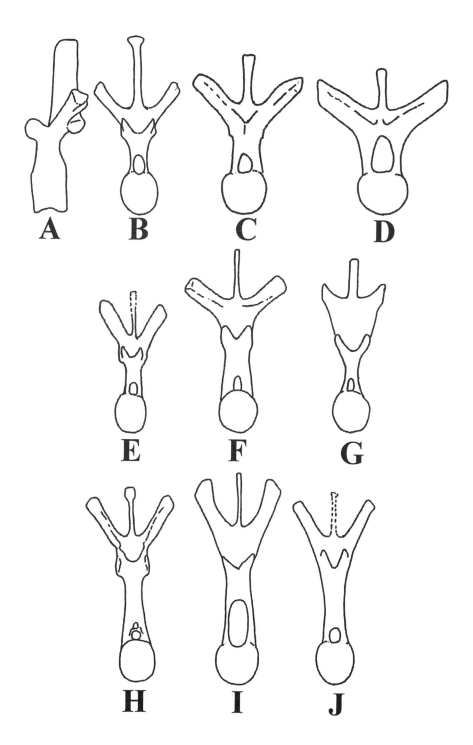

Figure 3.8. Comparison of posterior vertebrae. Low neural arch style: (A, B) Hesperosaurus in lateral and anterior views, (C) Dacentrurus, and (D) Huayangosaurus. Medium neural arch style: (E) Chialingosaurus, (F) Chunkingosaurus, and (G) Tuojiangosaurus. Tall neural arch style: (H) Stegosaurus, (I) Kentrosaurus, and (J) Wuerhosaurus (to scale).

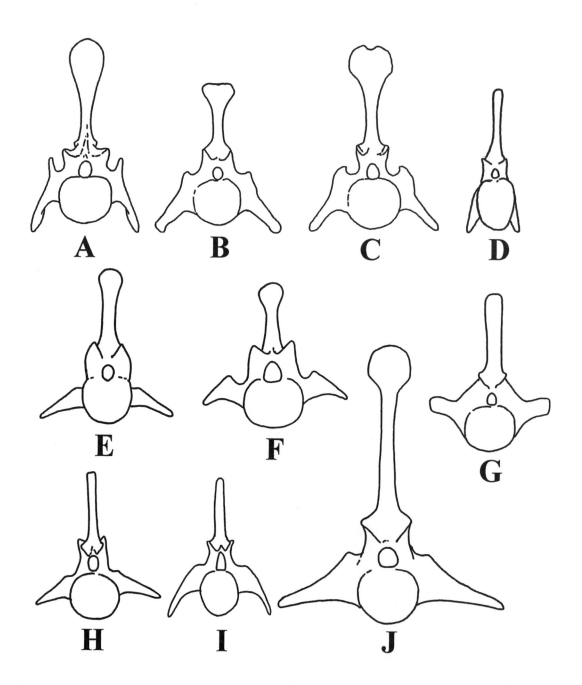

Figure 3.9. Comparison of anterior caudals in anterior view. (A) Hesperosaurus. *(B)* Stegosaurus stenops.
(C) S. ungulatus. *(D)* Lexovisaurus. *(E)* Kentrosaurus. *(F)* Dacentrurus. *(G)* Chialingosaurus. *(H)* Huayangosaurus.
(I) Chunkingosaurus. *(J)* Wuerhosaurus.

66 • Kenneth Carpenter, Clifford A. Miles, and Karen Cloward

Vertebrae. The vertebral column is almost complete (one caudal may have eroded away). There are 13 cervicals (including one transitional form based on the shape of the associated ribs; see below), 11 dorsals, a synsacrum of 2 dorsosacrals and 3 sacrals, and 45 caudals (Figs. 3.6–3.9).

Atlas and Axis. The neural arches are not coossified to the atlantal intercentrum, nor is the odontoid coossified to the axis (Figs. 3.6A, 3.7A). The atlantal intercentrum is almost twice as massive as that associated with a *S. stenops,* which has a similar-size occipital condyle (DMNH 33365). The neural spine of the axis is similar to that of *S. stenops,* except that it is twice as thick laterally. The anterior part of the centrum is low, so that the odontoid is located lower than in *S. stenops, Lexovisaurus,* or *Kentrosaurus.*

The cervicals have proportionally longer centra relative to both height and width than in other stegosaurs (Figs. 3.6B, C, 3.7E). Cervical 6 is representative of the midcervicals. The neural arch is vertical as in most stegosaurs, rather than sloped posteriorly as in *S. ungulatus*; it is slightly sloped anteriorly in *Huayangosaurus.* The neural spine is very low, being a slight swelling as in *Dacentrurus* and possibly *Lexovisaurus*; it is more prominent in other stegosaurs. The parapophyses on the side of the centra are situated as they are in *Dacentrurus* and *Lexovisaurus,* being more posteriorly positioned than in other stegosaurs. Furthermore, the parapophyses are situated on distinct pedestals. In addition, there is a wide, flattened ridge on the ventral surface of the centra in *Hesperosaurus* that connects the anterior and posterior articular faces (Fig. 3.6C); a similar ridge occurs in *Huayangosaurus taibaii.* In *Stegosaurus,* the ridge is narrow and rounded laterally.

Cervical 12 is representative of the posterior cervicals (Fig. 3.6D). The diapophysis is almost horizontal, whereas it is angled dorsally in *Stegosaurus.* As with the midcervicals, the parapophysis is situated almost midway between the anterior and posterior articular faces; it is closer to the anterior face in most other stegosaurs, such as *Stegosaurus.* Ventrally, the ridge connecting the anterior and posterior articular faces is broad, being about three quarters the length of the ridge.

A complete set of cervical ribs was found articulated to the cervical vertebrae. They are remarkable in that they do not project ventrally, as is often portrayed, but posterodorsally; only the 12th cervical rib curves ventrally. These ribs will be illustrated elsewhere.

The first dorsal is identified by the diapophysis located on the neural arch; it also has the first ventrally projecting long rib. The third dorsal represents the anterior dorsals (Fig. 3.6E). The centrum is longer than it is tall, as in *Chialingosaurus,* whereas the dimensions are equal, or nearly so, in all other stegosaurs, including *Huayangosaurus.* The prezygapophyses are situated just above the neural canal as in *Huayangosaurus,* whereas they are situated higher in most other stegosaurs. The prezygapophyses are long, projecting anteriorly well beyond the centrum.

The middle dorsals are represented by dorsal 6 (Fig. 3.6F, G). The centra height and length are about the same. The prezygapophyses are situated just above the neural canal as in a Bathonian stegosaur from

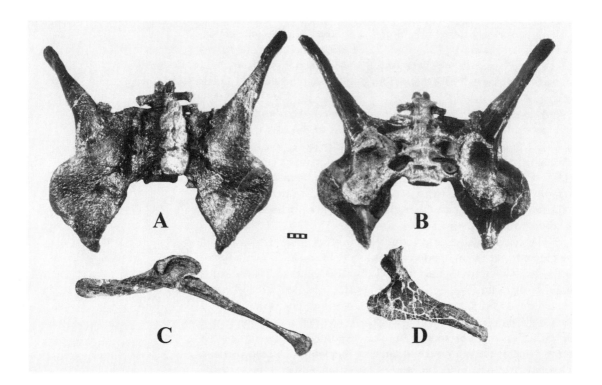

Figure 3.10. Pelvic elements of Hesperosaurus mjosi (HMNH 001). (A) Pelvis in dorsal view. (B) Pelvis in ventral view. (C) Left pubis in lateral view. (D) Left ischium in lateral view. Scale in centimeters.

England (?*Lexivosaurus vetustus* of Galton and Powell 1983), whereas in all other stegosaurs, they are situated higher. The posterior dorsals, exemplified by dorsal 11 (Figs. 3.6H, I, 3.8A, B), have a low pedicel onto which the prezygapophyses are situated, as in *Dacentrurus*; these are taller than in *Huayangosaurus*. Other stegosaurs can be divided into those with medium-height pedicles (*Chialingosaurus, Chunkingosaurus,* and *Tuojiangosaurus*) and those with tall pedicles or arches (*Stegosaurus, Kentrosaurus,* and *Wuerhosaurus*). The neural spine is very tall, unlike that of all other stegosaurs, possibly to compensate for the low neural pedicel.

Remarkably, ossified tendons were found associated with dorsal neural spines, but none with the caudals. Therefore, loss of these tendons occurred within the Stegosauria and cannot be used to define the taxon (e.g., Sereno 1986, 1997). A few ribs were coossified to their respective posterior dorsal vertebrae, indicating an old individual. Many of the dorsal and sacral neural spines are rugose because of partial ossification of the interspinous and supraspinous ligaments.

The three sacral vertebrae are coossified with the ilia (Figs. 3.10A, B, 3.11A). The sacrals are also fused together and to two dorsosacrals to form a short synsacrum of five vertebrae; *Kentrosaurus* has six vertebrae, including three dorsosacrals (Fig. 3.11E). The neural spines are coossified together with their tops expanded laterally. The last neural spine is as wide as the first caudal neural spine. The sacral ribs are coossified to both the sacrals and the ilia. The dorsal surfaces of the sacral ribs are expanded, forming a solid surface on the pelvis, a condition seen in other stegosaurids but not in *Huayangosaurus*.

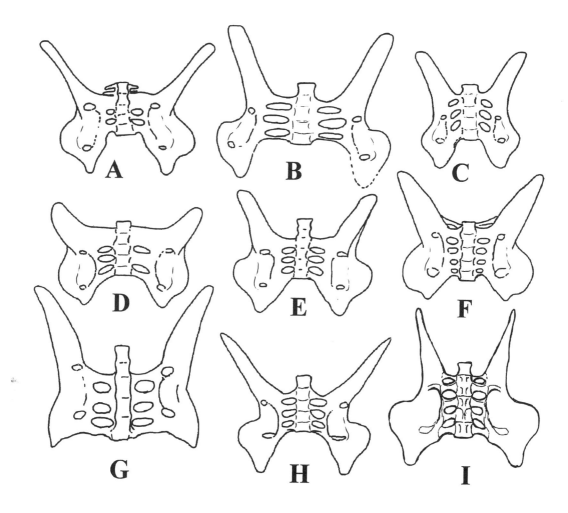

The caudal series is complete, or nearly so, including the very last caudal. It may be possible that one anterior caudal (possibly the third) is missing because of the abrupt transition between caudal 2 and the next following it, as preserved. In the cast of the mounted skeleton (Hayashibara Museum of Natural History), one caudal was inserted to make the transition more gradual. The anterior caudals are the most diagnostic segment of the tail, with characteristic neural spines and caudal ribs. The neural spine of *Hesperosaurus* is proportionally very tall in a manner similar to that in *Stegosaurus, Kentrosaurus,* and *Huayangosaurus,* but not to the extreme seen in *Wuerhosaurus* (Figs. 3.6J, K, 3.9). The top of the neural spine is uniquely teardrop-shaped in anterior view (Figs. 6K, 9A). The condition in other stegosaurs is either a nonexpanded top (*Lexovisaurus, Chialingosaurus, Huayangosaurus,* and *Chunkingosaurus*) or expanded, but not teardrop-shaped (*Stegosaurus, Kentrosaurus, Dacentrurus,* and *Wuerhosaurus*). A neural spine continues to caudal 33, after which it no longer projects dorsally of the postzygapophyses. The caudal rib or transverse process is cleatlike, with both a dorsally and ventrally projecting process to caudal 15.

Figure 3.11. Comparison of pelves in ventral view.
(A) Hesperosaurus.
(B) Lexovisaurus.
(C) Chungkingosaurus.
(D) Dacentrurus.
(E) Kentrosaurus.
(F) Monkonosaurus.
(G) Huayangosaurus.
(H) Wuerhosaurus.
(I) Stegosaurus.

From that point, the transverse process gradually decreases in size posteriorly until it is only a nubbin on caudal 20.

Appendicular Skeleton. Except for the scapula, nothing of the forelimb is known. The partial left scapula does not differ substantially from those of other stegosaurs. The pelvis is complete, with both ilia coossified to the sacrals (Fig. 3.10A, B). The ilia are widely divergent, unlike the condition in *Stegosaurus* and *Huayangosaurus* (Fig. 3.11). The short postacetabular process is more prominent than in *Huayangosaurus, Dacentrurus, Monkonosaurus,* and *Kentrosaurus.*

The pubis is long and slender (Figs. 3.10C, 3.12A), with a long, thin prepubic process. The postpubic process tapers distally with an expanded end as in *Wuerhosaurus* and *Huayangosaurus;* it is more parallel-sided without a distal expansion in *Stegosaurus.* The iliac process is short and low, whereas it is taller in most other stegosaurs, with the exception of *Huayangosaurus.* The ischium is closest to *Stegosaurus* in overall shape, except that the distal end is not tapered (Fig. 3.10D, 3.12).

Armor. The armor consists of 10 plates, two of which are incomplete, and four spikes (Fig. 3.13). The position of the plates can be estimated on the basis of their position relative to the vertebral column as found (Fig. 3.1). The cervical plates are considerably longer than they are tall (Fig. 3.13A, B), in marked contrast to *Stegosaurus,* where they are much taller than long. This orientation for *Hesperosaurus* is certain because the bases are greatly thickened. This thickening is asymmetrical, forming left-handed and right-handed plates, which correspond to their being on the left or right side of the midline. The two dorsal plates are incomplete (see, e.g., Fig. 3.13C), and the rest (four?) are believed to have eroded away. At least four caudal plates are preserved. The anterior plate is a parallelogram that is slightly taller than long, in marked contrast to this plate in *Stegosaurus,* in which it is triangular and taller than it is long. The more distal plates are longer than tall in a manner similar to those of the cervicals. If it is assumed that the anteriormost caudal plate somewhat reflects the shape of the dorsal plates, as it does in *Stegosaurus,* then it is probable that the dorsal plates were more parallelogram to rectangular (i.e., longer than tall) than triangular. The spikes are similar to those of *Stegosaurus* in that the anterior pair are larger than the posterior pair (Fig. 3.13F, G). In addition, the posterior spikes have a more angled base than do the anterior spikes so that they project more posteriorly.

Discussion

Recognition of *Hesperosaurus mjosi* increases the diversity of stegosaurs in the Morrison Formation. The other species are *Stegosaurus armatus, S. stenops, S. ungulatus, S. longispinus,* and *Hypsirophus discurus. Stegosaurus* is limited to Dinosaur Zone 2 and 3 of Turner and Peterson (1999), whereas *Hypsirophus* is limited to Dinosaur Zone 4. The stratigraphic position of *Hesperosaurus* is believed to be in Dinosaur Zone 1.

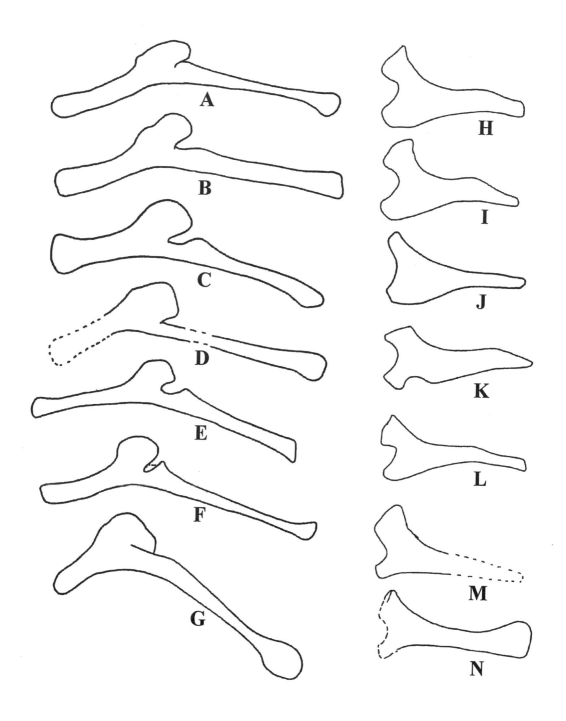

Figure 3.12. Comparison of the pubis in left lateral view: (A) Hesperosaurus, (B) Stegosaurus, (C) Dacentrurus, (D) Lexovisaurus, (E) Kentrosaurus, (F) Wuerhosaurus, *and* (G) Huayangosaurus. *Ischium in left lateral view:* (H) Hesperosaurus, (I) Stegosaurus, (J) Dacentrurus, (K) Lexovisaurus, (L) Kentrosaurus, (M) Chungkingosaurus, *and* (N) Wuerhosaurus *(to approximately the same length).*

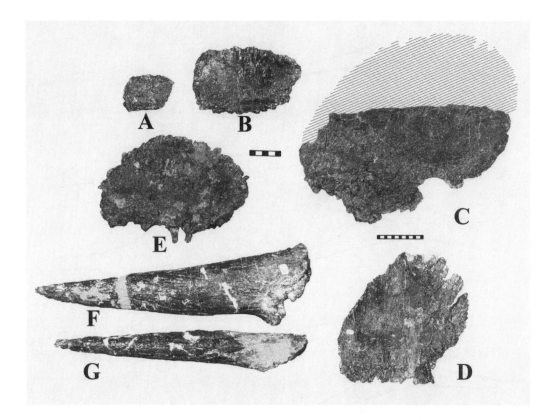

Figure 3.13. Representative armor of Hesperosaurus mjosi *(HMNH 001). (A, B) Cervical armor in left lateral views. (C) Partial dorsal plate. (D) ?Anterior caudal plate. (E) ?Midcaudal plate. (F) Left anterior spike in dorsal view. (G) Left posterior spike in dorsal view. Scales in centimeters. A, B, E, F, and G to scale; C and D to scale.*

Preliminary phyletic analysis suggests that *Hesperosaurus* is a primitive stegosaurid (Fig. 3.14; Carpenter and Miles 1997). The data matrix (Table 3.2) is incomplete and not exhaustive because many taxa are lacking crucial parts. Some taxa, such as *Monkonosaurus*, are not included because too many data are missing. The matrix was analyzed with PAUP version 4 (Swofford 2000), using the heuristic search command with the options nchuck = 1 and chuckscore = 1 (higher values up to 10 were tried and gave the same results). These options narrow the search to trees having the best score. The resultant tree had a length of 18, a consistency index of 0.67, a homoplasy index of 0.33, and a retention index of 0.70. The same data matrix analyzed with Hennig86 (Farris 1988), using both branch swapping and implicit enumeration, gave a similar tree, with a length of 18, a consistency index of 0.66, and a retention index of 0.70. Admittedly, these indexes are low and may be improved with the more rigorous analysis of the Stegosauria currently under preparation by Galton and Sereno (Galton, personal communication).

Conclusion

Next to sauropods, stegosaurs are the most abundant vertebrate in the Morrison Formation. Their diversity is high, with six species now known. The discovery of *Hesperosaurus* led us to reexamine most stegosaur specimens, and we conclude that the generic and specific

TABLE 3.2.
Data matrix for phyletic analysis (Fig. 3.14).

	1	2	3	4	5	6	7	8	9	10	11	12
Huayangosaurus taibai	0	0	0	0	0	0	0	0	0	0	0	0
Chialingosaurus kuani	?	?	?	?	?	1	?	0	?	1	1	?
Chungkingosaurus jiangbeiensis	?	1	0	?	?	1	1	0	?	1	0	0
Dacentrurus armatus	?	?	?	?	1	0	1	1	0	1	1	?
Hesperosaurus mjosi	0	?	0	0	1	0	1	1	0	1	?	1
Kentrosaurus aethiopicus	0	?	?	1	0	1	0	1	1	0	1	0
Lexovisaurus durobrivensis	0	?	?	1	0	1	0	1	1	0	1	0
Stegosaurus stenops	0	1	0	1	0	1	0	1	1	1	1	1
Stegosaurus ungulatus	1	?	?	1	0	1	0	1	1	1	0	1
Tuojiangosaurus multispinus	1	1	1	1	0	1	1	1	?	1	1	0
Wuerhosaurus homheni	?	?	?	?	?	1	?	1	0	1	?	1

1. Skull width > height.
2. Premaxilla edentulous.
3. Antorbital fenestra absent.
4. Skull roof flat in lateral view.
5. Midcervical spine nearly absent.
6. Height of prezygapophysis of mid- and posterior dorsals ≥ height of centrum.
7. Mid- and posterior dorsal neural spine taller than diapophysis.
8. Apex of proximal caudal neural spine laterally expanded.
9. Distal end of pubis not expanded.
10. Preacetabular blade of ilium vertical.
11. Anterior trochanter distinct (spike or ridge).
12. Cervical armor longer than tall.

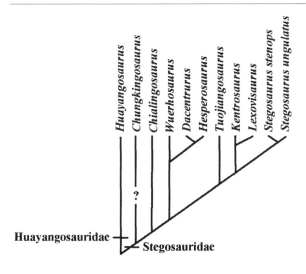

Figure 3.14. Preliminary phyletic tree of stegosaurs (see Table 3.2 for data matrix). The basal stegosaur Huayangosaurus *(see Sereno and Dong 1992) is used as the outgroup, allowing the Stegosauridae to be defined as all stegosaurs closer to* Stegosaurus *than to* Huayangosaurus. The position of Chungkingosaurus relative to Huayangosaurus is questionable because of missing data. Additional data may eventually move Chungkingosaurus up the phylogenetic tree.*

diversity is even greater. A detailed description of this material is in preparation.

Acknowledgments. Patrick McSherry discovered the holotype specimen, and we thank him for bringing the specimen to the attention of Tom Lindgren and Jeffrie Parker. The specimen was prepared by Western Paleontological Laboratory Inc. The Hayashibara Museum of Natural History gave us permission to describe the specimen. Access to specimens and archival information was made possible by Mary Ann Turner, Christine Chandler, and Jacques Gauthier (Yale Peabody Museum of Natural History); Mark Norell and Charlotte Holton (American Museum of Natural History); Robert Purdy (National Museum of Natural History); and Sandra Chandler and Angela Milner (Museum of Natural History, London). Peter Galton and Paul Sereno reviewed previous versions of the article, and their comments are greatly appreciated.

References Cited

Carpenter, K. 1998. Vertebrate biostratigraphy of the Morrison Formation near Cañon City, Colorado. In K. Carpenter, D. Chure, and J. I. Kirkland (eds.), *The Morrison Formation: An Interdisciplinary Study. Modern Geology* 23: 407–426.

Carpenter, K., and P. M. Galton. 2001. Othniel Charles Marsh and the myth of the eight-spiked *Stegosaurus*. In K. Carpenter (ed.), *The Armored Dinosaurs*. Bloomington: Indiana University Press [this volume, Chapter 4].

Carpenter, K., and C. A. Miles. 1997. New primitive stegosaur (Ornithischia) from the Upper Jurassic [abstract]. *Journal of Vertebrate Paleontology* 17(3, Suppl.): 35A.

Cope, E. D. 1878. A new opisthocoelous dinosaur. *American Naturalist* 12: 406.

Dong, Z. 1973. Dinosaurs from Wuerho. *Memoirs of the Institute of Vertebrate Paleontology and Paleoanthropology, Academia Sinica* 11: 45–52.

———. 1990. Stegosaurs of Asia. In K. Carpenter and P. Currie (eds.), *Dinosaur Systematics: Approaches and Perspectives,* pp. 255–268. New York: Cambridge University Press.

Dong, Z., S. Zhou, and Y. Zhang. 1983. The dinosaurian remains from Sichuan Basin, China. *Palaeontologica Sinica* 162: 1–145.

Farris, J. S. 1988. Hennig86 [software]. Diana Lipscomb, George Washington University [private distribution].

Galton, P. M. 1982. The postcranial anatomy of stegosaurian dinosaur *Kentrosaurus* from the Upper Jurassic of Tanzania, East Africa. *Geologica et Palaeontologica* 15: 139–160.

———. 1985. British plated dinosaurs (Ornithischia, Stegosauridae). *Journal of Vertebrate Paleontology* 5: 211–254.

———. 1988. Skull bones and endocranial casts of stegosaurian dinosaur *Kentrosaurus* Hennig, 1915, from the Upper Jurassic of Tanzania, East Africa. *Geologica et Palaeontologica* 22: 123–143.

———. 1990a. A partial skeleton of the stegosaurian dinosaur *Lexovisaurus* from the uppermost Lower Callovian (Middle Jurassic) of Normandy, France. *Geologica et Palaeontologica* 24: 185–199.

———. 1990b. Stegosauria. In D. Weishampel, P. Dodson, and H. Osmólska (eds.), *The Dinosauria*, pp. 435–455. Berkeley: University of California Press.

———. 1991. Postcranial remains of stegosaurian dinosaur *Dacentrurus* from Upper Jurassic of France and Portugal. *Geologica et Palaeontologica* 25: 299–327.

———. 2001. Endocranial casts of the plated dinosaur *Stegosaurus* (Upper Jurassic, Western USA): A complete undistorted cast and the original specimens of Othniel Charles Marsh. In K. Carpenter (ed.), *The Armored Dinosaurs*. Bloomington: Indiana University Press [this volume, Chapter 5].

Galton, P. M., R. Brun, and M. Rioult. 1980. Skeleton of the stegosaurian dinosaur *Lexivosaurus* from the lower part of Middle Callovian (Middle Jurassic) of Argences (Calvados), Normandy. *Bulletin Société Géologique Normandie et Amis Muséum du Havre* 67: 39–34.

Galton, P. M., and H. P. Powell. 1983. Stegosaurian dinosaurs from the Bathonian (Middle Jurassic) of England, the earliest record of the family Stegosauridae. *Geobios* 16: 219–229.

Marsh, O. C. 1877. A new order of extinct Reptilia (Stegosauria) from the Jurassic of the Rocky Mountains. *American Journal of Science,* Series 3, 14: 513–514.

Sereno, P. 1986. Phylogeny of the bird-hipped dinosaurs (order Ornithischia). *National Geographic Research* 2: 234–256.

———. 1997. The origin and evolution of dinosaurs. *Annual Review of Earth and Planetary Sciences* 25: 435–489.

Sereno, P., and Z. Dong. 1992. The skull of the basal stegosaur *Huayangosaurus* and a cladistic diagnosis of Stegosauria. *Journal of Vertebrate Paleontology* 12: 318–343.

Swofford, D. L. 2000. PAUP: Phylogenetic analysis using parsimony [software]. Version 4. Sunderland, Mass.: Sinauer Associates.

Turner, C. E., and F. Peterson. 1999. Biostratigraphy of dinosaurs in the Upper Jurassic Morrison Formation of the Western Interior, U.S.A. In D. Gillette (ed.), *Vertebrate Paleontology in Utah*. Utah Geological Survey Miscellaneous Publication 99-1: 77–114.

Zhou, S. 1984. The Middle Jurassic dinosaurian fauna from Dashanpu, Zigong, Sichuan. In: *Stegosaurs*, Vol. 2, pp. 1–52. Chengdu: Sichuan Scientific and Technical Publishing House.

4. Othniel Charles Marsh and the Myth of the Eight-Spiked *Stegosaurus*

Kenneth Carpenter and
Peter M. Galton

Abstract

Archival records are used to recreate Marsh's thinking in making the first skeletal reconstruction of *Stegosaurus*. This reconstruction shows a large, elephantine animal with a paired row of plates along its back, a single row on its tail, and four pairs of spikes on the end of its tail. This eight-spiked tail has long been considered a diagnostic feature of *Stegosaurus ungulatus,* but it is the result of an erroneous assumption on the part of Marsh and Richard Swann Lull, among others. The conclusion is based on the archival letters of Marsh's collectors, Arthur Lakes, William Reed, and Marshal Felch, as well as some of Marsh's own letters to his collectors. This correspondence is supplemented with quarry maps made by his collectors as required by Marsh. The conclusion from all the archival material is that Marsh's 1891 reconstruction is a composite of several individuals from different quarries and that no more than two pairs of spikes were ever found associated with distal caudals.

Introduction

The armor-plated *Stegosaurus* figures prominently in the history of dinosaur paleontology because of its bizarre appearance: large body; hind legs taller than the forelegs; tall, erect plates along the back and tail; and spikes at the end of the tail. *Stegosaurus* was named in 1877 by Othniel Charles Marsh, and most of the major features of our current view of *Stegosaurus* were already in place by 1891, when Marsh published the first skeletal reconstruction (Fig. 4.1). Subsequent ideas about the appearance of *Stegosaurus* are minor and mostly concern the arrangement of the plates along the back. These changing views were first summarized by Gilmore (1914) and more recently by Czerkas (1987; see also Carpenter 1998; Colbert 1962; Czerkas and Czerkas 1990).

Marsh meant for his skeletal reconstruction to give only a sense of what *Stegosaurus* was like because it was a composite based on several specimens: "The result . . . is believed to represent faithfully the main features of this remarkable reptile, as far as the skeleton and principal parts of the dermal armor are concerned," and "The peculiar group of extinct reptiles named by the writer, the Stegosauria, of which a typical example is represented in the present restoration" (Marsh 1891: 179, 181). Despite the composite nature of the skeletal reconstruction, Marsh referred to it as that of *Stegosaurus ungulatus*, noting "the four pairs of massive spines characteristic of the present species" (Marsh 1891: 180). When the skeleton of *S. ungulatus* was mounted at the Peabody Museum of Natural History at Yale University, New Haven, Connecticut, in 1910, Richard Swann Lull wrote, "The type of *Stegosaurus ungulatus* . . . shows four pairs [of spines] and no further evidence of the duplication of bone, so it is evident that they all belong to one individual" (Lull 1910b: 373). Lull also wrote that the mount "is a combination of two supplementary individuals . . . and contain but little duplicate bone" (Lull 1910b: 363). In actuality, several specimens were used based on a "map" of the skeleton made by Lull (Peabody Museum of Natural History collection) (Fig. 4.2, Table 4.1). Lull used many of the same bones Marsh had used for his skeletal reconstruction (see Table 4.1). The exceptions are the femur and tibia in Marsh's reconstruction that are based on YPM 1858 (see Ostrom and McIntosh 1966: pls. 47, 49), whereas Lull used YPM 1853 for the mount (see Ostrom and McIntosh 1966: pls. 46, 48). The mounted skeleton also differs in some respects from Marsh's reconstruction, notably in the elevated tail and the arrangement of the armor along the back. The skeleton was initially mounted with paired plates (Lull 1910a, 1910b), but when the mount was moved to its present location in 1924, it was remounted with alternating plates (Fig. 4.3; Lull, unpublished manuscript; Narendra 1990).

Marsh's reconstruction and the Peabody mount have both been used to justify the presence of eight terminal tail spikes in *Stegosaurus ungulatus*. This number of spikes has been used (as part or as the whole) to define this species by Marsh (1887, 1891), Lull (1910a, 1910b), Gilmore (1914), and more recently by Galton (1990, as *S. armatus*). In contrast, *S. stenops* is defined, in part, as having only two pairs of terminal tail spikes (Galton 1990; Gilmore 1914; Marsh 1887). But how accurate is the statement by Lull that the holotype of *S. ungulatus* had four pairs of spikes? Actually, Marsh (1880: 258) had written earlier, "Nine different spines of this character were recovered with this same skeleton [i.e., holotype of *S. ungulatus*]." This account by Marsh is supported by a letter of Arthur Lakes, who collected the specimen, and thus calls into question the eight-spiked species of *Stegosaurus*,

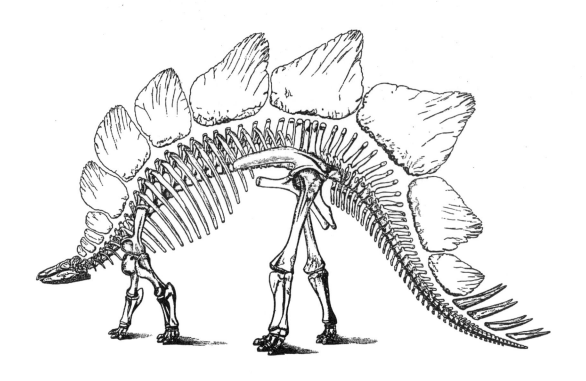

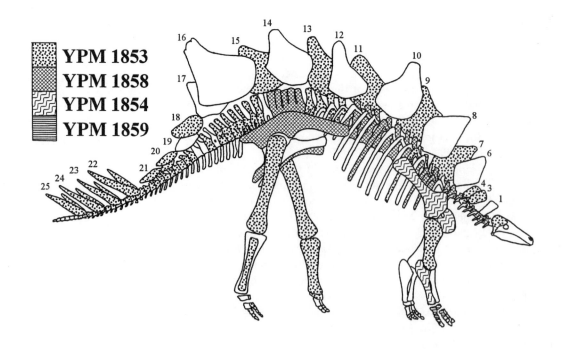

YPM 1853
YPM 1858
YPM 1854
YPM 1859

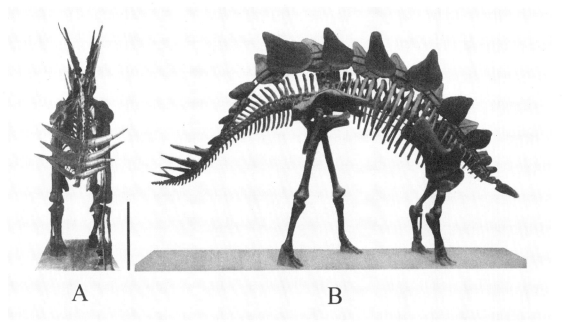

A   B

"the number which inclusive of others sent makes 9 or 10 in all" (Lakes, February 17, 1880).

As will be shown below, Marsh's ideas about *Stegosaurus* changed considerably over the years leading up to his 1891 skeletal reconstruction. These ideas changed as more and more specimens were discovered. Because of the importance of these specimens and the quarries where they were recovered, a short review is appropriate.

Institutional Abbreviations. AMNH: American Museum of Natural History, New York. USNM: National Museum of Natural History (formerly United States National Museum), Washington, D.C. YPM: Peabody Museum of Natural History, Yale University, New Haven, Connecticut.

Specimens and Their Quarries

We are fortunate that maps are available for most of the *Stegosaurus* specimens collected for Marsh. Marsh was adamant that all of his collectors write to him weekly informing him of their accomplishments, as can be inferred from Marshal Felch's response to a letter of Marsh: "I must apologize for not writing oftener—but as you did not mention about Weekly Reports—did not think it would be required" (Felch, May 25, 1883). Marsh also required drawings of the quarries: "Every quarry should be numbered as at Como, 1, 2, 3, etc. Sketches should be made of all bones in quarry before removal, so as to keep each skeleton by itself, if possible" (Marsh, letter to Frederick Brown, July 29, 1883). A historical survey of the excavations in Wyoming from 1877 to 1889, based on correspondence between Marsh and his collectors, is given by Ostrom and McIntosh (1966), and annotated selections from the field

Figure 4.1. (opposite page top) Skeletal reconstruction of Stegosaurus ungulatus *by Marsh (1891) showing four pairs of terminal caudal spikes.*

Figure 4.2. (opposite page bottom) Diagram showing the bones used in the mounted skeleton of Stegosaurus ungulatus *at the Peabody Museum of Natural History. Based on a color-coded map by Lull at the Peabody Museum of Natural History, probably made in 1924 when the skeleton was remounted because the plates are shown in an alternating arrangement; figures for bones, including numbered dermal plates and spikes, are given in Table 4.1. Length of mounted skeleton 19 feet, 5 inches (about 6 m) (Lull 1910b).*

Figure 4.3. (above) Skeleton of Stegosaurus ungulatus *as mounted at the Peabody Museum of Natural History. (A) Posterior view, modified from Lull (1910b). (B) Right lateral view of 1925 mount, modified from Lull (1929); note the alternating large plates, four pairs of terminal tail spikes, and the five low, elongate spines just anterior to the spikes.*

TABLE 4.1.

Bones used in the Peabody Museum of Natural History at Yale University (YPM) mounted skeleton of *Stegosaurus ungulatus* by Lull (1910b) (see Fig. 4.2).

Stegosaurus ungulatus (YPM 1853, holotype)

Posterior part of skull roof, braincase, quadrates (Galton 2001: fig. 5.4; replaced with cast of skull of USNM 4934 in the late 1980s)

Posterior dorsals (Ostrom and McIntosh 1966: pls. 15, 16)

Caudals (Ostrom and McIntosh 1966: pls. 25–27, 29)

Ribs

Right humerus (Ostrom and McIntosh 1966: pl. 33)

(Ischium *not* used, Ostrom and McIntosh 1966: pl. 43)

Both femora (Ostrom and McIntosh 1966: pl. 46)

Left tibia and fibula (Ostrom and McIntosh 1966: pl. 48)

Right fibula

Pes phalanges

Plates (see Fig. 4.2 for numbers): 3, 4, 6 (Ostrom and McIntosh 1966: ?pl. 62), 8, 10, 12,14 (Ostrom and McIntosh 1966: pl. 63), 17 (Ostrom and McIntosh 1966: pl. 60), 19 (Ostrom and McIntosh 1966: pl. 59, figs. 1–3), 20 (Ostrom and McIntosh 1966: pl. 59, figs. 4–6), 21 (Ostrom and McIntosh 1966: pl. 59, figs. 7, 8)

Spikes—paired (see Fig. 4.2 for numbers)—22 (Ostrom and McIntosh 1966: pl. 55), 23, 24 (Ostrom and McIntosh 1966: pl. 56, figs. 1–3), 25 (Ostrom and McIntosh 1966: pl. 56, figs. 4–6)

Stegosaurus duplex (YPM 1858—holotype)

Cervicals (Ostrom and McIntosh 1966: pl. 8)

Dorsals (Ostrom and McIntosh 1966: pls. 17, 18)

Dorsal ribs (Ostrom and McIntosh 1966: pl. 20)

Sacrum (Ostrom and McIntosh 1966: pls. 21, 45)

Distal caudals (Ostrom and McIntosh 1966: pl. 28)

Ilia (Ostrom and McIntosh 1966: pls. 42, 45)

Left pubis and ischium (Ostrom and McIntosh 1966: pl. 45)

(Femur, tibia and fibula *not* used, Ostrom and McIntosh 1966: pls. 47, 49)

Stegosaurus sulcatus (YPM 1854)

Right scapula and coracoid (cast) (Ostrom and McIntosh 1966: pl. 32)

Right ulna (Ostrom and McIntosh 1966: pl. 35)

Stegosaurus sulcatus (YPM 1859)

Right carpus (Ostrom and McIntosh 1966: pl. 36)

Right metacarpals and phalanges (Ostrom and McIntosh 1966: pls. 37–41)

Stegosaurus sp. (YPM)

Right radius, left carpus, metacarpals, phalanges

journals for 1877–1880 of Arthur Lakes are given by Kohl and McIntosh (1997).

Stegosaurus armatus *at Morrison*

Stegosaurus was named by Marsh in 1877 for bones recovered north of the town of Morrison, Colorado. The bones of what was to become the holotype of *Stegosaurus armatus* Marsh, 1877 (YPM 1850), were first mentioned in a letter by Lakes to Marsh (as "Saurian 5") on April 26, 1877, although no work was done for several months because of the hardness of the rock. Lakes had been contracted by Marsh to collect dinosaur bones near Morrison on his behalf and to send the fossils to him at the Peabody Museum of Natural History at what was then Yale College. The quarry is believed now to be part of a roadcut for the Alameda Parkway along the west side of Dinosaur Ridge (see Peterson and Turner 1998). This roadcut exposes most of the Morrison Formation, and it was designated the type section for the Morrison Formation by Waldschmidt and LeRoy (1944).

At the lower end of the roadcut are a series of lenticular channel sandstones that still contain abundant dinosaur bones. The sandstone can be matched lithologically with the matrix that still surrounds most of the bones of the holotype of *Stegosaurus armatus*. The bones were recovered from a yellowish to orange quartzitic sandstone, and many of the bones are only now being prepared at the Morrison Museum of Natural History, Morrison, Colorado. Most of the bones were poorly collected by modern standards, with the result that many contacts between bone fragments are now lost. Lakes broke and dynamited the sandstone into large pieces, and those showing bones were shipped to Marsh. Where possible, corresponding broken surfaces were marked with the same symbol (red painted lines, alphabetic letters, etc.) to facilitate reconstruction of the blocks at the museum. About 20 crates of blocks were eventually shipped to Marsh by Lakes and his crew from Saurian 5 (Kohl and McIntosh 1997: 166, 176). Marsh hired stonemasons as preparators to chisel the bones out, or at least to expose them for his inspection (see, e.g., Fig. 4.4C, D).

No quarry map exists for *Stegosaurus armatus*, apparently because Marsh had not yet established this procedure. However, this would have made no difference because the quarry was not conducive to mapping. As reported by Lakes,

> Will proceed in digging out several hundred weight of blocks of very hard & very difficultly worked sandstone containing many bones welded in the mass. The greatest difficulty lies in reducing these somewhat from half a ton weight to a hundred weight and so the rocks seams & cracks & flies in every direction & many hopeful looking bones have thus been broken to fragments. But it is hard to impossible to make piece match with piece as you purpose in such large fragments & such hard rock. (Lakes, June 27, 1877)

Later that same year, Samuel Williston was sent to Garden Park, Colorado, to assist Benjamin Mudge and Marshal Felch. In one of his letters, Williston sent a sketch of an articulated sauropod limb and

Figure 4.4. Stegosaurus armatus Marsh, 1877, YPM 1850, part of holotype from Saurian Quarry 5, Morrison, Colorado. (A) Partially prepared dorsal centrum. (B) Partial fragmentary neural spine and transverse process. (C) ?First caudal centrum and attached caudal rib (dorsal process missing). (D) Anterior caudals in right lateral view preserved in a sandstone block; note the absence of bifurcation of neural spines. (E, F) Anterior caudal in left lateral and anterior views (centrum length about 75 mm; anterior face median height 178 mm, width 180 mm); the simple caudal rib lacks a dorsal process. (G) Neural spines of anterior caudals in right lateral view preserved in sandstone block; weakly developed bifurcation appears as notch in top of neural spine. (H) Anterior middle caudal centra in two blocks with chevrons. (I) Posterior middle caudals as preserved. Sequence of caudals is C, E, D, G, H, I. All to same scale. Scales = 10 cm.

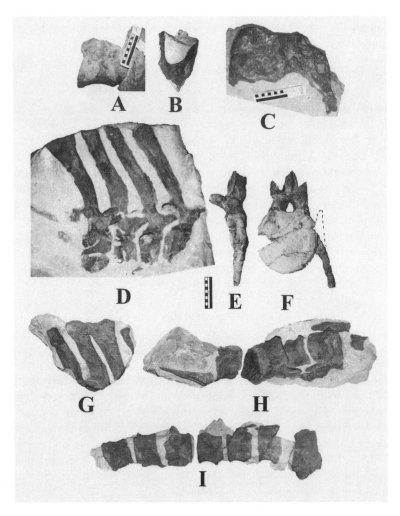

some articulated caudals (see McIntosh and Carpenter 1998: fig. 1). This sketch may have given Marsh the idea for having all of his collectors make quarry maps, so he could immediately see the relationship of the bones to each other. Often sketches of the bones in the field were all that Marsh's preparators had to guide them in piecing together broken fragments. The use of preservatives and plaster of Paris with burlap was still several years away.

Marsh (1879) originally included other bones with *Stegosaurus armatus*, such as jaw fragments of *Diplodocus* with teeth (Marsh 1880: pl. 6, figs. 2, 3), but these were soon shown to be incorrectly referred (Marsh 1884; see last section). The bones now include a dorsal centrum (Fig. 4.4A), a dorsal neural spine with transverse process (Fig. 4.4B), the ?first caudal with a vertically expanded caudal rib (Fig. 4.4C), several blocks containing an articulated series of vertebrae that represents the proximal half of the tail (Fig. 4.4D–I), a fragment of a femur (Fig. 4.5A), a fragment of the right ischium (Fig. 4.5B), and fragments of a large plate, which was the basis for the name *Stegosaurus armatus*

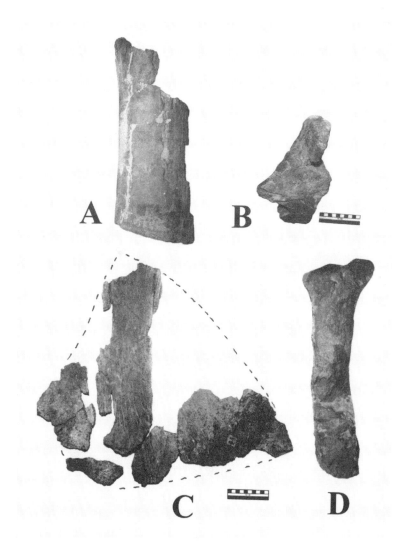

(Fig. 4.5C; tallest piece 414 mm). Spikes are not mentioned by Marsh, nor have any been found by us in the collection at the Peabody Museum. Other material cataloged as YPM 1850, which we have identified, includes a possible tibia of *Allosaurus* (Fig. 4.5D) and diplodocid limb bones. Analysis of these specimens indicates that not all the bones assigned to YPM 1850 are necessarily part of *S. armatus.*

Stegosaurus armatus has yet to be fully diagnosed, but we are unable to do so at present. We are awaiting the completion of preparations now underway by the Morrison Museum of Natural History. What has been identified so far suggests that only the posterior half of the animal was recovered; it is not known if any of the anterior parts were present at the time of burial. The genus is clearly valid on the basis of the tall neural spines with laterally expanded tops on the anterior caudals, the dorsoventrally expanded caudal ribs on the anterior caudals, anteroposteriorly short anterior caudal centra, and the very large dermal plate. Unfortunately, these characters also occur in the other

Figure 4.5. Stegosaurus armatus Marsh, 1877, YPM 1850, part of holotype from Saurian Quarry 5, Morrison, Colorado. (A) Femur section. (B) Fragment of right ischium in lateral view with part of iliac peduncle. (C) Fragments of a large plate; on the basis of other Stegosaurus specimens and the apparent large size of this plate, it may be from above the pelvic region. (D) Not all bones from Saurian Quarry 5 are of Stegosaurus armatus, *as indicated by this possible* Allosaurus tibia. All to same scale. Scales = 10 cm.

Stegosaurus taxa *S. ungulatus* and *S. stenops*. Until we can diagnose *S. armatus,* we prefer to restrict the name to the holotype specimen, although we are aware that it may supersede one of the other species.

Stegosaurus duplex *and Quarry 11*

Edward Ashley discovered the holotype of *Stegosaurus duplex* Marsh, 1887 (YPM 1858), on July 31, 1879, near the famous mammal quarry, Quarry 9 (Lakes, August 9, 1879; Reed, July 31, 1879; see map in Ostrom and McIntosh 1966). The site was worked sporadically (as were most of the Como Bluff quarries) by William Reed, aided by Frederick Brown and Ashley, until April 21, 1880, when it was deemed played out (Reed, January 25, 1879, April 21, 1880). The quarry is situated in mudstone and the preservation of the bone is very good.

Marsh originally identified the specimen as *Stegosaurus ungulatus*, although his reasons for doing so were never stated: "It was originally referred by the writer to *S. ungulatus* and the pelvic arch figured under that name" (Marsh 1887: 416). As Lull (1910b: 363) noted, "These two animals were of nearly equal size, coming from near-by localities of apparently the same geological level"; both came from about the same stratigraphic level, according to Lakes (Lakes, September 29, 1879). However, Marsh later decided that a distinct species was warranted because each sacral vertebrae supported its own sacral rib in *S. duplex* (Marsh 1887). In *S. ungulatus,* "the sacral ribs have shifted somewhat forward, so that they touch, also, the vertebra in front" (Marsh 1887: 416). We are puzzled why Marsh made such a claim, because the holotype of *S. ungulatus* lacks the sacrum. However, as Lull (1910b) noted, the intersegmental position for the ribs of this species may have been based on a referred sacrum of a juvenile individual (YPM 1857 from Quarry 13, identified as *S. ungulatus* by Marsh in a lithograph that he did not publish; see Ostrom and McIntosh 1966: pl. 22). Marsh also suggested the possibility that *S. duplex* had no dermal armor because none was found with the skeleton. Reed, however, had indicated that a *possible* spike was found on the slope below the quarry: "I send you two pieces of bone that I think belong to stegosaur spines. I found them in the wash below the quarry 50 yards [45.5 m]" (Reed, March 24, 1880). We have been unable to locate these fragments to determine their identity.

Lakes made a preliminary map of Quarry 11 on October 22, 1879, before extensive work was done. Reed mapped the site in greater detail as he excavated, and these maps appeared in several letters between February 17 and March 14, 1880. A composite map of the quarry based on all the maps is shown in Figure 4.6. The maps indicate a completely disarticulated skeleton that was widely scattered before burial by overbank deposits. The holotype includes a few cervicals and dorsals, ribs, midcaudals, femur, tibia, coossified fibula, astragalus and calcaneum, and a few phalanges (see Ostrom and McIntosh 1966: pls. 8, 17, 18, 20, 21, 28, 42, 45, 47, 49). Many of these bones were included in the composite mounted skeleton (Fig. 4.2, Table 4.1). The slight alignment of the larger bones east–west and of the vertebrae north–south suggests that the bones had been transported to the site at

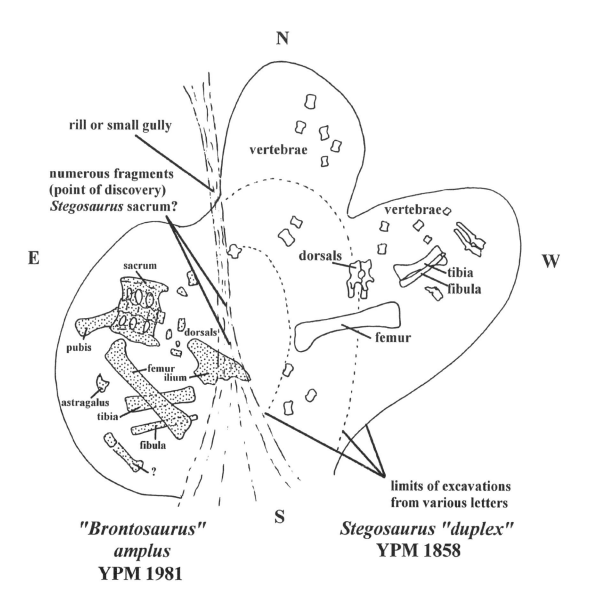

N

rill or small gully

numerous fragments
(point of discovery)
Stegosaurus sacrum?

vertebrae

vertebrae

dorsals

tibia
fibula

E

sacrum

dorsals

femur

pubis

femur
ilium

astragalus

tibia

fibula

?

W

S

limits of excavations
from various letters

"Brontosaurus"
amplus
YPM 1981

Stegosaurus "duplex"
YPM 1858

least a short distance before burial. This interpretation is supported by the absence of the very low profiled plates and spikes, which would not have been transported for any significant distance and may still be where the carcass initially lay.

Other specimens from the site include the holotype of *Apatosaurus amplus* and an unidentified mammal tooth (McIntosh 1995; Ostrom and McIntosh 1966).

Stegosaurus ungulatus *and Quarry 12*

Marsh named *Stegosaurus ungulatus* in 1879 for material (YPM 1853) collected at Quarry 12 near Robber's Roost, Como Bluff, Wyoming. The site is at the extreme west end of the Como anticline, near the

Figure 4.6. Composite quarry map of Quarry 11 based on maps by Reed and Lakes. The outline of the quarry is delineated around the bones. The holotype of Stegosaurus duplex (YPM 1858) lies scattered to the right of the small wash and the holotype of Apatosaurus amplus (YPM 1981) on the left. The dimensions of the quarry are not known because they were never given by Reed (modified from a sketch provided by J. McIntosh).

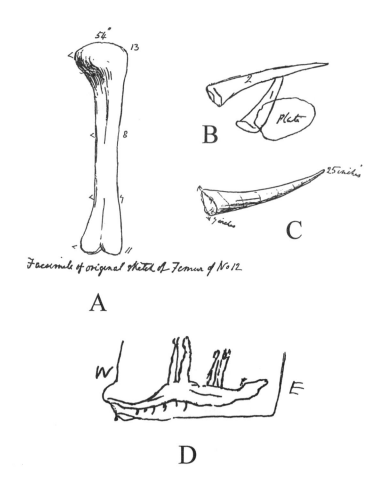

A

D

Figure 4.7. (A) Sketch of a femur for Stegosaurus ungulatus *(YPM 1853) made by Lakes (December 11, 1879) before its destruction during removal. Parts were subsequently collected and the bone restored. Measurements in inches. (B) Two spikes and a plate from Lakes's journal (February 17, 1879). (C) The "devil's tail," the first spike found, as recorded in a letter to Marsh (Lakes, December 27, 1879). (D) Lakes's attempted restoration of* Stegosaurus ungulatus *as it lay in Quarry 12 (Lakes, February 27, 1880); the spikes are arranged along the top of the tail.*

nose of the plunge where the beds dip steeply toward the north (Lakes, August 6, 1879, September 29, 1879; see Kohl and McIntosh 1997: 132, 136, 137). Lakes reported to Marsh that Quarry 12 was stratigraphically high in the Morrison: "Its position is about 40 to 50 feet [12–15 m] below the Dak[ota] sandstone and . . . about what we call the upper line of quarries" (Lakes, September 29, 1879). The site was discovered August 6, 1879, by Ashley (Reed, August 9, 1879), and it immediately produced a humerus and a much-shattered femur. At first only the ends of the femur were salvaged, but not before it had been sketched by Lakes (Fig. 4.7A). Later, in subsequent letters, Marsh requested of Lakes that all bone fragments of this and other bones be recovered from the quarry dump. Periodically, Lakes would mention in letters to Marsh that fragments of this bone were being shipped to him (e.g., letters of September 20, 1879, and December 27, 1879).Lakes continued to work Quarry 12 throughout the winter of 1879–1880 until March 20, 1880, when he left to join the faculty at the School of Mines in Golden, Colorado (today, the library there is named for Lakes). Lakes left his assistant, Richard Hallett, to enlarge the quarry in preparation for Lakes's possible return that summer, but he was never rehired

by Marsh to continue the work. Meanwhile, Marsh sent Reed a telegram on April 9 to investigate how the work by Hallett was progressing. As was typical of Reed, he antagonized Hallett when he arrived on the scene and immediately took over the work. Hallett, Ashley, and Brown resumed work at Quarry 12 for a few days, but Reed stopped all work after a cave-in of the quarry walls (Reed, April 12, 1880). The work was then shifted to other quarries but especially to the newly discovered, *Stegosaurus*-rich Quarry 13. However, Marsh did not forget Quarry 12, and six years later, Brown reopened the quarry with the assistance of William Beck and Henry Kessler (Beck, June 19, 1886). The quarry was abandoned for the last time in August 1886 (Beck, August 24, 1886).

Beck did collect nine pieces of ribs (cervicals, dorsals, or both), nine incomplete caudal vertebrae, and a large dermal plate (Fig. 4.8; maximum length, 655 mm; width, 415 mm; maximum thickness of base, 70 mm). This material was collected with U.S. Geological Survey funds when Marsh was their official vertebrate paleontologist; therefore, the material was transferred to the National Museum of Natural History after Marsh's death (see Gilmore 1914). The material, USNM 7414, is part of the holotype of *S. ungulatus* (or a syntype, if more than one individual is represented by the terminal tail spines).

Quarry 12 was difficult to excavate because the bone-bearing strata plunged at over 45°. The quarry was excavated as a long, deep, narrow trench along the strike of the beds (Fig. 4.9). Eventually it would be

Figure 4.8. Stegosaurus ungulatus *Marsh, 1879, from Quarry 12, Como Bluff, Wyoming, part of USNM 7414, part of holotype (or a syntype), large dermal plate in right lateral view.*

Othniel Charles Marsh and the Eight-Spiked *Stegosaurus* • 87

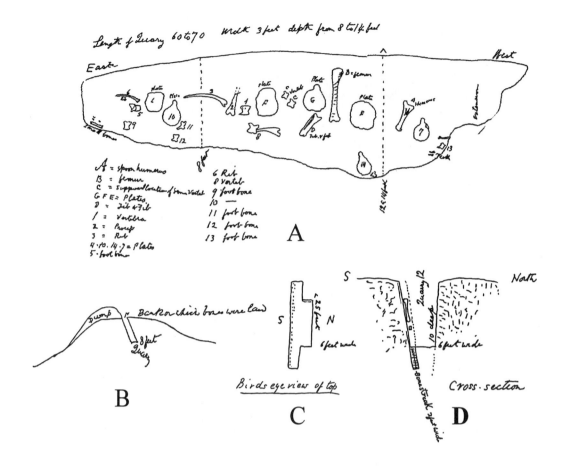

Figure 4.9. (A) Map of Quarry 12 summarizing work done between August 6 and October 29, 1879 (Lakes, November 1, 1879). (B) Cross section of Quarry 12 showing the position of the dump, slope of the quarry, and overhanging wall (Lakes, October 4, 1879). (C) Sketches by Lakes showing Quarry 12 in map view and in cross section with the overhanging wall removed (December 18, 1879). As the quarry was deepened, an unstable overhanging wall was recreated time and time again.

over 24 m (80 feet) long (including a 12-m ramp), 7.5 m (25 feet) deep, and over 3 m (10 feet) wide (Fig. 4.9A, C, D). As the quarry was deepened, the overhanging wall had to be removed (Fig. 4.9B, D), requiring that the dump be moved several times. The overhanging wall also proved to be unstable, resulting in numerous cave-ins, one of which nearly buried Brown and Ashley (Reed, April 12, 1880). Problems in the quarry worsened when the quarry reached the water table, so that there was much mud and standing water (Lakes, February 9, 1880). Lakes provided Marsh with a stratigraphic cross section of Quarry 12 (Lakes, January 17, 1880), which is reproduced in Figure 4.10. As can be seen in the figure, the hanging wall consisted of shale (technically a mudstone), which is notoriously unstable in vertical cuts.

Lakes was more of an artist than Reed, and this ability is reflected in his quarry maps and sketches, one of which is shown in Figure 4.9A. Lakes also made a final map of the site compiled from copies of all the smaller maps he made (Fig. 4.11). These smaller maps resulted from another of Marsh's requirements of his collectors: "Keep copies of all drawings so I can refer to them by number and date" (Marsh, letter to Brown, July 28, 1883).

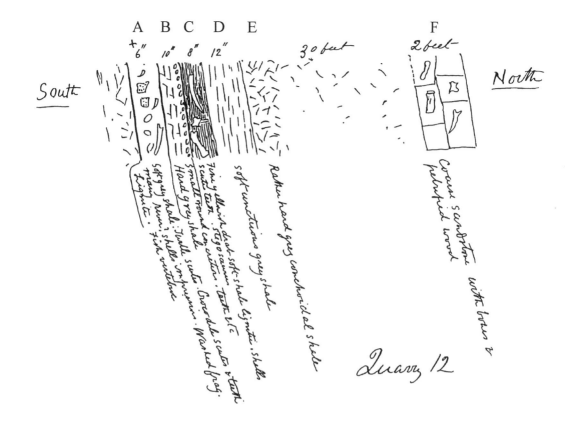

Figure 4.10. Cross section of Quarry 12 as drawn by Lakes (January 17, 1880). Annotations are provided for each layer. (A) "Soft grey shale: turtle scutes · crocodile scutes & teeth · many river? shells impressions · washed frag." At the base, "lignite · fish vertebrae." (B) "Small round concretions · teeth etc. Hard grey shale." (C) "Fine yellowish drab soft shale lignitic · shells scutes teeth · Stegosaurus." (D) "Soft unctuous grey shale." (E) "Rather hard grey concoidal shale."

As with Quarry 11, the bones were mostly scattered, although distinct clusters of bone occurred. Furthermore, plates and spikes coexist with the bones, demonstrating minimal transport of bones, possibly because the sediments (a carbonaceous shale or mudstone; Fig. 4.10) indicate a swampy site where water flow was minimal. The discovery of the first spike caused a great deal of perplexity:

> At the west end of the quarry we found a bunch of vertebrae in excellent preservation & with them a bone I am at a loss to name. It was shaped thus [see Fig 4.7C] something like a long horn running off to a sharp point. . . . It is not a rib and unless it is some portion of the pelvis or a process? or a horn on the snout as in Iguanodon or a spur on the foot. I do not know what to call it (My man calls it the devil's tail but he has no authority). (Lakes, December 27, 1877)

Spikes were found in four different areas (Fig. 4.11), with two clusters also containing distal caudals (I and III). One cluster (III) also contained four spikes in association with these distal caudals. That two clusters of distal caudals with spikes occur in the quarry suggests that two animals, not one, may be present. This interpretation is supported by the fact that the skeleton as mounted (Fig. 4.3) has two pairs of the moderately angled, broad-based anterior spines (e.g., Ostrom and McIntosh 1966: pls. 55, 56, figs. 1–3) and two pairs of the steeply angled,

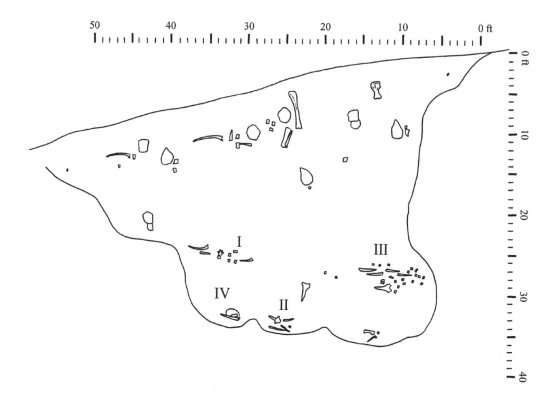

Figure 4.11. Summary map of Quarry 12 redrawn from a map made by Lakes. Spikes were found at bones denoted with Roman numerals.

narrow-based posterior spikes (e.g., Ostrom and McIntosh 1966: pl. 56, figs. 4–6). On the basis of articulated *Stegosaurus* tails (e.g., Carpenter et al. 2001; Marsh 1887: pl. 9; Ostrom and McIntosh 1966: pl. 54), the first, or anterior, pair of spikes typically have broad bases and the posterior pair have narrow, steeply angled bases. Thus, the presence of two pairs of broad-based spikes and two pairs of narrow, steeply angled spikes on the mounted skeleton of *S. ungulatus* indicates a mixture of two individuals. The absence of duplicate limb elements may be due to an incomplete overlap of two skeletons in the quarry. It is possible that reopening the quarry would reveal one or more additional skeletons. Occurrences of several *Stegosaurus* specimens together at one site are not rare and include the nearby Quarry 13, the Douglas Quarry at Dinosaur National Monument, Bone Cabin Quarry, and Felch Quarry 1.

The holotype (or probably syntypes) of *Stegosaurus ungulatus* (YPM 1853) consists of vertebrae from all regions except the sacrum (Lull 1910b), ribs, humerus, femora, tibia, fibulae, and 11 dermal plates and 8 dermal spines (see Ostrom and McIntosh 1966: pls. 15, 16, 25–27, 29, 33, 43, 46, 48, 55, 56, 59, 60, 62, 63), most of which were incorporated into the mounted skeleton (Fig. 4.2, Table 4.1; Lull 1910b). In addition, USNM 7414, which consists of a large dermal plate (Fig. 4.8) plus several ribs and cervical vertebrae, is probably part of the holotype (or syntypes).

Other isolated nonstegosaurian bones, especially teeth, were also

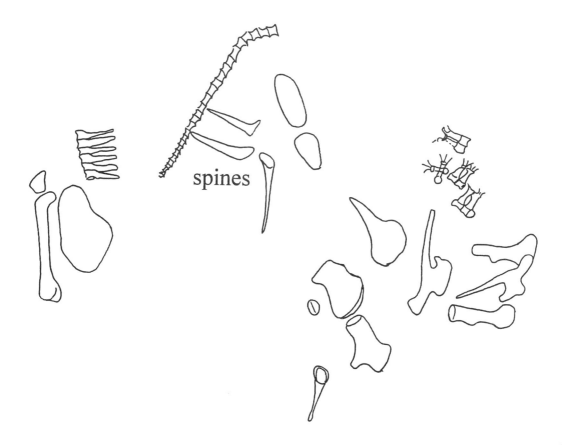

spines

recovered at Quarry 12, including bones from *Camarasaurus, Diplodocus, Allosaurus, Coelurus, Laosaurus, Goniopholis,* and cf. *Glyptops* (Lakes, January 17, 1880; Ostrom and McIntosh 1966).

Stegosaurus ungulatus S. stenops *and* S. sulcatus *from Quarry 13*

Quarry 13 has produced the greatest numbers of *Stegosaurus* specimens; the items taken from Quarry 13 are today split between the Peabody Museum of Natural History and the National Museum of Natural History (Smithsonian Institution). The site was first mentioned by Reed in a letter to Marsh on August 4, 1879. Although the quarry was worked by Reed (1880–1882) and Kenney (1883), only the maps of Brown were compiled into a single map by Gilmore (1914). Consequently, large areas of the map lack data. Fortunately, maps for some of these missing areas exist in the letters of Kenney and Reed, and one of them (Reed, May 19, 1880; Fig. 4.12) records the first articulated tail and associated spikes found there (Fig. 4.12). Over the next seven years, no fewer than five articulated tails and associated spikes would be collected (Reed and Brown, letters to Marsh, 1879–1887). Two of these sets were illustrated by Marsh (Fig. 4.13) for his planned monograph on the Stegosauria (Ostrom and McIntosh 1966: pl. 54), although only one set was ever published (Marsh 1887: pl. 9, 1896: pl. 51). Gilmore

Figure 4.12. Spikes (labeled as "spines" by Reed) associated with an articulated section of distal caudals from Quarry 13 (Reed, May 19, 1880).

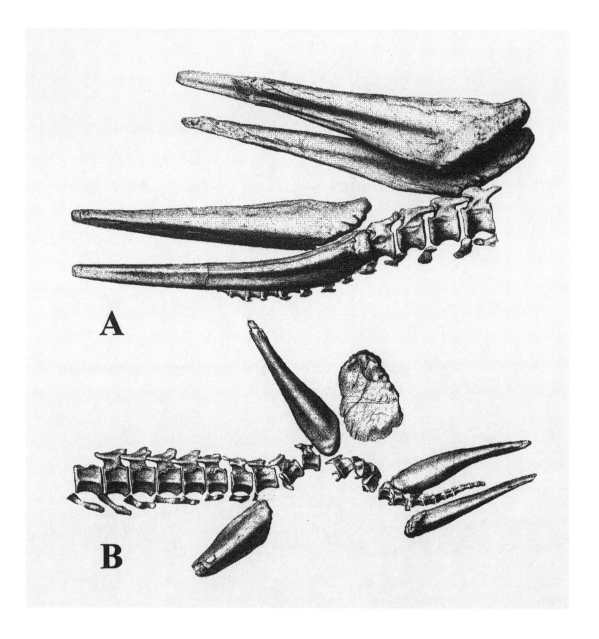

Figure 4.13. Two examples of spikes associated with distal caudals from Quarry 13. Marsh had numerous examples of two pairs of spikes associated with caudals when he presented his skeletal reconstruction of Stegosaurus ungulatus *in 1891.*

(1914: 8) refers a tail with four spikes from Quarry 13 to *S. ungulatus* (Fig. 4.14). This specimen is now incorporated into a mounted *Stegosaurus* skeleton (AMNH 650; archival records). Although no reasons were given by Gilmore for assigning the tail specimen to *S. ungulatus,* his identification is almost certainly correct because the deeply bifurcated neural spines on the anterior caudals and tall dorsal process on the caudal ribs are characteristic of *S. ungulatus.* Gilmore (1914: 4) also records *S. stenops* and *S. sulcatus* as coming from Quarry 13.

The sediments at Quarry 13 are a relatively soft, gray, fine- to medium-grain sandstone. It probably represents the distal end of a crevasse splay because of the fine nature of the sand. Unfortunately, no

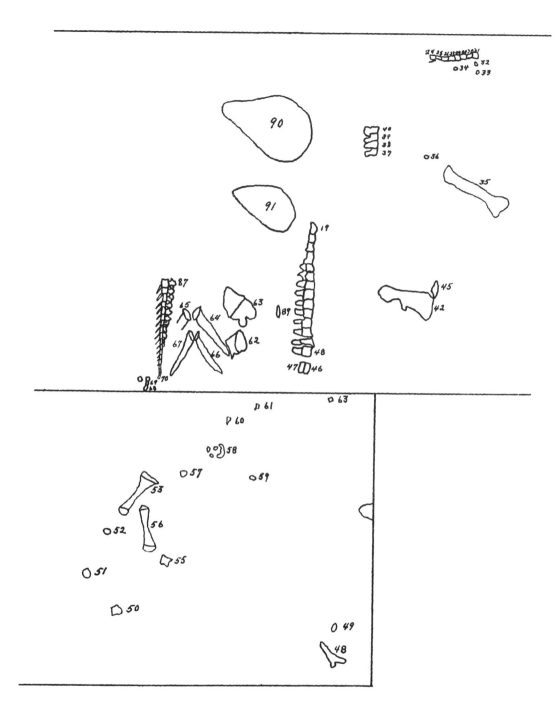

mention is made in any of the letters about the internal structure of the sediments, so characteristic features of crevasse splays, such as climbing ripples, are unknown. The presence of other taxa, most notably *Camptosaurus*, that is represented by several skeletons (see Gilmore 1909) and the wide range of disarticulation (isolated bones to partial articulated skeletons) do not support a catastrophic death assemblage. In-

Figure 4.14. Section of the map for Quarry 13 showing the tail referred by Gilmore to Stegosaurus ungulatus *(from Gilmore 1914). Only two pairs of spikes were present.*

stead, the assemblage appears to be partially attritional and partially noncatastrophic mass mortality (see Evanoff and Carpenter 1998).

Stegosaurus stenops *and Felch Quarry 1*

The holotype of *Stegosaurus stenops* Marsh, 1887 (USNM 4934), was collected by Marshal Felch between 1885 and 1887 at Garden Park north of Cañon City, Colorado (Evanoff and Carpenter 1998). The specimen was mentioned for the first time by Marshal Felch in a letter to Marsh on August 16, 1885 (as skeleton 11, a theropod; Felch did not correct the identification until October 27, 1885). The specimen was collected over a span of two years. The site of Felch Quarry 1 was originally discovered by Henry Felch in 1869 or 1870, but it was not until 1877 that the quarry was opened by Mudge (Evanoff and Carpenter 1998). Mudge excavated the site, with the help of Williston and Marshal Felch, from August 8 to October 27, 1877. The results were discouraging, so nothing further was done until the spring of 1882, when Marshal Felch reopened the quarry at Marsh's request (Felch, April 22, 1882). However, the quarry proved to be so productive that it was worked until 1889, making it one of the longest quarry operations for Marsh (Evanoff and Carpenter 1998).

Felch recorded the skeleton of *S. stenops* in his letters as it was being collected (Fig. 4.15) and later recorded the find in a composite quarry map. The skeleton was remarkably complete and articulated. The overlapping plates are illustrated in one of Felch's letters (December 9, 1885). Bones at the quarry occur partially as lag deposits in lenticular, coarse to conglomeratic sandstones, and partially in coarse-to medium-grained lateral accretion sediments (Carpenter 1999; Evanoff and Carpenter 1998). The site is the result of attritional and non-catastrophic mass mortality, with the articulated *S. stenops* skeleton possibly the result of an individual that died in a drought (Evanoff and Carpenter 1998).

Of the specimens used by Marsh (1891) in his skeletal reconstruction of *Stegosaurus*, *S. stenops* was the most complete, being an articulated skeleton (see Gilmore 1914). Remarkably, Marsh did not use this skeleton for the core of his reconstruction but only for the skull, throat ossicles, and the erect arrangement of the plates on the back and tail. This was because, for reasons known only to Marsh, little of the skeleton was prepared during Marsh's lifetime. After Marsh's death in 1899, the specimen was transferred from the Peabody Museum to the National Museum, where most of the preparation occurred (Gilmore 1914).

Marsh's Changing Views of *Stegosaurus*

The view of *Stegosaurus* as seen in the skeletal reconstruction by Marsh (1891) evolved slowly over the years as more and more specimens were discovered. These changes of how *Stegosaurus* was viewed by Marsh can be gleaned from his publications, beginning with the turtlelike view of *Stegosaurus armatus*. Marsh thought that the large plates found with *S. armatus* most closely resembled the shell parts of

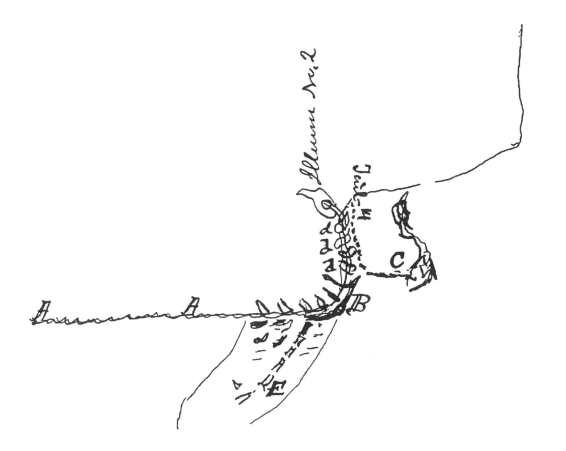

the Late Cretaceous sea turtle *Protostega*. Thus, his initial view of *Stegosaurus* was as a large turtle or turtlelike aquatic reptile with a long body encased in a carapace of large dermal plates (Marsh 1877: 513). It is to this covering, or "roof," formed by these large plates that the generic name refers (Marsh, 1879, 504). Because the carapace of a turtle is supported ventrally by neural spines, Marsh concluded that the tall neural spines of *Stegosaurus* supported the flat-lying plates along the back. Felch was probably influenced by Marsh's hypothesis when he speculated, many years later, about the position of the plates in *S. stenops* (from his letters, Felch is known to have received a reprint of Marsh's *Stegosaurus* paper). Felch noted that each plate in *S. stenops* was located over each neural spine and that the plates lay on the sides of the body (Felch, December 9, 1885). Felch elaborated his idea the following year: "Plates laid on over such a frame work—supported by those strong curved ribs would make a good design for an armor plated vessel. The only vulnerable point for attack would be on the top, where the spines could be placed to finish up" (Felch, March 25, 1886).

Marsh, however, had already abandoned the turtle shell–like hypothesis: "Some of them [i.e., plates] were so large and peculiar that their position is indicated by the structure of the anterior caudal vertebrae, whose enormous neural spines were especially adapted to support

Figure 4.15. Stegosaurus stenops *(USNM 4934) as it was recorded by Marshal Felch in a letter to Marsh (October 27, 1885). Annotations:*

A—A the front face of working ground
B—the corner where ilium lays
a—a a line of dermal plates—ribs—spine—vert. &c.
C—Sk. No. 4 [i.e., *Ceratosaurus nasicornis*]
The tail worked out right over four of the dermal plates and anterior portion of the ilium
E—the part worked out of No. 11.

them" (Marsh 1880: 258). In discussing a plate of *Stegosaurus ungulatus,* Marsh wrote, "The plate . . . was perhaps a dermo-neural spine, which stood erect over the caudal vertebrae" (Marsh 1880: 259, pl. 11, fig. 3). Other plates "are unsymmetrical, and their surfaces indicate that their position was on the back, arranged on each side of the medial line. There may have been several of these rows" (Marsh 1880: 259). Better evidence concerning the placement of the plates was not available until Felch discovered *S. stenops:*

> The upper portion of the neck . . . was protected by plates, arranged in pairs on either side. These plates increased in size farther back, and thus the trunk was shielded from injury. From the pelvic region backward, a series of huge plates stood upright along the median line, gradually diminishing in size to about the middle of the tail. (Marsh 1887: 415)

Marsh first mentions spikes (as spines) in his 1880 summary of *Stegosaurus.* He notes that "most of them are in pairs" (Marsh 1880: 258), apparently basing his conclusion on the mirrored similarities of spikes sent to him from Quarry 12. However, Marsh also wrote, "The position these various spines occupied in life is uncertain, as none of them were found in place with portions of the skeleton fitted to support them" (Marsh 1880: 258). A similar spine was found with *Omosaurus* (= *Dacentrurus*), which Owen (1875) suggested might be a carpal appendage. Although Marsh conceded the possibility that *Stegosaurus* was similarly equipped, he also expressed skepticism because "the number and variety of the spines found with one skeleton indicate that various others parts were equally well armed" (Marsh 1880: 258). Nevertheless, he also suggested that *Stegosaurus* could assume a tripodal posture and use the carpal spines as weapons: "The very short, powerful fore limbs . . . may have been well armed with spines, and thus used most effectively in defense" (Marsh 1880: 259). Evidence for the actual placement of these spikes was soon forthcoming in an illustration (Fig. 4.12) that accompanied a letter from Reed, showing an articulated segment of the posterior caudals associated with spikes (Reed, May 19, 1880). Marsh would later write, "The offensive weapons of this group were a series of huge spines arranged in pairs along the top of the distal portion of the tail, which was long and flexible" (Marsh 1887: 415). This description was accompanied by an illustration: "The position of these caudal spines with reference to the tail is indicated in the specimen figured . . . which shows the vertebrae, spines, and plate as found" (Fig. 4.13; Marsh 1887: 415).

As for the number of spikes, Marsh initially thought the spikes were sexually dimorphic: "It is possible that the difference between them [i.e., *S. armatus* and *S. ungulatus*] was only sexual, as spines were found with only one" (Marsh 1880: 259). Marsh later discarded this hypothesis once additional specimens were acquired:

> In *Stegosaurus ungulatus,* there were four pairs of these spines, diminishing in size backward. . . . In some other forms, there were three pairs, and in *S. stenops,* but two pairs have been found. . . . In one large species, which may be called *Stegosaurus sulcatus,*

there is at present evidence of only one pair of spines. . . . In one large specimen, of which the posterior half of the skeleton was secured, no trace of dermal armor of any kind was found. . . . This species represents a distinct species, which may be called *Stegosaurus duplex*. (Marsh 1887: 415–416)

Marsh gives no justification for the number of spikes found with each specimen, and in at least one case, *S. sulcatus,* there are in fact two pairs with the holotype (USNM 4937). Marsh (1887, 1896) does concede a possible taphonomic loss of the spikes for *S. duplex* but is skeptical. Sadly, Marsh never designated which specimen (or species) he thought had three pairs of spikes, and we are unable to locate such a specimen in the Marsh collections.

Marsh also mentioned the existence of other armor: "There were also, in the present species [i.e., *S. ungulatus*], four flat spines, which were probably in place below the tail, but as their position is somewhat in doubt, they are not in the present restoration" (Marsh 1891: 181). Marsh (1880: pl. 10, fig. 3) had previously illustrated this spine, which he calls a "flat dermal spine" in the caption. Lull mounted these "spines" (Ostrom and McIntosh 1966: pls. 59, 60) on the dorsal surface of the tail, just anterior to the spikes (Figs. 4.2, 4.3; Lull 1910b: figs. 4, 10, pl. 2).

As for the rest of the skeleton, Marsh (1879: 504) described the first partial skull of *Stegosaurus,* that of *S. ungulatus* (YPM 1853): "The skull is very small, and more lacertilian than in typical dinosaurs" (Marsh did not mention that he only had a partial skull). However, Marsh subsequently noted that "in its main features it agreed more nearly with that of the genus *Hatteria* [= *Sphenodon,* the Tuatara], from New Zealand, than with any other living lizard. The quadrates were fixed, and there was a quadrato-jugal arch" (Marsh 1880: 253). The cranial fragment consists of the braincase, with the postorbital, squamosal, and quadrate forming the upper temporal arch (figured only in dorsal outline in Marsh 1880: pl. 6, fig. 1; see Galton 2001: fig. 5.4B). However, this specimen is either now less complete (see Galton 2001: fig. 5.5) or Marsh deduced the presence of the lower temporal arch. Marsh finally received the first complete *Stegosaurus* skull from Felch in 1885, that of the holotype of *S. stenops* (USNM 4934). His description of this skull is remarkably brief, being a scant single page (Marsh 1887, repeated almost verbatim in 1896). Marsh (1880) earlier noted that an endocast of *Stegosaurus* (YPM 1853; see Galton 2001: fig. 5.4) indicated a more lizardlike, rather than birdlike, brain in overall appearance and indicated that it was unusually small for the body size.

Marsh's interpretation of the dentition of *Stegosaurus* was originally based on a series of associated but isolated maxillary teeth (YPM 1922; Kohl and McIntosh 1997: 22, 166; later made the holotype of *Diplodocus lacustris* Marsh 1884) and a partial jaw with cylindrical teeth (YPM 4677; Ostrom and McIntosh 1966: 370), both of which were supposedly found with *S. armatus* (Marsh 1880: pl. 6, figs. 4, 5, 1884: pl. 4, figs. 2, 3), but the jaw came from Peabody Museum of Natural History Quarry 8 at Como Bluff, Wyoming (Ostrom and McIntosh 1966: 370). Marsh thought the specimens were maxillae of *Stego-*

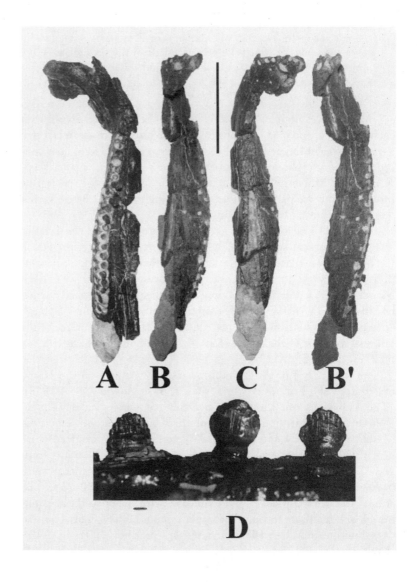

Figure 4.16. Stegosaurus *sp.*,
holotype material of the nomen
dubium Diracodon laticeps
Marsh, *1881b (YPM 1885), from*
Quarry 13. Views of a right
dentary. (A) Occlusal views.
(B, B') Stereoscopic pair of
medial view. (C) Lateral view.
(D) Lateral view of functional
dentary teeth 11, 9, 7.
Scale bars = 5 cm (A–C)
and 1 mm (D).

saurus and that the "entire dental series evidently formed a very weak
dentition, adapted to a herbivorous life" (Marsh 1880: 255). Marsh
was evidently still convinced that the jaw was that of *Stegosaurus* when
he named *Diracodon laticeps* Marsh, 1881b, for a "skull" that he
received from Quarry 13. Gilmore (1914: 108, 109) reidentified the
"skull" as consisting of two incomplete maxillae (YPM 1885), but
these are actually incomplete dentaries with a few replacement teeth
(Figs. 4.16, 4.17).

Marsh assumed, on the basis of the teeth and vertebrae (probably
of *Camptosaurus*), that *Diracodon* was related to *Laosaurus* (= *Dryo-
saurus*). Marsh did not recognize the jaw fragment that he referred to
as *Stegosaurus armatus* for what it was until he received the first complete *Diplodocus* skull from Felch in 1883. He then wrote, "This series
of teeth was found with the remains of *Stegosaurus,* and hence was at

first referred to that genus, as was also the specimen in figure 3 of the same plate. The teeth of *Stegosaurus* are now known to be of a different type, somewhat resembling those of *Scelidosaurus*" (Marsh 1884: 165, pl. 4; Figs. 4.2, 4.3). By this time (1884), Marsh had received several partial *Stegosaurus* skulls with teeth from Quarry 13 and undoubtedly recognized that the jaw of *Diracodon* was that of *Stegosaurus*. Nevertheless, he maintained *Diracodon* as a distinct genus on the basis of "the form of the skull [and] these specimens have in the fore foot the intermedium and ulnar bones separate, while in *Stegosaurus,* these carpals are firmly coossified" (Marsh 1887). Gilmore (1914) later showed that the lack of fusion of the carpal bones was a juvenile character.

Diracodon laticeps Marsh, 1881b, is a *nomen dubium,* but even if it was a valid taxon, as suggested by Bakker (1986: 227), it would be a junior synonym of *Stegosaurus* sp. Marsh never elaborated on the differences in the skulls of *Diracodon* and *Stegosaurus,* even in his 1896 summary of North American dinosaurs. When Marsh received the complete skull of *Stegosaurus stenops* (USNM 4934) from Felch, he noted of the teeth that "They are small, with compressed, fluted crowns, which are separated from the roots by a more or less distinct neck" (Marsh 1887: 414).

The limb fragments of *Stegosaurus* were originally thought by Marsh (1877) to indicate that the animal may have lived in an aquatic environment, probably because the bones were solid and therefore similar to the pachystosic bones—which counter the body's buoyancy in water—seen in some aquatic vertebrates. However, once Marsh received the complete limb material of *S. ungulatus,* he argued that *Stegosaurus* was bipedal: "The great disproportion in length between the fore and hind limbs . . . would imply that *Stegosaurus* was more or less bipedal in its movements on land" (Marsh 1880: 259). Still, he was unwilling to give up the aquatic model, adding that *Stegosaurus* was "probably more or less aquatic in habitat" (Marsh 1880: 259). In what might be considered an early biomechanical study, Marsh wrote of the forelimb,

Figure 4.17. Stegosaurus sp., holotype material of the nomen dubium Diracodon laticeps Marsh, 1881b (YPM 1885), *from Quarry 13. The fragments represent parts of a dentary. Scale in centimeters.*

The humerus and bones of the fore arm show clearly that this limb . . . was very powerful, and as it admitted of considerable rotation, it was doubtless used for other purposes than locomotion. . . . The massive posterior limbs, and huge tail doubtless formed a tripod on which the animal rested at times, while the fore limbs were used for prehension or defense. (Marsh 1881a: 170)

However, once Marsh made the skeletal reconstruction, he changed his mind, writing, "The head and neck, the massive forelimbs, and, in fact, the whole skeleton, indicate slow locomotion on all four feet" (Marsh 1891: 180), although he still maintained that *Stegosaurus* could assume a tripodal stance.

Conclusions

From the first description of *Stegosaurus* in 1879, Marsh's ideas about *Stegosaurus* underwent a considerable change, culminating in his 1891 skeletal reconstruction of *S. ungulatus*. Marsh had a wealth of specimens available to him when he made the reconstruction, including several distal tails with associated spikes (e.g., Fig. 4.13). In all instances, only four spikes were found with distal caudals. At present, the allegation that *S. ungulatus* had eight spikes is based on weak evidence, and it has not been supported by further discoveries. The genus *Stegosaurus* is apparently characterized by four distal caudal spikes, which is not variable among the different species.

Acknowledgments. For their assistance with the now divided Marsh collection of *Stegosaurus,* we thank Mary Ann Turner and Christine Chandler of the Peabody Museum of Natural History and Michael Brett-Surman (who kindly drew the attention of P.M.G. to USNM 7414; see Fig. 4.8) and Robert Purdy (who allowed us to use a set of the Marsh plates used to make Fig. 4.13) of the National Museum of Natural History. We also thank Bob O'Donnel at the Morrison Museum of Natural History (Morrison, Colorado) for preparing blocks of YPM 1950. Microfilm of the Marsh correspondence was made available by the Texas Tech University Library and the University of Delaware Library. Finally, we thank Jack McIntosh (Wesleyan University, Middletown, Connecticut) for review comments and for sharing his vast knowledge about the Marsh digs at Como Bluffs.

References Cited

Bakker, R. T. 1986. *The Dinosaur Heresies.* New York: William Morrow.
Carpenter, K. 1998. Armor of *Stegosaurus stenops,* and the taphonomic history of a new specimen from Garden Park, Colorado. In K. Carpenter, D. Chure, and J. I. Kirkland (eds.), *The Morrison Formation: An Interdisciplinary Study. Modern Geology* 23: 127–144.
———. 1999. The Cañon City dinosaur sites of Marsh and Cope. *Society of Vertebrate Paleontology Field Trip Guidebook,* 1–14.
Carpenter, K., C. A. Miles, and K. Cloward. 2001. New primitive stegosaur from the Morrison Formation, Wyoming. In K. Carpenter (ed.), *The Armored Dinosaurs.* Bloomington: Indiana University Press [this volume, Chapter 3].

Colbert, E. H. 1962. *Dinosaurs: Their Discovery and Their World.* London: Hutchinson Press.

Czerkas, S. A. 1987. A reevaluation of the plate arrangement on *Stegosaurus stenops.* In S. J. Czerkas and E. C. Olson (eds.) *Dinosaurs Past and Present,* Vol. 2, pp. 82–99. Seattle: University of Washington Press.

Czerkas, S. J., and S. A. Czerkas. 1990. *Dinosaurs: A Global World.* Limpsfield: Dragon's World.

Evanoff, E., and K. Carpenter. 1998. History, sedimentology, and taphonomy of Felch Quarry 1 and associated sandbodies, Morrison Formation, Garden Park, Colorado. In K. Carpenter, D. Chure, and J. I. Kirkland (eds.), *The Morrison Formation: An Interdisciplinary Study. Modern Geology* 23: 145–169.

Galton, P. M. 1990. Stegosauria. In D. Weishampel, P. Dodson, and H. Osmólska (eds.) *The Dinosauria,* pp. 435–455. Berkeley: University of California Press.

———. 2001. Endocranial casts of the plated dinosaur *Stegosaurus* (Upper Jurassic, Western USA): A complete undistorted cast and the original specimens of Othniel Charles Marsh. In K. Carpenter (ed.), *The Armored Dinosaurs.* Bloomington: Indiana University Press [this volume, Chapter 5].

Gilmore, C. W. 1909. Osteology of the Jurassic reptile *Camptosaurus,* with a revision of the species of the genus, and a description of two new species. *U.S. National Museum Proceedings* 36: 197–332.

———. 1914. Osteology of the armoured Dinosauria in the United States National Museum, with special reference to the genus *Stegosaurus. U.S. National Museum Bulletin* 89: 1–143.

Kohl, M. F., and J. S. McIntosh. 1997. *Discovering Dinosaurs in the Old West. The Field Journals of Arthur Lakes.* Washington, D.C.: Smithsonian Institution Press.

Lull, R. S. 1910a. The armor of *Stegosaurus. American Journal of Science,* Series 4, 29: 201–210.

———. 1910b. *Stegosaurus ungulatus* Marsh, recently mounted at the Peabody Museum of Yale University. *American Journal of Science,* Series 4, 30: 362–377.

———. 1929. *Organic Evolution.* New York: Macmillan.

Marsh, O. C. 1877. A new order of extinct Reptilia (Stegosauria) from the Jurassic of the Rocky Mountains. *American Journal of Science,* Series 3, 14: 513–514.

———. 1879. Notice of new Jurassic reptiles. *American Journal of Science,* Series 3, 18: 501–505.

———. 1880. Principal characters of American Jurassic dinosaurs, part III. *American Journal of Science,* Series 3, 19: 253–259.

———. 1881a. Principal characters of American Jurassic dinosaurs, part IV. Spinal chord, pelvis and limbs of *Stegosaurus. American Journal of Science,* Series 3, 21: 167–170.

———. 1881b. Principal characters of American Jurassic dinosaurs, part V. *American Journal of Science,* Series 3, 21: 417–423.

———. 1884. Principal characters of American Jurassic dinosaurs, part VII. On the Diplodocidae, a new family of the Sauropoda. *American Journal of Science,* Series 3, 27: 161–168.

———. 1887. Principal characters of American Jurassic dinosaurs, part IX. The skull and dermal armor of *Stegosaurus. American Journal of Science,* Series 3, 34: 413–417.

———. 1891. Restoration of *Stegosaurus. American Journal of Science,* Series 3, 42: 179–181.

———. 1896. The Dinosaurs of North America. *U.S. Geological and Geographical Survey, 16th Annual Report* 133–244.

McIntosh, J. S. 1995. Remarks on the North American sauropod *Apatosaurus* Marsh. In A. Sun and Y. Wang (eds.), *Sixth Symposium on Mesozoic Terrestrial Ecosystems and Biota, Short Papers,* pp. 119–123. Beijing: China Ocean Press.

McIntosh, J. S., and K. Carpenter. 1998. The holotype of *Diplodocus longus,* with comments on other specimens of the genus. In K. Carpenter, D. Chure, and J. I. Kirkland (eds.), *The Morrison Formation: An Interdisciplinary Study. Modern Geology* 23: 85–110.

Narendra, B. L. 1990. From the archives: The early years of the Great Hall, 1924–47. *Discovery (Yale Peabody Museum of Natural History)* 22: 18–23.

Ostrom, J., and J. S. McIntosh. 1966. *Marsh's Dinosaurs: The Collections from Como Bluff.* New Haven, Conn.: Yale University Press.

Owen, R. 1875. Monograph on the fossil Reptilia of the Mesozoic formations (parts 2 and 3) (*Bothriospndylus, Cetiosaurus, Omosaurus*). *Palaeontographical Society Monographs* 29: 15–94.

Peterson, F., and C. E. Turner. 1998. Stratigraphy of the Ralston Creek and Morrison Formations (Upper Jurassic) near Denver, Colorado. In K. Carpenter, D. Chure, and J. I. Kirkland (eds.), *The Morrison Formation: An Interdisciplinary Study. Modern Geology* 22: 3–38.

Waldschmidt, W., and L. W. LeRoy. 1944. Reconsideration of the Morrison Formation in the type area, Jefferson County, Colorado. *Geological Society of America Bulletin* 55: 1097–1114.

5. Endocranial Casts of the Plated Dinosaur *Stegosaurus* (Upper Jurassic, Western USA):

A Complete Undistorted Cast and the Original Specimens of Othniel Charles Marsh

PETER M. GALTON

Abstract

A complete endocranial cast from an undistorted braincase referred to *Stegosaurus ungulatus* is described and is shown to be similar to that of *Kentrosaurus* (Upper Jurassic, Tanzania). Because of differences between two endocranial casts figured by Marsh in the 1880s, doubts have been raised about their accuracy. The pituitary fossa was reinterpreted by Hopson as a cartilaginous filled gap between the bones and the fossa relocated more anteriorly. Reexamination of braincases and endocasts shows that the endocast of the holotype of *S. ungulatus* is incomplete and that the original lateral view, although accurate, did not indicate the reconstructed parts. The accurately figured complete endocast of a referred specimen of *S. stenops,* which represents a large individual as shown by the postcranial skeleton, is from a badly distorted braincase. Comparisons with the adjacent cranial foramina show that Marsh correctly identified the pituitary fossa on this braincase.

Introduction

The plated dinosaur *Stegosaurus* Marsh, 1877 (suborder Stegosauria Marsh, 1877; family Stegosauridae Marsh, 1880b), is only known from the Morrison Formation (Upper Jurassic, Kimmeridgian to early Tithonian; Kowallis et al. 1998) of the Western Interior, United States. The holotype postcranial material of *Stegosaurus armatus* Marsh, 1877, from Morrison, Colorado, is figured for the first time by Carpenter and Galton (2001). Marsh (1880b, 1880c, 1896) provided a dorsal outline view of the partial skull of the holotype (YPM 1853) of *Stegosaurus ungulatus* Marsh, 1879, from Como Bluff, Wyoming. This partial skull was incorporated into the mounted skeleton (see Lull 1910, 1912). The complete skull of the holotype (USNM 4934) of *Stegosaurus stenops* Marsh, 1887, from Garden Park, Colorado, was illustrated in three views and briefly described by Marsh (1887, 1896, 1897). Marsh also had prepared, but did not publish, lithographic plates of the skull of USNM 4934 in six views, including one of the medianly sectioned skull, plus views of the posterior part of a skull (USNM 4936) from Garden Park that he referred to *Stegosaurus armatus*. Copies of these lithographic plates (along with others of *Stegosaurus* postcrania and sauropods) were eventually distributed to libraries and research institutions around the world (see Osborn 1931) and were finally published in a smaller format by Ostrom and McIntosh (1966, 2000).

Gilmore (1914) provided a more detailed description of the skull of *Stegosaurus* by use of these specimens, some of Marsh's lithographic plates, and additional material from Como Bluff. Huene (1914) also briefly described the complete skull (USNM 4934), providing four views as preserved (three with some changes in bone identifications in Huene 1956), plus described some additional braincases, including CM 106 from Sheep Creek, Wyoming. McIntosh (1981) published four unlabeled photographs of CM 106, the undistorted posterior part of a skull. The cranial anatomy of *Stegosaurus* was reviewed by Galton (1990), who illustrated the skull of USNM 4934 and the braincase of CM 106; Sereno and Dong (1992) also illustrated the skull of USNM 4934 and made comparisons with that of the basal stegosaur *Huayangosaurus* (Middle Jurassic, China).

The endocranial casts of *Stegosaurus* have figured prominently in discussions about dinosaurian brains, but as Hopson (1979) noted, published knowledge of stegosaurian endocasts is limited to those of *Stegosaurus* and *Kentrosaurus* (Upper Jurassic, Tanzania; Hennig 1924; Janensch 1936). The endocast of *Stegosaurus ungulatus* (YPM 1853) was illustrated and briefly described by Marsh (1880b, 1880c, 1896), who also figured it in lateral view (Marsh 1881, 1897), but the nerves were not labeled, an omission that was rectified by Huene (1907–1908: fig. 328.5). The Marsh lithographs of the braincase and endocast of *Stegosaurus armatus* (USNM 4936) were published by Gilmore (1914: 43–44, 132, pl. 10), who described them as *S. stenops*, with the identification as *S. armatus*(?) in the caption and on the plate. Gilmore (1914: pl. 8) also published the Marsh lithograph of the sectioned skull of *S. stenops* (USNM 4934; all lithographs published by Ostrom and McIntosh 1966: 248–255, 2000). Lull (1910) published a photograph of a reconstructed endocast of USNM 4936 (note that dimensions are mistakenly given in meters, not centimeters; correct as centimeters in Lull 1912), Huene (1914: pl. 10, fig. 4) labeled the openings visible in the medial view of this braincase, and Edinger (1929: fig. 99) copied Marsh's figures of the endocasts of YPM 1853 and USNM 4936.

By use of figures of an endocast of *Kentrosaurus* given by Hennig (1924) for comparison, Hopson (1979: fig. 20) reidentified many of the cranial nerves and relocated the pituitary fossa on the endocast of *Stegosaurus armatus* (USNM 4936). However, Galton (1990) noted that Hopson (1979) was misled by the distortion of this braincase. This became apparent when comparisons were made between the endocasts of USNM 4936, *Kentrosaurus* (Galton 1988), and, in particular, the undistorted braincase and endocast of CM 106 (briefly described by Galton 1990).

To correct the problems of past descriptions of *Stegosaurus* endocasts, new illustrations and descriptions are presented, and because of the uncertainty of the species involved, the postcranial bones of USNM 4936 are illustrated and briefly described for the first time. The braincase of YPM 1853 was removed from the mounted skeleton and replaced with a cast of the skull of USNM 4934. The encasing plaster and the matrix remaining in the ear region were removed mechanically. The undistorted and complete braincase of CM 106 was also prepared free of matrix. Latex endocranial casts were prepared from these and other braincases by use of the method outlined by Radinsky (1968).

Institutional Abbreviations. CM: Carnegie Museum of Natural History, Pittsburgh, Pennsylvania. USNM: National Museum of Natural History (formerly United States National Museum), Washington, D.C. YPM: Peabody Museum of Natural History, Yale University, New Haven, Connecticut.

Anatomical Abbreviations. aa, anterior ampulla. ac, anterior circular canal. aw, anterior wall of pituitary space. ba, basilar artery. bo, basioccipital. bpt, basipterygoid process. bs, basisphenoid. bt, basisphenoid tubera. ca, cartilage. car, cartilage-filled space of Hopson (1979). cb, cerebellar region. cc, common carotid artery. ce, cerebrum. de, ductus endolymphaticus. ds, dorsum sellae of Hopson (1979). f, frontal. fj, foramen jugular for internal jugular vein. fl, flocular lobe. fla, foramen lacerum anterior for vena cerebralis anterior. flp, foramen lacerum posterior for cranial nerves IX and X, ?XI. fm, fissure mitotica (= fj + flp). fo, fenestra ovalis. fom, fenestra ovalis + fissure mitotica. ha, horizontal ampulla. hc, horizontal circular canal. ic, internal carotid artery in Vidian canal. ie, inner ear. l, lagena. ls, laterosphenoid. m, medulla oblongata. ns, sutural surface for nasal. o, orbit. oc, occipital condyle. ol, olfactory lobe. on, optic nerve of Marsh (1880b, 1880c, 1896). op, optic process of Marsh (1880b, 1880c, 1896), probably mostly cartilage. os, orbitosphenoid. ot, olfactory tract. p, parietal. pa, posterior ampulla. pas, parasphenoid. pc, posterior semicircular canal. pf, pituitary foramen. pfs, sutural surface for prefrontal. pit, pituitary space or fossa of Hopson (1979). po, posterior process of postorbital (for squamosal). pp, paroccipital process. pr, prootic. prs, presphenoid. ps, pituitary space or fossa. s, superior temporal fenestra. so, supraoccipital. sq, squamosal. su, superior utriculus. vc, Vidian canal (for internal carotid artery). vca, vena cerebralis anterior. vcm, vena cerebralis media. vcp, vena cerebralis posterior. vp, ventral process of postorbital (for jugal). 2–5, metatarsals II–V. Cranial nerves indicated by Roman numerals. VIIIa, anterior ramus of VIII. VIIIp, posterior ramus of VIII.

Material

CM 106 *(Figs. 5.1, 5.2, 5.7G).* Partial skull of *Stegosaurus ungulatus* (referral based on proportional shortness of suboccipital in ventral view, see below under USNM 4936), collected by O. A. Peterson and C. W. Gilmore in 1900 from Quarry D, Sheep Creek, Albany County, Wyoming. Includes right prefrontal and supraorbitals, plus both frontals, postorbitals, squamosals, braincase, right quadrate, posterior part of both nasals, anterior part of right dentary, and various other pieces of skull and mandible (Galton, in preparation). The posterior part of the skull was illustrated in occipital view in Huene (1914: pl. 10, fig. 2), in four unlabeled photographs by McIntosh (1981: fig. 18), and a lateroventral view of the braincase and lateral views of endocranial cast in Galton (1990: fig. 21.2G–I).

USNM 4934 *(Fig. 5.12G, H).* Almost complete skeleton as a slab mount (see Gilmore 1914, 1918) that was collected by M. P. Felch in 1886 from YPM Quarry 1 (see Evanoff and Carpenter 1998), Garden Park near Cañon City, Fremont County, Colorado. It is the holotype of *Stegosaurus stenops* Marsh, 1887, and includes the complete skull that was illustrated by Marsh (1887: pl. 6, 1896: pl. 43, 1897: fig. 52), Huene (1914: pl. 10, fig. 1a–d, 1956: fig. 565h, i, k), Gilmore (1914: fig. 1, pls. 5–9, pl. 19, fig. 1), Ostrom and McIntosh (1966: 249, 251, 253), Galton (1990: fig. 21.2A–D), and Sereno and Dong (1992: fig. 10). Marsh's median view of the sectioned skull (Fig. 5.12G) shows details of the inside of the braincase that are now inaccessible in the reassembled skull (Fig. 5.12H).

USNM 4936 *(Figs. 5.8–5.11, 5.12A–F).* Partial skeleton (number 8) collected by M. P. Felch in 1884 from YPM Quarry 1, Garden Park, Colorado. The specimen includes the posterior part of a distorted skull (with frontals, postorbitals, squamosals, and braincase), and the endocranial cast was figured by Gilmore (1914: pl. 10), Ostrom and McIntosh (1966: 255) (Fig. 5.8), Lapparent and Lavocat (1955: fig. 137), and Hopson (1979: fig. 20) (Fig. 5.12A). A left medial view of the sectioned braincase (Figs. 5.9D, 5.12F) was given by Huene (1914: pl. 10, fig. 4), and the endocast was given by Lull (1910, 1912; a reconstruction) and Edinger (1929: fig. 99a, b). The braincase was referred to *Stegosaurus armatus* by Marsh (in Gilmore 1914), to *S. stenops* by Gilmore (1914: 43), to *Stegosaurus* sp. by Huene (1914), and to *S. armatus*(?) and *S. armatus* by Ostrom and McIntosh (1966: 63, 254). However, no reasons were given for these identifications, and the associated postcranial skeleton was not described. It includes a posterior cervical vertebra (Fig. 5.10A, B; length of centrum 100 mm, another

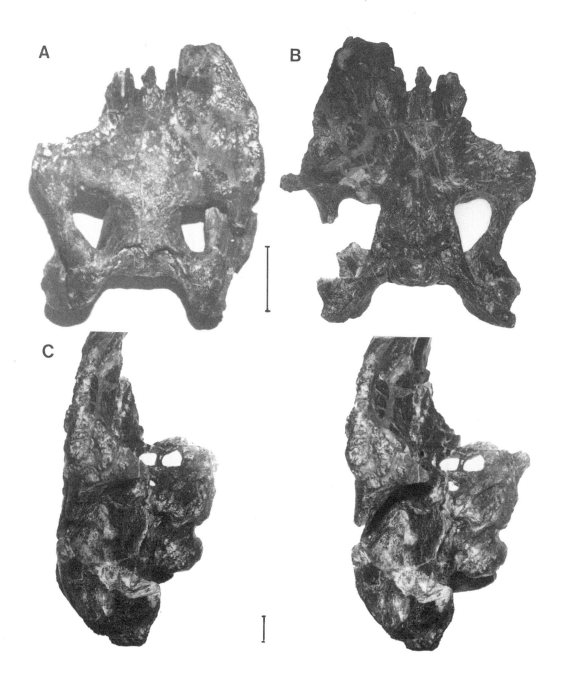

centrum 95 mm), six dorsal vertebrae (Fig. 5.10C; lengths of centra range from 98 to 108 mm) and several incomplete cervical and dorsal ribs (Fig. 5.11F–J), an almost complete anterior caudal vertebra (Fig. 5.10D–G), the neural spines of four other anterior caudals (Fig. 5.10 H–K), plus 10 incomplete caudals from the proximal half of the tail, a left ilium with the attached sacrum that was sectioned medially (Figs. 5.10M–O, 5.11K, L); separate right ilium represented by the anterior

Figure 5.1. Stegosaurus ungulatus from Upper Jurassic of Sheep Creek, Wyoming, CM 106. Braincase and adjacent parts of skull in (A) dorsal, (B) ventral, and (C) right lateral (see Fig. 5.2D) views. Scale bars = 5 cm (A, B) and 1 cm (C).

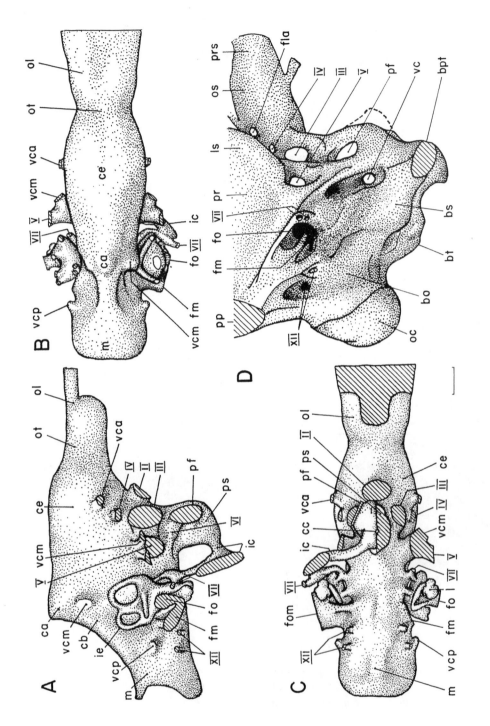

Figure 5.2. Stegosaurus ungulatus from Sheep Creek, Wyoming, CM 106. Endocranial cast in (A) lateral view with inner ear restored (cf. Fig. 5.7G); (B) dorsal view with right inner ear restored and left as preserved; and (C) ventral view. (D) Right side of braincase in ventrolateral view to show foramina (see Fig. 5.1C). From Galton (in preparation). Scale bar = 1 cm.

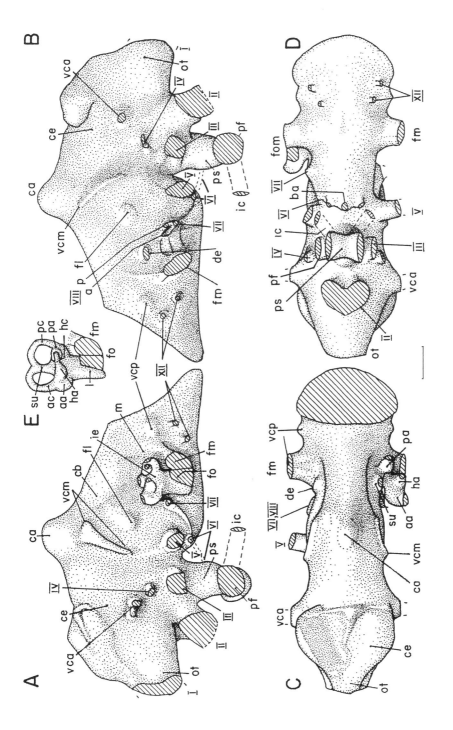

Figure 5.3. *Kentrosaurus aethiopicus* from Upper Jurassic of Tanzania. Endocranial cast as preserved in (A) left lateral, (B) right lateral, (C) dorsal, and (D) ventral views. (E) Restoration of inner ear. From Galton (1988). Scale bar = 1 cm.

process and acetabular region; incomplete right ischium (Fig. 5.11M), the left humerus (Fig. 5.11A, B; length 555 mm), the left hindlimb with femur (Fig. 5.10P, Q; length 1190 mm), tibia with attached astragalus (Fig. 5.11C, D; length 725 mm), fibula (Fig. 5.11E; length 612 mm), an incomplete pes (Fig. 5.10R, S), and four dermal plates (Fig. 5.11N–Q), one lacking the base (Fig. 5.10L, maximum length 550 mm).

The identity of USNM 4936, and hence its braincase, is based on the sacrum and anterior caudals. The sacrum of the holotype of *Stegosaurus duplex* (= *S. ungulatus*) has four pairs of massive sacral ribs originating from the posterior four vertebrae of the synsacrum with the ribs of dorsosacrals 1 and 2 free (Gilmore 1914: fig. 54; Ostrom and McIntosh 1966: 275). In the holotype of *S. stenops*, the rib of dorsosacral 1 forms a slender, sheetlike additional sacral rib (Gilmore 1914: fig. 23), as also appears to be the case in USNM 4936 (Figs. 5.10N, 5.11K). This sacral difference was noted for *Stegosaurus* by Galton (1982a), but it is not clear if this is a species difference; it is interpreted as a sexual dimorphism in the Upper Jurassic stegosaurs *Kentrosaurus aethiopicus* (Tanzania, quarry St, three sacra with four sacral ribs, four with five ribs, Galton 1982b; latter probably female, see Galton 1999) and in *Dacentrurus armatus* (western Europe, Galton 1991). Referral of USNM 4936 to *Stegosaurus stenops* may be supported by the convex top to the neural spine of the ?first caudal vertebra (Fig. 5.10D, G; cf. caudal 2 of *S. stenops,* Ostrom and McIntosh 1966: 295, figs. 3, 5), rather than being forked as in more distal caudals (Fig. 5.10I, K); in *S. duplex,* the top of the neural spine of the last sacral vertebra is gently concave (Ostrom and McIntosh 1966: fig. 21), so this was presumably the case for the first caudal vertebra. The pes is more complete than that figured by Gilmore (1914: fig. 52) in having two ungual phalanges rather than one. Medially, on the posterior surface of metatarsal IV, there is a vestigial metatarsal V (in the same position as in the ornithopod *Hypsilophodon,* Galton 1974: fig. 58B), the broader end of which is proximally situated (Fig. 5.10S) so the figure of an isolated metatarsal V (as IV) in Gilmore (1914: fig. 53) is shown upside down.

In ventral view, the basioccipital of USNM 4936 (region posterior to basisphenoid tubera) is relatively long, with a relatively short basisphenoid (Figs. 5.8B, 5.9B), as is also the case in USNM 4934 (Fig. 5.12G; see Galton 1990: fig. 21.2C; Gilmore 1914: pl. 7; Ostrom and McIntosh 1966: 253, fig. 2; Sereno and Dong 1992: fig. 10D), USNM 2274 (Gilmore 1914: fig. 10), and USNM 6645 (Galton, in preparation; Gilmore 1914: fig. 5). This similarity supports the referral of USNM 4936 to *Stegosaurus stenops,* rather than to *S. ungulatus,* because YPM 1853 has a relatively short basioccipital and a long basisphenoid (Fig. 5.5B; also in CM 106, Fig. 5.1B). The short basioccipital may represent the plesiomorphic condition because it is also present in *Kentrosaurus* (Upper Jurassic, Tanzania; Galton 1988: fig. 1D, pl. 1, figs. 2, 7, 9), *Huayangosaurus* (Middle Jurassic, China; Sereno and Dong 1992: figs. 5D, 6D), the basal thyreophoran *Emausaurus* (Lower Jurassic, Germany; Haubold 1990: fig. 3), and the basal ornithischian *Lesothosaurus* (Sereno 1991: figs. 11D, 12D). However, an elongate basioccipital is not an autapomorphy for *Stegosaurus stenops* because

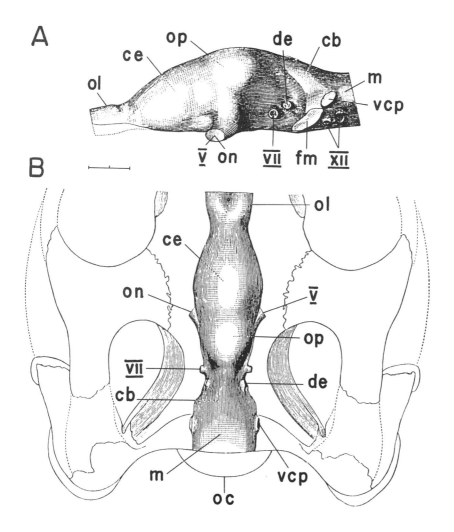

A

ol · ce · op · de · cb · m · vcp · v̄ · on · VII · fm · XII

B

ol · ce · on · v̄ · op · VII · de · cb · m · vcp · oc

it is also present in the stegosaur *Tuojiangosaurus* (Upper Jurassic, China; Dong 1990: fig. 19.9B; Dong et al. 1983: figs. 77, 81).

YPM 1853 (Figs. 5.4–5.6, 5.7A–F). Partial skeleton (see Carpenter and Galton 2001; Lull 1910, 1912; Marsh 1896; Ostrom and McIntosh 1966, 2000) collected by A. Lakes in 1879 from YPM Quarry 12, Como Bluff, Carbon County, Wyoming (see Ostrom and McIntosh 1966, 2000). It is the holotype of *Stegosaurus ungulatus* Marsh, 1879, and includes the posterior part of the skull with partial postorbitals, squamosals, quadrates, parts of ?supraorbitals, and the right ?lacrimal. An outline of the cranium was given in dorsal view with the endocranial cast by Marsh (1880b: pl. 6, figs. 1, 2, 1880c: figs. A, B, 1896: pl. 44, figs. 3, 4). Views of the endocast only were given by Marsh (1881: pl. 6, fig. 1, 1897: fig. 54), Huene (1907–1908: fig. 328.5), and Edinger (1929: fig. 99c, d).

Figure 5.4. Stegosaurus ungulatus *Marsh, 1879, holotype YPM 1853 from Upper Jurassic of YPM Quarry 12, Como Bluff, Wyoming. Partial endocranial cast in (A) lateral view and (B) dorsal view with outline of posterior part of skull; from Marsh (1896), with new labels. Scale bar = 2 cm.

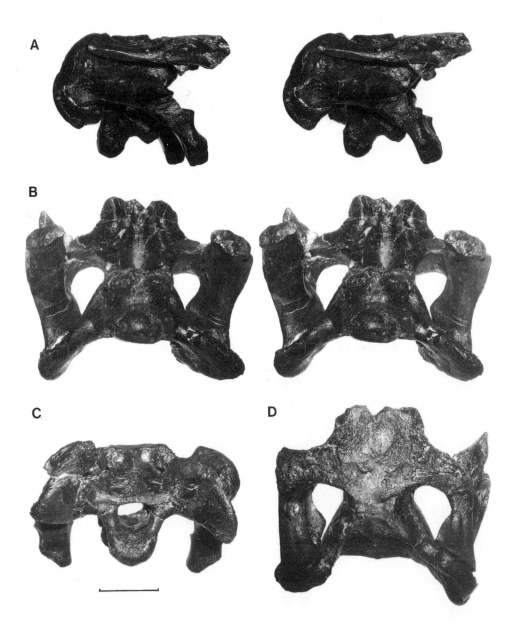

A

B

C

D

Figure 5.5. Stegosaurus ungulatus
Marsh, 1879, holotype YPM
1853 from YPM Quarry 12,
Como Bluff, Wyoming. Incom-
plete cranium in (A) right lateral,
(B) ventral, (C) posterior, and
(D) dorsal views. Scale bar = 5 cm.

Description

The description of the endocranial cast assumes that it approximately represents the form of the brain because, as Hopson (1979: 78)
noted, the brains of dinosaurs "appear to have molded the cranial
cavity to a greater extent than is usual in reptiles." *Kentrosaurus* is the
only other stegosaur for which endocasts are available, and these consist of one complete but slightly distorted cast (Fig. 5.3) and three
partial casts (Galton 1988). In most respects, the endocasts of *Stegosau-*

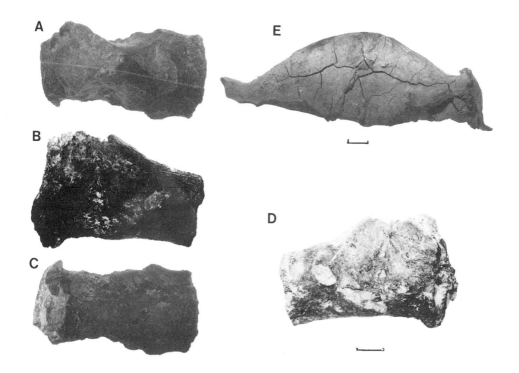

A

E

B

D

C

rus and *Kentrosaurus* are very similar (Figs. 5.2A–C, 5.3, 5.7, 5.12). The brain of *Stegosaurus* is relatively short and deep, with strong cerebral and pontine flexures and a steeply inclined posterior edge, when compared with those of ornithopod dinosaurs (Galton 1989). Because of the prominent flexures, many of the regions of the brain can only be recognized by landmarks.

Figure 5.6. Stegosaurus ungulatus Marsh, 1879, holotype YPM 1853 from YPM Quarry 12, Como Bluff, Wyoming. Partial endocranial casts. Natural endocranial cast in (A) dorsal, (B) left lateral, (C) ventral, and (D) right lateral (see Figs. 5.4A, 5.7C) views. (E) Endocranial cast made in the 1880s in left lateral view. Scale bar = 1 cm.

Brain

Telencephalon. The cerebral hemispheres form the widest part of the brain, but dorsally, they are not as well differentiated as in *Kentrosaurus* (Figs. 5.2A, B, 5.3A–C). These hemispheres taper anteriorly to short olfactory tracts and then widen again into small olfactory bulbs (Fig. 5.2B).

Diencephalon. The endocasts have a small dorsal projection (ca, Fig. 5.2A, B) that probably represents the unossified space between the top of the supraoccipital and the overlying parietal, as in some ornithopods, and in life, it was probably occupied by cartilage (Galton 1989). The optic nerve (II) arose laterally from this region, which ventrally occupied part of the pituitary space (ps) within the sella turcica. The walls of the sella turcica were ossified except for a laterally facing pituitary foramen (pf, Fig. 5.2A, D), as in *Kentrosaurus* (Fig. 5.3A, B).

Mesencephalon. The extent of this region is uncertain because there are no dorsal optic lobes and the exact points of origin of the oculomo-

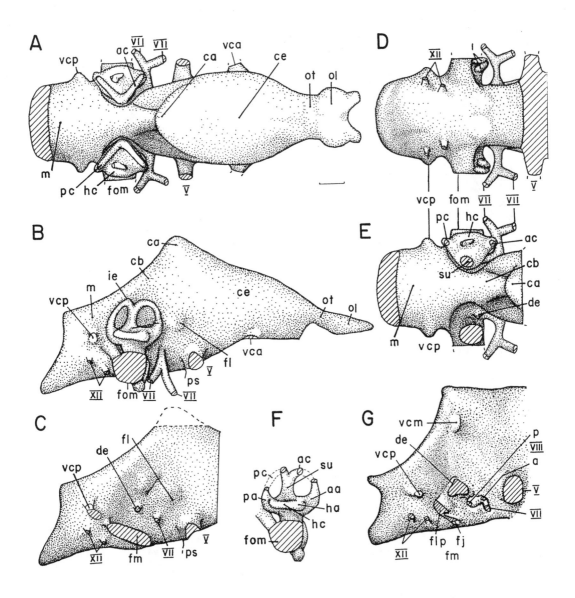

Figure 5.7. Stegosaurus ungulatus *Marsh, 1879. (A–F) Holotype YPM 1853 from YPM Quarry 12, Como Bluff, Wyoming. Partial endocranial cast. Restored endocast with inner ear in (A) dorsal and (B) right lateral views. (C) Posterior part of endocranial cast without ear region in right lateral view (see Figs. 5.4B, 5.6B, D). Posterior part of endocranial cast with inner ear as preserved in (D) dorsal and (E) ventral views. (F) Cast of right inner ear as preserved in lateral view. (G) CM 106 from Sheep Creek, Wyoming, posterior part of endocranial cast in right lateral view with inner ear removed (cf. Fig. 5.2A). From Galton (in preparation). Scale bar = 1 cm.*

tor and trochlear nerves (III, IV) cannot be determined. Gilmore (1914: 43) drew attention to the large size of the optic lobes, but as Edinger (1975: 66) noted, these are "actually the dorsad non-neural evagination of the cavity (see Zangerl 1960)."

Metencephalon. There is no cerebellar expansion in the dorsal region of the metencephalon. The floccular lobes of the cerebellum (fl, Figs. 5.3A, B, 5.7B, C, G) are interpreted from a slight concavity, the fossa subarcuata, in the medial wall of the prootic and supraoccipital. The trigeminal nerve (V) originates from this region of the brain stem, the posterior part of which has transversely constricted side walls to accommodate the inner ears (Figs. 5.2B, 5.3C, 5.7E). With the adjacent part of the myelencephalic walls, this area is the narrowest part of the brain.

Myelencephalon. Cranial nerves VI to XII and the inner ear originate from the ventral part of the myelencephalon, the widest part of which is slightly posterior to the vena cerebralis posterior (Figs. 5.2A–C, 5.3C).

Cranial Nerves

Olfactory Nerve (I). The short olfactory tracts narrow anteriorly from the cerebrum and widen again into small olfactory bulbs (Fig. 5.2A–C).

Optic Nerve (II). The combined optic foramina form a large oval to subcircular opening that opens anteroventrally and laterally (Fig. 5.2A, C). In *Kentrosaurus,* this opening is heart-shaped (Fig. 5.3D) or triangular in outline.

Oculomotor Nerve (III). Each oculomotor nerve exits just posterior to the optic foramen, and the oculomotor foramen is proportionally larger in CM 106 than it is in *Kentrosaurus* (Figs. 5.2A, 5.3A, B). The foramina identified by Huene (1914) for this nerve and for IV are located too posteriorly on the braincase to belong to these nerves (Fig. 5.9F; cf. Figs. 5.2A, D, 5.12C).

Trochlear Nerve (IV). The trochlear foramen is a small opening above the oculomotor foramen.

Trigeminal Nerve (V). The foramen prooticum is immediately posterodorsal to the sella turcica, and part of this large foramen was probably occupied by the large sensory trigeminal or Gasserian ganglion, which is close to the root of the trigeminal nerve. The dorsal margin of the opening has notches anteriorly and posteriorly that probably indicate the routes of the ramus ophthalmicus (V_1, with vena cerebralis medius) and the rami maxillaris and mandibularis (V_2, V_3), respectively, as in *Kentrosaurus.*

Abducens Nerve (VI). The abducens foramen either passes anteroventrally through the floor of the braincase into the dorsal part of the sella turcica (cf. Fig. 5.3A, B), or, if the posterodorsal wall of the sella turcica is lower, the route is indicated by a groove on the medial surface of the side wall of the sella turcica (Fig. 5.2A). A similar variation in the route of this nerve occurs in *Kentrosaurus.* This nerve presumably exited the pituitary space via the large pituitary foramen.

Facialis Nerve (VII). Medially, the facialis foramen is small and close to the acoustic foramina. The canal forks within the bone to give two posterolateral openings to the outside for the ramus hyomandibularis (Fig. 5.2D; only one exit in *Kentrosaurus*) and a ventral one into the Vidian canal for the ramus palatinus (Fig. 5.2A), which enters the sella turcica to join the internal carotid artery and presumably exited this space with the abducens nerve through the pituitary foramen.

Acoustic Nerve (VIII). Immediately above the facialis nerve are two other small openings; the three openings are connected together by a groove that appears as a ridge on the cast (Fig. 5.7G). The other two openings are for the anterior (vestibular, VIIIa) and posterior (acoustic, VIIIp) branches of the acoustic nerve.

Glossopharyngeal, Vagus, and Accessory Nerves (IX, X, XI). These nerves exited through the upper part of the fenestra mitotica. In life, there was presumably a cartilage bar separating this part (foramen lacerum posterior) from the more ventral jugular foramen for the vena jugularis internus (Fig. 5.2A).

Hypoglossus Nerve (XII). The two foramina for this nerve differ in size, with the anterior ramus smaller and anteroventral to the posterior ramus (Figs. 5.2A, 5.7C).

Inner Ear

Casts of the inner ear include most of the bony labyrinth, except for the semicircular canals, the more central parts of which are still filled with matrix. However, enough is indicated (Figs. 5.2B, 5.7E, F) so that a reasonable reconstruction is possible (Figs. 5.2A, B, 5.7A, B). The horizontal (or lateral) canal was much shorter than the subequal vertical ones in YPM 1853 (Fig. 5.7B), as it is in *Kentrosaurus* (Fig. 5.3E). In CM 106, the posterior canal is shorter than the anterior one (Fig. 5.2A). Swellings at the base of each canal, the ampullae, are not as prominent as in *Kentrosaurus* (Figs. 5.2A, 5.3E, 5.7A, B, F). By analogy with lizards, the utriculus was a roughly Y-shaped system of tubes, the ends of which connected with the semicircular canals. Only the base of the superior utriculus, from which the medial ends of the vertical canals originate, is preserved. The sacculus is poorly developed, whereas in lizards it is much enlarged. The lagena is straight and short. The fenestra ovalis is poorly delimited from the fenestra mitotica in YPM 1853 because of the incompleteness of the crista interfenestralis. This crista is more complete in CM 106 (Fig. 5.2D), in which it funnels into a relatively small fenestra ovalis, the margin of which is preserved except for the unossified ventral part. The ductus endolymphaticus is represented by an opening posterodorsal to the facialis foramen; it is small in YPM 1853 (Fig. 5.7C) and larger in CM 106 (Fig. 5.7G), where this opening was less well ossified.

Blood Vessels

Arteries. The internal carotid artery enters the lateral opening of the Vidian canal (Fig. 5.2D) and is joined by the palatine ramus of the facialis nerve (Fig. 5.2A, C) with which it enters the posteroventral part of the pituitary space (Fig. 5.2A). A palatine artery presumably branched off the internal carotid arteries, passed upward in the pituitary space, and united as a single basilar artery that passed into the cranial cavity through the median notch in the posterodorsal wall of the sella turcica.

Veins. A small opening anterodorsal to the trochlear foramen (Fig. 5.2D) was identified by Janensch (1936) for *Kentrosaurus* as the fenestra epioptica for the vena cerebralis anterior (Figs. 5.2A, C, 5.3A, B, D). A prominent groove on the medial surface of the supraoccipital and the adjacent part of the prootic was for the vena cerebralis media. The vena capitis dorsalis drained blood from the occipital musculature, and its entrance is visible externally in the suture between the parietal and the supraoccipital (Fig. 5.5C; Gilmore 1914: figs. 4, 5). An opening on the lateral part of the same suture (Fig. 5.1C; Gilmore 1914: fig. 10) was for the vena parietalis. The connections between the vena parietalis, vena capitis dorsalis, and vena capitis media are visible in a sectioned braincase of *Kentrosaurus* (Galton 1988: fig. 4K, pl. 4, figs. 6–8). Ventrally, the vena cerebralis media leaves the braincase in the anterodorsal part of the trigeminal foramen (Fig. 5.2A, D; separate foramen in braincase of *Kentrosaurus*). A pit opening medially above the hypoglossal foramina is a remnant of the vena cerebralis posterior (Figs. 5.2A, 5.7B, C). The major route of blood drainage through the posterior side wall of the braincase was by the vena jugularis internus. It exited through the jugular foramen, the ventral part of the fenestra mitotica, which in life was probably separated by cartilage from the more dorsal foramen lacerum posterior (Figs. 5.2A, D, 5.7G).

Discussion

The first endocranial casts of *Stegosaurus* to be illustrated were those of *S. ungulatus* (YPM 1853) by Marsh (1880b, 1880c, 1896) (Fig. 5.4) and a referred specimen (USNM 4936) of *S. stenops* by Marsh (as *S. armatus* in Gilmore 1914; in Ostrom and McIntosh 1966, 2000; copied by Hopson 1979) (Figs. 5.8, 5.12A). Edinger (1962: 72) noted that Marsh "never had so complete an endocast specimen as is seen in his figures," and with particular reference to the braincase and endocast of YPM 1853, Edinger (1975: 104) noted that "no such specimen as figured is now in Marsh's collection at Yale University; there is a different, unnumbered cast [USNM 4936] with, e.g. the pituitary, but showing only one of the six posterior nerve canals of Marsh's figure." Hopson (1979: 109) noted that Marsh's "endocasts were composites, based on several specimens" and that his figures of the endocasts of YPM 1853 and USNM 4936 (Figs. 5.4, 5.8D, E) are sufficiently different "to cast doubt on the accuracy of both specimens." Such doubts were reasonable, given the demonstration by Edinger (1951) that the figures of the complete endocranial casts of the toothed birds *Hesperornis* and *Ichthyornis* (Upper Cretaceous, western United States) in Marsh (1880a) are actually reconstructions based on very little cranial material. However, in the case of *Stegosaurus*, the braincase and endocasts (natural and artificial) of YPM 1853 still exist. Marsh did not figure composites involving more than one specimen; the endocasts were accurately illustrated, and the discrepancies result from differences in

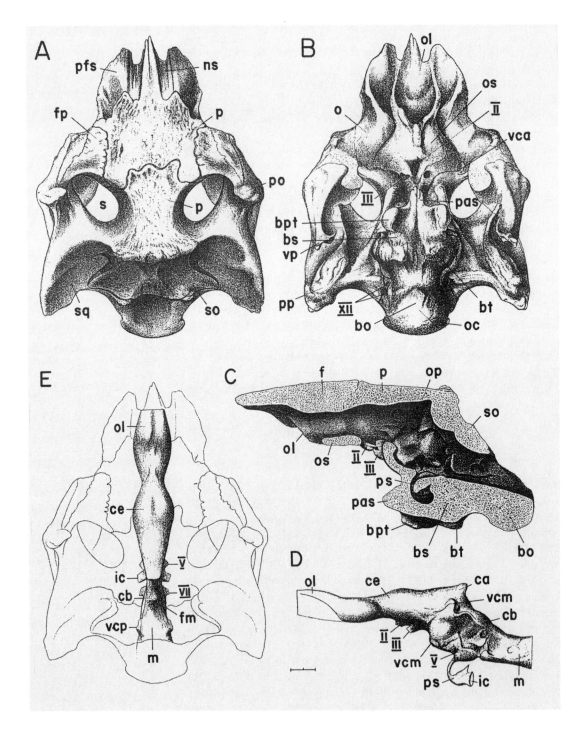

Figure 5.8. Stegosaurus stenops *from Upper Jurassic of Garden Park, Colorado, USNM 4936. Cranium in (A) dorsal and (B) ventral views (see Fig. 5.9A, B); (C) right wall of sectioned braincase in medial view. Endocranial cast in (D) left lateral and (E) dorsal views (with outline of cranium from A). C and D combine information from both sides (see discussion for details, cf. Fig. 5.9D, E). From a lithographic plate prepared under the direction of O. C. Marsh in the 1880s (with new labels). Scale bar = 2 cm.*

preservation (viz., the incompleteness of YPM 1853 and the postmortem distortion of USNM 4936).

Until recently, the incomplete braincase of YPM 1853 (Fig. 5.5) was mounted on the skeleton at the Peabody Museum. Edinger would have seen this mounted skeleton on exhibit, but probably because of the color matching and the inevitable coat of dust, she did not realize that the posterior part of the reconstructed skull was bone rather than plaster. The illustrations (Fig. 5.4) given by Marsh (1880b, 1880c, 1896) are clearly based on two endocasts from this braincase in the Peabody Museum collection: an old endocast for the "complete" outline and a less complete natural endocast for the "foramina" (Fig. 5.6), most of which were not identified in the original illustration (Fig. 4.4A). The Peabody Museum collection was not properly reorganized until the early 1960s, which would explain why Edinger failed to locate these endocasts. Marsh (1880b, 1880c, 1896; figure copied by Edinger 1929: fig. 99c) failed to indicate that some of the outline in lateral view was restored (cf. Figs. 5.4A, 5.6B, D, E, 5.7B, C; middle part of dorsal edge and all of anterior half of ventral edge missing). Only Huene (1907–1908: fig. 328.5) provided identifications, some of them incorrect, for all the foramina on the endocast of YPM 1853. The trigeminal nerve was misidentified as the optic nerve (V, on; Fig. 5.4A) by Marsh (1880b, 1880c, 1896) and by Huene (1907–1908) as the pituitary fossa (hypophysis), the actual location of which is indicated by a slight downturning of the ventral margin adjacent to the trigeminal foramen (ps, Figs. 5.6B–D, 5.7B, C). Huene (1907–1908) also misidentified the foramen for the facial nerve as that for the trigeminal nerve, the opening for the ductus endolymphaticus as that for the facialis nerve, and the protuberance for the vena cerebralis posterior as the endolymphatic sac (VII, de, vcp; Figs. 5.4A, 5.7C). He correctly identified the foramen for the internal carotid artery entering the pituitary fossa, the fissura mitotica, and the two rami of the hypoglossal nerve (vc, fm, XII, Figs. 5.4A, 5.7C).

Hopson (1979: 109) notes that the figure of the braincase of USNM 4936 (Fig. 5.12F) in Huene (1914) contains "misinterpretations of the position of the pituitary fossa and various nerve foramina," but this is misleading because only a few of the foramina were incorrectly identified. He also notes that everyone (but he cites no references) who has discussed the pituitary fossa of this endocast (Figs. 5.8C, D, 5.12A)

has commented on the posterior position and unusual rearward curvature of the pituitary fossa. Comparison with *Kentrosaurus,* or any other reptile for that matter, indicates that the pit in the floor of the braincase [car, Fig. 5.12A] cannot be the pituitary fossa as it lies below the medulla oblongata well behind the level of both the pontine flexure and the bony structure identified here as the dorsum sellae [ds, Fig. 5.12A]. Instead, the pit probably represents an unossified zone within the basicranium. The braincase appears to pertain to a subadult individual in which cartilaginous areas between the bones are represented by ridges on the endocast; in such a specimen the floor of the cranial cavity may have been poorly ossified. Correct placement of the pituitary fossa [pit, Fig. 5.12A] anterior to the dorsum sellae and below the for-

A   B

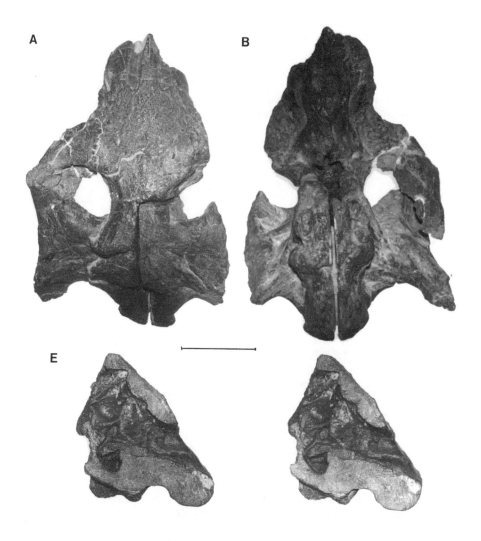

E

Figure 5.9. (This page and opposite page) Stegosaurus stenops *from Garden Park, Colorado, USNM 4936. Complete cranium in (A) dorsal, (B) ventral, and (C) left lateral views (A, B cf. Fig. 5.8A, B). (D) Right lateral view anteriorly, posteriorly left wall of sectioned braincase in medial view (cf. Fig. 5.12F). (E) Right wall of posteriorly sectioned braincase in medial view (cf. Fig. 5.8C). Scale bar = 5 cm.*

amen of the oculomotor nerve [V, Fig. 5.12A] results in a reptilian endocast of more conventional appearance. Hopson (1979: 109)

From the previously unfigured lateral view (Fig. 5.9C), it is obvious that the partial skull of USNM 4936 was badly damaged during preservation. The postorbital is rotated through 90° so the posterior process for the squamosal is directed dorsally and the anteroventral process for the jugal is directed posteroventrally (po, Vp, Figs. 5.8A, B, 5.9A–C). The braincase of USNM 4936 (Fig. 5.9C) has been compressed dorsoventrally, and although not obviously indicated by breakage, the dorsal part has been sheared by plastic distortion anteriorly relative to the floor (Fig. 5.9A, B, D). The amount of anterior shear can be assessed from the fact that although the superior edge of the supraoccipital superior to the foramen magnum is slightly incomplete, it is 35 mm anterior to the posterior edge of the occipital condyle (Figs. 5.8C, 5.9D, E, 5.12F), rather than almost directly over it, as in USNM 4934 (Fig. 12G). This is not a species difference because both braincases are

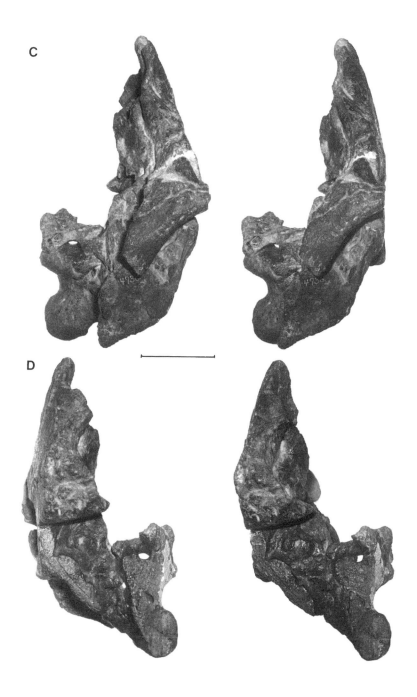

referred to *Stegosaurus stenops,* and in addition, the situation in CM 106 (*S. ungulatus*) is similar to that of USNM 4934. The distortion of the braincase is reflected in the form of the endocranial cavity and the resulting endocast (Figs. 5.8C, D, 5.9D, E, 5.12C, E, F), when compared with the undistorted braincases and endocasts of CM 106 (Figs. 5.1, 5.2), USNM 4934 (Fig. 5.12G, H), other specimens of *Stegosaurus stenops* and *S. ungulatus* (Galton, in preparation), and those of *Kentro-*

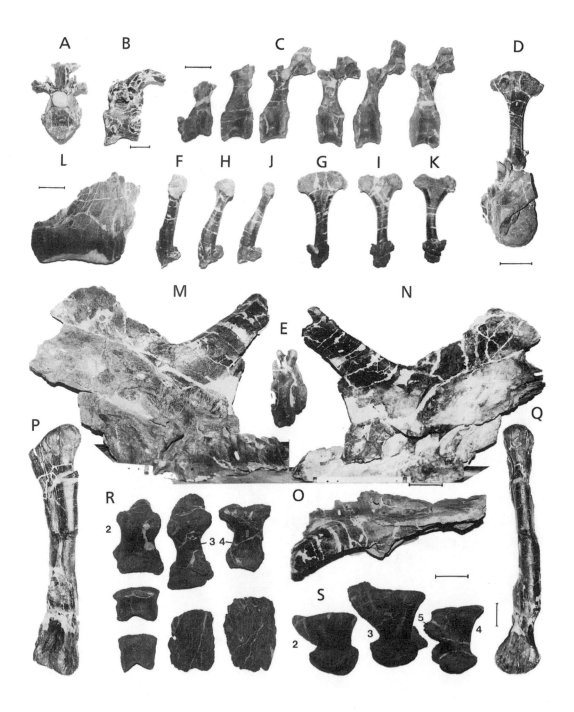

Figure 5.10. Stegosaurus stenops *from Garden Park, Colorado, USNM 4936. Posterior cervical vertebra in*
(A) anterior and (B) left lateral views. (C) Six dorsal vertebrae in left lateral view. Anterior caudal vertebra, ?first,
(D) complete vertebra in anterior view; (E) centrum and neural arch with prezygapophysis in left lateral view; neural
spine and postzygapophysis in (F) left lateral and (G) posterior views. Neural spine and postzygapophyses of anterior
caudal vertebrae, (H, I) ?third and (J, K) ?fifth in (H, J) left lateral and (I–K) posterior views. (L) Incomplete dermal
plate in lateral view. Left ilium and left half of medially sectioned sacrum in (M) dorsal; (N) ventral; and (O) left
lateral views (see Fig. 5.11K, L). Left femur in (P) posterior and (Q) medial views. (R) Incomplete left pes in anterior
(dorsal) view. (S) Metatarsals II to V in medial view. Scale bars = 5 cm (A, B, R, S) and 10 cm (C–Q).

122 • Peter M. Galton

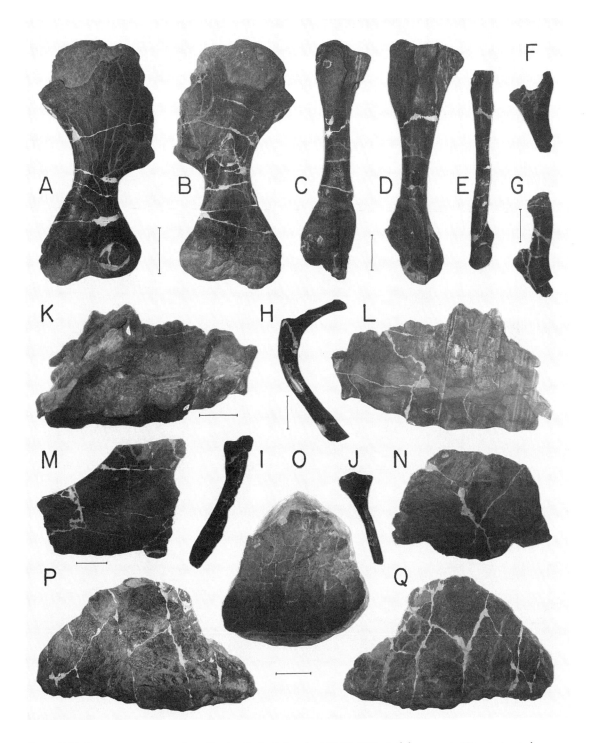

Figure 5.11. Stegosaurus stenops *from Garden Park, Colorado, USNM 4936. Left humerus in (A) anterior and (B) posterior views. Left tibia in (C) anterior and (D) lateral views. (E) Left fibula in lateral view. Left ribs in anterodorsal view: (F) midcervical; (G, H) posterior cervicals; (I, J) posterior dorsals. Right half of medially sectioned sacrum in (K) lateral and (L) medial views (see Fig. 5.10M–O). (M) Proximal part of right ischium in lateral view. Dermal plates in lateral view: (N, O) medium-sized plates; (P, Q) large dermal plate in opposite lateral views. Scale bars = 5 cm (M) and 10 cm (A–L, N–Q).*

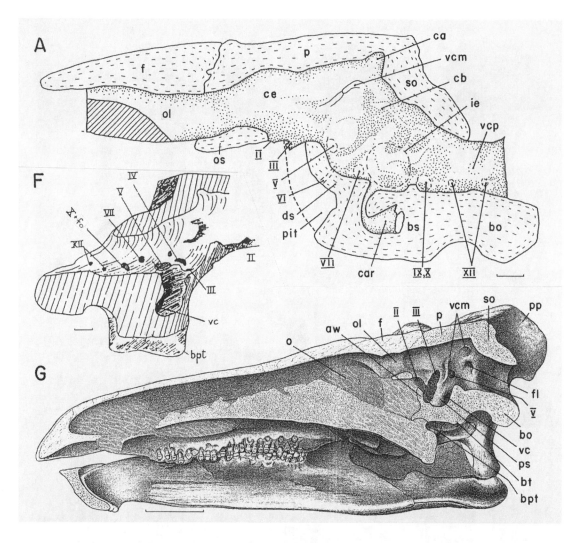

Figure 5.12. (This page and the opposite page) Stegosaurus stenops Marsh, 1887, from Garden Park, Colorado. (A–F) USNM 4936. (A) Left lateral view of endocranial cast in sectioned braincase to show identifications of Hopson (1979: fig. 20A, B) (figure combines information from both sides; cf. Fig. 5.9D, E; see discussion for details and for identifications of car, ds, pit). Complete endocranial cast in (B) dorsal; (C) left lateral; and (D) ventral views. (E) Partial endocranial cast in right lateral view. (F) Sectioned left wall of braincase in medial view (cf. Fig. 5.9D), from Huene (1914) (III and IV incorrectly placed; should be close to II; see Figs. 5.2A, D, 5.8B). (G, H) Holotype USNM 4934. (G) Medial view of sectioned right half of skull, from lithographic plate prepared under the direction of O. C. Marsh in the 1880s, with new labels. (H) Partial endocranial cast in left lateral view. Scale bars = 1 cm (A–F, H) and 5 cm (G).

saurus (Fig. 5.3; Galton 1988). This explains why the medulla oblongata of USNM 4936 is over the structure long interpreted as the pituitary fossa (Fig. 5.12C, E; car Fig. 5.12A).

Hopson (1979) (Fig. 5.12A) followed Marsh (in Gilmore 1914), who used the complete left side of the braincase of USNM 4936 in reverse (Figs. 5.9D, 5.12F) for the overall outline and the posterior half of the right side (Fig. 5.9E) for the details. This mixture of sides was unfortunate, because the more laterally placed right sagittal section of the pituitary fossa has a peculiar shape (Figs. 5.9E, 5.12E) that is not

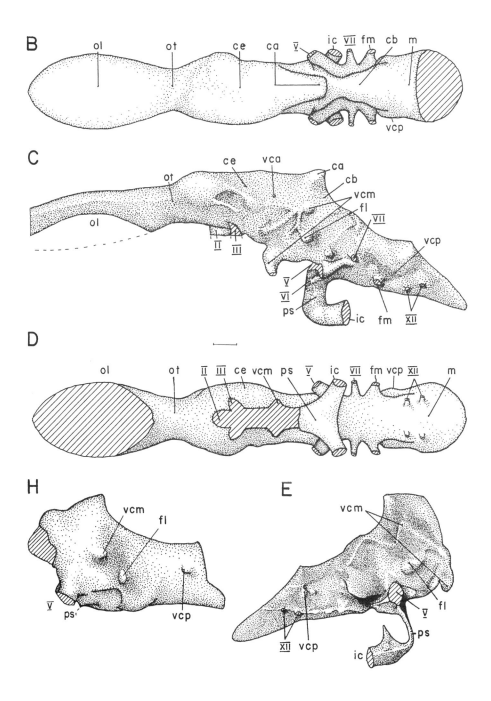

present on the slightly more medially placed section of the left side (Figs. 5.9D, 5.12C). The result is a reconstructed figure for the right side of the complete sectioned braincase (Fig. 5.8C), the "cast" of which was then reversed to give the complete endocast (Fig. 5.8D; cf. actual endocasts, Fig. 5.12C, E). However, Marsh's illustrations of the braincase and endocranial cast of USNM 4936 are accurate (cf. Figs. 5.8, 5.9, 5.12B, C, E).

The actual partial right side of the braincase (Fig. 5.9E) is difficult to interpret because more ventrally, no foramina open onto the lateral surface; it is difficult to identify the resulting structures on the endocast (Fig. 5.12E). However, such openings are present on the complete left side (Figs. 5.9C, D, 5.12F), and these foramina compare quite closely with those of the undistorted braincase of CM 106 (Fig. 5.2D), allowing the structures on the endocast to be correctly identified (Fig. 5.2A; cf. Fig. 5.12C). The two posterior openings are for the two branches of the hypoglossal nerve (XII). The next opening, which exits just anterior to the crista tuberalis, is part of either the fenestra ovalis (X + fo, Fig. 5.12F) or the fenestra mitotica (fm, Fig. 5.12C). Anterior to this is a conical indentation that represents the medial part of the facial foramen for the facial nerve (VII). The next opening, which is incompletely bordered by bone anteriorly, is the trigeminal foramen for the trigeminal nerve (V). The cavity immediately ventral to the trigeminal foramen, the hypophysial or pituitary fossa (ps, Fig. 5.12C, D, F; car, Fig. 5.12A), has a diagonally inclined groove on the upper part of the side wall for the abducens nerve (VI, Fig. 5.12C), as in CM 106 (Fig. 5.2A). The Vidian canal for the internal carotid artery opens from the posterolateral part of the pituitary fossa (vc, ic, ps, Fig. 5.12C, D, F) and exits close to the base of the basipterygoid tubera (Fig. 5.9C) as in CM 106 (Figs. 5.1C, 5.2A, C, D) and *Kentrosaurus* (Fig. 5.3A, C, D). This side of the pituitary fossa (Figs. 5.9D, 5.12C, F) lacks the unusual rearward curvature shown by the most lateral part of the opposite wall of this cavity, the one preserved on the right side (Figs. 5.9E, 5.12E).

USNM 4936 does not represent a subadult animal because the ends of the long bones are extremely rugose (Figs. 5.10P, Q, 5.11A–E), and as indicated by the maximum lengths of these bones (humerus, 555 mm; femur, 1190 mm; tibia with attached astragalus, 725 mm; and fibula, 612 mm), this represents a very large individual. The structure interpreted as the dorsum sellae (ds, Fig. 5.12A) by Hopson (1979) is only present on the right side, and it is the ossified anterior wall of the pituitary space (also visible in medial section of USNM 4934; see aw, Fig. 5.12G). Consequently, I conclude that the cavity immediately ventral to the trigeminal foramen in USNM 4936 is the pituitary space or fossa (Fig. 5.12C–E), as identified by Marsh (in Gilmore 1914), rather than an unossified space in the floor of the braincase as identified by Hopson (1979).

Conclusions

The complete endocranial cast of *Stegosaurus ungulatus* from the undistorted braincase of CM 106 (Figs. 5.1, 5.2), which includes most of the inner ear, is similar to that of *Kentrosaurus* (Fig. 5.3; Upper Jurassic, Tanzania). Two endocranial casts, YPM 1853 and USNM 4936, were accurately figured by Marsh in the 1880s. However, the endocast of *S. ungulatus* Marsh, 1879 (YPM 1853), is incomplete and the original lateral view of given by Marsh (1880b, 1880c, 1896), although accurate (Fig. 5.4A), does not indicate the reconstructed parts (cf. Figs. 5.6, 5.7B, C). The complete endocast of a referred specimen of *S. stenops* (USNM 4936, Fig. 5.8D, E), is from a distorted skull (Figs.

5.8A–C, 5.9, 5.12F) of an adult individual (as shown by the postcranial skeleton, Figs. 5.10, 5.11), but comparisons with the adjacent cranial foramina show that Marsh (in Gilmore 1914) correctly identified the pituitary fossa on this endocast (Fig. 5.8C, D), as did Huene (1914) (Fig. 5.12F). Consequently, the reidentification by Hopson (1979) of the pituitary fossa as a cartilage filled space between the bones of a subadult individual, with the pituitary fossa situated more anteriorly (car, pit, Fig. 5.12A), is incorrect.

Acknowledgments. I thank the following people for their assistance while I was studying specimens at their respective institutions, for the loan of important specimens for a very extended period of time (Carnegie Museum, National Museum of Natural History), and for permission for further preparation (Carnegie Museum, Peabody Museum): D. S. Berman (Carnegie Museum); N. Hotton III, R. Purdy, and M. Brett-Surman (National Museum of Natural History; also R. Chapman for hand-carrying the cranial material back to the National Museum of Natural History); and J. H. Ostrom and M. A. Turner (Peabody Museum). I am especially grateful to John Ostrom for permitting the substitution of a cast of the skull of USNM 4934 for that on the Peabody Museum mounted skeleton. This braincase was skillfully freed from plaster and further prepared by Robert Allen (Peabody Museum; now at University of Florida, Gainesville), who also made the substitute cast from an old mold of USNM 4934 (kindly supplied by R. Purdy) and did the further preparation of CM 106. I thank the late Jim Jensen (Brigham Young University, Provo) for giving me the original Marsh lithographic plate that was used to prepare Figure 5.8, and I thank an anonymous reviewer and especially Kenneth Carpenter, who kindly invited me to submit a paper, for their constructive comments, some of which prompted me to try and determine the species for USNM 4936. This research was partly supported by National Science Foundation grants DEB 77-24088 and BSR 85-00342.

References Cited

Carpenter, K., and P. M. Galton. 2001. Othniel Charles Marsh and the myth of the eight-spiked *Stegosaurus*. In K. Carpenter (ed.), *The Armored Dinosaurs*. Bloomington: Indiana University Press [this volume, Chapter 4].

Dong Z. 1990. Stegosaurs in Asia. In K. Carpenter and P. J. Currie (eds.), *Dinosaur Systematics: Approaches and Perspectives,* pp. 255–268. New York: Cambridge University Press.

Dong Z., S. Zhou, and Y. Zhang 1983. The dinosaur remains from Sichuan Basin, China [in Chinese with English summary]. *Palaeontologica Sinica* 162: 1–166.

Edinger, T. 1929. Die Fossilen Gerhirne. *Ergebnisse der Anatomie und Entwicklungsgeschichte* 28: 1–249.

———. 1951. The brains of the Odontognathidae. *Evolution* 5: 6–25.

———. 1962. Anthropocentric misconceptions in paleoneurology. *Rudolf Virchow Medical Society of New York, Proceedings* 19 (1960): 56–107.

———. 1975. Paleoneurology 1804–1966: An annotated bibliography. *Advances in Anatomy, Embryology and Cell Biology* 49: 1–258.

Evanoff, E., and K. Carpenter. 1998. History, sedimentology, and taphonomy of Felch Quarry 1 and associated sandbodies, Morrison Formation, Garden Park, Colorado. In K. Carpenter, D. Chure, and J. I. Kirkland (eds.), *The Morrison Formation: An Interdisciplinary Study. Modern Geology* 22: 145–169.

Galton, P. M. 1974. The ornithischian dinosaur *Hypsilophodon* from the Wealden of the Isle of Wight. *British Museum (Natural History), Bulletin, Geology* 25: 1–152c.

———. 1982a. Juveniles of the stegosaurian dinosaur *Stegosaurus* from the Upper Jurassic of North America. *Journal of Vertebrate Paleontology* 2: 47–62.

———. 1982b. The postcranial anatomy of the stegosaurian dinosaur *Kentrosaurus* from the Upper Jurassic of Tanzania, East Africa. *Geologica et Palaeontologica* 15: 139–160.

———. 1988. Skull bones and endocranial casts of stegosaurian dinosaur *Kentrosaurus* Hennig, 1915 from Upper Jurassic of Tanzania, East Africa. *Geologica et Palaeontologica* 22: 123–143.

———. 1989. Crania and endocranial casts from ornithopod dinosaurs of the families Dryosauridae and Hypsilophodontidae (Reptilia: Ornithischia). *Geologica et Palaeontologica* 23: 217–239.

———. 1990. Stegosauria. In D. B. Weishampel, P. Dodson, and H. Osmólska (eds.), *The Dinosauria*, pp. 435–455. Berkeley: University of California Press.

———. 1991. Postcranial remains of stegosaurian dinosaur *Dacentrurus* from Upper Jurassic of France and Portugal. *Geologica et Palaeontologica* 25: 299–327.

———. 1999. Sacra, sex, *Sellosaurus* (Saurischia: Sauropodomorpha; Upper Triassic, Germany)—Or why the character "two sacral vertebrae" is plesiomorphic for Dinosauria. *Neues Jahrbuch für Mineralogie, Geologie und Paläontologie, Abhandlungen* 213: 19–55.

Gilmore, C. W. 1914. Osteology of the armored dinosaurs in the United States National Museum, with special reference to the genus *Stegosaurus. United States National Museum, Bulletin* 89: 1–136.

———. 1918. A newly mounted skeleton of the armored dinosaur *Stegosaurus stenops* in the United States National Museum. *United States National Museum, Proceedings* 54: 383–396.

Haubold, H. 1990. Ein neuer Dinosaurier (Ornithischia, Thyreophora) aus dem unteren Jura des Nördlichen Mitteleuropa. *Revue de Paléobiologie* 9: 149–177.

Hennig, E. 1924. *Kentrurosaurus aethiopicus,* die Stegosaurier-funde von Tendaguru, Deutsch-Ostafrika. *Palaeontographica, Supplement* 71(1): 103–254.

Hopson, J. A. 1979. Paleoneurology. In C. Gans, R. G. Northcutt, and P. Ulinski (eds.), *Biology of the Reptilia,* Vol. 9, pp. 39–146. New York: Academic Press.

Huene, F. 1907–1908. Die Dinosaurier der europäischen Triasformation mit Berucksichtigung der aussereuropäischen Vorkommnise. *Geologische und Palaeontologische Abhandlungen, Supplement* 1: 1–419.

———. 1914. Über die Zweistammigkeit der Dinosaurier, mit Beitragen zur Kenntnis einiger Schädel. *Neues Jahrbuch für Mineralogie, Geologie und Paläontologie (Bell.-Bd.)* 37: 577–589.

———. 1956. *Paläontologie und Phylogenie der Niederen Tetrapoden.* Jena: Fischer.

Janensch, W. 1936. Über Bahnen von Hirnvenen bei Saurischiern und Ornithiern, sowie einigen anderen fossilen und rezenten Reptilien. *Palaeontologische Zeitschrift* 18: 181–198.

Kowallis, B. J., E. H. Christiansen, A. L. Deino, F. Peterson, C. E. Turner, M. J. Kunk, and J. D. Obradovich. 1998. The age of the Morrison Formation. In K. Carpenter, D. Chure, and J. I. Kirkland (eds.), *The Morrison Formation: An Interdisciplinary Study. Modern Geology* 22: 235–260.

Lapparent, A. F. de, and R. Lavocat. 1955. Dinosauriens. In J. Piveteau (ed.), *Traité de Paléontologie,* Vol. 5, pp. 785–962. Paris: Masson et Cie.

Lull, R. S. 1910. *Stegosaurus ungulatus* Marsh, recently mounted at the Peabody Museum of Yale University. *American Journal of Science,* Series 4, 30: 361–376.

———. 1912. The armored dinosaur *Stegosaurus ungulatus,* recently restored at Yale University. *Verhandlungen des internationalen Zoologen-Kongresses Graz 1910* 8: 672–681.

Marsh, O. C. 1877. New order of extinct Reptilia (Stegosauria) from the Jurassic of the Rocky Mountains. *American Journal of Science,* Series 3, 14: 513–514.

———. 1879. Notice of new Jurassic reptiles. *American Journal of Science,* Series 3, 18: 501–505.

———. 1880a. Odontornithes: A monograph on the extinct toothed birds of North America. *United States Geological Exploration of the Fortieth Parallel, Report* 7: 1–201.

———. 1880b. Principal characters of American Jurassic dinosaurs. Part III. *American Journal of Science,* Series 3, 19: 253–259.

———. 1880c. Die Stegosaurier. *Kosmos* 7: 213–215.

———. 1881. Principal characters of American Jurassic dinosaurs. Part IV. Spinal cord, pelvis and limbs of *Stegosaurus. American Journal of Science,* Series 3, 21: 167–170.

———. 1887. Principal characters of American Jurassic dinosaurs. Part IX. The skull and dermal armor of *Stegosaurus. American Journal of Science,* Series 3, 34: 413–417.

———. 1896. Dinosaurs of North America. *United States Geological Survey, 16th Annual Report* 1894–1895: 133–244.

———. 1897. Vertebrate fossils of the Denver Basin. *United States Geological Survey, Monograph* 27: 473–527.

McIntosh, J. S. 1981. Annotated catalogue of the dinosaurs in the collections of Carnegie Museum of Natural History. *Carnegie Museum of Natural History, Bulletin* 18: 1–65.

Osborn, H. F. 1931. United States Geological Survey unpublished lithographic plates on vertebrate fossils for distribution. *Science* 74: 43–44.

Ostrom, J. H., and J. S. McIntosh. 1966. *Marsh's Dinosaurs: The Collections from Como Bluff.* New Haven, Conn.: Yale University Press.

———. 2000. *Marsh's Dinosaurs: The Collections from Como Bluff,* 2nd ed. New Haven, Conn.: Yale University Press.

Radinsky, L. B. 1968. A new approach to mammalian cranial analysis, illustrated by examples of prosimian primates. *Journal of Morphology* 124: 167–180.

Sereno, P. C. 1991. *Lesothosaurus,* "fabrosaurids," and the early evolution of Ornithischia. *Journal of Vertebrate Paleontology* 11: 168–197.

Sereno, P. C., and Dong, Z. 1992. The skull of the basal stegosaur *Huayangosaurus taibaii* and a cladistic diagnosis of Stegosauria. *Journal of Vertebrate Paleontology* 12: 318–343.

Zangerl, R. 1960. The vertebrate fauna of the Selma Formation of Alabama. Part V. An advanced chelonid sea turtle. *Fieldiana, Geological Memoirs* 3: 281–312.

6. Possible Stegosaur Dermal Armor from the Lower Cretaceous of Southern England

WILLIAM T. BLOWS

Abstract

Some dermal bones, unlike those of the better known polacanthine dinosaurs, are described. These may be from stegosaurs from the British Wealden, which are poorly known and for which no previous dermal remains are recorded. It is not possible to refer these specimens to the few European Lower Cretaceous stegosaur taxa that are named. Spanish Wealden stegosaur material has no armor comparative to that of the British specimens. Five distinctive characters suggesting a possible stegosaurian origin separate the armor from that of specimens assigned to the Ankylosauria. These characters could form the basis of a standardization in stegosaur dermal armor descriptions.

Introduction

Lower Cretaceous stegosaurs in England are rare, and descriptions are based on fragmentary specimens (Blows 1998). *Regnosaurus northamptoni* Mantell, 1848, is based on a right mandibular fragment with teeth from the Mantell collection (BMNH 2422) and was found in the Wealden (Valanginian) of the Tilgate Forest area, near Cuckfield, Sussex. This specimen was redescribed by Barrett and Upchurch (1995) as a stegosaur. *Craterosaurus pottonensis* Seeley, 1874, was based on an incomplete neural arch of a dorsal vertebra (SMC B.28814), which was recovered from Aptian strata of Bedfordshire, but was probably reworked from the earlier Valanginian. This specimen is discussed by Galton (1981).

A Wealden stegosaur caudal spine is described from Valencia, Spain (Casanovas-Cladellas et al. 1995b), but no Lower Cretaceous stegosaurian dermal armor has been formally described from Britain. Four dermal armor specimens from the Lower Cretaceous of England (BMNH 40458, BMNH 39533, MIWG 1191a, and MIWG 5307) are listed by Olshevsky (personal communication) as being stegosaurian because they are atypical of the better known ankylosaurs *Hylaeosaurus* and *Polacanthus*. Of these, BMNH 40458 is a water-worn ossicle referable to *Polacanthus*; the others are discussed below.

Lydekker (1888: 191) listed a number of specimens in the Natural History Museum, London, of dermal armor as "generically undetermined specimens" from the Lower Cretaceous of the Isle of Wight and Hastings. Some of these are now recognizable as *Polacanthus* (i.e., BMNH R133 in part, BMNH R202 in part, and BMNH R643, BMNH 40458, BMNH 37713, and BMNH 37714). Others could be stegosaurian (e.g., BMNH R15950 and BMNH 15951, formerly BMNH R202 and BMNH 36515–17, respectively). One bone, BMNH R9533, was listed and described by Lydekker (1888: 191) as "a dermal spine. This specimen has not the compressed form characteristic of *Polacanthus* and *Hylaeosaurus*. Presented by E. Backhouse, 1866." This specimen could not be found in the collections in 1998. However, another spine, BMNH R1875 (described below), not previously listed by anyone, was found in the collection and could be from a stegosaur.

BMNH R202 consists of at least six bones from the Fox collection. Six were located, but it is not clear if a seventh label relates to these six or to another bone that was not located. Four of the six are *Polacanthus*, but two may be stegosaurian and have been renumbered as BMNH R15950 and BMNH R15951; they are described below.

Institutional Abbreviations. BMNH: Natural History Museum (formerly British Museum [Natural History]), London. MIWG: Museum of Isle of Wight Geology, Sandown, UK. SMC: Sedgwick Museum of Geology, Cambridge. USNM: National Museum of Natural History (formerly United States National Museum), Washington, D.C.

Description

BMNH R1875 (Fig. 6.1) is a moderately large, conical, dermal spine from the Wealden beds (Valanginian or Hauterivian) of Hastings, Sussex, and is part of the Beckles collection. The bone surface is poorly preserved. A strong bend in the bone appears to be natural (Fig. 6.1B). The roughly circular base (Fig. 6.1C) has a low central ridge extending from one edge to the other, as indicated by the broken line. The spine itself is relatively thick but flattened on both sides, and it has rounded edges.

BMNH R15950 (previously part of BMNH R202; Fig. 6.2) is a nearly complete specimen from the Isle of Wight in the Fox collection. It is a short, conelike spine with a flattened, slightly concave base and a rounded posterior edge. The base shape is broad (Fig. 6.2C) and

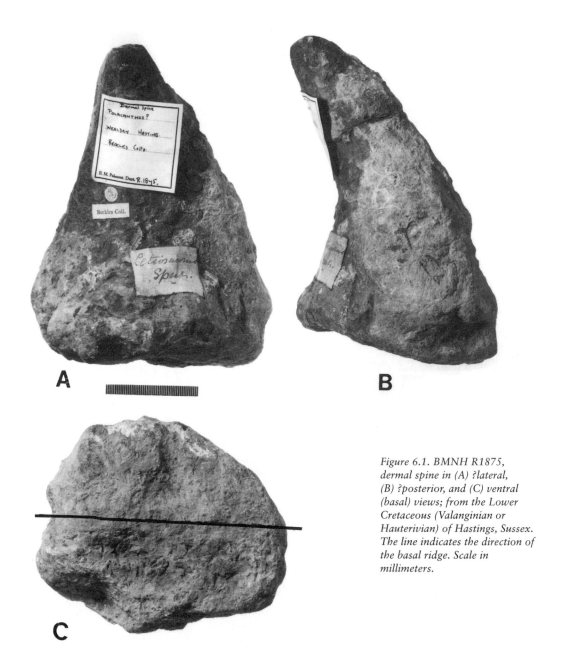

Figure 6.1. BMNH R1875, dermal spine in (A) ?lateral, (B) ?posterior, and (C) ventral (basal) views; from the Lower Cretaceous (Valanginian or Hauterivian) of Hastings, Sussex. The line indicates the direction of the basal ridge. Scale in millimeters.

slightly expanded laterally, especially at an area that inserted into the dermis (Fig. 6.2B, similar to the Museum of Isle of Wight Geology specimens described below). A second expanded area for dermal insertion on the opposite side probably existed but is now lost.

BMNH R15951 (previously part of BMNH R202, Fig. 6.3) is the basal half of a small spine from the Isle of Wight in the Fox collection. The nearly flat base is oval (apart from damage) and has a slightly raised ridge extending centrally across the greater length (Fig. 6.3C). The spine is flattened on both sides, and the apex is missing. One edge curves

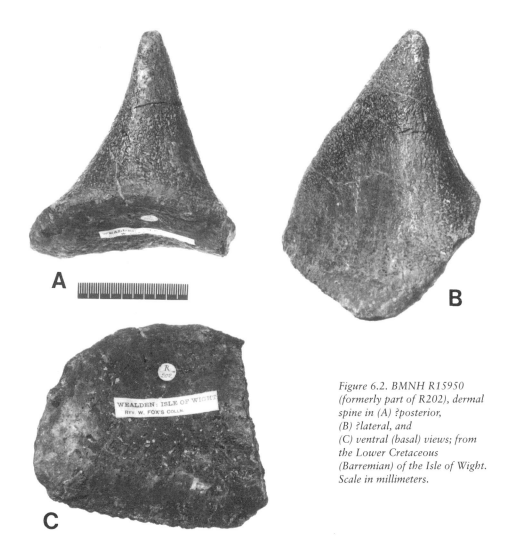

Figure 6.2. BMNH R15950
(formerly part of R202), dermal
spine in (A) ?posterior,
(B) ?lateral, and
(C) ventral (basal) views; from
the Lower Cretaceous
(Barremian) of the Isle of Wight.
Scale in millimeters.

sharply to the side as it approaches the base (Fig. 6.3B) and has a
shallow groove behind it.

BMNH 36515 and 36516 (Fig. 6.4) are two dermal spines that
superficially do not appear to be bone. However, microscopic examination shows a surface of fragmented compact and cancellous bone with
crystal and iron pyrite deposits. The bases of the spines show the natural vascular openings associated with dermal spines. Both of these
spines are small and conical, and they have oval bases. The side faces
of the spines are slightly flattened, and the edges are very rounded.
BMNH 36517 consists of two bone fragments associated with these
spines. They have broken or cut surfaces with a similar microscopic
appearance.

MIWG 5307 (Figs. 6.5, 6.6) was described by Blows (1987) as a
possible variation of the polacanthine presacral spine Type A, but was
designated as Type B. It differs from the polacanthine Type A spines in

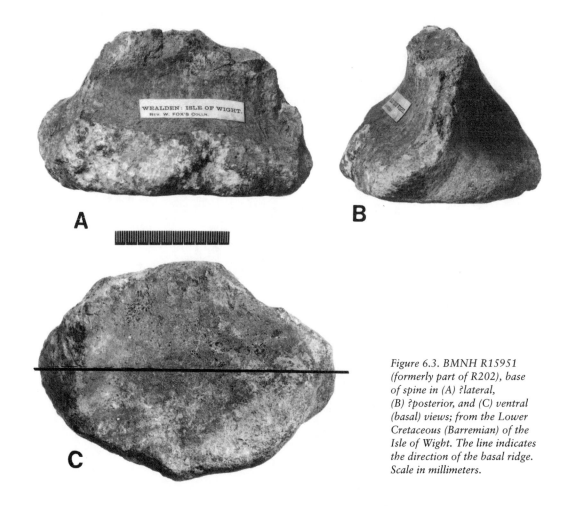

WEALDEN: ISLE OF WIGHT.
REV. W. FOX'S COLLN.

A

B

C

Figure 6.3. BMNH R15951 (formerly part of R202), base of spine in (A) ?lateral, (B) ?posterior, and (C) ventral (basal) views; from the Lower Cretaceous (Barremian) of the Isle of Wight. The line indicates the direction of the basal ridge. Scale in millimeters.

that "the dorsal keel is flat on both sides and the edges are straight, or very gently curved, twisting slightly or not at all. The base has two distinct areas of dermal attachment, a prominent medial anterior process and lateral posterior process. Both areas are separated from the dorsal keel by a step in the bone. There is no hooklike feature in Type B spines" (Blows 1987: 566). Blows also noted that

> the existence of Type B spines as purely isolated elements causes problems. They strongly suggest a similar position as Type A spines, but their absence in skeletons remains unexplained. No Type B spine can therefore be directly attributed to *Polacanthus* on the basis of known material. However, with absence of other nodosaurs identified from the Isle of Wight Wealden strata, tentative assignation of Type B spines to *Polacanthus* is made on the grounds of possible sexual dimorphism. (Blows 1987: 572)

However, he did not consider a possible stegosaur origin for the Type B spines at that time.

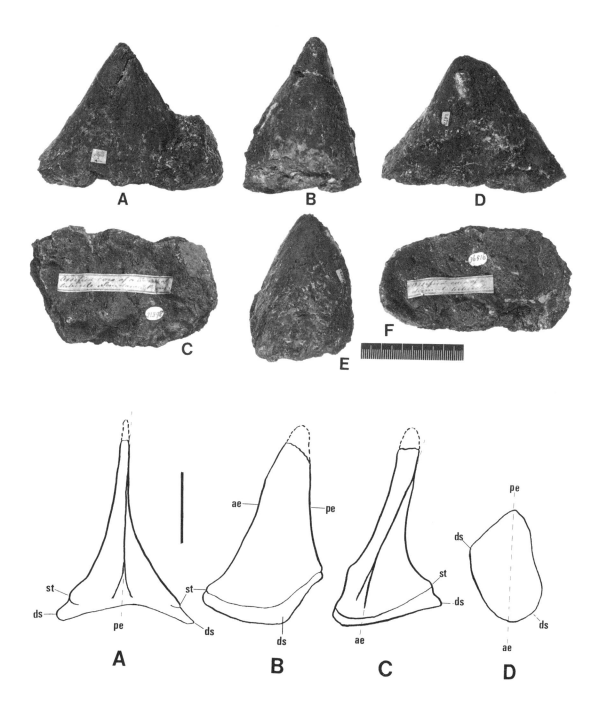

Figure 6.4. (top) BMNH 36515 (A–C) and BMNH 36516 (D–F) dermal spines in (A–D) ?lateral, (B, E) ?posterior, and (C, F) ventral (basal) views; from the Lower Cretaceous (Barremian) of the Isle of Wight. Scale in millimeters.

Figure 6.5. (bottom) MIWG 5307, dermal spine in (A) posterior, (B) ?lateral, (C) anterior, and (D) ventral (basal) views, from the Lower Cretaceous (Barremian) of the Isle of Wight. ae = anterior edge; ds = dermal spur; pe = posterior edge; st = step in the bone. Scale bar = 100 mm. Redrawn from Blows (1987).

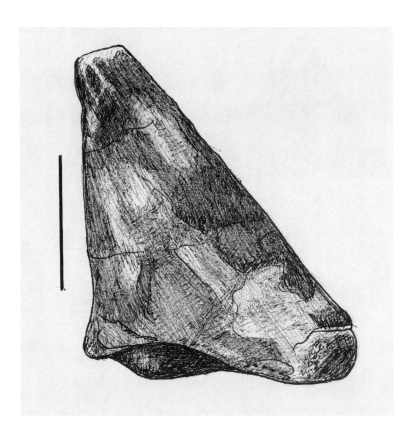

Figure 6.6. MIWG 5307, dermal
spine in ?medial view, from the
Lower Cretaceous (Barremian)
of the Isle of Wight.
Scale bar = 100 mm.

MIWG 1191a is a large, nearly complete dermal spine taken from the Lower Cretaceous Barremian beds of Sedmore Point, Isle of Wight. For many years, this specimen was labeled as "*Hylaeosaurus,*" but it is now known that *Hylaeosaurus* does not occur in the Barremian fauna (Blows 1998). It was subsequently referred to the ankylosaur *Polacanthus,* but, as with MIWG 5307, Blows (1987) pointed out differences from the usual type of polacanthine presacral spine (Type A) and included it as another Type B. As discussed for MIWG 5307 above, these differences may mean that MIWG 1191a could have been of stegosaurian origin (see character 4 in Discussion). If so, it is a rare specimen, but unfortunately, it had been sectioned both horizontally and vertically, and in 1996, only about one third of the bone was located. It is now unavailable for either full description or illustration. Neither MIWG 5307 nor MIWG 1191a appears to have a basal ridge, but if correctly assigned to the Stegosauria, these spines, together with the few others described above, would be among the few Barremian stegosaurian remains described from Britain and from Europe.

Discussion

Comparison of these spines with the armor of Jurassic and Lower Cretaceous stegosaurs is difficult. Stegosaurs generally had bilateral rows of plates close to the sagittal plane over the anterior part of the body, with a continuation of these rows as spines over the posterior part of the body. Armor may have existed outside of these two rows, but the evidence for this is weak. The loss of lateral rows of armor occurred in all stegosaurs more advanced than the Huayangosauridae (Sereno and Dong 1992). The exception are the parascapular spines of the Kentrosauridae and the throat ossicles of *Stegosaurus.* There appears to be no well-illustrated review of stegosaurian armor in the literature, which makes comparisons difficult, especially comparisons with the excellent Chinese specimens.

The following classification of the Stegosauria is based on that of Olshevsky (personal communication). The primitive group, the Huayangosauridae, primarily the Chinese Middle Jurassic stegosaur *Huayangosaurus,* had a double row of plates and pointed spines close to the sagittal plane and at least one lateral row of armor on each side (Sereno and Dong 1992). Most of the large, medial elements are flattened on the lateral–medial faces, and some have laterally expanded bases (Dong 1992; Dong and Milner 1988). Sereno and Dong (1992) described ovate, keeled scutes and considered these lateral rows of armor to be a primitive character for *Huayangosaurus.* Barrett and Upchurch (1995) considered the jaw of *Regnosaurus northamptoni* to compare closely with that of *Huayangosaurus,* and *Regnosaurus* may therefore be a member of this primitive family. Such an assignment is made difficult by the difference in age and location of the two taxa (*Huayangosaurus* from the Middle Jurassic of China and *Regnosaurus* from the Lower Cretaceous of Britain) and the lack of comparative material for *Regnosaurus.* If, however, *Regnosaurus* is closely assigned to *Huayangosaurus,* it too may have had a lateral row of armor on each side, possibly similar to those described above.

The Dacentruridae—that is, the Jurassic *Dacentrurus*—has large tail spines similar to those of *Stegosaurus* (Galton 1985, 1991). Casanovas-Cladellas et al. (1995b) describe a similar large caudal spine from the Wealden beds of Valencia, Spain, and from the same general locality, Casanovas-Cladellas et al. (1995a) tentatively refer some postcranial endoskeletal remains to *Dacentrurus,* which, if correctly assigned, extends this taxon into the Lower Cretaceous.

The Kentrosauridae, represented by the East African Upper Jurassic stegosaur *Kentrosaurus,* has spines that are large and elongate, with expanded bases, as well as long parascapular spines (Galton 1982). The armor of the European Jurassic *Lexovisaurus* is poorly known but is similar to that of *Kentrosaurus* in that both have a gradation of armor morphology from anterior plates to posterior spines and parascapular spines (Galton 1985; Galton et al. 1980; Sereno and Dong 1992). *Chungkingosaurus,* from the Upper Jurassic of China, has triangular, flattened, platelike spines, without laterally expanded bases, over the neck and trunk. The tail spines have a central rodlike thickening with some basal expansion (Dong 1992). The Chinese Jurassic stegosaur *Tuojiangosaurus* has large, elongate spines with expanded bases that occupied the last two thirds of the back and most of the tail. The smallest pair of spines, closest in size to the spines of the specimens discussed here, was just anterior to the caudal spikes. They differ from the British specimens by the lateral expansion of the bases. The spines have flattened lateral–medial faces (Dong 1992, and examination of a cast skeleton in London of BMNH R12158).

In the most advanced group of the Stegosauridae, exemplified by the American Upper Jurassic *Stegosaurus,* the spines are large, with angled and sometimes expanded bases that do not compare well with the much smaller Wealden specimens. In the Lower Cretaceous Chinese stegosaurid *Wuerhosaurus,* the armor is mostly unknown, except that the presacral plates are considerably shorter in height than their Jurassic counterparts (Dong and Milner 1988: 65).

Generally, the armor of Jurassic stegosaurs exceeds the size of most of the spines described here, and as with *Wuerhosaurus,* this may indicate an overall reduction in the size of the medial rows of armor from Jurassic to Cretaceous stegosaurs. Blows (2001) points out a gradual transition in the function of thyreophoran dermal armor over the latter half of the Mesozoic, and medial row plate size reduction in early Cretaceous stegosaurs, such as *Wuerhosaurus,* and perhaps the spines discussed here, if not from lateral rows, could be seen as part of that trend.

The major characters that appear to separate the British armor specimens from their polacanthine equivalent and also put them closer to a stegosaurian origin are as follows.

Character 1. Dorsal keel of the spine is flattened from side to side— that is, it is more platelike than the thicker polacanthine spines, which often have one face expanded into three surfaces. They cannot be assigned to polacanthine caudal plates, which have deeply hollowed, highly characteristic bases (Blows 1987).

Character 2. Dorsal keel edges from the base up are rounded and thicker, as opposed to sharper edges found in polacanthine specimens. Stegosaur armor has a mix of sharp and rounded edges within the various elements, but the only rounded edges found in *Polacanthus* armor occur toward the peak of the very largest spines.

Character 3. The edges are straighter than the polacanthine type, in which the edges twist to varying amounts (up to 90° in larger spines). All the stegosaurian armor I have seen has straight edges.

Character 4. The spines are elongated anteroposteriorly, which provides an oval shape to the base. This is the case in many stegosaurs where the armor elements are stretched longitudinally but remain relatively narrow transversely. The two Museum of Isle of Wight Geology specimens described above have broad bases (MIWG 1191a more so than MIWG 5307), but the dorsal keel fits characters 1, 2, and 3, so they may therefore represent a different Barremian stegosaurian taxon or armor that occurred outside the main bilateral rows.

Character 5. A low, centrally placed ridge extends across the solid, flattened base in several specimens. This ridge is also seen in other stegosaur spines, such as *Stegosaurus sulcatus* Marsh (USNM 4937) and *Dacentrurus armatus* (BMNH 46320 and BMNH 46322, the holotype of *Omosaurus hastiger* Owen, 1877). The ridge is not known in ankylosaur specimens and therefore may be known only in the Stegosauria. The function of this ridge was discussed by Lull (1910: 209), who states, "Some of the larger spines, notably . . . *Stegosaurus sulcatus*, have the base divided by an asymmetrically placed longitudinal ridge . . . into two facets which seem to have borne against the neural process and centrum of the vertebra," and he mentions this feature in *Dacentrurus*. Because this basal ridge is a characteristic of caudal spines in *Stegosaurus sulcatus* and *Dacentrurus armatus*, it may help place the Wealden specimens with basal ridges on the tail region of the animal. Flat or slightly excavated, solid bases may indicate a presacral position, with larger spines probably emanating from the trunk area.

The features listed here currently preclude the inclusion of these specimens within the Polacanthinae (unless new armor types that can be attributed to *Polacanthus* armor are found). Their tentative assignment to the Stegosauria is made on the grounds that they have a solid base (some centrally ridged), combined with a flattened, straight-edged, platelike quality of the dorsal keel. Identification of BMNH R15950 is problematic. It may be a previously unknown spine type of the Polacanthinae (e.g., a cervical spine) because the base is similar to the presacral spines of *Polacanthus*, or it could be stegosaurian because it has characters 2 and 3. It cannot currently be assigned to either thyreophoran group with any confidence.

Conclusion

The current names proposed for the British Cretaceous stegosaurs, *Regnosaurus northamptoni* and *Craterosaurus pottonensis*, should be retained for the holotypes only, because they are not comparable with each other or with the dermal bones described here. For this reason, attempts to synonymize these taxa are unjustified. Proposing new names for the dermal specimens is also inappropriate because their designation as stegosaurian is only tentative, and new taxa should not be based on isolated dermal bones. Therefore, these specimens are better assigned to ?Stegosauria *incertae sedis*.

Descriptions of stegosaur armor need to be standardized to include at least the five characteristics listed above. This standardization would allow direct comparisons between armor of the various taxonomic groups and provide a broader understanding of the role of dermal bones in stegosaurian systematics and evolution.

Acknowledgments. I thank Sandra Chapman (Natural History Museum, London) for assistance with the specimens and Dr. Angela Milner for access to the collections. I am also grateful to Dr. Peter Galton and Dr. Kenneth Carpenter for their valuable comments on the paper. The photographs are courtesy of the photographic unit of the Natural History Museum, London.

References Cited

Barrett, P. M., and P. Upchurch. 1995. *Regnosaurus northamptoni*, a stegosaurian dinosaur from the Lower Cretaceous of Southern England. *Geological Magazine* 132: 213–222.

Blows, W. T. 1987. The armored dinosaur *Polacanthus foxi* from the Lower Cretaceous of the Isle of Wight. *Palaeontology* 30: 557–580.

————. 1998. A review of Lower and Middle Cretaceous dinosaurs of England. In S. G. Lucas, J. I. Kirkland, and J. W. Estep (eds.), *Lower and Middle Cretaceous Terrestrial Ecosystems. New Mexico Museum of Natural History and Science Bulletin* 14: 29–38.

————. 2001. Dermal armor of the polacanthine dinosaurs. In K. Carpenter (ed.), *The Armored Dinosaurs*. Bloomington: Indiana University Press [this volume, Chapter 17].

Casanovas-Cladellas, M. L., J. V. Santafe-Llopis, and X. Pereda Suberbiola. 1995a. Nuevo material de estegosaurios en el Cretacico Inferior de Valencia (Aras de Alpuente, Localidad de Losilla I). *Paleontologia I Evolucio* 28–29: 269–274.

Casanovas-Cladellas, M. L., J. V. Santafe-Llopis, J. Pereda-Suberbiola, and C. Santisteban-Bove. 1995b. Presencia, por primera vez en Espana de dinosaurios estegosaurios Cretacico inferior de Aldea de Losilla, Valencia. *Revista-Espanola de Paleontologia* 10: 83–89.

Dong, Z. 1992. Stegosaurs of Asia. In K. Carpenter and P. J. Currie (eds.), *Dinosaur Systematics: Approaches and Perspectives,* pp. 255–268. Cambridge: Cambridge University Press.

Dong, Z., and A. Milner. 1988. *Dinosaurs from China.* London: British Museum (Natural History) and China Ocean Press.

Galton, P. M. 1981. *Craterosaurus pottonensis* Seeley, a stegosaurian dinosaur from the Lower Cretaceous of England, and a review of Cretaceous stegosaurs. *Neues Jahrbuch für Geologie und Paläontologie Abhandlungen* 161: 28–46.

————. 1982. The postcranial anatomy of stegosaurian dinosaur *Kentrosaurus* from the Upper Jurassic of Tanzania, East Africa. *Geologica et Palaeontologica* 15: 139–159.

————. 1985. British plated dinosaurs (Ornithischia, Stegosauridae). *Journal of Vertebrate Paleontology* 5: 211–254.

————. 1991. Postcranial remains of stegosaurian dinosaur *Dacentrurus* from Upper Jurassic of France and Portugal. *Geologica et Palaeontologica* 25: 299–327.

Galton, P. M., R. Brun, and M. Rioult. 1980. Skeleton of the stegosaurian dinosaur *Lexovisaurus* from the lower part of Middle Callovian (Middle Jurassic) of Argences (Calvados), Normandy. *Bulletin Société Géologique Normandie et Amis Muséum du Havre* 67: 39–60.

Lull, R. S. 1910. The armor of *Stegosaurus. American Journal of Science* 29: 201–210.

Lydekker, R. 1888. *Catalogue of the Fossil Reptilia and Amphibia in the British Museum. Part 1. Containing the Orders Ornithosauria, Crocodilia, Dinosauria, Squamata, Rhynchocephalia and Pterosauria.* London: British Museum.

Mantell, G. A. 1848. On the structure of the jaws and teeth of the *Iguanodon. Philosophical Transactions of the Royal Society (London)* 138: 183–202.

Owen, R. 1877. Monograph of the fossil Reptilia of the Mesozoic formations (*Omosaurus,* continued). *Palaeontographical Society Monograph* 33: 1–19.

Seeley, H. G. 1874. On the base of a large lacertilian cranium from the Potton Sands, presumably dinosaurian. *Quarterly Journal of the Geological Society* 30: 690–692.

Sereno, P., and Dong Z. 1992. The skull of the basal stegosaur *Huayangosaurus taibaii* and a cladistic diagnosis of Stegosauria. *Journal of Vertebrate Paleontology* 12: 318–343.

7. Posttraumatic Chronic Osteomyelitis in *Stegosaurus* Dermal Spikes

Lorrie A. McWhinney,
Bruce M. Rothschild, and
Kenneth Carpenter

Abstract

Trauma-related injuries affecting the bony structure can be seen in the fossil record as a unique opportunity to view certain life and death events. Evidence of trauma-induced bone fractures in the skeletal remains of animals that survived the trauma shows alterations in the bony structure that are seen as endosteal new bone formation (callus). The alterations seen that are the result of bone infections produce various degrees of bone resorption, bone formation, and necrosis. No callus or other bony alterations due to infection occurs postmortem. Although infectious disease resulting from trauma is well documented in the fossil record, infection in dermal armor is rare. In a sample of 51 *Stegosaurus* dermal tail spikes, 5 (approximately 10%) show trauma. Two of these spikes exhibit the effects of posttraumatic chronic osteomyelitis. Another spike either exhibits a contiguous spread of the disease or represents a separate injury with secondary infection. The pathologic spikes showed remodeled bone growth, indicating that each *Stegosaurus* survived after a traumatic event occurred that resulted in broken spikes. The outer cortical surface exhibits deposits of extraneous bone as well as prominent areas of filigree bone. Drainage areas (sinus cavities) for the evacuation of pyogenic material are noted at or near the broken ends of two spikes. The disease exhibited in two of three individuals was a direct response to the traumatic open fractures. Open fractures have a higher rate of disturbances in the healing process because of the increased frequency of infection. The open fracture allowed bacteria to enter the wound, resulting in posttraumatic chronic osteomyelitis.

Introduction

Although the pathogenesis of osteomyelitis is a complicated process, it is regarded as an inflammation of the bone and marrow (Resnick and Niwayama 1988). Osteomyelitis may be either pyogenic (pus-producing) or nonpyogenic (Rothschild and Martin 1993) and can affect any bone. Infection in bone is a result of the interplay between the pathogen, the susceptible host, and any predisposing environmental factors (Jauregui and Senour 1995). Once the infection begins, bacterial colonization is produced within an exopolysaccharide biofilm (glycocalyx), which adheres to devitalized bone (Gristina et al. 1990). In cases of chronic osteomyelitis, there is a prolonged microbial infection of the bone (McCarthy 1996) that is extremely difficult and sometimes impossible to eradicate, even with antibiotics. *Staphylococcus aureus* is the most common bacteria to cause bone infection because of its association with trauma and tissue destruction; however, other microorganisms, such as *Pneumococcus, Streptococcus, Pseudomonas,* and *Escherichia coli,* as well as several forms of fungi and viruses, have also been reported as etiologic agents (Gupta and Frenkel 1995). Complications from osteomyelitis can include chronic osteomyelitis, septicemia (blood poisoning), metastatic infection (most often seen in children), endocarditis, amyloidosis, and acute pyogenic arthritis. Depending on its etiology, osteomyelitis can sometimes be distinguished between endogenous and exogenous forms. Hematogenous osteomyelitis develops via the bloodstream, whereas contiguous osteomyelitis develops when the infection of the bone is a result of direct implantation (e.g., open fracture) that allows the microbial infection to have direct access to the bone. Open fractures have a higher rate of disturbances in the healing process, the result of the increased frequency of infection, which corresponds to posttraumatic osteomyelitis.

Not all fractures, however, will become infected. When a fracture occurs, the bone goes through a series of events that morphologically alter its appearance. When a bone fractures, the alteration in its structure is seen as endosteal new bone formation (callus). Although the majority of fractures will heal, wounds that remain open have a greater chance for contamination (Resnick and Niwayama 1988). Besides infection, there are several different factors that can affect the healing process of the fracture. These include the degree of trauma or bone loss, the type of bone involved (cancellous or cortical bone), or an underlying pathologic process (e.g., metabolic neoplasm or other disorders). In humans, approximately 7–10 days is required to see the early antemortem reparative changes in bone (Aufderheide and Rodriguez-Martin 1998), and a minimum of two weeks' postfracture healing is required to recognize alterations in bone (Mann and Murphy 1990). Rhinelander et al. (1968) noted that the earliest sign of callus development is in the third week after injury on the basis of a study of dogs with limited fracture immobilization.

Wolff's law hypothesizes that every change in the form or function of a bone is followed by adaptive changes in its internal architecture and its external shape (Rubin et al. 1994). Trauma-induced injuries affecting the bony structure can be seen in the fossil record as unique opportunities to view these adaptive changes and how they affected certain life and death events. Taphonomic evidence can indicate whether trauma to bone occurred during life or after death. The alterations seen in all bone infections produce varying degrees of bone resorption, bone formation (e.g., involucrum), and necrosis (bone death). In chronic osteomyelitis, this change can be recognized by several different characteristics that include drainage sinus tracks, bony sequestra, reactive new bone having a filigree appearance on the bone surface, and, internally, a disorganized trabecular pattern. On rare occasions, chronic osteomyelitis produces a well-circumscribed osteolytic (bone resorbing) lesion, which is surrounded by a rim of reactive bone, called a Brodie abscess. New bone formation is a response of the periosteum being elevated from the cortex by the infectious process. Active infection destroys bone faster than the bone can rebuild, causing the newly formed bone to have an immature, fibrous (filigree) appearance and disorganized trabeculae.

Increased intraosseous pressure can also interfere with the normal blood flow. In attempting to relieve the internal pressure, drainage tracks in the bone find routes from the medullary cavity through the Haversian systems and Volkmann canals to the outer cortex for the pyogenic material to ooze out. When the blood supply is cut off by thrombosis of the nutrient artery, the bone becomes avascular, resulting in bone death. In cases of untreated osteomyelitis, the obliteration of the medullary cavity is common.

The results of the infectious process may produce bone that is visually quite different from normal bone. In cases of chronic osteomyelitis, the bone may have a swollen appearance, with rugose, extraneous bone deposits located in the infected site, as well as irregular-shaped holes at the drainage sites. No callus or other bony alterations due to infection can form after death. Different processes such as aneurysmal bone cysts, malignant and benign tumors, and fibrous dysplasia can be ruled out as a primary diagnosis because of the bony characteristics associated with osteomyelitis.

Although secondary infectious disease of the bone resulting from trauma is well documented (Moodie 1923; Rothschild and Martin 1993), it is not commonly observed in the fossil record. These rare events seem to be biased toward predominantly bipedal, rather than quadrupedal, posture, or they were incompatible with survival (Rothschild and Tanke 1992). Bennett (1989) reported on pterosaurs that exhibited deformed bones, apparently resulting from injury and infection, with subsequent repair. Moodie (1926) reported an infected fracture in a hadrosaur manus. Moodie (1921) also described a *Dimetro-*

don skeleton having a transverse fracture with sinus cavities due to chronic osteomyelitis near the base of the dorsal vertebral spine. This spine showed a roughened and swollen appearance due to callus formation around the fracture site and from the ensuing infection. In cross section, highly developed sinuses were seen, which in life were filled with pus. Moodie (1923) described the histology of the pathologic bone as having a few, small, and widely scattered lacunae with short and unbranched canaliculi throughout the osteoid substrate.

Marshall et al. (1998) reported a healing fracture with secondary osteomyelitis in the second phalanx of the second pedal digit of *Deinonychus antirrhopus*. One of the only known examples of a Brodie abscess has been reported on the left tibia of a partial, articulated hypsilophodont skeleton (Gross et al. 1993). A fracture with associated secondary infection was noted by Moodie (1923) in the caudal vertebrae of an *Apatosaurus*. This specimen has been rediagnosed by Rothschild and Berman (1991) as ossification of ligaments spanning consecutive centra, or diffuse idiopathic skeletal hyperostosis (known as DISH), and not a pathologic response to injury.

Most examples of osteomyelitis in the fossil record occur in the endoskeleton. Gilmore (1914) briefly mentioned broken tail spikes in *Stegosaurus,* but he did not illustrate them. The recent discovery of another example of a broken tail spike associated with a nearly complete *Stegosaurus stenops* (Carpenter 1998) has led us to reexamine Gilmore's specimens and to survey collections that contain *Stegosaurus* tail spikes for additional pathologies.

Institutional Abbreviations. DMNH: Denver Museum of Natural History, Denver, Colorado. USNM: National Museum of Natural History (formerly United States National Museum), Washington, D.C.

Materials and Methods

A survey of *Stegosaurus* dermal tail spikes from the Morrison Formation (Upper Jurassic) was conducted in the collections of the American Museum of Natural History, the Yale Peabody Museum, the Denver Museum of Natural History, and the Smithsonian Institution. The material from the National Museum of Natural History included the bulk of the material from Quarry 13, Como Bluff, Wyoming, and represents the largest portion of this analysis. A total of 51 spikes were examined, of which 5 (approximately 10%) show trauma, and 2 of these also exhibit posttraumatic chronic osteomyelitis. One additional spike exhibits either an early contiguous spread of the disease (an infection that is in contact with or in close proximity to the original infected site) or represents a separate injury with secondary infection.

A pair of pathologic spikes occurs with an adult *Stegosaurus stenops* (DMNH 2818) found by Bryan Small in 1992 at Garden Park, Colorado. The nearly articulated skeleton was found on its left side in lenticular gray lacustrine mudstone with numerous disarticulated nonstegosaurian vertebrate elements. The third pathologic dermal spike is that of an adult *S. ungulatus* (USNM 6646). Originally described by Gilmore (1914), this isolated spike is from Quarry 13, 12.9 km (8 miles) east of Como, Wyoming, and was found with associated disarticulated elements of *Stegosaurus*. The bones were found in a layer of sandy clay between bands of brown and green marl that may be interpreted as a lacustrine environment. The total length of the *S. ungulatus* (USNM 6646) pathologic right dermal tail spike is 38.0 cm. In comparison, the total length of a normal *S. stenops* (DMNH 1483) left anterior dermal tail spike is 57 cm.

Evidence for osteomyelitis in the 51 spikes was first determined by visual examination. Three of those displaying osteomyelitis were also examined with computed tomography (CT). They included the two spikes from DMNH 2818 and one spike from USNM 6646. They were also compared with CT scans of a normal *S. stenops* tail spike (DMNH 1483). CT scans of DMNH 2818 and DMNH 1483 were performed by use of 1.5-mm axial and sagittal cuts, whereas those of USNM 6646 were performed by use of 3-mm axial and sagittal cuts.

Thin sections were made of a right anterior pathologic spike (DMNH 2818) and the normal spike (DMNH 1483). Transverse and longitudinal thin sections of the normal cortical bone (DMNH 1483) were taken approximately 21 cm from the base. Thin sections of the pathologic cortical bone (DMNH 2818) were taken from two sites. One section was taken 9.3 cm from the base on the anterior margin, and the other sample was taken 6.9 cm from the distal end on the posterior margin of the spike. No thin sections of USNM 6646 were obtained for this study.

Results

The tail spikes of adult *S. stenops* and *S. ungulatus* project in a posterolateral position to the caudal vertebral column (Carpenter 1998). The spikes have oblique, rugose bases that are not expanded and an elongated shaft tapering to a point. In cross section, the spikes are subcircular at approximately the midshaft. The total length of the spikes vary, with the anterior pair always larger than the posterior pair. In life, the spikes probably were covered with a horny sheath and embedded by their bases in thick skin. Gilmore (1914) observed that the anterior spikes were apparently more deeply embedded than the posterior pair and that the anterior spikes were more prone to injury and fracture than the posterior pair. The longitudinal grooves that run along the outer cortical layer represent vascular impression. The nutrient foramen was located on the medial side of the base, with the nutrient artery extending up the interior of the shaft, as determined by CT scans. The internal bony structure was found to be cancellous in cross section, which, according to Gilmore, implied that the spikes were lightweight in life.

Macroscopic Analysis

In contrast, an adult *S. stenops* (DMNH 2818) exhibits the effects of posttraumatic chronic osteomyelitis on the right anterior dermal tail spike. The left anterior dermal tail spike exhibits either an early contiguous spread of the disease or represents a separate injury with sec-

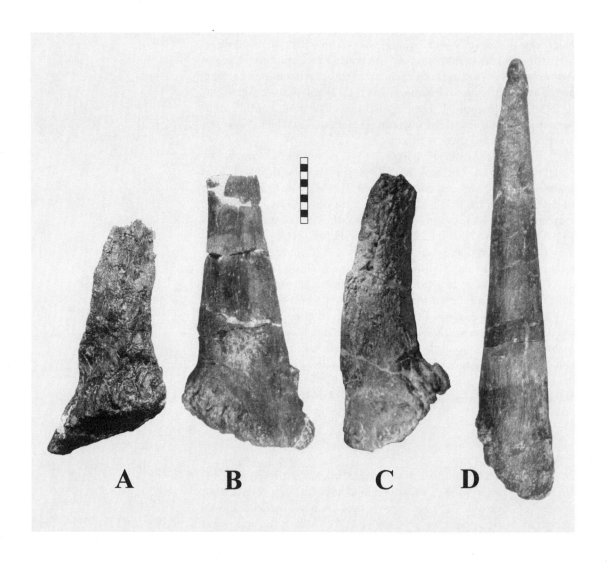

Figure 7.1. Pathologic dermal tail spikes (A–C), with a normal tail spike (D) for comparison. (A) Dorsal view of Stegosaurus stenops *(DMNH 2818) right anterior spike. (B) Dorsal view of* S. stenops *(DMNH 2818) left anterior spike. (C) Ventral view of* S. ungulatus *(USNM 6646) right anterior spike. (D) Dorsal view of* S. stenops *(DMNH 1483) anterior spike.*

ondary infection. The total length of the pathologic right anterior dermal spike is 29.2 cm, which is 78% of the corresponding pathologic left anterior dermal spike (37.5 cm), and about half the length of a normal anterior spike from a same-size adult (57.0 cm; DMNH 1483). The right dermal tail spike of *S. ungulatus* (USNM 6646) also exhibited visual signs of osteomyelitis (Fig. 7.1). The total length of the *S. ungulatus* (USNM 6646) pathologic right dermal spike is 38.0 cm.

The pathologic right anterior dermal spike of DMNH 2818 was found lying in a posterolateral position across the corresponding pathologic left anterior dermal spike. This position resulted in the dorsal distal end of the right spike to mold over the dorsal surface of the left spike during compaction of the matrix sediments (Carpenter 1998). The right spike has an open fracture at the distal end of the spike, with some callus formation around the opening. It has been dorsoventrally flattened into a wedge shape, with thinning of the bone starting from the proximal rugose base and decreasing to a feather edge at the broken

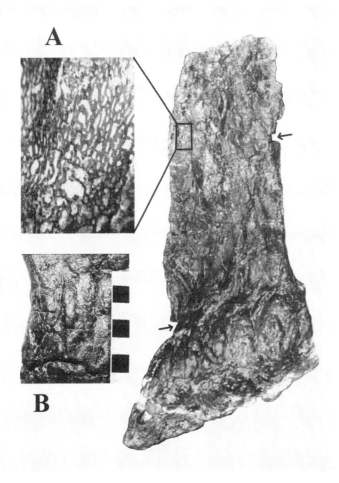

distal end. This crushing produced sharply defined edges along the anterior and posterior margins. The external surface of the pathologic right spike has a waxy or eburnated appearance and is marked by a pronounced network of ridges and furrows along the length of the spike on the ventral side (Fig. 7.2). The furrows do not correspond to grooves for blood vessels feeding the horny sheath. Instead, the ridges most likely indicate areas of reactive new bone formation or involucrum. The involucrum, representing a layer of living bone, formed around the necrotic bone, eventually merging with the parent bone. The furrows in the involucrum are perforated by sinus tracks, where pyogenic material could escape from the medullary cavity. The dorsal side of the spike is curved, slightly twisted obliquely, and lacks the extraneous bone formation found on the ventral side.

Areas of bone with a filigree appearance (Fig. 7.2) occur at the dorsoposterior and ventroposterior distal ends of the spike. Subperiosteal cavities resulting in a pitted surface (Fig. 7.3) are observed most notably near the proximal ventral end of the specimen, as well as on the dorsoposterior distal end. With loss of the intraosseous space, cortical

Figure 7.2. The pathologic characteristics of Stegosaurus stenops *right anterior dermal tail spike (DMNH 2818). (A) Close-up in dorsal view of bone with filigree appearance, which is outlined in the box. (B) Close-up in ventral view of reactive new bone formation (involucrum). Arrows denote areas of bone that were removed for thin-section analysis.*

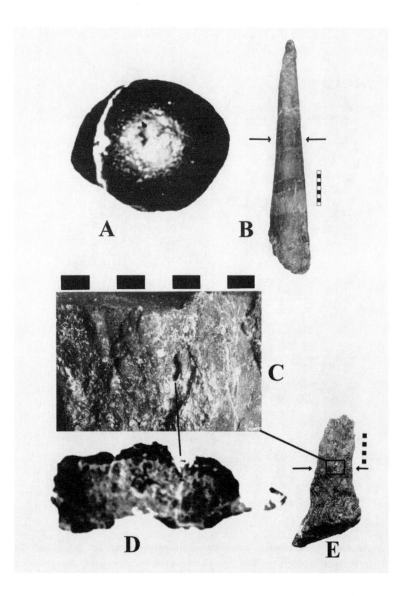

Figure 7.3. Comparison of normal and pathologic bone structure using computed tomography (CT) scans and photographic images. (A) CT scan of Stegosaurus stenops *(DMNH 1483) showing normal bone structure. The fracture line through the CT slice is an artifact. (B) Photographic image of S. stenops (DMNH 1483). Arrows point to CT slice shown in Figure 7.3A. (C) Close-up image of drainage site located on the dorsal side of the right anterior dermal tail spike S. stenops (DMNH 2818) in association with (D) CT scan of pathologic bone structure and (E) photographic image.*

bone lies on top of cortical bone at the distal end of the broken spike. The spike appears to have periosteal reaction throughout, which would indicate generalized infection. The nutrient foramen is not visible in the base of the spike, possibly due to compaction by the sediments or due to disease. The abnormal internal structure of the bone is seen in cross section along a break at approximately the middle of the spike. When compared with the normal dermal spike (DMNH 1483), the pathologic spike showed no evidence of a normal medullary cavity or cancellous bone. The absence of normal bone structure was confirmed by the CT images (Fig. 7.3), which showed disorganization of the trabecular framework and remnants of a medullary cavity.

The corresponding left anterior tail spike of DMNH 2818 also exhibits the visible signs of an early infectious process. The tip of the

drainage site

drainage site

bone spicule

Figure 7.4. Close-up image of the proximal end of Stegosaurus stenops *(DMNH 2818) left anterior dermal tail spike showing pathologic characteristics.*

spike was unfortunately lost during the excavation of the skeleton. In cross section, along a break at approximately one third of its length, the spike appears to have a normal internal structure in comparison to the corresponding right anterior spike and the normal spike (DMNH 1483). The nutrient foramen is seen piercing the base, and its duct is seen in CT images as extending into the spike. The duct extends 3 cm into the interior of the spike before bifurcating and terminates approximately 4.5 cm from the base. The widest margin of the nutrient foramen is 4 mm.

A spiculated, periosteal reaction occurs 9 cm from the base, on its anterior margin (Fig. 7.4). The spicule measures 2 cm in length. Different processes may cause a spiculated appearance in bone, including osteomyelitis, hemangiomas, anemia, trauma, and several different types of localized primary malignant tumors and bone metastases. Because this spicule is relatively short and squat, it is considered benign, as opposed to malignant spicules, which tend to form long and slender structures (Greenfield 1986). The formation of spicules is due to disturbance in the reparative stage that follows periosteal elevation. The spicule is formed along stretched periosteal vessels and extensions of Sharpey fibers.

There is a single 3.2- by 0.2-cm depression and a ridge of extraneous bone that runs slightly distal and medial to the spicule on the dorsal side. This most likely corresponds to an area of involucrum. Periosteal reaction starts 10.2 cm from the base of the spike and extends 12.8 cm. It has a filigree appearance, with a single drainage site on the dorsal side, medial to the depression. A smaller drainage site is located near the base of the spike, close to the posterior margin. Although this may represent a nutrient foramen, it was not observed in the other specimens in the study. The pathology seen on the left spike is minimal when compared with the right spike and may represent either an early contiguous spread of the disease or a separate injury from the right spike with secondary infection. The skeleton was otherwise unaffected by the infectious process observed in the pathologic dermal spikes.

USNM 6646 was described by Gilmore (1914) as being curved and thickened along the shaft and in having the top broken off, with subsequent healing (callus). We reexamined this spike and found it to have characteristics similar to those of the pathologic spikes of DMNH 2818 (Fig. 7.6). The spike has an open fracture at the distal end, with callus formation around the opening. Although the outer cortical surface has a roughened, pitted, and swollen appearance due to callus formation and disease, the spike maintained a comparatively normal contour. Because the spike was not crushed, it shows more definition in the drainage cavities than the right pathologic spike of DMNH 2818. There are numerous drainage sites in USNM 6646, the majority located around the tip of the fractured distal end of the spike. The drainage cavities are irregularly shaped, tapering shafts. These holes in the bone do not have the appearance of those made by bone-boring organisms, which would have nontapering shafts or a punched-out appearance (Rogers 1992). A network of sinus tracks interconnect the drainage sites. We noted filigree periosteal reaction on both the ventral and dorsal sides of the spike. The nutrient foramen can be seen piercing the base. Normal vascular scarring is not apparent. The base of the spike is not abnormally expanded.

Analysis of the CT scans of USNM 6646 showed the outer cortical margin to be irregular in appearance, with remodeled bone growth. From the base of the spike to approximately 20 cm from the base, there appears to be a normal trabecular pattern and medullary cavity. Beyond this level, there is an abrupt change within the medullary cavity. An irregular, ovoid-shaped radiolucent mass (Fig. 7.6), 3.5 by 2.0 cm, is seen right of the midline, extending dorsally and ventrally through the medullary cavity into the remodeled cortical bone. The margins of the radiolucent mass are poorly defined. The mass is located near one of the drainage sites seen by visual examination on the surface. Within this mass are small, circular densities that lack trabecular structure and may represent normal structures (e.g., vascular canals). Some of the densities probably represent sinus cavities. The mass may represent a focal area of destruction due to infection. Analysis of the CT scans also revealed that the nutrient artery entered the diaphysis at an oblique angle from the nutrient foramen, then progressed medially up and terminating approximately 13 cm from the base.

Microscopic Analysis

Thin sections were made of the pathologic right anterior spike of DMNH 2818 as well as the normal left anterior spike of DMNH 1483 (Fig. 7.5). The transverse and longitudinal thin sections of the normal spike, viewed at 40× magnification, revealed a normal Haversian system that is very dense in appearance. Primary osteons (primary compact bone) can be seen in the cortical zone layer that transitions to secondary osteons (secondary compact bone) in the endosteal zone seen

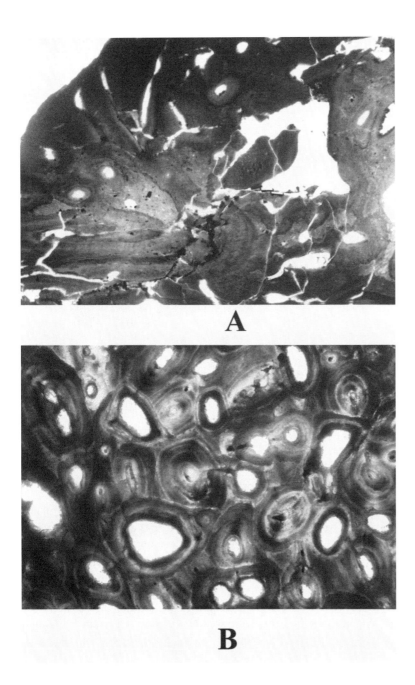

A

B

in transverse thin sections. The secondary osteons are of various ages, with cementing or reversal lines and interstitial remnants. Lacunae and canaliculi are observed. Howship lacunae, which indicate areas of bone resorption, are also visible. Haversian canals can be seen in the longitudinal thin sections. The bone tissue is a highly vascularized azonal fibrolamellar type. As reported by Reid (1990), the occurrence of this type of bone tissue indicates a rapid rate of bone deposition that does not imply ectothermy or endothermy.

Figure 7.5. Thin sections of the pathologic (top) Stegosaurus stenops (DMNH 2818) right anterior dermal tail spike and normal (bottom) S. stenops (DMNH 1483) left anterior dermal tail spike using 40× magnification.

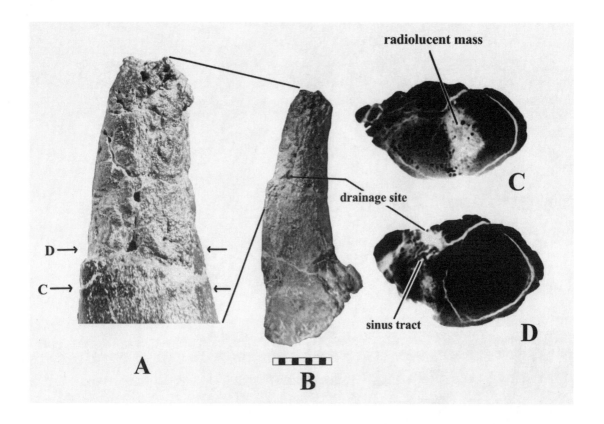

Figure 7.6. Pathologic structure of Stegosaurus ungulatus *(USNM 6646) using computed tomography (CT) scans and plain imaging. (A) Close-up in ventral view of the distal end of the dermal tail spike. (B) Ventral view of dermal tail spike with drainage site. (C) CT image of radiolucent mass correlated with the location of the mass on the close-up plain imaging. (D) CT image of drainage site and sinus tract correlated with the location on the close-up photographic imaging.*

In contrast, the pathologic spike DMNH 2818 at 40× magnification showed abnormal microstructure in the transverse samples (transverse and longitudinal samples were difficult to cut because of poor permineralization). The thin sections revealed an absence of dense Haversian system, and only small and few, irregularly spaced osteons (Fig. 7.5). The lumens of the remaining osteons are uniform in their small size, suggesting that the secondary osteons represent mostly old osteons when compared with the normal thin sections at the same magnification. Distinct cementing or reversal lines and interstitial remnants were not observed in these samples. At 100× magnification, more of the abnormal microstructure was observed in one of the transverse samples. No other microstructure could be identified in the transverse samples.

Discussion

The taphonomic context of DMNH 2818 indicated that the animal may have died in a drought (Carpenter 1998). In the Late Jurassic, droughts occurred throughout the Western Interior, including the Garden Park area (Demko and Parrish 1998; Evanoff and Carpenter 1998). The habitat of the *Stegosaurus* included areas farther away from water sources (Galton 1997), although Dodson et al. (1980) reports several different environments. Skeletal remains are frequently located in channel sands, but in the case of *S. stenops* (DMNH 2818), the skeleton was found in lacustrine mudstone. During times of increased stress due to drought conditions, animals tend to stay close to a water source (Behrensmeyer and Boaz 1980; Cornfield 1973). Death in a drought is not usually from the lack of water but from malnutrition and other associated problems (Carpenter 1987). Often there is increased death during droughts, with some carcasses left untouched by scavengers. Lack of scavenging on this specimen would also support the notion that death was drought related. Drought conditions and the infection in the spike were probably factors in the animal's death.

Marsh (1880) suggested that the armor of the *Stegosaurus* was used for offensive and defensive behavior, although Gilmore (1914) thought the primary usage was for display. Since then, intraspecific behavior in dinosaurs has been suggested by Molnar (1977) and interspecific behavior by Galton (1990) and Carpenter (1998). Molnar (1977) commented on three different types of intraspecific display and the possibility of such behavior among dinosaurs. These included courtship display, combat-related display (threat or dominance), and combat. Horned antelopes engage in intraspecific encounters between males that range from low-intensity play fights to fierce conflict (Allsopp 1978). Such encounters may result in severe injuries (Ritter and Bednekoff 1995). As noted by Spassov (1979), *Stegosaurus* armor was ideally arranged for lateral display of agonistic behavior. In quadrupedal ornithischians, there is evidence of caudal involvement in combat or display (Molnar 1977). Although the tails of dinosaurs with ossified tendons would have had a significant reduction of lateral motion, the tail of *Stegosaurus* lacks ossified tendons and therefore still had the ability to deliver damaging blows (Carpenter 1998). In modern reptiles, the use of the tail for combat and combat-related display is widely known (Carpenter 1961; Holland 1915; Milstead 1970; Pooley and Gans 1976). The usage of the dermal armor by *Stegosaurus* is demonstrated by the breakage and subsequent healing, infection, or both of the spikes. The spikes were likely viewed by predators as the main obstacle between them and their next meal.

Conclusions

On the bases of visual examination and analysis of CT scans of the *Stegosaurus* dermal tail spikes, the right anterior dermal spike of *S. stenops* (DMNH 2818) and right dermal spike of *S. ungulatus* (USNM 6646) suffered from posttraumatic chronic osteomyelitis. Analysis of the thin sections taken from *S. stenops* (DMNH 2818) supports a pathologic condition due to the abnormal microstructure. This disease originated in a microbial infection in the open fractured spikes, causing an exogenous form of osteomyelitis, which then progressed into a chronic state. The left anterior dermal spike of *S. stenops* (DMNH 2818) exhibits either an early contiguous spread of the disease or represents a separate injury with secondary infection.

The open fracture, with secondary infection, indicates that both stegosaurs survived the initial trauma of the broken spikes, because it takes approximately two to three weeks to recognize antemortem reparative changes in bone. In the chronic infectious state, there can be periods of remission and intermittent flare-ups over a period of months to years. Though the disease might not have been initially fatal, once a pyogenic infection takes hold, it is nearly impossible to eradicate. Eventually, however, other physiological and environmental factors (e.g., drought-induced stress) might have compounded the affliction, debilitating the animals and leading to their deaths.

Gilmore (1914) thought the purpose of the spikes of *Stegosaurus* was primarily to make the animal appear formidable; he did not believe that the spikes were actively used as a defensive weapon. We suggest, as evidenced by trauma, infection, or both sustained in the spikes, that the primary purpose of the *Stegosaurus* dermal tail spikes was to actively use them in defense and offense posturing in interspecific and intraspecific combat.

Finally, analysis of the pathologic spikes indicates that the horny sheath, which is believed to have covered the bone spike, must not have extended very far beyond the tip of the spike. Otherwise, the tip of the horny sheath would have broken off, rather than the tip of the bone spike. As we will show elsewhere, the sheath was probably less than 1 cm beyond the bone spike (Carpenter, McWhinney, and Sanders, in preparation).

Acknowledgments. We thank Mary Ann Turner (Yale Peabody Museum) and Robert Purdy (National Museum of Natural History) for access to the *Stegosaurus* specimens in their care. We also thank all the staff in the Radiology Department at Kaiser Permanente, especially radiologists Thomas Barsch, M.D., Margaret Montana, M.D., Richard Obregon, M.D., and Darwin Kuhlmann, M.D., and radiological technologists Kim Hardy and Ann Canavan, for CT scans, and Beth Carpenter and Erin Erskine for radiographic analysis. We thank Rick Spurlock, M.D., Department of Pathology, Kaiser Permanente, for his comments and interpretations. L.A.M. would also like to extend special thanks to Pamela Isaacs, D.O., radiologist, Department of Radiology, University Hospital, for support and friendship; Kail Goplin, pathologist assistant, Department of Pathology, St. Joseph Hospital, for photomicroscopy; and Ruth Barrett, histologist, Department of Histology, St. Joseph Hospital, for technical support and friendship.

References Cited

Allsopp, R. 1978. Social biology of bushbuck (*Tragelaphus scriptus* Pallus 1776) in the Nairobi National Park, Kenya. *East Africa Wildlife Journal* 16: 153–165.

Aufderheide, A. C., and C. Rodriguez-Martin. 1998. *The Cambridge Encyclopedia of Human Paleopathology.* Cambridge: Cambridge University Press.

Behrensmeyer, A., and D. Boaz. 1980. The recent bones of Amboseli National Park, Kenya, in relation to East African paleoecology. In A. Behrensmeyer and A. Hill (eds.), *Fossils in the Making,* pp. 72–92. Chicago: University of Chicago Press.

Bennett, S. C. 1989. Pathologies of the large pterodactyloid pterosaurs *Ornithocheirus* and *Pteranodon* [abstract]. *Journal of Vertebrate Paleontology* 9: 13A.

Carpenter, C. C. 1961. Patterns of social behavior in the desert iguana, *Dipsosaurus doralis. Copeia* 1961: 396–405.

Carpenter, K. 1987. Paleoecological significance of droughts during the Late Cretaceous of the Western Interior. In P. Currie and E. Koster (eds.), *Fourth Symposium on Mesozoic Terrestrial Ecosystems, Short Papers. Occasional Papers of the Tyrrell Museum of Palaeontology* 3: 42–47.

————. 1998. Armor of *Stegosaurus stenops*, and the taphonomic history of a new specimen from Garden Park, Colorado. In K. Carpenter, J. Kirkland, and D. Chure (eds.), *Interdisciplinary Study of the Morrison Formation, Upper Jurassic. Modern Geology* 23: 127–144.

Cornfield, T. 1973. Elephant mortality in Tsavo National Park, Kenya. *East Africa Wildlife Journal* 11: 339–368.

Demko, T. M., and J. Parrish. 1998. Paleoclimate setting of the Upper Jurassic Morrison Formation. In K. Carpenter, J. Kirkland, and D. Chure (eds.), *Interdisciplinary Study of the Morrison Formation, Upper Jurassic. Modern Geology* 22: 283–296.

Dodson, P., A. K. Behrensmeyer, R. T. Bakker, and J. S. McIntosh. 1980. Taphonomy and paleoecology of the Upper Jurassic Morrison Formation. *Paleobiology* 6: 208–232.

Evanoff, E., and K. Carpenter. 1998. History sedimentology and taphonomy of Fetch Quarry and associated sandbody in the Morrison Formation near Garden Park, Fremont County, Colorado. In K. Carpenter, J. Kirkland, and D. Chure (eds.), *Interdisciplinary Study of the Morrison Formation, Upper Jurassic. Modern Geology* 22: 145–169.

Galton, P. 1990. Stegosauria. In D. B. Weishampel, P. Dodson, and H. Osmólska (eds.), *The Dinosaria*, pp. 427–455. Berkeley: University of California Press.

————. 1997. Stegosaurs. In J. O. Farlow and M. K. Brett-Surman (eds.), *The Complete Dinosaur*, pp. 291–306. Bloomington: Indiana University Press.

Gilmore, C. 1914. Osteology of the armored dinosauria in the United States National Museum, with special reference to the *Stegosaurus. U.S. National Museum Bulletin* 89: 1–136.

Greenfield, G. B. 1986. *Radiology of Bone Diseases*. Philadelphia: Lippincott.

Gristina, A. G., P. T. Naylor, and Q. N. Myrvik. 1990. Musculoskeletal infection, microbial adhesion, and antibiotic resistance. *Infectious Disease Clinics of North America* 4: 391–406.

Gross, D., T. H. Rich, and P. Vickers-Rich. 1993. Dinosaur bone infection. *National Geographic Research and Exploration* 9(3): 286–293.

Gupta, M., and L. D. Frenkel. 1995. Acute osteomyelitis. In L. E. Jauregui (ed.), *Diagnosis and Management of Bone Infection*, Vol. 16, pp. 1–21. New York: Marcel Dekker.

Holland, W. J. 1915. Heads and tails, a few notes relating to the structure of sauropod dinosaurs. *Annals of the Carnegie Museum* 9: 273–278.

Jauregui, L. E., and C. L. Senour. 1995. Chronic osteomyelitis. In L. E. Jauregui (ed.), *Diagnosis and Management of Bone Infection*, Vol. 16, pp. 37–49. New York: Marcel Dekker.

Mann, R. W., and S. P. Murphy. 1990. *Regional Atlas of Bone Disease*. Illinois: C. C. Thomas.

Marsh, O. C. 1880. Principal characteristics of American Jurassic dinosaurs, part III. *American Journal of Science* 19: 253–259.

Marshall, C. D., D. Brinkman, R. Lau, and K. Bowman. 1998. Fracture and osteomyelitis in PII of the second pedal digit of *Deinoychus antir-*

rhopus (Ostrom) an Early Cretaceous "raptor" dinosaur [abstract]. In *Palaeontological Association Annual Meeting Abstracts* (n.p.).

McCarthy, E. F. 1996. *Differential Diagnosis in Pathology Bone and Joint Disorders.* New York: Igaku-Schoin Medical Publishers.

Milstead, W. W. 1970. Late summer behavior of the lizards *Sceloporus merriami* and *Urosaurus ornatus* in the field. *Herpetologica* 26: 343–354.

Molnar, R. E. 1977. Analogies in the evolution of combat and display structures in ornithopods and ungulates. *Evolutionary Theory* 3: 165–190.

Moodie, R. L. 1921. Osteomyelitis in the Permian. *Science* 53: 333.

———. 1923. *Paleopathology.* Chicago: University of Chicago Press.

———. 1926. Excess callus following fracture of the fore foot in a Cretaceous dinosaur. *Annals of Medical History* 8: 73–77.

Pooley, A. C., and C. Gans. 1976. The Nile crocodile. *Scientific American* 234: 114–124.

Reid, R. E. H. 1990. Zonal "growth rings" in dinosaurs. *Modern Geology* 11: 133–154.

Resnick, D., and G. Niwayama. 1988. Osteomyelitis, septic arthritis, and soft tissue infection: The mechanisms and situations. In *Diagnosis of Bone and Joint Disorders,* pp. 2525–2618. Philadelphia: W. B. Saunders.

Rhinelander, F. W., R. S. Phillips, W. M. Steel, and J. C. Beer. 1968. Microangiography in bone healing. II. Displaced closed fractures. *Journal of Bone and Joint Surgery* 50A: 643–663.

Ritter, R. C., and P. A. Bednekoff. 1995. Dry season water, female movements and male territoriality in springbok: Preliminary evidence of waterhole-directed sexual selection. *African Journal of Ecology* 33: 395–404.

Rogers, R. R. 1992. Non-marine borings in dinosaur bones from the Upper Cretaceous Two Medicine Formation, Northwestern Montana. *Journal of Vertebrate Paleontology* 12: 528–531.

Rothschild, B. M., and D. S. Berman. 1991. Fusion of Caudal Vertebrae in Late Jurassic Sauropods. *Journal of Vertebrate Paleontology* 11: 29–36.

Rothschild, B. M, and L. D. Martin. 1993. *Paleopathology Disease in the Fossil Record.* Boca Raton, Fla.: CRC Press.

Rothschild, B. M., and D. Tanke. 1992. Paleopathology insights to lifestyle and health in prehistory. *Geoscience Canada* 19: 73–82.

Rubin, C., G. Friedlaender, and J. M. Lane. 1994. Physical and environmental influences on bone formation. *Bone Formation and Repair.* Brighton, Conn.: AAOS.

Spassov, N. B. 1979. Sexual selection and the evolution of horn-like structures of ceratopsian dinosaurs. *Paleontology Stratigraphy Lithology* 11: 37–48.

Part III
Ankylosauria

8. South American Ankylosaurs

RODOLFO A. CORIA AND
LEONARDO SALGADO

Abstract

Many isolated bones recovered from Upper Cretaceous beds of Patagonia have been recently identified as belonging to ankylosaurs. The femur has a hemispherical femoral head, the fourth trochanter is absent or reduced, the greater and anterior trochanters are fused, and small, conical, ventrally concave dermal plates are present. Other features, such as the retention of a proximally positioned fourth trochanter, suggest that the material has closer affinities to the nodosaurids than to the ankylosaurids. All remains described here come from the same geological levels as hadrosaurs, suggesting that the South American ankylosaurs were part of the same faunal interchange with North America that occurred near the end of the Cretaceous.

Introduction

In contrast with the rich saurischian dinosaur faunas known from South America, the ornithischian faunas have remained scarce and poorly recorded. Among them, ornithopods are by far the best known ornithischian dinosaurs (Coria 1999; Coria and Salgado 1996a, 1996b). The presence of ankylosaurs in South America was first claimed for *Loricosaurus scutatus* (Huene 1929) on the basis of several dermal plates. These elements were later identified as dermal scutes of the titanosaur sauropod *Saltasaurus loricatus* (Bonaparte and Powell 1980).

The first unquestionable report of armored dinosaurs in South America was the description of a single femur and a few dermal scutes with distinctive ankylosaur features (Coria 1994; Salgado and Coria

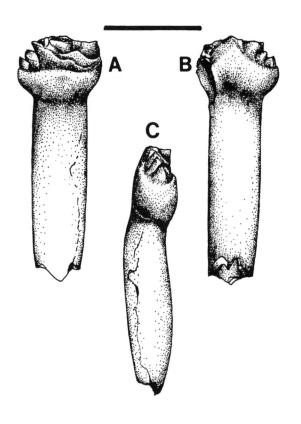

Figure 8.1. Ankylosaur tooth in
(A) lingual, (B) labial, and
(C) posterior views.
Scale bar = 5 mm.

1996). All the material comes from the Allen Formation (Upper Cretaceous) at Salitral Moreno, Río Negro Province, Argentina. They were
found associated with titanosaur and hadrosaur remains, although
ankylosaurs are much scarcer. Ankylosaurs are the most poorly represented dinosaurs from this continent, and their phylogenetic relationships have yet to be determined. We summarize all the ankylosaur
material from this locality, including newly collected specimens.

Institutional Abbreviations. MPCA-Pv: Museo Provincial Carlos
Ameghino, Cipolleti, Argentina. MPCA-SM: Museo Provincial Carlos
Ameghino, Salitral Moreno Collection, Cipolleti, Argentina.

Specimens

Material. MPCA-SM 1, right femur; MPCA-Pv 68/69/70, three
posterior dorsal vertebrae; MPCA-Pv 71, caudal vertebrae; MPCA-Pv
72/73, two caudal centra; MPCA-Pv 41/42/43/74/75/76, six conical
dermal plates; MPCA-Pv 78, two fused dermal plates; and MPCA-Pv
77, one tooth.

Horizon. Sandy Member, Allen Formation, Campanian-Maastrichtian, Upper Cretaceous (Powell 1986).

Locality. Salitral Moreno, 40 km south of General Roca, Río Negro Province, Argentina.

Description

Tooth

The single tooth (MPCA-Pv 77) has a crown 4 mm tall and a root 10 mm long (Fig. 8.1). The crown is laterally compressed and leaf-shaped, as in all ankylosaurs (Fig. 8.1C). The root is 3 mm wide, and the crown is 5 mm anteroposteriorly. A conspicuous neck separates the crown and root. The crown has cingula on both lingual and labial surfaces, which are asymmetrically placed as in nodosaurids (Carpenter 1990; Coombs 1990; Coombs and Maryańska 1990). On one side of the crown, the cingulum is straight and aligned with the anteroposterior tooth axis (Fig. 8.1A). On the opposite side, the cingulum arches toward the crown apex (Fig. 8.1B). The wear surface of the crown is eroded. The apex is not preserved, although two marginal denticles are present.

Vertebrae

Dorsals

Three dorsal vertebrae have been collected. Their centra are long, higher than they are broad, and have a rounded cross section, as in most ankylosaurs (Coombs and Maryańska 1990) (Fig. 8.2A'–C'). The articular surfaces are amphyplatian (Fig. 8.2A–C). The neural arch is high and has a high and narrow neural canal (Fig. 8.2A, C). The transverse processes, although not preserved completely, seem to have been upwardly oriented (Fig. 8.2A, A'). Laterally, the neural arch is excavated by a triangular fossa, bordered by two thick infraparapophysial lamina (Fig. 8.2A'). The prezygapophyses are fused ventrally to form the typical ankylosaur U-shaped structure (Coombs and Maryańska 1990) enclosing a broad prespinal basin (Fig. 8.2A, B). The neural spine is broken and bears a shallow prespinal laminae.

Caudal

The material includes a complete caudal vertebra with an amphicelous centrum (Fig. 8.3B). The articular surfaces are roughly circular but are slightly wider than high (Fig. 8.3). The centrum is very short anteroposteriorly and has a deep ventral groove. The facets for the chevrons are more developed on the posterior edge of the centrum. The transverse processes are long, thin, flattened slightly dorsoventrally, and projected laterally (Fig. 8.3B). The high neural arch is directed slightly backward. It is formed mainly by the neural spine, which is anteroposteriorly short, and flattened laterally. The zygapophyses are not developed. The neural canal is high and very narrow. Two other caudal centra found lack their neural arches (Fig. 8.3A, C). Ventrally, they lack grooves.

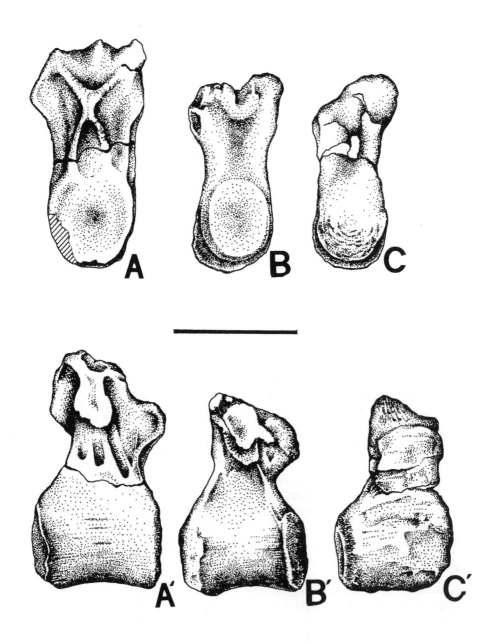

Figure 8.2. Ankylosaur dorsal vertebrae in (A–C) anterior view; A'–C') same vertebrae in lateral view. Scale bar = 5 cm.

Appendicular Skeleton

Femur

The femur is relatively short, stout, and straight, and both the proximal and distal ends are transversely expanded (Fig. 8.4B). The femoral head is large and hemispherical and projects dorsomedially (Fig. 8.4A, C). A distinct neck with a pronounced notch between the

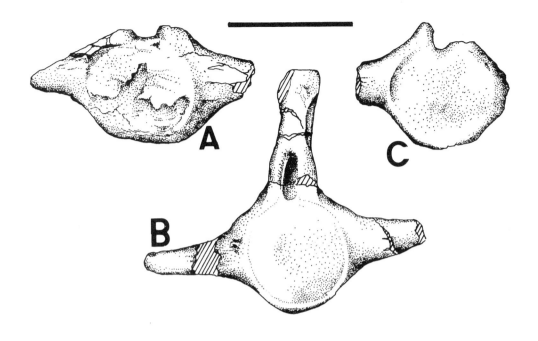

Figure 8.3. Ankylosaur caudal vertebra in anterior view. Scale bar = 5 cm.

femoral head and the trochanteric area is present. The greater and anterior trochanters form a blunt prominence, as in most nodosaurids, including *Sauropelta* (Coombs and Maryańka 1990; Ostrom 1970) (Fig. 8.4A, C). They are positioned low in relation to the femoral head, as in the nodosaurids *Polacanthus* (Pereda Suberbiola 1991) and *Edmontonia* (Russell 1940). The fourth trochanter is reduced to a low ridge located proximally on the posterolateral side of the femur, as in most nodosaurids. The shaft is anteroposteriorly compressed, bearing two longitudinal, slightly ventromedially oriented ridges on its anterior side, which suggest an adult ontogenic stage for the individual (Fig. 8.4A). Distally, the condyles are enlarged (Fig. 8.4F). The medial condyle has a wide, convex articular surface for the tibia. The strongly projected lateral condyle has the condylid displaced toward the anterior and posterior intercondylar grooves.

Dermal Plates

There are two different morphological types among the plates collected. The first type is conical and roughly elliptical (Fig. 8.5). The ventral side is concave, with a smooth surface. The spike is blunt and inclines slightly laterally and off center. The lateral surface of the spike has a complex system of small foramina, sometimes connected by shallow radial grooves. The height of the spike is less than the long axis of the plate.

The second type of plate is characterized by an acute spike and a

South American Ankylosaurs • 163

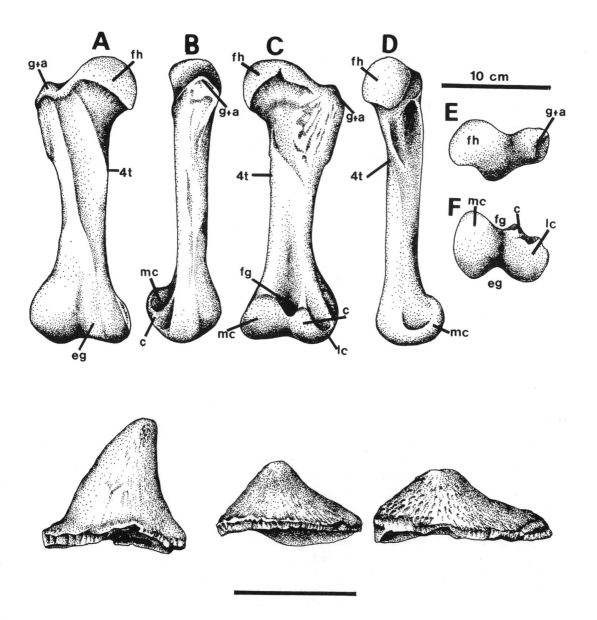

Figure 8.4. (top)
Right femur in (A) anterior,
(B) lateral, (C) posterior,
(D) medial, (E) proximal, and
(F) distal views. Scale bar = 5 cm
(from Salgado and Coria 1996).

Figure 8.5. (bottom) Ankylosaur
conical dermal plates in lateral
view. Scale bar = 5 cm.

sharp anterior edge (MPCA-Pv 78, two fused dermal plates; Fig. 8.6). The spike inclines well outward and is higher and more massive than the other type of plate. It also has a thicker base. The ventral side has a rugose surface. The sides of the spike have similar foramina and grooves as the first type of plate. Some plates have double, parallel spikes.

Discussion

All the material described here was collected from an area no larger than 50 m². Nevertheless, we cannot determine if the material belongs

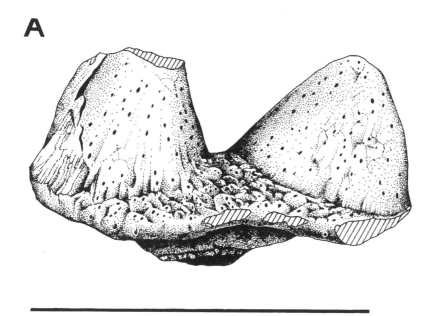

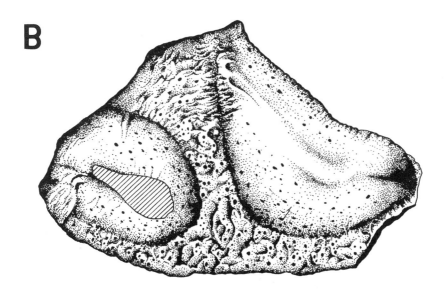

to a single animal or to several because the bones were found in fluvial paleochannels associated with titanosaur and hadrosaur bones. Although quite fragmentary, the material is diagnostically that of an ankylosaur, especially the leaf-shaped, laterally compressed tooth crown with a cingulum, U-shaped dorsal prezygapophyses, the upwardly directed dorsal transverse processes, the fused greater and anterior trochanters of the femur, the hemispherical femoral head (suggesting a

Figure 8.6. Ankylosaur dermal plates fused in (A) lateral and (B) dorsal views. Scale bar = 10 cm.

completely closed acetabulum), and the dermal scutes with sharp keels and concave ventral surfaces. The tooth suggests an affinity with the nodosaurids because of the presence of the asymmetrically placed cingula on both sides of a relatively broad crown (Coombs 1990).

Salgado and Coria (1996) proposed a nodosaurid affinity for the femur (MPCA-SM 1) because of the plesiomorphic position of the fourth trochanter placed proximally. The fourth trochanter is apomorphically placed on the distal half of the femur in ankylosaurids (Sereno 1986). Also, all specimens described here are closer to expected nodosaurid sizes than to ankylosaurids. However, the evidence available so far does not include any anatomical feature based on derived characters.

Besides ankylosaurs, the record of South American thyreophorans has increased recently with the reports of a possible stegosaur (Bonaparte 1996). This material from the Lower Cretaceous of Neuquen Group suggests the presence of endemic armored dinosaurs in Gondwana. Despite the equally parsimonious hypothesis of indigenous thyreophoran dinosaurs in South America (see Gasparini et al. 1996), the ankylosaur remains here seem to us to be more likely associated with the faunal interchange between North America and South America in the Campanian-Maastrichtian (Bonaparte 1986). At present, three different hadrosaur forms have been recorded, *Secernosaurus koerneri* (Brett-Surman 1979), *Kritosaurus australis* (Bonaparte et al. 1984), and an undescribed lambeosaur (Powell 1987). This lambeosaur comes from the same locality as the ankylosaurs described here. The association of the ankylosaurs with hadrosaurs strongly suggests that North American dinosaurs reached southern latitudes by the end of the Cretaceous. Furthermore, the nodosaurids reported from Antarctica (Gasparini et al. 1987) could be also the result of the same Late Cretaceous dinosaur dispersion into the southern continents.

Acknowledgments. We thank the technicians from Museo de la Universidad Nacional del Comahue for their collaboration in fieldwork and Juan Carlos Muñoz for allowing us to study the specimens under his care. Illustrations were made by Aldo Beroisa. We are also grateful to Dr. Philip Currie for his comments on early drafts and to Dr. R. E. Molnar and Dr. K. Carpenter for critical reviews. Research was supported by Dirección Provincial de Cultura of Neuquén Province, Neuquén, Argentina; Museo Carmen Funes, Neuquén, Argentina; and University of Comahue, Buenos Aires, Argentina.

References Cited

Bonaparte, J. F. 1986. History of the terrestrial Cretaceous vertebrates of Gondwana. *Actas IV Congreso Argentino de Paleontología y Bioestratigrafía* 2: 63–95.

———. 1996. Cretaceous tetrapods of Argentina. In G. Arratia (ed.), *Contributions of Southern South America to Vertebrate Paleontology. Münchner Geowissenschaftliche Abhandlungen* 30: 73–130.

Bonaparte, J. F., M. R. Franchi, J. E. Powell, and E. C. Sepulveda. 1984. La Formación Los Alamitos (Campaniano-Maastrichtiano) del sudeste de Río Negro, con descripción de *Kritosaurus australis* nov. sp. (Hadrosauridae). Significación paleobiogeográfica de los vertebrados. *Revista de la Asociación Geológica Argentina* 39: 284–299.

Bonaparte, J. F., and J. E. Powell. 1980. A continental assemblage of tetrapods from the Upper Cretaceous beds of El Brete, northwestern Argentina. *Mémoires de la Société Géologique de France,* n.s., 139: 19–28.

Brett-Surman, M. K. 1979. Phylogeny and palaeobiogeography of hadrosaurian dinosaurs. *Nature* 277: 560–562.

Carpenter, K. 1990. Ankylosaur systematics: Example using *Panoplosaurus* and *Edmontonia* (Ankylosauria: Nodosauridae). In K. Carpenter and P. J. Currie (eds.), *Dinosaur Systematics: Perspectives and Approaches,* pp. 281–298. New York: Cambridge University Press.

Coombs, W. P., Jr. 1990. Teeth and taxonomy in ankylosaurs. In K. Carpenter and P. J. Currie (eds.), *Dinosaur Systematics: Perspectives and Approaches,* pp. 269–279. New York: Cambridge University Press.

Coombs, W. P., Jr., and T. Maryańska. 1990. Ankylosauria. In D. B. Weishampel, P. Dodson, and H. Osmólska (eds.), *The Dinosauria,* pp. 456–483. Berkeley: University of California Press.

Coria, R. A. 1994. Sobre la presencia de dinosaurios ornitisquios acorazados en Sudamérica [abstract]. *Ameghiniana* 31(4): 398.

———. 1999. Ornithopod dinosaurs from the Neuquén Group, Patagonia, Argentina: Phylogeny and biostratigraphy. In Y. Tomida, T. H. Rich, and P. Vickers-Rich (eds.), *Proceedings of the Second Gondwanan Dinosaur Symposium. National Science Museum Monographs* 15: 47–60.

Coria, R. A., and L. Salgado. 1996a. A basal iguanodontian (Ornithischia: Ornithopoda) from the Late Cretaceous of South America. *Journal of Vertebrate Paleontology* 16: 445–457.

———. 1996b. "*Loncosaurus argentinus*" Ameghino, 1899 (Ornithischia, Ornithopoda): A revised description with comments on its phylogenetic relationships. *Ameghiniana* 33(4): 373–376.

Gasparini, Z., E. Olivero, R. Scasso, and C. Rinaldi. 1987. Un ankylosaurio (Reptilia, Ornithischia) campaniano en el continente antártico. *Anais X Congreso Brasilero de Paleontologia,* pp. 131–141. Río de Janeiro: Sociedade Brasileira de Paleontologia.

Gasparini, Z., X. Pereda Suberbiola, and R. E. Molnar. 1996. New data on the ankylosaurian dinosaur from the Late Cretaceous of the Antarctica peninsula. In F. E. Novas and R. E. Molnar (eds.), *Proceedings of the Gondwanan Dinosaur Symposium. Memoirs of the Queensland Museum* 39: 583–594.

Huene, F. von. 1929. Los saurisquios y ornitisquios del Cretácico Argentino. *Anales del Museo de La Plata,* Serie 2: 1–196.

Ostrom, J. H. 1970. Stratigraphy and paleontology of the Cloverly Formation (Lower Cretaceous) of the Bighorn Basin area. Wyoming and Montana. *Bulletin of the Peabody Museum of Natural History* 35: 1–234.

Pereda Suberbiola, X. 1991. Nouvelle evidence d'une connexion terrestre entre Europe et Amérique du Nord au Crétacé inférieur: *Hoplitosaurus* synonyme de *Polacanthus* (Ornithischia: Ankylosauria). *Comptes Rendus de la Academie des Sciences* 313: 971–976.

Powell, J. E. 1986. *Revisión de los Titanosáuridos de América del Sur.* Ph.D. diss. Universidad Nacional de Tucumán, Argentina.

———. 1987. Hallazgo de un dinosaurio hadrosáurido (Ornithischia, Ornithopoda) en la Formación Allen (Cretácico Superior) de Salitral

Moreno, Provincia de Río Negro, Argentina. *X Congreso Geológico Argentino. Actas* 3: 149–152.

Russell, L. S. 1940. *Edmontonia rugosidens* (Gilmore), an armored dinosaur from Belly River Series of Alberta. *University of Toronto Studies Geological Series* 43: 1–27.

Salgado, L., and R. A. Coria. 1996. First evidence of an ankylosaur (Dinosauria. Ornithischia) in South America. *Ameghiniana* 33(4): 367–371.

Sereno, P. C. 1986. Phylogeny of the bird-hipped dinosaurs (order Ornithischia). *National Geographic Research* 2: 234–256.

9. Skull of the Polacanthid Ankylosaur *Hylaeosaurus armatus* Mantell, 1833, from the Lower Cretaceous of England

KENNETH CARPENTER

Abstract

The holotype of the polacanthid ankylosaur *Hylaeosaurus armatus* includes a partial, highly damaged skull that has never been described in detail. It consists of the rear portion, which is exposed ventrally, and includes the left quadrate articulated to its paroccipital process, part of the left quadratojugal, right postorbital horn, and possibly the rear portions of both lower jaws. The postorbital horn is triangular and small and is comparable with that of the polacanthid *Gargoyleosaurus*. The quadrate has a prominent rim along its anterior lateral edge, a feature also seen in the primitive nodosaurid *Animantarx*.

Introduction

The armored dinosaur *Hylaeosaurus armatus* was one of the earliest named dinosaurs. It was named by Gideon Mantell in 1833, only eight years after the naming of *Iguanodon* and nine years after the naming of *Megalosaurus*. The holotype, BMNH R3775, consisted of the anterior portion of an articulated skeleton and was briefly described by Mantell (1833) and in more detail by Owen (1857) and Pereda-Suberbiola (1993). To date, however, the fragmentary skull associated with the specimen has not been described.

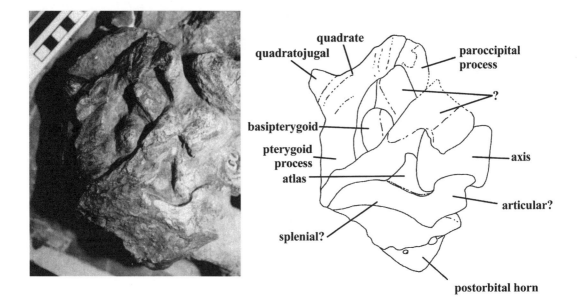

Figure 9.1. Photograph of the skull of Hylaeosaurus *in close-up (left), and an outline drawing of the skull with parts labeled (right). The postorbital horn is not seen very well in the photograph because of parallax.*

Institutional Abbreviation. BMNH: Natural History Museum (formerly British Museum [Natural History]), London.

Description

Only the rear portion of the skull remains and is seen in ventral view (Fig. 9.1). The fragment is 18.5 cm wide and 12.2 cm long. The surface of some of the bones is damaged, making it difficult to delineate their margins, especially because the matrix and bones are presently nearly the same color. The axis and left quadrate provide key anatomical reference points to aid in the identification of the surrounding elements.

The left quadrate is visible in lateral view, with the exposed portion 10.7 cm long. The quadrate is partially twisted so that a portion of the posterior side faces laterally. This surface is damaged, but it can be seen to articulate with the paroccipital process. No suture is seen separating the quadrate head from the paroccipital process, implying that the two elements are coossified. The distal, or condylar, end is overlain by extensively damaged bone or bones that may be the rear portion of the right lower jaw. The quadrate is arched laterally, as it is in the polacanthid *Gargoyleosaurus* (see Carpenter 2001 for a definition of the Polacanthidae). A rim along the lateral edge of the quadrate, near the quadratojugal, makes the shaft appear to have a groove along its back side. The extent of this rim, however, cannot be determined because of damage. This feature is seen in *Animantarx ramaljonesi* (Carpenter et al. 1999) and *Stegosaurus,* and it probably marks the anterior edge of the tympanic membrane. The quadrate shaft is 2.8 cm in lateromedial width, where it is least damaged near the quadratojugal.

The quadratojugal is identified on the basis of its anatomical position relative to the quadrate. It is short, measuring 1.4 cm long and 2.4 cm deep. The quadratojugal is small in most ankylosaurs and braces the quadrate anterolaterally with the jugal. The anterior edge is damaged, but only a little is probably missing on the basis of the proportions of *Gargoyleosaurus* (personal observation). The quadratojugal is coossified higher on the quadrate shaft than it is in the nodosaurid *Edmontonia,* were it is located near the distal condyle. The high position is characteristic of the polacanthid *Gargoyleosaurus* and the primitive nodosaurids *Pawpawsaurus* and *Animantarx,* but not the advanced nodosaurid *Edmontonia.*

The only part of the paroccipital process that is visible is its distal end, where it articulates with the quadrate head. It is at least 2.8 cm deep. The quadrate head is damaged, but no suture is visible, suggesting that the head was fused, at least in part, to the paroccipital process. The medial part of the paroccipital process is covered by damaged bone or bones that may be part of the left lower jaw. The damage is so severe, however, that this identification is questionable. This damaged bone is in contact with another bone anteriorly and is separated from it posteriorly by the axis. This bone is also severely damaged, but it may be the rear portion of the right mandible on the basis of what appears to be the medial process of the articular. If this object is the right mandible, it has shifted back almost 10 cm from the estimated position of the distal end of the quadrate.

Anterior to the paroccipital process is a somewhat bulbous bone protruding posterior to the quadrate and lateral to the questionable left mandible. Little detail can be seen, but on the basis of its topographic position, is probably the left basitubera. If this identification is correct, it is large for an ankylosaur.

The right postorbital horn is seen in lateral view. The surface is damaged, but it still retains its triangular shape. It is 5.5 cm wide at its base and projects laterally about 3.2 cm. The horn is proportional to that seen in the polacanthids *Gargoyleosaurus* and *Gastonia.* There are two small depressions on the surface near where the horn joins the rest of the skull. Unfortunately, because the surrounding bone is damaged, it is not certain if these depressions are pathologic or not.

Discussion

The skull to the holotype of *Hylaeosaurus armatus* is incomplete and much of the surface is damaged, making comparisons with other, more complete ankylosaur skulls difficult. The identifications offered here, with the exception of the quadrate, are tentative and may be changed if the skull fragment is prepared free of the block containing the skeleton.

The similarities between the skulls of *Gargoyleosaurus* and *Hylaeosaurus* enumerated above indicate a close affinity between the two taxa and support the inclusion of *Hylaeosaurus* into the Polacanthidae (see Carpenter 2001).

Acknowledgments. My thanks to Sandra Chapman and Angela

Milner (Natural History Museum, London) for access to the ankylosaur specimens in their care. Thanks also to the review comments by
Xabier Pereda Suberbiola.

References Cited

Carpenter, K. 2001. Phylogenetic analysis of the Ankylosauria. In K. Car-
    penter (ed.), *The Armored Dinosaurs*. Bloomington: Indiana Univer-
    sity Press [this volume, Chapter 21].
Carpenter, K., J. Kirkland, D. Burge, and J. Bird. 1999. Ankylosaurs (Dino-
    sauria: Ornithischia) of the Cedar Mountain Formation, Utah, and
    their stratigraphic distribution. In D. Gillette (ed.), *Vertebrate Paleon-
    tology in Utah*. *Utah Geological Survey Miscellaneous Publication*
    99-1: 244–251.
Mantell, G. 1833. *The Geology of the South-East of England*. London:
    Longman.
Owen, R. 1857. Monograph of the fossil Reptilia of the Wealden and
    Purbeck Formations. Part IV. *Hylaeosaurus*. *Palaeontological Society
    of London Monograph* 10: 8–26.
Pereda-Suberbiola, J. 1993. *Hylaeosaurus, Polacanthus,* and the systemat-
    ics and stratigraphy of Wealden armored dinosaurs. *Geological Maga-
    zine* 130: 767–781.

10. Reappraisal of the Nodosaurid Ankylosaur *Struthiosaurus austriacus* Bunzel from the Upper Cretaceous Gosau Beds of Austria

XABIER PEREDA SUBERBIOLA AND
PETER M. GALTON

Abstract

The Campanian continental coal-bearing Gosau Beds of Muth-mannsdorf (Lower Austria) have yielded a rich vertebrate assemblage. The fauna, which contain dinosaurs, crocodilians, pterosaurs, turtles, lizards, and champsosaurids, is dominated by nodosaurid ankylosaurs. The fossil remains were mostly recovered during the 19th century but have received little attention since the 1920s. A systematic revision of the ankylosaurian material suggests the occurrence of a single species, *Struthiosaurus austriacus* Bunzel. At least three partial skeletons, representing different growth stages, are known in the assemblage. Bone differences are interpreted as the result of both ontogenetic changes and individual variation. The taxa *Crataeomus pawlowitschii, C. lepido-phorus, Danubiosaurus anceps, Pleuropeltus suessii,* and *Rhadino-saurus alcimus* are regarded as junior synonyms of *Struthiosaurus aus-triacus*. *Struthiosaurus* is conservative within the Nodosauridae. The primitive phylogenetic position and small size of *Struthiosaurus* (esti-mated body length 3 m in adults) may be related to the insular endemic

Figure 10.1. Location map showing the Muthmannsdorf locality, in the Wiener Neustadt region, Austria.

evolution of European ankylosaurs during the Late Cretaceous. *Struthiosaurus* is interpreted as a pedomorphic dwarf ankylosaur.

Introduction

The continental coal-bearing series of the Gosau Beds of Austria have yielded one of the most important vertebrate assemblages from the Late Cretaceous of Europe. In 1859, Suess found shells of freshwater molluscs in the dumps of the Feldering Lignite Mine in the Neue Welt area, near the town of Wiener Neustadt (Niederösterreich). Shortly after that, Stoliczka discovered an iguanodontidlike tooth in coal in the same general area. Another isolated tooth, described as crocodilian by Stoliczka (1860), was found in shale at Russbachtal, near the town of Gosau, on the boundary between the Oberösterreich and Salzburg. During the 1860s, Suess and his colleague Pawlowitsch gathered abundant reptilian remains from a thin, marly bed associated with a lignite layer near the village of Muthmannsdorf, about 45 km south of Vienna (Fig. 10.1). Bunzel (1870a, 1870b) mentioned the discovery of a vertebrate assemblage in two short papers and named *Struthiosaurus* ("ostrich reptile") on the basis of a fragmentary skull, which resembled that of a birdlike dinosaur (Bunzel 1870a). Later, Bunzel (1871) described most of the specimens recovered by Suess and Pawlowitsch and listed crocodilians, dinosaurs, lizards, and turtles. In the same paper, Bunzel (1871) referred to the birdlike braincase as *Struthiosaurus austriacus* and erected *Danubiosaurus anceps* as a new species of lizard.

TABLE 10.1.

Different systematic identifications of the ankylosaurian material from the Gosau Beds of Muthmannsdorf, Austria.

	Bunzel 1871	Seeley 1881	Nopcsa 1902	Nopcsa 1915–1923	Nopcsa 1929
Lizards:	Danubiosaurus anceps Lacerta sp.	Crataeomus (part) + Pleuropeltus (part)			
Crocodilians:	"Crocodili ambigui" Crocodilus sp.	Dinosaurs: Crataeomus pawlowitschii Crataeomus lepidophorus			
Dinosaurs:	Scelidosaurus sp. Iguanodon mantelli Struthiosaurus austriacus	Struthiosaurus austriacus Rhadinosaurus alcimus Hoplosaurus ischyrus Pleuropeltus suessii	Dinosaurs: S. austriacus H. ischyrus Turtles: Leipsanosaurus noricus	S. austriacus	S. austriacus

TABLE 10.2.

Fossil bones and different systematic identifications of the ankylosaurian material from the Gosau Beds of Muthmannsdorf, Austria.

	Bunzel (1871)	Seeley (1881)	Nopcsa (1929)	Molnar (1980)	This chapter
Skull (braincase)	*Struthiosaurus austriacus* (pl. 5, figs. 1–6)	*Struthiosaurus austriacus* (pl. 27, figs. 5, 6)	*Struthiosaurus austriacus*		*S. austriacus* (Fig. 10.2A–E)
Orbital roof		*Pleuropeltus suessii* (pl. 28, figs. 8, 9)	*S. austriacus*		(Fig. 10.2F–I)
Quadrate			*S. austriacus*		(Fig. 10.2J–L)
Lower jaw		*Crataeomus* sp. (pl. 27, figs. 9, 10)	*S. austriacus* (fig. 4)		(Fig. 10.2O–S)
Teeth		*Crataeomus* sp. (pl. 27, figs. 11–16)	*S. austriacus* (fig. 5)		(Fig. 10.2M–N)
Cervical vertebrae	"*Crocodili ambigui*" (pl. 2, figs. 9, 10)	*Crataeomus pawlowitschii*	*S. austriacus*		(Fig. 10.3A–D)
Dorsal vertebrae	"*Crocodili ambigui*" (pl. 2, figs. 1–3, pl. 7, fig. 24; pl. 1, figs. 24–26)	*C. pawlowitschii* (pl. 30, fig. 3)	*S. austriacus*	*Crateomus pawlowitschii* (table 1)	*Crataeomus pawlowitschii* (Fig. 10.3F–H)
		C. lepidophorus (pl. 30, fig. 5)	"	*S. austriacus* = *C. lepidophorus*	= *C. lepidophorus*
Caudal vertebrae	"*Crocodili ambigui*" (pl. 2, figs. 4–8, pl. 8, figs. 1, 7, 8)	*C. pawlowitschii* (pl. 30, fig. 4)	*S. austriacus*		(Fig. 10.3J–J')
	Scelidosaurus sp. (pl. 4, figs. 6–9, pl. 8, figs. 7, 8, 16)	"	"		
Cervical rib	*Crocodilus* sp. (pl. 1, fig. 27)	*C. lepidophorus*	*S. austriacus*		Fig. 10.3E
Dorsal ribs	*Lacerta* sp. (pl. 3, figs. 5, 6)	*C. pawlowitschii* (pl. 31, fig. 12)	*S. austriacus*		(Fig. 10.3I)
	Dinosaur indet. (pl. 8, figs. 14, 15)	*C. lepidophorus* (pl. 27, figs. 17, 18)	"		
Scapulocoracoid	*Danubiosaurus anceps* (pl. 5, figs. 7–9)	*C. pawlowitschii* (fig. A, p. 656)	*S. austriacus*	*Crataeomus*	(Fig. 10.4A–E)
		C. lepidophorus (fig. B, p. 656)	"	*Struthiosaurus*	(Fig. 10.4F–H)

TABLE 10.2. (cont.)

	Bunzel (1871)	Seeley (1881)	Nopcsa (1929)	Molnar (1980)	This chapter
Humerus		*C. lepidophorus* (pl. 29, figs. 1–3)	*S. austriacus*		(Fig. 10.4I–K)
Ulna			*S. austriacus*		(Fig. 10.4L)
Radius		*C. lepidophorus* (pl. 27, fig. 19)	*S. austriacus*		(Fig. 10.4M–N)
Ilium	*Damubiosaurus anceps* (pl. 6, figs. 4, 5)	*Pleuropeltus suessii* (pl. 30, fig. 15)	*S. austriacus*		(Fig. 10. 6A–E)
Ischium	"*Crocodili ambigui*" (pl. 3, figs. 12, 13)	*Crataeomus* sp. (pl. 27, fig. 20)	*S. austriacus*		(Fig. 10.6F–H)
Femur	"*Crocodili ambigui*" (pl. 3, figs. 2–4)	*C. pawlowitschii* (pl. 31, fig. 1)	*S. austriacus*		(Fig. 10.6I–M)
		C. lepidophorus (pl. 31, figs. 4, 5)	"		(Fig. 10.7A–L)
Tibia		*C. pawlowitschii* (pl. 31, fig. 2)	*S. austriacus*		(Fig. 10.7M–N)
Fibula	"*Crocodili ambigui*" (pl. 3, fig. 1)	*Rhadinosaurus alcimus* (pl. 31, figs. 6, 7)	*S. austriacus*		(Fig. 10.7O–R)
Metatarsal	"*Crocodili ambigui*" (pl. 4, figs. 11, 12)	*C. lepidophorus*	*S. austriacus*		(Fig. 10.7S)
Phalanx	*Scelidosaurus* sp. (pl. 4, figs. 4, 5)	*C. lepidophorus* (pl. 29, fig. 6)	*S. austriacus*		(Fig. 10.7T)
Dermal armor	*Scelidosaurus* sp. (pl. 7, figs. 20, 21, pl. 8, figs. 10–12)	*C. pawlowitschii* (pl. 28, figs. 2–4)	*S. austriacus* (pl. 4, figs. 1–15)		(Fig. 10.8A–W)
	Damubiosaurus anceps (pl. 5, fig. 10)	*C. lepidophorus* (pl. 30, fig. 2, pl. 31, fig. 3; pl. 28, fig. 5)	"		

Between 1870 and 1880, the Muthmannsdorf site yielded new reptile bones. Other fossil remains were recovered from a calcareous clay at an indeterminate neighboring locality (Seeley 1881). Seeley spent a month studying the reptilian collection kept in Vienna and subsequently published a memoir on the Gosau fauna (Seeley 1881). He modified some of Bunzel's identifications. The skull of *Struthiosaurus austriacus* was regarded as that of a scelidosaurian dinosaur. Other pieces, mostly postcranial bones, were attributed to several new armored dinosaurs, *Crataeomus pawlowitschii*, *C. lepidophorus*, and *Hoplosaurus ischyrus*, and the small-size *Rhadinosaurus alcimus* (Tables 10.1, 10.2). Moreover, Seeley (1881) described crocodiles, turtles (*Pleuropeltus suessii*), lizards, and pterosaurs.

Baron F. Nopcsa (1902, 1926, 1929) reviewed the Muthmannsdorf fauna and discussed the status and affinities of *Struthiosaurus*. He regarded *Struthiosaurus* as an acanthopholidid ankylosaur (now included in the ankylosaurian family Nodosauridae *sensu* Coombs, 1978b). Nopcsa (1918) named *Leipsanosaurus noricus* from an isolated tooth from Frankenhof, near Piesting, but he later regarded it as a junior synonym of *Struthiosaurus austriacus* (Nopcsa 1923). In 1929, Nopcsa gave an illustrated description of *Struthiosaurus* on the basis of the Austrian material and additional remains from the Maastrichtian Hateg Basin of Transylvania (*Struthiosaurus transylvanicus* Nopcsa, 1915). Since Nopcsa, no further dinosaur remains have been described from the coal-bearing Gosau Beds of Muthmannsdorf.

The ankylosaurian material from the Gosau-Schichten of Austria is kept in Vienna, and here we provide an illustrated description of the more interesting specimens. Previous descriptions of the skull (holotype) of *S. austriacus* and a discussion of its systematic position were given by Pereda Suberbiola and Galton (1992, 1994). The phylogenetic relationships of *Struthiosaurus* were also discussed.

Institutional Abbreviations. BMNH: Natural History Museum (formerly British Museum [Natural History]), London. NHM: Naturhistorisches Museum, Vienna. PIUW: Paläontologisches Institut der Universität Vienna, Vienna.

Figure 10.2. (opposite page) Skull, lower jaw, and teeth of Struthiosaurus austriacus *from the Gosau Beds of Muthmannsdorf, Austria. (A–E) PIUW 2349/6 (C4c), holotype, partial skull in dorsal, ventral, posterior, anterior, and left lateral views. (F–I) PIUW 2349/17 and uncataloged (B6a, b), two fragments of the ?supraorbital skull roof in dorsal and ventral views. (J–L) PIUW 2349/uncataloged (C4b), distal articular condyle of a right quadrate in posteromedial, anterolateral, and ventral views. (M, N) PIUW 2349/105a, b, two teeth in ?lingual and ?labial views. (O, P) PIUW 2349/ uncataloged (B5b), anterior end of a right dentary in medial and lateral views. (Q–S) PIUW 2349/ 5 (C4a), right dentary in medial, lateral, and dorsal views. Scale bars = 2 cm (A–L, O–S) and 1 cm (M, N).*

Systematic Paleontology
Dinosauria Owen, 1842
Ornithischia Seeley, 1888
Thyreophora Nopcsa, 1915
Suborder Ankylosauria Osborn, 1923
Family Nodosauridae Marsh, 1890
Genus *Struthiosaurus* Bunzel, 1870

Crataeomus *Seeley, 1881*
Danubiosaurus *Bunzel, 1871*
Pleuropeltus *Seeley, 1881*
Rhadinosaurus *Seeley, 1881* partim

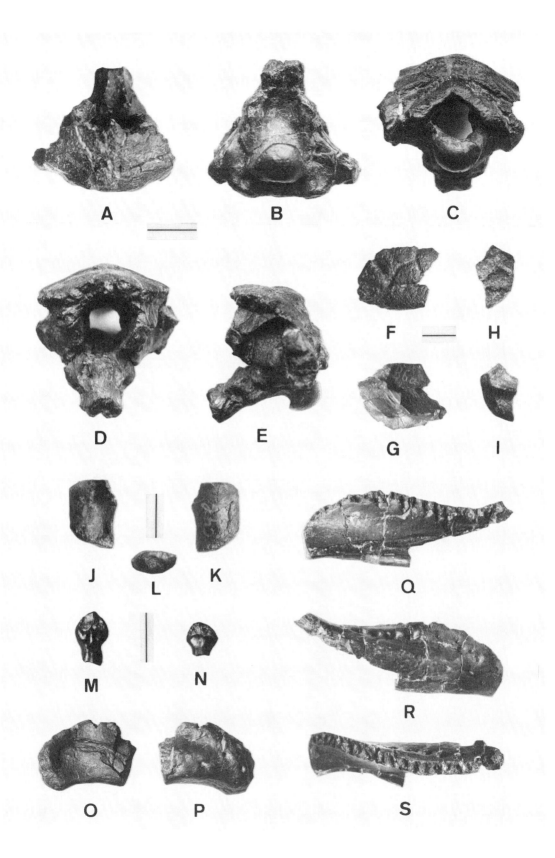

Type Species. Struthiosaurus austriacus Bunzel, 1871.

Diagnosis. As for *Struthiosaurus austriacus* Bunzel, 1871, given below.

Known Distribution. Campanian to Maastrichtian of Europe.

Struthiosaurus austriacus *Bunzel, 1871*
Crataeomus pawlowitschii *Seeley, 1881*
Crataeomus lepidophorus *Seeley, 1881*
Danubiosaurus anceps *Bunzel, 1871*
Pleuropeltus suessii *Seeley, 1881*
Rhadinosaurus alcimus *Seeley, 1881* partim

Type Specimen. PIUW 2349/6, a fragmentary cranium (Fig. 10.2A–E; Bunzel 1871: pl. 5, figs. 1–6; Seeley 1881: pl. 27, figs. 5, 6; Pereda Suberbiola and Galton 1992: fig. 1, 1994: figs. 1, 2).

Type Locality and Horizon. Muthmannsdorf, Wiener Neustadt area, about 45 km south of Vienna, Lower Austria; continental coal-bearing series of the Gosau Formation, Grünbach Basin, Lower Campanian (Brix and Plöchinger 1988).

Referred Specimens (the numbers in parentheses correspond to Nopcsa's 1929 catalog): PIUW 2349/17 and uncataloged, two fragments of orbital? skull roof with associated armor (B6a, b); PIUW 2349/uncataloged, distal articular condyle of right quadrate (C4b); PIUW 2349/5, right dentary (C4a); PIUW 2349/uncataloged, anterior ends of left and right mandibles (B5a, b); PIUW 2349/7–9, 39, 105a, b, and uncataloged, about 18 isolated teeth (C4d); PIUW 2349/uncataloged, two cervical vertebrae (B7a); PIUW 2349/uncataloged, cervical rib (B8a); PIUW 2349/24, 37, 110, and uncataloged, seven dorsal vertebrae (A3c, B7b, c, C3a, b); PIUW 2349/10 and uncataloged, about 20 dorsal ribs (B8b–j, C5a–f); PIUW 2349/25 and uncataloged, 22 caudal vertebrae with hemal arches (A3d, e, A3g–t, B7d–g, C3c); PIUW 2349/1, right scapula (C1); PIUW 2349/uncataloged, left scapulocoracoid (A2) and right scapula (B9); PIUW 2349/18, 19, left and right humeri (B2a, b); PIUW 2349/11 and uncataloged, left and right radii (B4a, b); PIUW 2349/uncataloged, ulna? (B3b); PIUW 2349/uncataloged, left and right ilia (B1a, b); PIUW 2349/41, left ischium?; PIUW 2349/31, 32, left and right small femora (C2a, b); PIUW 2349/29 and uncataloged, left and right femora (A5a, b); PIUW 2349/30, left tibia? (B3a); PIUW 2349/34 and uncataloged, left and right fibulae; PIUW 2349/uncataloged, metapodial and ungual phalanx (A6); PIUW 2349/13, 14, and uncataloged, parts of three cervical half-rings (B10a–c); PIUW 2349/15 and uncataloged, two hornlike spikes (A1a); PIUW 2349/uncataloged, eight caudal plates (A1b, A1d, B11a, b, B12a, b, C), and about 20 low-keeled scutes and ossicles (A1c, B13a–h, B14a–i, C6a, b); PIUW 2349/20, indeterminate bone (Figs. 10.2–10.8; Bunzel 1871: pls. 1–6; Nopcsa 1929: pl. 4; Seeley 1881: pls. 27–31). Casts of some specimens are kept in London (Natural History Museum), Lyon (Faculté des Sciences, Université Claude Bernard), and Pittsburgh, Pennsylvania (Carnegie Museum of Natural History; McIntosh 1981). For measurements, see Table 10.3.

TABLE 10.3.

Measurements of the holotype and referred specimens of *Struthiosaurus austriacus*. The numbers in parentheses correspond to Nopcsa's (1929) catalog.

Skull fragment PIUW 2349/6 (C4c)

Preserved length	+54 mm
Transverse diameter of occipital condyle	21 mm
Greatest diameter of foramen magnum	17 mm
Height from base of occipital condyle to skull roof	56 mm

Right lower jaw PIUW 2349/5 (C4a)

Preserved length	+87 mm

Left lower jaw, symphysis PIUW 2349/(B5b)

Preserved length	+40 mm

Cervical vertebra PIUW 2349/(B7a), centrum length	57 mm
/(no number), length	58 mm
Dorsal vertebra PIUW 2349/24 (B7b), centrum length	61 mm
/(B7c)	49 mm
/(A3c)	Incomplete
/110 (C3a)	14 mm
/(C3b)	Incomplete
Caudal vertebra PIUW 2349/A3d centrum length	43 mm
/A3	+41 mm
/A3g	44 mm
/A3h	47 mm
/A3i	45 mm
/A3j	47 mm
/A3k	49 mm
/A3l	49 mm
/A3m	Incomplete
/A3n	47 mm
/A3o	Incomplete
/A3p	44 mm
/A3q	Incomplete
/A3r	45 mm
/A3s	Incomplete
/A3t$_1$	34 mm
/A3t$_2$	34 mm
/B7d	42 mm
/B7e	43 mm
/B7f	41 mm
/B7g	42 mm
/C3c	31 mm

TABLE 10.3. *(cont.)*

Right scapula PIUW 2349/1 (C1)

Maximum length	205 mm
Maximum width	+90 mm

Left scapula PIUW 2349/uncataloged (A2)

Maximum length	+310 mm
Maximum width	+85 mm

Left scapula PIUW 2349/uncataloged (B9)

Maximum length	+270 mm
Maximum width	+90 mm

Left humerus PIUW 2349/18 (B2a), maximum length	+185 mm
Right humerus PIUW 2349/19 (B2b), maximum length	+190 mm
Ulna? PIUW 2349/uncataloged (B3b), maximum length	+195 mm
Radius PIUW 2349/uncataloged (B4a), maximum length	+155 mm
Radius PIUW 2349/11 (B4b), maximum length	+160 mm
Right ilium PIUW 2349/uncataloged (B1a), maximum length	+470 mm
Left ilium PIUW 2349/41 (B1b), maximum length	+280 mm
Left? ischium PIUW 2349/41, maximum length	+125 mm
Right femur PIUW 2349/29 (A5a), maximum length	+265 mm
Left femur PIUW 2349/uncataloged (A5b), maximum length	+180 mm

Right femur PIUW 2349/31 (C2a)

Maximum length	265 mm
Maximum distal width	76 mm

Left femur PIUW 2349/32 (C2b)

Maximum length	255 mm
Maximum distal width	78 mm

Left tibia? PIUW 2349/30 (B3a), maximum length	+210 mm
Left fibula PIUW 2349/34, maximum length	+150 mm
Right fibula PIUW 2349/15, maximum length	+120 mm
Metapodial PIUW 2349/uncataloged, maximum length	+70 mm
Ungual phalanx PIUW 2349/uncat. (A6), maximum length	35 mm
Cervical half-ring PIUW 2349/14 (B10a), maximum width	90 mm
Cervical half-ring PIUW 2349/13 (B10b), maximum width	Incomplete
Cervical half-ring PIUW 2349/uncataloged (B10c), maximum width	195 mm
Dorsal spine PIUW 2349/15 (A1a), maximum height	185 mm
Sacral? plate PIUW 2349/uncataloged (A1c), maximum length	110 mm

Caudal plate PIUW 2349/uncataloged (B12a)

Antero-postererior length	95 mm
Maximum width	135 mm

Caudal plate PIUW 2349/uncataloged (B12b?)

Antero-postererior length	90 mm

TABLE 10.3. *(cont.)*

Maximum width	140 mm
Caudal plate PIUW 2349/uncataloged (A1b?)	
Antero-postererior length	130 mm
Maximum width	50 mm
Caudal plate PIUW 2349/uncataloged (C?)	
Antero-postererior length	170 mm
Maxmum width	60 mm
Caudal plate PIUW 2349/uncataloged (B11a?)	
Antero-postererior length	120 mm
Maximum width	45 mm
Caudal plate PIUW 2349/uncataloged (B11b)	
Antero-postererior length	120 mm
Maximum width	50 mm
Caudal plate PIUW 2349/uncataloged (A?)	
Antero-postererior length	80 mm
Maximum width	40 mm
Caudal plate PIUW 2349/uncataloged (?)	
Antero-postererior length	70 mm
Maximum width	40 mm

Additional Material Tentatively Referred to as S. austriacus. PIUW 2349/uncataloged, a set of fragmentary bones, including parts of vertebrae, ribs, limb bones, and scutes; type of *Hoplosaurus ischyrus* Seeley, 1881. These remains are embedded in a calcareous clay and come from an imprecise locality near Wiener Neustadt (Seeley 1881: 681). NHM 1861.I.46, a single tooth from Frankenhof, near Piesting; type of *Leipsanosaurus noricus* Nopcsa, 1918. This tooth was referred to *Struthiosaurus austriacus* by Nopcsa (1923).

Emended Diagnosis. Small-size nodosaurid ankylosaur (total length about 3 m in adults); relatively high, narrow occiput compared with that of more derived nodosaurids such as *Edmontonia* and *Panoplosaurus*; basisphenoid projecting ventrally; nearly oval, almost symmetrical distal articular condyle of quadrate; elongated cervical vertebrae, centra longer than wide; hook-shaped acromion process centrally on scapula; lesser (anterior) trochanter forms a ridge; three cervical half-rings from fusion of spines, small ossicles, and oval scutes; at least a pair of hornlike spikes on trunk; tall, hollow-based triangular plates on the tail.

Description and Comparisons

Skull

Cranium and Cranial Scutes (Fig. 10.2A–L). The cranium of *Struthiosaurus austriacus* (PIUW 2349/6) has been described in detail elsewhere (Pereda Suberbiola and Galton 1992, 1994) and compared with the skull of *S. transylvanicus* (BMNH R4966; see Nopcsa 1929; Pereda Suberbiola and Galton 1997). The cranium fragment preserves the posterior part of the skull roof, most of the occiput, and part of the basicranium (Fig. 10.2A–E; Bunzel 1871: pl. 5, figs. 1–6; Seeley 1881: pl. 27, figs. 5, 6). The sutures between the bones are mostly obliterated, and the dermal scutes are coossified with the skull roof. The basipterygoid processes consist of a pair of rugose stubs, and the paroccipital processes are visible in dorsal view, as occurs in nodosaurids (Coombs and Maryańska 1990; Sereno 1986). Moreover, the hemispherical occipital condyle is formed exclusively by the basioccipital; it is set off from the braincase by a distinct neck that is angled about 55° posteroventrally, as in most nodosaurids (Coombs and Maryańska 1990; Lee 1996; Pereda Suberbiola and Galton 1994; Sereno 1986). The basisphenoid projects below the level of the basioccipital in occipital view. This feature, also seen in *S. transylvanicus,* serves to characterize *Struthiosaurus* (Pereda Suberbiola and Galton 1992, 1994).

Two small, isolated, fragmentary cranial remains with armor (PIUW 2349/17 and uncataloged) are tentatively regarded as parts of the orbital roof (Pereda Suberbiola and Galton 1994: fig. 4E–H). They were originally described as postfrontal bones of the chelonian *Pleuropeltus suessii* (Seeley 1881: pl. 28, figs. 8, 9) and later referred to as prefrontals of *Struthiosaurus austriacus* (Nopcsa 1929). Dorsally, both specimens preserve small irregular scutes arranged in rows, presumably parallel to the border of the orbit (Fig. 10.2F–I). Each series is composed of three (or four) scutes approximately as wide as the distinct scute along the posterior edge of the cranium piece. The ventral surface is smooth and is crossed by a sharp, sinuous ridge. *S. austriacus* probably had several rows of small cranial scutes on each side of the supraorbitals and a large plate centrally between the orbits (Pereda Suberbiola and Galton 1994). A similar scute pattern is known in the nodosaurids *Panoplosaurus* and *Edmontonia* (Carpenter 1990). In *S. austriacus,* an anteroposteriorly narrow scute occurs along the rear border of the skull roof, as in the nodosaurids *Panoplosaurus* and *Pawpawsaurus* (Lee 1996).

As noted by Nopcsa (1929), the distal end of an isolated right quadrate ramus is known (PIUW 2349/uncataloged). The specimen strongly resembles the quadrate condyles of *Struthiosaurus transylvanicus* (Nopcsa 1929; Pereda Suberbiola and Galton 1994, 1997). The quadrate ramus is narrow and flat transversely (Fig. 10.2J–L). The sutural surface between the quadrate and the quadratojugal is visible as in *S. transylvanicus*; this character is unusual in adult ankylosaurs but occurs in juveniles. The articular face is nearly oval and almost symmetrical, with the medial portion slightly more robust than the lateral one, unlike that of typical ankylosaurs.

The cranium of *S. transylvanicus* is more complete than that of *S. austriacus* and shows some additional features (Pereda Suberbiola and Galton 1994, 1997). The skull roof is highly domed. The pterygoid closes the space between the palate and braincase. A postocular shelf closes the orbital cavity posteromedially. The lateral temporal fenestra is visible in lateral view and, reconstructed, was probably oval and relatively elongate. This shape is more similar to that seen in *Sauropelta, Silvisaurus,* and *Pawpawsaurus* (Carpenter and Kirkland 1998; Lee 1996) than that of *Panoplosaurus* and *Edmontonia* (Carpenter 1990). Moreover, the occiput of *Struthiosaurus* is relatively narrow and high for a nodosaurid, with the tip of the paroccipital processes being short and much curved posteriorly, so it resembles that of early nodosaurids such as *Silvisaurus* (Pereda Suberbiola and Galton 1994).

Endocranial Cast. A latex endocranial cast was prepared from the partial braincase of *Struthiosaurus austriacus.* Pereda Suberbiola and Galton (1994: fig. 5) described and compared it with the endocast of *S. transylvanicus* (BMNH R4966), the latter being the first to be described from an ankylosaur (Nopcsa 1929). The general form of the endocranial cast of *Struthiosaurus* is roughly similar to that of the nodosaurid cf. *Polacanthus* (Norman and Faiers 1996), mainly in having expanded cerebral lobes and cerebellum, which are more prominent than those of the ankylosaurid *Euoplocephalus* (Coombs 1978a; Hopson 1979). Moreover, the endocranial cast of *Struthiosaurus* has a sharp flexure, as in cf. *Polacanthus* and *Gargoyleosaurus* (Carpenter et al. 1998), rather than the moderate flexure seen in *Euoplocephalus.*

Lower Jaw (Fig. 10.2O–S). Jaw remains include a small right dentary and the anterior end of the left and right mandibles of larger size. This material was originally described as *Crataeomus* sp. (Seeley 1881: pl. 27, figs. 9, 10) and later referred to as *Struthiosaurus austriacus* (Nopcsa 1929). Seeley (1881) suggested the occurrence of two different species of *Crataeomus* on the basis of three mandibular characters: the form of the ventral surface of the lower jaw; the extension of the tooth row; and the thickness of the anterior end of the dentary. However, the available material is too fragmentary to support this idea. The differences are more probably due to both ontogenetic changes and preservational differences rather than specific features.

The small dentary (PIUW 2349/5) is broken anteriorly near the symphysis, and posteriorly just anterior to the coronoid process. Nineteen or 20 alveoli are present. The tooth row forms a sinuous curve that extends from the mesial process of the coronoid process to the proximal end of the lower jaw, near the symphysis. A similar pattern is known in early nodosaurids, such as *Sarcolestes* (Galton 1983), *Sauropelta* (Ostrom 1970), and *Gargoyleosaurus* (Carpenter et al. 1998). The dentary of *Struthiosaurus* is short compared with those of other nodosaurids (Pereda Suberbiola et al. 1995). The splenial is missing, and a narrow Meckelian canal is open medially. This groove tapers anteriorly to below the first alveolus. The lateral side of the dentary is traversed by a ridge that separates a dorsal smooth surface, pierced by four large foramina, from a ventral textured surface. There is no evidence of a scute coossified to the lateral side of the lower jaw, unlike typical adult ankylosaurs (Coombs and Maryańska 1990; Galton 1983) and *Sceli-dosaurus* (Norman 1984). The absence of a dermal scute fused to the mandible may have an ontogenetic significance, but we cannot exclude the possibility that this is an artifact of preservation.

The anterior end of a comparatively larger right dentary (PIUW 2349/uncataloged) preserves a scooplike symphysis, but no alveoli are preserved. A V-shaped suture is visible ventromedially that is interpreted as the contact between the dentary and the splenial. If this is correct, the absence of the splenial in the small dentary is an immature feature (see growth stages below).

Teeth (Fig. 10.2M–N). Seeley (1881: pl. 27, figs. 11–16) described nine isolated teeth and referred them to *Crataeomus* sp. In fact, 18 teeth are present at the University of Vienna (PIUW 2349/7–9, 39, 105a, b, and uncataloged). The roots of most of the teeth are larger than the diameter of the alveoli preserved on the small dentary and probably came from a larger lower jaw (*contra* Nopcsa 1929). As Seeley (1881) noted, the surface of the crowns is damaged, with the enamel partially or totally missing. Consequently, the main tooth features are lacking. The two best preserved teeth (PIUW 2349/105a, b) show a conspicuous basal cingulum on each side, significantly higher on one side than the other, as is common in nodosaurids (Coombs and Maryańska 1990). Seeley (1881) listed several differences between the teeth on the basis of the shape and extension of the cingulum, the denticle count, and the presence or absence of vertical ridges on the crown. According to him, these differences resulted from either a difference in position in the jaw or were of specific importance (Seeley 1881: 641). The tooth pattern seen in ankylosaurs is primitive for Ornithischia (Sereno 1986). Recent studies on the systematic value of ankylosaurian teeth indicate that there are several sources of variation (Coombs 1990). Thus, an intraspecific variation is likely to explain the tooth differences of Seeley (1881). Following Nopcsa (1929), we provisionally refer the teeth to *Struthiosaurus austriacus*.

Nopcsa (1918) reported an isolated tooth (NHM 1861.I.46) from Frankenhof, near Piesting, a site close to Muthmannsdorf (Fig. 10.1). This specimen, purchased by the Museum of Vienna in 1861, is the first record of a dinosaur in Austria and one of the first discoveries of reptiles from the Gosau Beds. This tooth was originally described as the type of *Leipsanosaurus noricus* (Nopcsa 1918) but was later referred to *Struthiosaurus austriacus* (Nopcsa 1923, 1929). The crown is asymmetrical, leaf-shaped, and labiolingually compressed, with about 10 marginal denticles. A double-faced cingulum is developed at the base of the crown. On the crown surface, there are no marked vertical ridges or grooves that are confluent with the marginal denticles, unlike those seen in nodosaurids such as *Edmontonia* (Carpenter 1990; Coombs 1990; Coombs and Maryańska 1990).

Axial Skeleton

The axial skeleton consists of fragmentary cervical, dorsal, and caudal vertebrae, with associated ribs (Fig. 10.3). The centra are amphiplatyan to slightly amphicoelous. All the material was described as

Crataeomus pawlowitschii by Seeley (1881), except for a small dorsal vertebra and some ribs that he referred to *C. lepidophorus*. Nopcsa (1929) assigned all the vertebral remains from Muthmannsdorf to *Struthiosaurus austriacus*. More recently, Molnar (1980) and Carpenter (personal communication) have supported the idea that two distinct ankylosaurs are represented in the assemblage. Molnar (1980) listed a number of dorsal vertebral differences between *Crataeomus* and *Struthiosaurus*; these differences are discussed below. The material of *Hoplosaurus ischyrus* described by Seeley (1881) includes fragmentary sacral and caudal remains.

Cervical Vertebrae and Rib (Fig. 10.3A–E). The cervical region is represented by two vertebrae and an isolated rib (PIUW 2349/uncataloged). These remains were described in detail by Seeley (1881), but additional comments can now be made. The diapophyses of the most complete vertebra are robust and extend laterally to form a horizontal platelike structure (Bunzel 1871: pl. 2, figs. 9, 10). The neural spine is broken but has a square-shaped base. The Muthmannsdorf vertebrae resemble that of *S. transylvanicus* (BMNH R4966) in having a very elongate centrum and well-expanded transverse processes, although the centra of *S. austriacus* are not as long (Nopcsa 1929; Pereda Suberbiola and Galton 1997). The elongated cervical vertebrae of *Struthiosaurus* suggest that it was a long-necked ankylosaur (Pereda-Suberbiola 1992). Elongate cervicals are also known in *Stegopelta* (Carpenter and Kirkland 1998).

The only known cervical rib is Y-shaped, and it bears an oval capitulum and a fragmentary round tuberculum (Bunzel 1871: pl. 1, fig. 27; Seeley 1881). Cervical ribs were apparently not fused to the vertebrae.

Dorsal Vertebrae and Ribs (Fig. 10.3F–I). Three incomplete and four fragmentary dorsals are present. Two of them (PIUW 2349/110 and uncataloged) were referred to *Crataeomus lepidophorus,* and the other vertebrae, comparatively larger in size (PIUW 2349/24 and uncataloged), was referred to *C. pawlowitschii* by Seeley (1881).

Molnar (1980: table 1) listed several differences between the dorsal vertebrae of *Crataeomus pawlowitschii* and those of *Struthiosaurus austriacus* (= *Crataeomus lepidophorus*): the form of the centrum, presence of a ventral keel, orientation of neural spine, size of neural canal, and location of parapophyses. Carpenter (personal communication) agrees and suggests the following additional taxonomic differences: compared with *Danubiosaurus anceps* (= *Crataeomus pawlowitschii*), the dorsals of *Struthiosaurus austriacus* seem to have a lower neural arch, the neural canal is not as tall, and the centrum is proportionally longer relative to centrum height. These conclusions of Molnar and Carpenter are based on the three dorsal vertebrae illustrated by Bunzel (1871: pl. 1, figs. 24–26, pl. 2, figs. 1–3, pl. 7, fig. 24) and Seeley (1881: pl. 30, figs. 3, 5). However, these differences are probably insufficient to support a viable distinction between *S. austriacus* and a second taxon. They could be explained by different positions in the dorsal series, preservational differences, or both. Moreover, size differences could result in ontogenetic changes. The small dorsal vertebra (PIUW 2349/

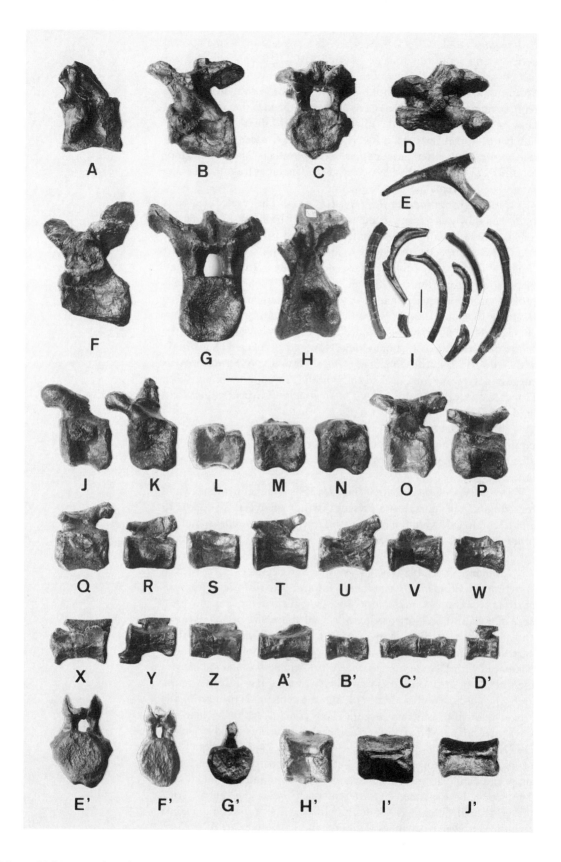

110), assigned to *C. lepidophorus* by Seeley (1881), shows a suture between the centrum and the neural arch and a large neural canal. It probably belongs to an immature individual (called "C" by Nopcsa 1929). The form of the centra of the larger vertebrae of *C. pawlowit-schii* suggests that the anterior dorsals were long. There are no posterior dorsals in the Muthmmansdorf assemblage, but some are known for *Struthiosaurus transylvanicus* (BMNH R4966, R3848 *partim*). The dorsal centra are cylindrical. The most striking feature of the dorsals is the presence of tall neural arch pedicles and a high neural canal equal to three quarters of the centrum height. This morphology of the posterior dorsals of *Struthiosaurus* is considered distinctive among ankylosaurs (Pereda Suberbiola and Galton 1997).

About 20 fragmentary dorsal ribs are known (PIUW 2349/37, uncataloged) (Fig. 10.3I; Bunzel 1871: pl. 8, fig. 15; Seeley 1881: pl. 27, figs. 17, 18). The proximal portion is T-shaped in cross section, as is common in ankylosaurs (Coombs 1978b). The distal half of the shaft shows a roughened area (the "uncinate process" of Brown 1908). There is no evidence of fusion of the ribs to the dorsal vertebrae (but this character is present in other dorsal vertebrae of *Struthiosaurus*; see Pereda Suberbiola 1999).

Sacral Vertebrae. No synsacral remains are known in the Muth-mannsdorf assemblage. Seeley (1881) described two portions of sacrum among the material of *Hoplosaurus ischyrus*. One fragment consists of two fused sacral vertebrae. The material is badly preserved and does not give any useful information.

Caudal Vertebrae and Hemal Arches (Fig. 10.3J, J'). Seeley (1881) listed 18 caudal vertebrae that he referred to *Crataeomus pawlowit-schii,* but 22 are actually preserved (PIUW 2349/25 and uncataloged). Some of the vertebrae have been partially reconstructed (presumably by Suess; see Seeley 1881) with black plaster to resemble the bone. Most of the caudal vertebrae come from the middle and distal portions of the tail, but they do not form a continuous series. Contrary to Seeley's opinion, several individuals could be represented. Nopcsa (1929) assigned these remains to three distinct individuals of *Struthiosaurus austriacus*.

The caudal centra have roughly hexagonal articular surfaces and show longitudinal ridges laterally, as well as a ventral groove (Fig. 10.3J, J'; Bunzel 1871: pl. 2, figs. 4–8, pl. 4, figs. 6–9, pl. 8, figs. 1, 7, 8, 15; Seeley 1881: pl. 30, fig. 4). Notochordal protuberances are occasionally present on the articular surfaces. Unlike typical nodosaurids, such as *Sauropelta* (Carpenter 1984; Ostrom 1970) and *Nodosaurus* (Lull 1921), transverse processes are still present in the middle and distal caudals. The chevrons may be coossified intervertebrally with the centra in the distal caudals. As noted by Nopcsa (1929), the most posterior caudals are fused to each other (probably pathological).

Appendicular Skeleton

The appendicular skeleton is represented by bones from the shoulder girdle (scapulae and coracoid), forelimb (humeri, radii, and ulna?),

Figure 10.3. (opposite page) Vertebrae and ribs of Struthiosaurus austriacus from the Gosau Beds of Muthmannsdorf, Austria. (A) PIUW 2349/uncataloged, cervical vertebra in left lateral view. (B–D) PIUW 2349/ uncataloged (B7a), cervical vertebra in left lateral, posterior, and dorsal views. (E) PIUW 2349/uncataloged (B8a), right cervical rib in anterior view. (F) PIUW 2349/uncataloged (B7a), dorsal vertebra in left lateral view. (G, H) PIUW 2349/ uncataloged (B7b), dorsal vertebra in anterior and left lateral views. (I) PIUW 2349/10 and uncataloged (B8c–e, B8g, B8l, C5c), left and right dorsal ribs in posterior view. (J, J') PIUW 2349/uncataloged (A3d, A3e, B7e, B7f, B7d, A3g, A3h, A3i, B7g, A3j, A3l, A3k, A3m, A3o, A3n, A3p, A3q, A3r, A3s, A3t$_{1-2}$, C3c), caudal vertebrae in lateral (J–D'), anterior (E', F', cf. J, O), posterior (G', cf. U) and ventral (H'–J', cf. O, Q, X) views. Scale bar = 5 cm.

pelvic girdle (ilia and ischium?) and hindlimb (femora, tibia?, and fibulae). A metapodial and a phalanx are also known.

Shoulder Girdle (Figs. 10.4A–H, 10.5). Seeley (1881: fig. on p. 656) referred two scapulae to *Crataeomus pawlowitschii*, a smaller scapula and a coracoid to *C. lepidophorus,* and fragments of a scapula and a coracoid to *Hoplosaurus ischyrus.* Earlier, Bunzel (1871: pl. 5, figs. 7–9, pl. 6, figs. 1–3) described the two scapulae of *C. pawlowitschii* as part of the type of *Danubiosaurus anceps.* Nopcsa (1929) assigned three scapulae and a fragmentary coracoid to *Struthiosaurus austriacus* (PIUW 2349/1 and uncataloged). These scapulae suggest the occurrence of at least three individuals at Muthmannsdorf (Nopcsa 1929). The coracoid referred to *C. lepidophorus* by Seeley (1881) is actually a large dermal scute, and the material of *H. ischyrus* is too fragmentary for an accurate description.

Following Seeley (1881), Molnar (1980) distinguished *Crataeomus* (or *Danubiosaurus* according to the priority's rule; Molnar and Frey 1987) from *Struthiosaurus* on the basis of the general form and position of the acromial process. Carpenter (personal communication) supports this hypothesis and notes that the scapula of *Danubiosaurus anceps* from Austria looks like that of *Struthiosaurus transylvanicus* from Romania.

According to Seeley (1881), the scapular differences between *Crataeomus pawlowitschii* and *C. lepidophorus* involved the greater convexity of the borders in the former and the flatness of the shaft in the latter. In fact, the larger scapula is somewhat deformed by twisting; this distinction is probably an artifact. If the two scapulae are arranged with the glenoid planes parallel to each other, then there is no significant difference in the convexity of the borders (Fig. 10.5). On the other hand, the blade of the larger scapula is more convex than that of the small specimen, as noted by Seeley (1881). This curvature is also present in the scapula of *Struthiosaurus transylvanicus.* On the basis of size differences, the flatness of the small scapula (*C. lepidophorus*) may be due to ontogenetic changes rather than a specific feature.

The scapular spine extends obliquely from the anterior edge to the glenoid cavity and ends in a prominent knoblike acromion, as is typical for nodosaurids (Coombs 1978b, 1978c; Galton 1983). This process is complete and hook-shaped in the scapula of *C. pawlowitschii,* but only the proximal part is preserved in *C. lepidophorus,* giving the impression that it terminates in a rounded ridge (Seeley 1881). A well-defined hooklike acromion also occurs in the scapula of *S. transylvanicus* and suggests that this character is autopomorphic for *Struthiosaurus* (Pereda Suberbiola and Galton 1997). Moreover, the acromion process is centrally placed, well above the level of the glenoid cavity. Seeley (1881) noted that the acromion of *C. lepidophorus* is placed near the glenoid, but the scapula is damaged in this area, so it could have been extended beyond the broken part. In the larger scapula of *Crataeomus* and that of *S. transylvanicus,* the placement of the acromion process seems to be far dorsal to the glenoid (Carpenter, personal communication), but the sutural surface for the coracoid is preserved.

The scapulae have a prespinous fossa anterior to the scapular

Figure 10.4. (opposite page) Shoulder girdle and forelimb bones of Struthiosaurus austriacus *from the Gosau Beds of Muthmannsdorf, Austria. (A–C) PIUW 2349/uncataloged (A2), left scapulocoracoid in lateral, anterior, and medial views. (D, E) PIUW 2349/ uncataloged (B9), right scapula in medial and lateral views. (F–H) PIUW 2349/1 (C1), right scapula in lateral, anterior, and medial views. (I–K) PIUW 2349/ 18, 19 (B2a, b), right and left humerus in anterior (I, K) and dorsal (J) views. (L) PIUW 2349/ uncataloged (B3b), ?ulna in lateral view; (M, N) PIUW 2349/ 11 (B4a, b) left and right radii in medial view. Scale bar = 5 cm.*

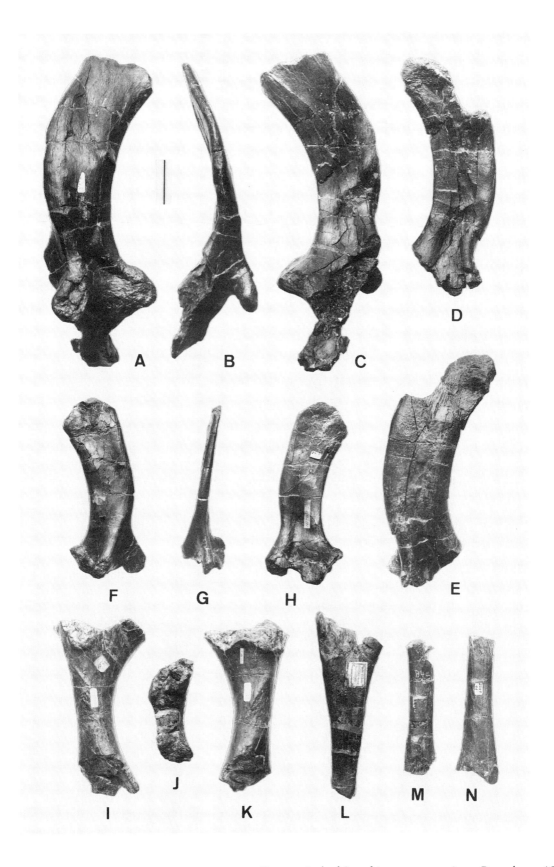

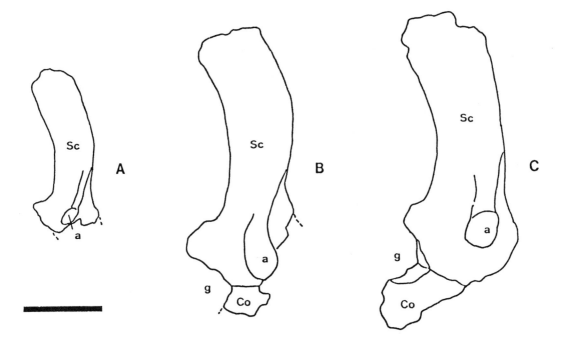

Figure 10.5. Shoulder girdles of
Struthiosaurus *in lateral view.*
(A, B) S. austriacus *(PIUW 2349/*
uncataloged C1 and A2) from
Muthmannsdorf, Austria (B
drawn reverse). (C) S.
transylvanicus *(BMNH R4966)*
from Hateg Basin, Romania. a,
acromion process; Co, coracoid;
g, glenoid; Sc, scapula.
Scale bar = 10 cm.

spine, as in advanced nodosaurids (Coombs 1978b, 1978c; Coombs and Maryańska 1990). The area for the M. supracoracoideus is restricted to the anteroventral part of the scapula, and it appears to be considerably reduced relative to that of *Sauropelta* and *Anoplosaurus* (Galton 1983: fig. 4). Moreover, *Struthiosaurus* differs from *Anoplosaurus* in having a reduced area for the M. scapulohumeralis anterior (Galton 1983; Pereda Suberbiola and Barrett 1999).

As interpreted here, the differences between the scapulae of *Danubiosaurus anceps* (= *Crataeomus pawlowitschii*) and *Struthiosaurus austriacus* (= *C. lepidophorus*) do not support a taxonomic distinction, but the differences are probably due to preservation, ontogenetic changes, individual variation, or a combination of these. Following Nopcsa (1929), the shoulder remains of three nodosaurid individuals are referred to *S. austriacus*.

A coracoid fragment is fused to the larger scapula of *Crateomus pawlowitschii*, but it seems that the specimen was restored (presumably by Nopcsa), because only the scapula was illustrated by Bunzel (1871) and Seeley (1881). Coombs (1978b) interpreted the scapula and coracoid of *Struthiosaurus* as separate bones. The partial skeleton of *S. transylvanicus* (BMNH R4966), regarded as that of an adult (Pereda Suberbiola and Galton 1994, 1997), preserves the coracoid fused to the scapula, but there is still evidence of a suture between the bones.

Forelimb (Fig. 10.4I–N). Forelimb bones consist of two fragmentary humeri, an ulna?, and two radii (PIUW 2349/11, 18, 19, 30, and uncataloged). These bones were originally interpreted as two pairs of complementary humeri and tibiae of *Crataemous lepidophorus* and a tibia of *C. pawlowitschii* (Seeley 1881: pl. 27, fig. 19, pl. 29, figs. 1–3). The proximal and distal ends of the bones are broken. They could be from one (called "B" by Nopcsa 1929) or several individuals. The deltopectoral crest of the humerus is mostly missing, but it does not extend to the midpoint of the shaft, as in *Polacanthus* (Pereda Suberbiola 1994) and *Gastonia* (Kirkland 1998). The shaft of the ulna? and radii are more slender than in *Gastonia, Mymoorapelta* (Kirkland and Carpenter 1994; Kirkland et al. 1998), and *Edmontonia* (Carpenter 1990).

Seeley (1881: pl. 29, fig. 4) described the distal half of a left humerus (PIUW 2349/20; A4 in Nopcsa's catalog) and referred it to *Crataeomus pawlowitschii*. Nopcsa (1929) interpreted the bone as a tibial fragment. Unfortunately, the piece is too fragmentary for an exact identification (see Fig. 10.6N).

Pelvic Girdle (Fig. 10.6A–H). The pelvic girdle includes parts of two ilia and an ischium? (PIUW 2349/41 and uncataloged). One ilium was originally described as that of the lizard *Danubiosaurus anceps* by Bunzel (1871: pl. 6, figs. 4, 5), but later reidentified, with the other ilium, as costal plates of the turtle *Pleuropeltus suessii* by Seeley (1881: pl. 30, fig. 5). Both ilia are fragmentary; the right is more complete than the left. They are damaged by crushing and show transverse fractures that are developed parallel to each other. These artifacts lead Seeley to think that the bones were chelonian. The iliac blade is a platelike structure, slightly convex dorsally above the acetabulum. As preserved, the preacetabular process of the right ilium is long. The left ilium preserves part of the dorsal wall of the (apparently closed) acetabulum. The dorsal surface of the ilia is smooth, and there is no trace of dermal armor coossified to it.

A fragmentary bone, described by Seeley (1881: pl. 27, fig. 30; see also Bunzel 1871: pl. 3, figs. 12, 13) as a fibula of *Crataeomus*, may actually be an ischium (PIUW 2349/41). The specimen is long, broad anteriorly, and slender distally, but it lacks the proximal and distal ends. It looks much like a nodosaurid ischium but does not show a sharp flexion as in *Sauropelta* (Coombs 1978b; Ostrom 1970), *Polacanthus* (Pereda-Suberbiola 1994), and *Gastonia* (Kirkland 1998).

Hindlimb (Figs. 10.6I–M, 10.7A–T). Two complete small femora, a left and a right, were described by Seeley (1881: pl. 31, figs. 4, 5) as *Crataeomus lepidophorus* (PIUW 2349/31, 32). The bones are well preserved but crushed anteroposteriorly. The shaft is straight in the right femur, but in the left, it has a sigmoidal curve in lateral view due to deformation (Fig. 10.7A–L). The head is slightly delimited from the shaft and projects anteromedially, as in nodosaurids (Coombs 1978b, 1979). The shaft is comparatively more slender than that of other nodosaurids (see Carpenter et al. 1995: fig. 13; Pereda Suberbiola and Barrett 1999: fig. 7). The lesser trochanter (anterior trochanter of Carpenter and Kirkland 1998) is not obliterated by fusion as in *Sauropelta* (Ostrom 1970) and *Edmontonia* (Carpenter 1990), but there is no well-developed cleft or groove separating the lesser and greater trochanters, as in *Cryptodraco* and *Hoplitosaurus* (Galton 1983). In fact, the anterior trochanter is a prominent ridge visible in anterior and lateral views,

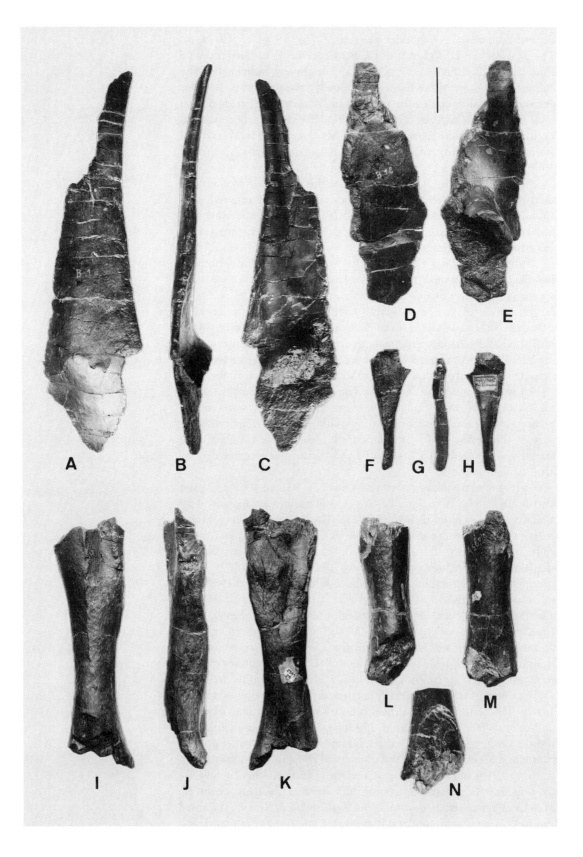

a condition comparable to that of *Polacanthus* (as interpreted by Carpenter and Kirkland 1998). The fourth trochanter is moderately well developed and located proximally near the midlength of the femur. The distal condyles are well developed and separated anteriorly by a shallow intercondylar groove (seen in left femur). The ratio of the maximum distal width relative to the total length of the femur is about 0.3, a low value among ankylosaurs (Galton 1983; Pereda-Suberbiola 1994).

Seeley (1881: pl. 31, fig. 1; see also Bunzel 1871: pl. 3, figs. 2–4) assigned two femora to *Crataeomus pawlowitschii* (PIUW 2349/29 and uncataloged). These bones are fragmentary and lack the articular ends (Fig. 10.6I–M). As noted by Seeley (1881), the best preserved right femur looks roughly like the femora described above, but it is larger and has well-marked anterior ridges for muscular attachment.

A tibia referred to *Crataeomus pawlowitschii* by Seeley (1881: pl. 31, fig. 2) is broken proximally and distally (PIUW 2349/30). The bone was later considered to be an ulna by Nopcsa (1929). The shaft is slender, twisted, and well expanded distally (Fig. 10.7M, N).

Two fragmentary fibulae (PIUW 2349/34) were identified by Seeley (1881: pl. 31, fig. 6; see also Bunzel 1871: pl. 3, fig. 1) as the femora of the crocodilian *Rhadinosaurus alcimus*. The fibular shaft is slender, compressed, and almost straight (Fig. 10.7O–R). It bears a rugose scar on the proximal half of the shaft, as in some ankylosaurs (Coombs 1979).

An isolated metapodial? and an ungual phalanx (PIUW 2349/uncataloged; Fig. 10.7S, T) are the only known autopodial bones (see Bunzel 1871: pl. 4, figs. 4, 5, 11, 12; Seeley 1881: pl. 29, fig. 6). They were referred to *Crataeomus lepidophorus* by Seeley (1881) and regarded as a second or third metatarsal and a second or third pedal ungual phalanx, respectively. On the basis of the narrow proximal end, the phalanx could belong to an immature individual.

Postcranial Armor

Seeley (1881: 650) listed 50 dermal plates and a large number of fragments of armor, which he referred to as *Crataeomus*, although he noted that "it is quite possible that the remains may have belonged to more than one species." Nopcsa (1929) recorded about 30 elements and referred them to three different individuals of *Struthiosaurus austriacus*. Because details of the excavation are not known, it is difficult to say whether the armor of one or of several individuals is represented in the assemblage. Because the dermal elements do not exhibit conclusive evidence for the presence of more than one species, Nopcsa's proposition is followed.

Five types of armor are known at Muthmannsdorf: low-keeled scutes, spines, hornlike spikes, triangular hollow-based plates, and small ossicles (PIUW 2349/13–15 and uncataloged collection). Cervical, dorsal, sacral?, and caudal armor is known (Fig. 10.8A–W; Bunzel 1871: pl. 4, figs. 1, 2, pl. 5, fig. 10, pl. 7, figs. 20, 21, pl. 8, figs. 10–13; Nopcsa 1929: pl. 4; Seeley 1881: pl. 28, figs. 2–4, pl. 30, fig. 2, pl. 31, figs. 3, 11).

Figure 10.6. (opposite page) Pelvic girdle and hindlimb bones of Struthiosaurus austriacus *from the Gosau Beds of Muthmannsdorf, Austria. (A–C) PIUW 2349/uncataloged (B1a), right ilium in dorsal, lateral, and ventral views. (D, E) PIUW 2349/uncataloged (B1b), left ilium in dorsal and ventral views. (F–H) PIUW 2349/41, ?left ischium in lateral, dorsal, and medial views. (I–K) PIUW 2349/29 (A5a), right femur in anterior, medial, and posterior views. (L, M) PIUW 2349/uncataloged (A5b), left femur in anterior and posterior views. (N) PIUW 2349/20 (A4), indeterminate bone. Scale bar = 5 cm.*

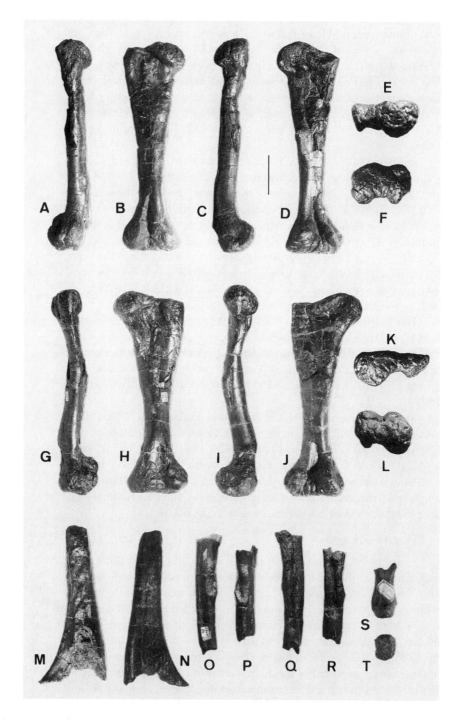

Figure 10.7. Hindlimb bones of Struthiosaurus austriacus *from the Gosau Beds of Muthmannsdorf, Austria.*
(A–F) PIUW 2349/31 (C2a), right femur in lateral, anterior, medial, posterior, proximal, and distal views.
(G–L) PIUW 2349/32 (C2b), left femur in lateral, anterior, medial, posterior, proximal, and distal views.
(M, N) PIUW 2349/uncataloged (B3a), left ?tibia in posterior and anterior views. (O–R) PIUW 2349/34 and
uncataloged, left and right fibulae in medial and lateral views. (S) PIUW 2349/uncataloged, metatarsal. (T) PIUW
2349/uncataloged (A6), ungual phalanx in dorsal view. Scale bar = 5 cm.

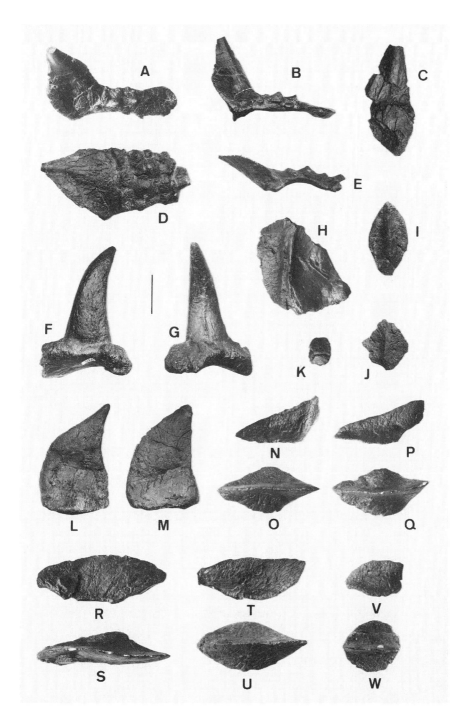

Figure 10.8. Dermal armor of Struthiosaurus austriacus *from the Gosau Beds of Muthmannsdorf, Austria.*
(A, B) PIUW 2349/14 (B10a), right cervical half-ring in dorsal and anterior views. (C) PIUW 2349/uncataloged (B10b), cervical spine in dorsal view. (D, E) PIUW 2349/13 (B10c), right (?posterior) cervical half-ring in dorsal and anterior views. (F, G) PIUW 2349/15 (A1a), ?dorsal spike in lateral and posterior views. (H) PIUW 2349/uncataloged (A1c), dorsal or ?sacral scute (partially reconstructed) in dorsal view. (I, J) PIUW 2349/uncataloged, low-keeled scutes in dorsal view. (K) PIUW 2349/uncataloged, scute in dorsal view. (L, M) PIUW 2349/uncataloged (B12a, b?), ?anterior caudal plates (partially reconstructed) in lateral view. (N–W) PIUW 2349/uncataloged (B11a, b, A1b [partially reconstructed], A1d, C6), caudal high-keeled plates in left lateral and dorsal views. Scale bar = 5 cm.

The cervical armor consists of three half-rings, one complete and two partial (PIUW 2349/13, 14, and uncataloged). Each half-ring is composed of three types of fused elements: spines, small ossicles, and oval scutes (Fig. 10.8A–E). The spines are asymmetrical and blunt. They are dorsoventrally compressed, with a sharp keel on the anterior edge and a ventral rim. The smaller (presumed anterior) spines projected dorsolaterally and slightly posteriorly, whereas the larger (posterior) spines were almost horizontal (*contra* Nopcsa 1929). Medially, the anterior half-ring has a suboval scute that is wider than long and spoon-shaped. Between the lateral spines and the medial scutes is a mosaic band of small, irregular to polygonal ossicles. Similar cervical half-rings have been found in the Upper Cretaceous of the Laño Quarry, Iberian Peninsula (Pereda Suberbiola et al. 1995).

A pair of hornlike spikes (PIUW 2349/15 and uncataloged) is known, one complete and the other fragmentary (Fig. 10.8F, G). The best preserved specimen has a large, concave base and bears a conical spine that is curved toward the tip. A sharp keel is present on one (posterior?) side of the spike, and a ridge is present on the lateral side. The position of these spikes on the body is unknown, but they could be situated on the shoulder region.

Dorsal armor is presumably represented by 10 low-keeled oval scutes and small, flat ossicles (PIUW 2349/uncataloged; Fig. 10.8I–K). A subcircular, keeled plate was regarded as a coracoid of *C. lepidophorus* by Seeley (1881). On the basis of its large size and flat base, it may be part of the dorsal or sacral armor (Fig. 10.8H). A sacral shield of fused armor is known in *Polacanthus* (Blows 1987; Hulke 1888; Pereda-Suberbiola 1994), *Mymoorapelta* (Kirkland and Carpenter 1994; Kirkland et al. 1998), *Gastonia* (Kirkland 1998), and *Gargoyleosaurus* (Carpenter, personal communication). In addition, portions of sacral armor provisionally referred to *Stuthiosaurus* have been found in the Iberian Peninsula (Pereda Suberbiola 1999).

The armor regarded as caudal consists of a number of asymmetrical, triangular plates (PIUW 2349/uncataloged). Some plates are higher than long, laterally compressed, and hollow-based. They bear a spine oriented posteriorly (Fig. 10.8L, M). The distal edge is not as concave as in *Polacanthus* (Blows 1987; Pereda-Suberbiola 1994), *Mymoorapelta* (Kirkland and Carpenter 1994; Kirkland et al. 1998), and *Gastonia* (Kirkland 1998). Moreover, the apical spine is more pointed than in the North American "polacanthines." Other armor elements of Muthmannsdorf include low, rooflike plates and small, conical scutes (Fig. 10.8N–W).

Discussion

Phylogenetics

I. *Struthiosaurus* is referred to the Ankylosauria on the basis of the following synapomorphies (see list of characters in Coombs 1978b; Coombs and Maryańska 1990; Sereno 1986): cranium low, rear of skull wider than high; supratemporal fenestra closed; pterygoid closes the passage between the space above the palate and that below the braincase; postocular shelf encloses the orbital cavity; fusion of dermal ossifications to the dorsal skull roof; atlar neural arches joined dorsally; ilium rotated into horizontal plane; acetabulum closed; and extensive armor composed of several kinds of dermal plates.

II. Within the Ankylosauria, *Struthiosaurus* shows diagnostic features of the Nodosauridae as follows (see Coombs and Maryan´ska 1990; Lee 1996; Sereno 1986): hemispherical occipital condyle separated from the ventral braincase by a constricted neck; paroccipital processes posteroventrally directed and visible in dorsal view; basipterygoid processes consist of a pair of rounded, rugose stubs; scapular spine displaced ventrally toward glenoid; three transverse rows of cervical armor; tall conical spikes; and keeled plates only slightly excavated ventrally.

III. *Struthiosaurus* is characterized by the following combination of characters:

1. The basisphenoid projects ventrally below the general level of the basioccipital in posterior view. If the fragmentary braincase of *S. austriacus* is arranged with the posterior part of the roof skull at 20° relative to the horizontal and the occipital condyle is angled at approximately 55° posteroventrally (following the model of the nodosaurid *Panoplosaurus mirus*), then the main part of the ventral surface of the basisphenoid is directed anteroventrally at about 20° (Pereda Suberbiola and Galton 1994: fig. 6). The floor of the braincase, however, is horizontal or nearly horizontal in other well-known ankylosaurs—for example, the nodosaurids *Silvisaurus* (cast BMNH R11189) and *Panoplosaurus* (cast in Natural History Museum, London) and the ankylosaurid *Euoplocephalus* (BMNH R4947). The ventral projecting shape of the basisphenoid bone has been reported in some prosauropods, sauropods, and theropods, but it is not present in typical ornithischian basicrania (see discussion in Pereda Suberbiola and Galton 1994). However, the ankylosaur *Gastonia* has elongate basisphenoid processes that form a gliding joint with the unfused pterygoids (Kirkland 1998). The functional significance of this feature is not accurately understood.

2. Nearly oval, symmetrical, distal articular condyle of quadrate. The mandibular condyle of the quadrate is peanut-shaped in most ankylosaurs and stegosaurs (Coombs and Maryańska 1990; Galton 1990; Sereno 1986). In nodosaurids, such as *Panoplosaurus* and *Edmontonia,* the medial portion of the quadrate condyle is much more robust than the lateral portion, and viewed from below, the articular end is roughly triangular in outline (Gilmore 1930). In contrast, it is almost symmetrical in *Struthiosaurus*, with the medial condyle only a little wider than the lateral condyle, so in ventral view, it is nearly oval.

3. Very elongate, longer than wide, cervical centra. The cervical centra of *Struthiosaurus* differ considerably from those of most ankylosaurs in having a much elongated centrum. In typical ankylosaurs, such as the nodosaurids *Sauropelta* (Ostrom 1970), *Silvisaurus* (Carpenter and Kirkland 1998; Eaton 1960), *Edmontonia* (Gilmore 1930), *Niobrarasaurus* (Carpenter et al. 1995), *Polacanthus* (Blows 1987), *Hylaeosaurus* (Pereda-Suberbiola 1993), and the ankylosaurids *Sai-*

chania (Maryańska 1977) and *Ankylosaurus* (Brown 1908), the cervical centra are wider than long, but the contrary appears to be the case in *Struthiosaurus* and *Stegopelta* (Carpenter and Kirkland 1998: fig. 22).

4. Hooklike acromion process, placed centrally on the scapula. Some nodosaurids such as *Mymoorapelta* (Kirkland and Carpenter 1994), *Gastonia* (Kirkland 1998), *Hylaeosaurus*, and *Polacanthus* (Blows 1996; Pereda-Suberbiola 1993, 1994), possess a bladelike scapular spine. In more derived nodosaurids, the scapular spine ends in a prominent knoblike acromion process (Coombs 1978b, 1978c; Coombs and Maryańska 1990). As suggested by Carpenter et al. (1995), both the shape and position of the acromion process are variable among the Nodosauridae and may be useful systematically. *Sauropelta* and *Anoplosaurus* have a thumblike acromion process; it is located posteriorly near the edge of the glenoid in *Sauropelta* (Coombs 1978b; Ostrom 1970), whereas in *Anoplosaurus*, the acromion is placed centrally (Pereda Suberbiola and Barrett 1999; Seeley 1879). Coombs (1995) described *Texasetes* as having a finger- or pronglike acromion directed toward the innermost extent of the glenoid, but this process is broken and was probably longer than has been thought (Carpenter and Kirkland 1998). *Panoplosaurus* possesses a handle-shaped acromion process that is centrally situated and projects obliquely anteriorly (Russell 1940; Sternberg 1921). In *Struthiosaurus*, the acromion of both *S. austriacus* and *S. transylvanicus* is centrally located, with the distal end projected anteriorly to form a hooklike process. This morphology seems to be unique among ankylosaurs.

5. Distinctive postcranial armor. The armor of *S. austriacus* includes cervical plates composed of spines, a band of coossified small ossicles, and oval scutes; and dorsal hornlike spikes and tall triangular hollow-based plates, presumably on the tail.

IV. *Struthiosaurus* is conservative within the Nodosauridae in retaining the following primitive characters:

1. Occiput relatively narrow and high, with the tip of the paroccipital processes short and strongly curved posteriorly. A similar pattern in known in *Silvisaurus* (Eaton 1960), *Sauropelta* (Carpenter and Kirkland 1998), *Pawpawsaurus* (Lee 1996), and *Gastonia* (Kirkland 1998). On the other hand, *Panoplosaurus* and *Edmontonia* (Carpenter 1990; Gilmore 1930; Russell 1940) have a very low occiput, with the paroccipital processes comparatively longer and more slender, and not projecting as much posteriorly (Pereda Suberbiola and Galton 1994: fig. 7)

2. Narrow and relatively elongate infratemporal fenestra (after reconstruction). This morphology is intermediate between that of *Silvisaurus* (with an oval infratemporal fenestra) and *Panoplosaurus* and *Edmontonia* (with a very narrow fenestra) (Pereda Suberbiola and Galton 1994). The size and shape of the infratemporal fenestra of *Struthiosaurus* seems roughly similar to those of *Pawpawsaurus* (Lee 1996), *Sauropelta* (Carpenter and Kirkland 1998), and *Gastonia* (Kirkland 1998).

3. Long tooth row extending from the coronoid process to the anterior end of the dentary, near the symphysis. Besides *Struthiosaurus*,

this pattern is known in *Sarcolestes* (Galton 1983), *Sauropelta* (Ostrom 1970), *Anoplosaurus* (Pereda Suberbiola and Barrett 1999), and *Gargoyleosaurus* (Carpenter et al. 1998). In contrast, the anterior end of the dentary is edentulous in *Panoplosaurus* (Russell 1940), *Edmontonia* (Sternberg 1928), and *Silvisaurus* (Carpenter and Kirkland 1998).

4. Atlas and axis separate. Among nodosaurids, the atlas and axis are separate in *Struthiosaurus, Sauropelta* (Ostrom 1970), *Polacanthus* (Blows 1987), and probably *Hylaeosaurus* (Pereda-Suberbiola 1993). The atlas and axis are fused in *Panoplosaurus* (Russell 1940; Sternberg 1921) and *Edmontonia* (Gilmore 1930).

5. Lesser (anterior) trochanter of femur not completely obliterated by fusion. Primitive nodosaurids such as *Cryptodraco* and *Hoplitosaurus* retain a prominent lesser trochanter separated from the greater trochanter by a deep cleft (Galton 1983). The lesser trochanter is represented by a distinct ridge in the more derived *Polacanthus* (Pereda-Suberbiola 1994; see also Carpenter and Kirkland 1998) and *Anoplosaurus* (Pereda Suberbiola and Barrett 1999; as *Acanthopholis* in Galton 1983). The lesser trochanter is reduced to a slight crest in *Sauropelta* and completely obliterated in derived nodosaurids (Coombs 1979).

V. *Struthiosaurus* is clearly more primitive than *Edmontonia* and *Panoplosaurus* in retaining several symplesiomorphies (see above). The relationship of *Struthiosaurus* to other nodosaurids is problematic, mainly because of the incompleteness of its skull. *Struthiosaurus* is more derived than *Mymoorapelta*, but probably not as derived as *Sauropelta*. With regard to European ankylosaurs, *Struthiosaurus* is more derived than the Wealden nodosaurids *Polacanthus* and *Hylaeosaurus* in having a prominent acromion process and a prespinous fossa on the scapula. It seems to be more derived than *Anoplosaurus* in having a further reduced area of origin for the M. scapulohumeralis anterior on the scapula and a more reduced lesser (anterior) trochanter on the femur.

VI. Finally, *Struthiosaurus* apparently shares some characters with the polacanthines. The Polacanthinae, which consists of *Polacanthus, Hoplitosaurus, Gastonia, Mymoorapelta,* and possibly *Hylaeosaurus,* has been defined by Kirkland (1998) as the sister group to most of the Ankylosauridae on the basis of a preliminary cladistic analysis. Armor of typical polacanthines has large, erect dorsal shoulder spines, a sacral shield of fused armor, and asymmetrical, hollow-based, triangular caudal plates (Kirkland 1998). Curiously, these features are also present in *Struthiosaurus* (Pereda Suberbiola 1999). In contrast, the scapula of *Struthiosaurus* clearly differs from that of polacanthines and strongly resembles that of derived Nodosauridae. This casts doubt on the validity of the Polacanthinae as a clade (Carpenter 1998). The oldest representative of the Polacanthinae is *Mymoorapelta maysi* from the Upper Jurassic of North America (Kirkland and Carpenter 1994; Kirkland et al. 1998). This species is mainly characterized by a very plesiomorphic ilium (preacetabular process curves lateroventrally and is not horizontal as in all other known ankylosaurs; the acetabulum is partially closed). The Upper Jurassic–Lower Cretaceous distribution and the primitive position of most of the polacanthines suggest that they are early ankylosaurs. However, more information is needed to resolve the question of the position of polacanthine taxa within the Ankylosauria. In any case, Kirkland's (1998) conclusion that the early radiation of the armored dinosaurs was more complicated than previously thought is worth considering.

Species Identity

Struthiosaurus was one of the last ankylosaurs and ranges geochronologically from the Campanian to the Maastrichtian (Pereda-Suberbiola 1992). The genus was originally described in central Europe, including the type species *S. austriacus* in Muthmannsdorf (Lower Austria) and *S. transylvanicus* in the Sânpetru area of Transylvania (now Romania) (see Nopcsa 1915, 1923, 1929). Additional *Struthiosaurus* remains have been recently recovered in southwestern Europe sites, such as the Laño Quarry in the Iberian Peninsula (Pereda Suberbiola 1999; Pereda Suberbiola et al. 1995) and, tentatively, in Languedoc in southern France (Pereda-Suberbiola 1993). These discoveries indicate a large geographical distribution for *Struthiosaurus*.

The specific diversity within the genus *Struthiosaurus* is still unresolved. Nopcsa (1929) suggested the occurrence of two different species of *Struthiosaurus* in Austria and Transylvania and established a distinct genus, *Rhodanosaurus,* for the French material. *Rhodanosaurus ludgunensis* is nondiagnostic, and this species should be regarded as Nodosauridae *nomen dubium* (Coombs and Maryańska 1990; Pereda-Suberbiola 1993). The material may belong to *Struthiosaurus*.

Struthiosaurus austriacus apparently differs from *S. transylvanicus* in having a less ventrally projected basisphenoid, cervical vertebrae comparatively less elongated, and dorsal vertebrae with a lower neural arch, pedicles, and neural canal. These differences may have a specific significance, but the available material is probably inadequate to resolve the question of the relationship of *S. austriacus* to *S. transylvanicus* (Pereda Suberbiola and Galton, 1994, 1997). The fossil remains referred to as *Struthiosaurus* in Austria, Transylvania, and the Iberian Peninsula exhibit a great variation in shape. The significance of this osteological variability is currently uncertain. It may reflect several sources of variation (individual differences, ontogenetic changes, or even sexual dimorphism) within *S. austriacus*; or it may reflect the occurrence of two or more species of *Struthiosaurus* (see further discussion in Pereda Suberbiola 1999; Pereda Suberbiola et al. 1995).

Seeley (1881), and more recently Molnar (1980) and Carpenter (personal communication), have supported the presence of two different ankylosaurs in the Gosau Beds of Austria. This interpretation is based on differences in the dorsal vertebrae and scapulae. Moreover, the two nodosaurids at Muthmannsdorf can apparently be separated by size: the larger bones belong to *Danubiosaurus anceps* Bunzel, 1871 (= *Crataeomus pawlowitschii* Seeley, 1881), and the smaller bones belong to *Struthiosaurus austriacus* Bunzel, 1871 (= *Crataeomus lepidophorus* Seeley, 1881).

Nopcsa (1929), however, considered that several individuals of a single nodosaurid species were present at Muthmannsdorf and referred all the material to *Struthiosaurus austriacus*. At least three individuals, one of small size (including the type specimen) and two larger specimens, are present. There are size differences between the bones, which suggests ontogeny in immature and adult individuals. Additional differences between the bones may be due to individual variation or preservation.

The distinct placement of the acromion process on the scapulae of *Crataeomus* may be taxonomically significant, but the importance of this feature is debatable. *Struthiosaurus* shows a considerable amount of variation. Thus, we prefer to adopt a conservative approach, and following Nopcsa (1929), we provisionally refer all the ankylosaurian remains found at Muthmannsdorf to *Struthiosaurus austriacus*.

Growth Stages

The skull of *Struthiosaurus austriacus* is slightly smaller than that of *S. transylvanicus* (approximate ratio 0.83:1). Qualitative differences —for example, the lack of fusion of the paroccipital process, dorsal end of the quadrate and squamosal, the open suture of quadrate–quadratojugal, and the weak posterodorsal development of the skull roof armor—suggest that the type specimen of *S. austriacus* belongs to an immature, probably subadult individual (Pereda Suberbiola and Galton 1992, 1994).

The absence of a splenial and the lack of fusion of dermal ossification to the lower jaw seen in *S. austriacus* may also be ontogenetically significant (Pereda Suberbiola et al. 1995). The splenial is commonly firmly fused to the dentary in adult ankylosaurs, and the Meckelian groove is hidden medially (Galton 1983). The fusion of a dermal scute to the ventrolateral side of the mandible has been regarded as diagnostic for the Ankylosauria (Galton 1983; Sereno 1986). However, the lower jaw of presumably immature specimens, such as the small dentaries of *Struthiosaurus austriacus* and *Anoplosaurus curtonotus* (Seeley 1879; Pereda Suberbiola and Barrett 1999), lack a splenial, and the Meckelian canal remains open. There is no evidence of an ossification on the lateral side of these mandibles.

On the basis of the juvenile features listed by Galton (1982a, 1982b) for stegosaurs and listed by Coombs (1986) for ankylosaurs, postcranial characters for *S. austriacus* that we tentatively regard as immature include the following: presence of a suture between the vertebral centrum and the neural arch (obliterated in adults); ungual phalanx widest at about one third down the length (proximally in adults); and surface of large bones with neither deep rugosities nor strong muscular insertions (contrary to full-size adults).

Galton (1982a, 1982b, 1990) considered the lesser trochanter to be a distinct process in the femora of juvenile and subadult stegosaurs, but it becomes completely obliterated in adults. A distinct lesser (anterior) trochanter is absent or reduced in adult individuals of derived ankylosaurs (Coombs and Maryańska 1990). Primitive nodosaurids, such as *Cryptodraco* (Galton 1983), *Polacanthus* (Pereda-Suberbiola 1994), and *Anoplosaurus* (Pereda Suberbiola and Barrett 1999), retain a distinct lesser trochanter. It has been suggested that this character reflects slight locomotor differences within ankylosaurs (Galton 1983).

Juvenile and subadult ankylosaurs are not abundant in the fossil record. Among ankylosaurids, juvenile *Pinacosaurus* are known from Mongolia (Jerzykiewicz et al. 1993; Maryańska 1977), and a juvenile or subadult *Euoplocephalus* has been described from North America (Coombs 1986). Immature nodosaurids are represented by a partial skeleton of *Anoplosaurus curtonotus* in the Cambridge Greensand of England (Pereda Suberbiola and Barrett 1999), a partial skeleton from the Paw Paw Formation of Texas (Jacobs et al. 1994), and isolated bones from the Niobrara Formation of Kansas (Carpenter et al. 1995). Other records consist of isolated teeth of baby or juvenile ankylosaurs obtained by screen-washing the sediment (Antunes and Sigogneau-Russell 1991; Carpenter 1982; Pereda Suberbiola 1999).

Evolutionary Observations

Struthiosaurus was a small ankylosaur, with adult individuals having a total length of about 3 m and an estimated body weight of about 300 kg (Pereda-Suberbiola 1992). It is much smaller than the nodosaurids *Edmontonia* and *Panoplosaurus* from the Late Cretaceous of western North America, which reached 6–7 m long (Carpenter 1990; Coombs and Maryańska 1990). In spite of its late chronostratigraphic distribution, *Struthiosaurus* is about the same size as the enigmatic ankylosaur *Minmi* from the Early Cretaceous of Australia (Molnar 1996) and as the early ankylosaurs *Mymoorapelta* (Kirkland and Carpenter 1994), *Dracopelta* (Galton 1980), and *Tianchisaurus* (Dong 1993). *Struthiosaurus* is about the same size as the basal thyreophoran *Scelidosaurus* from the Early Jurassic of England (Coombs et al. 1990).

As early as 1923, Nopcsa recognized that *Struthiosaurus* was much smaller than its contemporary relatives in North America, and he claimed that the Transylvanian dinosaurs included relict, dwarf forms. This hypothesis is now widely followed (Pereda Suberbiola 1996; Weishampel et al. 1991). Nopcsa regarded the small size of *Struthiosaurus* and other dinosaurs as a consequence of insular evolution in the European archipelago during the Late Cretaceous. The body-size effects (mainly heterochronic processes) via island habitation in Transylvanian dinosaurs have been reevaluated by D. Weishampel and collaborators. Weishampel et al. (1993) noted that dwarfism in the hadrosaurid *Telmatosaurus transsylvanicus* may involve heterocronic alterations of growth processes. Similarly, Jianu and Weishampel (1998) have suggested that the "juvenile" morphology of the sauropod *Magyarosaurus dacus* may constitute dwarfing by heterocronic pedomorphosis. Other insular, apparently dwarf dinosaurs also exhibit similar heterocronic features. Molnar (1996) observed juvenilelike features (e.g., lack of fusion between the cranial bones and coracoid not fused to the scapula) in a mature or almost mature skeleton of *Minmi* and interpreted them as plesiomorphic. In *Struthiosaurus*, the lack of fusion between the quadratojugal and jugal, the unusual shortness of the dentary, and the presence of a sutural surface between the scapula and coracoid look like immature features, but they occur in adult individuals. Preliminarily, these features suggest that *Stuthiosaurus* was a pedomorphic dwarf.

Size changes in dwarf vertebrates are generally accompanied by both allometric and locomotor changes. Weishampel et al. (1993) hypothesized that *Telmatosaurus* may have different hindlimb kinematics than other hadrosaurids. In regard to ankylosaurs, Molnar and Frey (1987) speculated on differences in the locomotory system between *Minmi* and well-known forms and suggested that it may have had more cursorial habits than other ankylosaurs. Similarly, the elongated cervical vertebrae of *Struthiosaurus* are probably an adaptation and suggest a long neck (Pereda-Suberbiola 1992). However, the available evidence is inconclusive as to whether this specialization involved an altered feeding strategy or not.

Summary

A revision is given of the ankylosaurian remains from the Late Cretaceous (early Campanian) Gosau Beds of Muthmannsdorf, Lower Austria. The Gosau vertebrate fauna is largely dominated by ankylosaurs. More than a hundred bones from at least three skeletons of different size are recognized. Following Nopcsa (1929), all the remains (including lower jaws, teeth, vertebrae, ribs, limb bones, and armor elements) are referred to *Struthiosaurus austriacus* Bunzel (holotype: a fragmentary skull). Differences between the bones are interpreted as the result of ontogenetic changes and individual variation rather than viable taxonomic characters. As a result, the taxa *Crataeomus lepidophorus, Crataeomus pawlowitschii, Danubiosaurus anceps, Pleuropeltus suessii,* and *Rhadinosaurus alcimus* (in part) are regarded as junior synonyms of *Stuthiosaurus austriacus*.

Struthiosaurus possesses typical ankylosaurian synapomorphies and can be referred to the Nodosauridae. It is characterized by the following diagnostic features: ventrally projected basisphenoid; almost symmetrical distal articular condyle of quadrate; cervical centra longer than wide; hooklike acromion process on the scapula; ridgelike lesser (anterior) trochanter on the femur; and a distinctive dermal armor, mainly composed of cervical rows provided with spines, small ossicles, and scutes fused together, dorsal hornlike spikes, and tall, hollow-based, triangular caudal plates.

Struthiosaurus is more derived than other known European ankylosaurs, but it seems much more conservative than the derived North American nodosaurids *Edmontonia* and *Panoplosaurus*. It retains a number of primitive characters within nodosaurids, including occiput relatively narrow and high; relatively elongate infratemporal fenestra; tooth row extending near the symphysis; atlas and axis separate; and lesser (anterior) trochanter of femur not completely obliterated by fusion.

Struthiosaurus has a large geographical distribution (central and southern Europe) and ranges chronostratigraphically from the lower Campanian to the Maastrichtian. The significance of the strong morphological variability seen in *Struthiosaurus* is not yet understood, but the occurrence of several species is not excluded.

Finally, *Struthiosaurus* was an endemic, presumably pedomorphic dwarf ankylosaur. Heterochronic alterations of growth processes could explain the juvenilelike features retained by mature individuals of *S. austriacus*.

Acknowledgments. We thank Prof. Dr. G. Rabeder and Mag. Dr. K. L. Rauscher (Paläontologisches Institut der Universität, Vienna), Drs. H. A. Kollman and H. Summesberger (Naturhistorisches Museum, Vienna), and Miss S. Chapman and Dr. A. C. Milner (Natural History Museum, London) for their assistance during the study of dinosaurian material in their care. We are appreciative of the constructive criticisms of both Dr. K. Carpenter (Denver) and Dr. D. Norman (Cambridge) whose suggestions have improved this article. Photos are by X.P.B., C. Abrial (Université Pierre et Marie Curie, Paris), and D. Serrette (Muséum National d'Histoire Naturelle, Paris).

References Cited

Antunes, M., and D. Sigogneau-Russell. 1991. Nouvelles données sur les dinosaures du Crétacé supérieur du Portugal. *Comptes Rendus de l'Académie des Sciences Paris* 313: 113–119.

Blows, W. T. 1987. The armored dinosaur *Polacanthus foxi* from the Lower Cretaceous of the Isle of Wight. *Palaeontology* 30: 557–580.

———. 1996. A new species of *Polacanthus* (Ornithischia: Ankylosauria) from the Lower Cretaceous of Sussex, England. *Geological Magazine* 133: 671–682.

Brix, F., and B. Plöchinger. 1988. Erläuterungen zu Blatt 76 Wiener Neustadt. *Geologische Karte der Republik Österreich 1:50000.* Vienna: Geologische Bundesanstalt.

Brown, B. 1908. The Ankylosauridae, a new family of armored dinosaurs from the Upper Cretaceous. *Bulletin of the American Museum of Natural History* 24: 187–201.

Bunzel, E. 1870a. Notice of a fragment of a reptilian skull from the Upper Cretaceous of Grünbach. *Quarterly Journal of the Geological Society of London* 26: 394.

———. 1870b. Ueber die Reptilien-Fauna der Kreide-schichten von Grünbach. *Verhandlungen Geologische Reichsanstalt* 180: 80.

———. 1871. Die Reptilfauna der Gosau-Formation in der Neuen Welt bei Wiener-Neustadt. *Abhandlungen Geologische Reichsanstalt* 5: 1–18.

Carpenter, K. 1982. Baby dinosaurs from the Late Cretaceous Lance and Hell Creek formations and a description of a new species of theropod. *Contributions to Geology, University of Wyoming* 20: 123–134.

———. 1984. Skeletal reconstruction and life restoration of *Sauropelta* (Ankylosauria: Nodosauridae) from the Cretaceous of North America. *Canadian Journal of Earth Sciences* 21: 1491–1498.

———. 1990. Ankylosaur systematics: Example using *Panoplosaurus* and *Edmontonia*. In K. Carpenter and P. J. Currie (eds.), *Dinosaur Systematics: Approaches and Perspectives,* pp. 281–198. Cambridge: Cambridge University Press.

———. 1998. Ankylosaur odds and ends [abstract]. *Journal of Vertebrate Paleontology* 17(3, Suppl.): 31A.

Carpenter, K., and J. I. Kirkland. 1998. Review of Lower and Middle

Cretaceous ankylosaurs from North America. In S. G. Lucas, J. I. Kirkland, and J. W. Estep (eds.), *Lower and Middle Cretaceous Terrestrial Ecosystems. New Mexico Museum of Natural History and Science Bulletin* 14: 249–270.

Carpenter, K., D. Dilkes, and D. B. Weishampel. 1995. The dinosaurs of the Niobrara Chalk Formation (Upper Cretaceous, Kansas). *Journal of Vertebrate Paleontology* 15: 275–297.

Carpenter, K., C. Miles, and K. Cloward. 1998. Skull of a Jurassic ankylosaur (Dinosauria). *Nature* 393: 782–783.

Coombs, W. P., Jr. 1978a. An endocranial cast of *Euoplocephalus* (Reptilia, Ornithischia). *Palaeontographica* A 161: 176–182.

———. 1978b. The families of the ornithischian dinosaur order Ankylosauria. *Palaeontology* 21: 143–170.

———. 1978c. Forelimb muscles of the Ankylosauria (Reptilia, Ornithischia). *Journal of Paleontology* 52: 642–658.

———. 1979. Osteology and myology of the hindlimb in the Ankylosauria (Reptilia, Ornithischia). *Journal of Paleontology* 53: 666–684.

———. 1986. A juvenile ankylosaur referable to the genus *Euoplocephalus* (Reptilia, Ornithischia). *Journal of Vertebrate Paleontology* 6: 162–173.

———. 1990. Teeth and taxonomy in ankylosaurs. In K. Carpenter, and P. J. Currie (eds.). *Dinosaur Systematics: Perspectives and Approaches,* pp. 269–279. Cambridge: Cambridge University Press.

———. 1995. A new nodosaurid ankylosaur (Dinosauria: Ornithischia) from the Lower Cretaceous of Texas. *Journal of Vertebrate Paleontology* 15: 298–312.

Coombs, W. P., Jr., and T. Maryańska. 1990. Ankylosauria. In D. B. Weishampel, P. Dodson, and H. Osmólska (eds.), *The Dinosauria,* pp. 456–483. Berkeley: University of California Press.

Coombs, W. P., Jr., D. B. Weishampel, and L. M. Witmer. 1990. Basal Thyreophora. In D. B. Weishampel, P. Dodson, and H. Osmólska (eds.), *The Dinosauria,* pp. 427–434. Berkeley: University of California Press.

Dong, Z. 1993. An ankylosaur (ornithischian dinosaur) from the Middle Jurassic of the Junggar Basin, China. *Vertebrata PalAsiatica* 10: 257–266.

Eaton, T. H., Jr. 1960. A new armored dinosaur from the Cretaceous of Kansas. *University of Kansas Paleontological Contributions* 25(8): 1–24.

Galton, P. M. 1980. Partial skeleton of *Dracopelta zbyszewskii* n. gen. and n. sp., an ankylosaurian dinosaur from the Upper Jurassic of Portugal. *Geobios* 13: 451–457.

———. 1982a. Juveniles of the stegosaurian dinosaur *Stegosaurus* from the Upper Jurassic of North America. *Journal of Vertebrate Paleontology* 2: 47–62.

———. 1982b. The postcranial anatomy of the stegosaurian dinosaur *Kentrosaurus* from the Upper Jurassic of Tanzania, east Africa. *Geologica et Palaeontologica* 15: 139–160.

———. 1983. Armored dinosaurs (Ornithischia: Ankylosauria) from the Middle and Upper Jurassic of Europe. *Palaeontographica* A 182: 1–25.

———. 1990. Stegosauria. In D. B. Weishampel, P. Dodson, and H. Osmólska (eds.), *The Dinosauria,* pp. 435–455. Berkeley: University of California Press.

Gilmore, C. W. 1930. On dinosaurian reptiles from the Two Medicine

Formation of Montana. *Proceedings of the United States National Museum* 77: 1–39.

Hopson, J. A. 1979. Paleoneurology. In C. Gans (ed.), *The Biology of the Reptilia,* Vol. 9, pp. 39–146. New York: Academic Press.

Hulke, J. W. 1888. Supplemental note on *Polacanthus foxii,* describing the dorsal shield and some parts of the endoskeleton, imperfectly known in 1881. *Philosophical Transactions of the Royal Society of London* 178: 169–172.

Jacobs, L. L., D. A. Winkler, P. A. Murry, and J. M. Maurice. 1994. A nodosaurid scuteling from the Texas shore of the Western Interior Seaway. In K. Carpenter, K. F. Hirsch, and J. R. Horner (eds.), *Dinosaur Eggs and Babies,* pp. 337–346. Cambridge: Cambridge University Press.

Jerzykiewicz, T., P. J. Currie, D. A. Eberth, P. A. Johnston, E. H. Koster, and J.-J. Zeng. 1993. Djadokhta Formation correlative strata in Chinese Inner Mongolia: An overview of the stratigraphy, sedimentary geology, and paleontology and comparisons with the type locality in the pre-Altai Gobi. *Canadian Journal of Earth Sciences* 30: 2180–2195.

Jianu, C.-M., and D. B. Weishampel. 1998. The smallest of the largest: A new look at possible dwarfing in sauropods. In J. W. M. Jagt, P. H. Lambers, E. W. A. Mulder, and A. S. Schulp (eds.), *Third European Workshop on Vertebrate Palaeontology Abstracts:* 36. Maastricht, The Netherlands: Naturhistorisch Museum Maastricht.

Kirkland, J. I. 1998. A polacanthine ankylosaur (Ornithischia: Dinosauria) from the Early Cretaceous (Barremian) of eastern Utah. In S. G. Lucas, J. I. Kirkland, and J. W. Estep (eds.), *Lower and Middle Cretaceous Terrestrial Ecosystems. New Mexico Museum of Natural History and Science Bulletin* 14: 271–281.

Kirkland, J. I., and K. Carpenter. 1994. North American's first pre-Cretaceous ankylosaur (Dinosauria) from the Upper Cretaceous Morrison Formation of western Colorado. *Brigham Young University Geology Studies* 40: 25–42.

Kirkland, J. I., K. Carpenter, A. P. Hunt, and R. D. Scheetz. 1998. Ankylosaur (Dinosauria) specimens from the Upper Jurassic Morrison Formation. *Modern Geology* 23: 145–177.

Lee, Y.-M. 1996. A new nodosaurid ankylosaur (Dinosauria: Ornithischia) from the Paw Paw Formation (late Albian) of Texas. *Journal of Vertebrate Paleontology* 16: 232–245.

Lull, R. S. 1921. The Cretaceous armored dinosaur *Nodosaurus textilis* Marsh. *American Journal of Science,* Series 5, 1: 97–126.

Maryańska, T. 1977. Ankylosauridae (Dinosauria) from Mongolia. *Paleontologia Polonica* 37: 85–151.

McIntosh, J. S. 1981. Annotated catalogue of the dinosaurs (Reptilia, Archosauria) in the collections of the Carnegie Museum of Natural History. *Bulletin of the Carnegie Museum of Natural History* 18: 1–67.

Molnar, R. E. 1980. An ankylosaur (Ornithischia: Reptilia) from the Lower Cretaceous of southern Queensland. *Memoirs of the Queensland Museum* 20: 77–87.

———. 1996. Preliminary report on a new ankylosaur from the Early Cretaceous of Queensland, Australia. In F. A. Novas and R. E. Molnar (eds.), *Proceedings of the Gondwanan Symposium. Memoirs of the Queensland Museum* 39: 653–668.

Molnar, R. E., and E. Frey. 1987. The paravertebral elements of the Australian ankylosaur *Minmi* (Reptilia: Ornithischia, Cretaceous). *Neues Jahrbuch für Geologie und Paläontologie Abhandlungen* 175: 19–37.

Nopcsa, F. 1902. Notizen über cretacische Dinosaurier. 1. Zur systematischen Stellung von *Struthiosaurus (Crataeomus)*. *Sintzungberichte Akademie der Wissenchaften Mathematisch-naturwissenchafliche Klasse* 111: 93–103.

———. 1915. Die Dinosaurier der Siebenbürgischen Landesteile Ungarns. *Mitteilungen aus dem Jahrbuche der königlich ungarischen Geologischen Reichsanstalt* 23: 1–26.

———. 1918. *Leipsanosaurus* n. gen. in neuer Thyreophore aus der Gosau. *Födtani Közlöny* 48: 324–328.

———. 1923. On the geological importance of the primitive reptilian fauna in the uppermost Cretaceous of Hungary; with a description of a new tortoise (*Kallokibotion*). *Quarterly Journal of the Geological Society of London* 79: 100–116.

———. 1926. Die Reptilien der Gosau in neuer Beleuchtung. *Centralblatt für Mineralogie, Geologie und Paläontologie* B 14: 520–523.

———. 1929. Dinosaurierreste aus Siebenbürgen. V. *Geologica Hungarica* 4: 1–76.

Norman, D. B. 1984. A systematic reappraisal of the reptile order Ornithischia. In W. E. Reif and F. Westphal (eds.), *Third Symposium on Mesozoic Terrestrial Ecosystems Short Papers*, pp. 157–162. Tubingen: Attempto Verlag.

Norman, D. B., and T. Faiers. 1996. On the first partial skull of an ankylosaurian dinosaur from the Lower Cretaceous of the Isle of Wight, southern England. *Geological Magazine* 133: 299–310.

Ostrom, J. H. 1970. Stratigraphy and paleontology of the Cloverly Formation (Lower Cretaceous) of the Bighorn Basin area, Wyoming and Montana. *Bulletin of the Peabody Museum of Natural History* 35: 1–234.

Pereda-Suberbiola, J. 1992. A revised census of European Late Cretaceous nodosaurids (Ornithischia: Ankylosauria): Last occurrence and possible extinction scenarios. *Terra Nova* 4: 641–648.

———. 1993. Armored dinosaurs from the Late Cretaceous of southern France: A review. *Revue de Paléobiologie* vol. spéc. 7: 163–172.

———. 1994. *Polacanthus* (Ornithischia, Ankylosauria), a transatlantic armored dinosaur from the Early Cretaceous of Europe and North America. *Palaeontographica Abteilung A* 232: 133–159.

Pereda Suberbiola, J., and P. M. Galton. 1992. On the taxonomic status of the dinosaur *Struthiosaurus austriacus* from the Late Cretaceous of Austria. *Comptes Rendus de l'Académie des Sciences Paris* 315: 1275–1280.

———. 1994. A revision of the cranial features of the dinosaur *Struthiosaurus austriacus* Bunzel (Ornithischia: Ankylosauria) from the Late Cretaceous of Europe. *Neues Jarhbuch für Geologie und Paläontologie Abhandlungen* 191: 173–200.

———. 1997. Armored dinosaurs from the Late Cretaceous of Transylvania. In *Proceedings of the Mesozoic Vertebrate Faunes of Central Europe Symposium. Sargetia Scienta naturae*, Vol. 17, pp. 203–217. Deva: Museul Civilizatiei Dacice si Romane.

Pereda Suberbiola, X. 1996. La contribución del Barón Nopcsa al estudio de las fauna de vertebrados continentales del Cretácico final de Europa. *Gaia* 13: 43–66.

———. 1999. Ankylosaurian dinosaur remains from the Upper Cretaceous of Laño (Iberian Peninsula). *Estudios del Museo de Ciencias Naturales de Alava* 14(Núm. esp. 1): 273–288.

Pereda Suberbiola, X., H. Astibia, and E. Buffetaut. 1995. New remains of the armored dinosaur *Struthiosaurus* from the Late Cretaceous of the Iberian peninsula (Laño locality, Basque-Cantabric Basin). *Bulletin de la Société géologique de France* 166: 207–211.

Pereda Suberbiola, X., and P. M. Barrett. 1999. A systematic review of ankylosaurian dinosaur remains from the Albian–Cenomanian of England. *Special Papers in Palaeontology* 60: 177–208.

Russell, L. S. 1940. *Edmontonia rugosidens* (Gilmore), an armored dinosaur from the Belly River Series of Alberta. *University of Toronto Studies Geological Series* 43: 3–28.

Seeley, H. G. 1879. On the Dinosauria of the Cambridge Greensand. *Quarterly Journal of the Geological Society of London* 35: 591–635.

———. 1881. The reptile fauna of the Gosau Formation preserved in the Geological Museum of the University of Vienna. *Quarterly Journal of the Geological Society of London* 37: 620–704.

Sereno, P. C. 1986. Phylogeny of the bird-hipped dinosaurs (order Ornithischia). *National Geographic Research* 2: 234–256.

Sternberg, C. M. 1921. A supplementary study of *Panoplosaurus mirus*. *Transactions of the Royal Society of Canada*, Series 3, 15: 93–102.

———. 1928. A new armored dinosaur from the Edmonton Formation of Alberta. *Transactions of the Royal Society of Canada*, Series 3, 22: 93–106.

Stoliczka, F. 1860. Über eine der Kreideformation angehörige Süswasserbildung in den nordöstlichen Alpen. *Sitzungberichte der Mathematisch-naturwissenchaftlichen Classe der Kaiserlichen Akademi der Wissenschaften* 38: 482–496.

Weishampel, D. B., D. Grigorescu, and D. B. Norman. 1991. The dinosaurs of Transylvania: Island biogeography in the Late Cretaceous. *National Geographic Research and Exploration* 7: 68–87.

Weishampel, D. B., D. B. Norman, and D. Grigorescu, D. 1993. *Telmatosaurus transsylvanicus* from the Late Cretaceous of Romania: The most basal hadrosaurid dinosaur. *Palaeontology* 36: 361–385.

11. Disarticulated Skull of a New Primitive Ankylosaurid from the Lower Cretaceous of Eastern Utah

Kenneth Carpenter,
James I. Kirkland,
Don Burge, and
John Bird

Abstract

A new genus and species of ankylosaurid is described. The specimen was taken from the top of the Ruby Ranch Member of the Lower Cretaceous Cedar Mountain Formation of eastern Utah. Two specimens were recovered from a single quarry and represent large individuals (estimated skull length 60 cm). Skulls are known for both specimens, although one is mostly disarticulated. This skull provides the first description of disarticulated skull elements for an ankylosaur. Some elements (e.g., lachrymal) resemble their counterparts in *Stegosaurus*, whereas others differ considerably (e.g., prefrontal). The other skull is not disarticulated, although it is about the same size as the disarticulated skull. The new genus has premaxillary teeth, a plesiomorphy for ankylosaurids not previously reported. The ornamentation of the skull surface is through periosteal osteogenesis, not through coossification of armor to the skull surface.

Introduction

The skull bones of ankylosaurs have long been a mystery because thin sheets of dermal armor were thought to be coossified to the surface. The discovery of a juvenile *Pinacosaurus grangeri* without the covering armor provided the first glimpse of the underlying bones (Maryańska 1971, 1977). More recently, Molnar (1996) described the skull of *Minmi* sp., which also revealed many of the skull bones. Together, these two skulls provide most of what we know about the structure of ankylosaur skulls. Two new skulls are described below, but one is unique in that it is mostly disarticulated, giving the first glimpse of individual skull bones. The specimens are one of several newly discovered ankylosaur taxa from the Lower Cretaceous Cedar Mountain Formation of Utah (Carpenter and Kirkland 1998; Carpenter et al. 1999; Kirkland 1998).

Lower Cretaceous ankylosaurs are diverse (Table 11.1), with 18 species known (including one unnamed new genus and species of giant nodosaurid). Four of these more recent taxa have been found in the Cedar Mountain Formation (Barremian?, lower Cenomanian) of eastern Utah (Carpenter and Kirkland 1998; Kirkland et al. 1997). The Cedar Mountain Formation has proven to be the richest formation anywhere in the world for Lower Cretaceous ankylosaurs (Carpenter et al. 1999), and ankylosaurs are numerically dominant among all the dinosaurs found so far from this formation.

The most recently collected ankylosaur, a *Shamosaurus*-like form from the top of the Ruby Ranch Member (Aptian–Albian) of the Cedar Mountain Formation, was found southeast of Price, Carbon County, Utah, on the north end of the San Rafael Swell (Figs. 11.1, 11.2). The type locality near the Price River is known as the CEM (pronounced "kim") site and was discovered by Evan Hall and Sue Ann Bilbey of the Vernal Field House of Natural History while they were visiting a College of Eastern Utah (CEU) Prehistoric Museum excavation in the vicinity. The CEU Prehistoric Museum has located a number of other important sites at the same stratigraphic horizon in the area (Price River [PR]-2, PR-3, etc.) that have produced the remains of brachiosaurid sauropods, huge nodosaurids, additional ankylosaurid remains, and elements identifiable as coming from the large theropod *Acrocanthosaurus*. The specimen of *Tenontosaurus* from PR-1 reported by Kirkland et al. (1997: 77) as being excavated from the base of the Cedar Mountain Formation is from a conglomeratic sandstone in the overlying Mussentuchit Member that caps the ridge above the CEM, PR-1, and PR-2 localities (Fig. 11.2). This realization, plus the identification of isolated teeth collected by the University of Oklahoma as cf. *Tenontosaurus* sp. (Kirkland et al. 1997), indicates that *Tenontosaurus* ranges stratigraphically from the Aptian up into the basal Cenomanian.

The holotype skull of the new ankylosaurid specimen shares features with *Shamosaurus scutatus* from the Lower Cretaceous of Mongolia, as discussed below. This occurrence of a shamosaurlike ankylosaur in Utah adds to the growing body of evidence for faunal exchange

TABLE 11.1.
Lower Cretaceous ankylosaur taxa.

Taxa	Horizon	Age	Major references
Animantarx ramaljonesi	Mussentuchit Member, Cedar Mountain Formation, Utah	Albian–Cenomanian	Carpenter et al. 1999
Anoplosaurus curtonotus	Gault? Formation or Cambridge Greensand, England	Albian	Pereda Suberbiola and Barrett 1999
Cedarpelta bilbeyhallorum	Mussentuchit Member, Cedar Mountain Formation, Utah	Albian–Cenomanian	This chapter
Gastonia burgei	Yellow Cat Member, Cedar Mountain Formation, Utah	Barremian?	Kirkland 1998
Hoplitosaurus marshi	Lakota Formation, South Dakota	Barremian	Carpenter and Kirkland 1998
Hylaeosaurus armatus	Hastings Beds; Grinstead Clay, England	Upper Valanginian	Mantell 1833; Pereda Suberbiola 1993
Minmi paravertebra	Bungil Formation, Allaru Formation Queensland, Australia	Aptian	Molnar 1980, 1996
Nodosaurid n. g., n. sp	Ruby Ranch Member, Cedar Mountain Formation, Utah	Albian	Carpenter and Kirkland, in preparation
Pawpawsaurus campbelli	Paw Paw Formation, Texas	Late Albian	Carpenter and Kirkland 1998; Lee 1996
Polacanthus foxii	Wessex Formation, Vectis Formation, Isle of Wight, England	Barremian	Blows 1987; Hulke 1881; Pereda Suberbiola 1993, 1994
Polacanthus rudgwickensis	Wessex Formation, Sussex, England	Barremian	Blows 1996
Priconodon crassus	Arundel Formation, Maryland	Aptian–Albian boundary	Carpenter and Kirkland 1998
Sauropelta edwardsorum	Cloverly Formation, Wyoming and Montana	Aptian (see May et al. 1995)	Carpenter and Kirkland 1998; Ostrom 1970
Shamosaurus scutatus	Dzun Bayn Formation, Mongolia	Lower Cretaceous	Tumanova 1985, 1987
Shamosaur n. g., n. sp.	Zhonggou? Formation, China	Lower Cretaceous	Currie, personal communication
Silvisaurus condrayi	Terra Cotta Member, Dakota Group, Kansas	Albian	Carpenter and Kirkland 1998; Eaton 1960
Stegopelta landerensis	Belle Fourche Member, Frontier Formation, Wyoming	Latest Albian or earliest Cenomanian	Carpenter and Kirkland 1998; Moodie 1911
Texasetes pleurohalio	Paw Paw Formation, Texas	Late Albian	Carpenter and Kirkland 1998; Coombs 1995

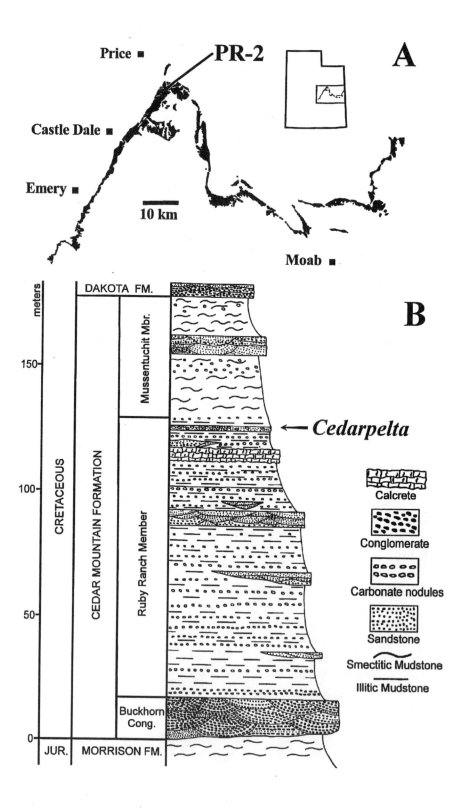

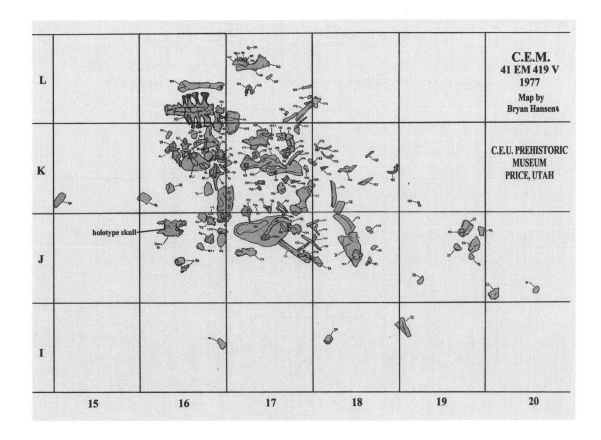

C.E.M.
41 EM 419 V
1977
Map by
Bryan Hansen

C.E.U. PREHISTORIC
MUSEUM
PRICE, UTAH

holotype skull

between Asia and North America during the latest Albian (Cifelli et al. 1997; Kirkland et al. 1997, 1998; Russell 1993).

The holotype skull is incomplete anterior to the left orbit and most of the right side, and the squamosal region is damaged; the incomplete paratype skull is disarticulated. Most of the description is of the disarticulated skull, with supplemental comments based on the articulated skull. One of the most remarkable features of the two new skulls is that the cranial "armor" is restricted to the individual elements and does not cross sutural boundaries. The rugose texture, which is typically identified as cranial armor, suggests elaboration of the individual elements, rather than the coossification of osteoderms to the bone surface (see Vickaryous et al. 2001). For this reason, the term "ornamentation" is preferred to "cranial armor." The ornamentation is discussed further below.

Institutional Abbreviation. CEUM: College of Eastern Utah Prehistoric Museum, Price, Utah.

Systematic Paleontology
Order Ankylosauria
Family Ankylosauridae

Amended Diagnosis. Ankylosaurid with narrow snout; orbit at midpoint of skull length; supraorbital domed above anterior rim of

Figure 11.1. (opposite page) (A) Outcrop belt of Cedar Mountain Formation showing the location for the Price River localities, including the CEM site. (B) Stratigraphic position of Cedarpelta bilbeyhallorum *in the Cedar Mountain Formation.*

Figure 11.2. (above) Quarry map of the CEM site showing the distribution of bones of the two Cedarpelta bilbeyhallorum *specimens, including the holotype skull. Grid in 1-m squares.*

orbit; occipital condyle hemispherical and set on neck; skull ornamentation formed from remodeling of surface, not coossification of dermal armor; absence of pyramidal boss at posterior corners of skull; interpterygoid fenestra narrow; pterygoid adjacent to interpterygoid fenestra extending posteriorly over basioccipital and is elongate compared with skull size.

Cedarpelta n. g.

Holotype. CEUM 12360, partial skull missing the snout and lower jaws; possible rear portion of parietal.

Type Locality. Price River 1 (PR-1) locality (CEM site), CEUM 42EM352U, CEU90–2, uppermost Ruby Ranch Member of the Cedar Mountain Formation, Carbon County, Utah.

Diagnosis. As for the species.

C. bilbeyhallorum n. sp.

Diagnosis. Premaxilla with short rostrum anterior to nasal process; paired premaxillae parallel-sided, not divergent posterolaterally as in other ankylosaurids; cutting edge of beak confined to anteriormost portion of premaxillae; six alveoli in premaxilla; quadrate sloped posteriorly; quadrate head not coossified with paroccipital process as in *Shamosaurus*; occipital condyle on a long neck as in nodosaurids and projecting horizontally, not obliquely downward as in all other known ankylosaurids and nodosaurids; pterygoids anteroposteriorly elongated and vaulted in nodosaurid fashion; well-developed trochlearlike process along the lateral edge of pterygoid; basitubera as a very large, ventral projecting wedge; large oval process for adductor tendon on medial side of coronoid process; ischium with large knob on medial side near pubic peduncle.

Etymology. Species named for Sue Ann Bilbey and Evan Hall, who discovered the locality.

Paratype Material. Disarticulated skull: including left premaxilla CEUM 10405; left nasal fragment CEUM 10410; right prefrontal CEUM 10421; right lachrymal CEUM 10560; right postorbital CEUM 10352; jugal fragment CEUM 10598; left frontal CEUM 10325; parietal CEUM 10332; right squamosal CEUM 10345; left quadrate with attached quadrojugal CEUM 10417; right quadratojugal CEUM 10561; braincase CEUM 10267; left surangular CEUM 10270; left angular CEUM 10529. Vertebrae: cervical centrum CEUM 11288; dorsal centra CEUM 10258, CEUM 10409, CEUM 10442, CEUM 10360; synsacrum of two dorsals, four sacrals, and one caudal CEUM 12163; first? caudal CEUM 10258; anterior caudals CEUM 10258, CEUM 10387, CEUM 10366; midcaudals CEUM 10255, CEUM 10257, CEUM 10260, CEUM 10261, CEUM 10262, CEUM 10349, CEUM 10400, CEUM 10412; posterior caudals CEUM 10404, CEUM 10407. Appendicular: right partial humerus CEUM 10258; left ulna CEUM 10425; left ischium CEUM 10266; partial right ischium CEUM 10537; fragment of right ilium CEUM10375; cervical ribs CEUM 10248, CEUM 10445; metacarpals CEUM 10254, CEUM 10356, CEUM 10430, CEUM 10449, CEUM 10984; phalanges CEUM 10247, CEUM 9970; unguals CEUM 9922, CEUM 10253. Armor: keeled plates CEUM 10526, CEUM 10359, CEUM 10394, CEUM 10431, CEUM 10459, CEUM 10248; compressed conical plates CEUM 10359, CEUM 9960, CEUM 9962, CEUM 10548, CEUM 10414, CEUM 10441; flat osteoderm CEUM 10338.

Description

Skull

The holotype skull consists of a rear portion that has separated from the preorbital region along the frontal and maxillary sutures (Fig. 11.3). The right side and dorsoposterior portions of the skull are damaged; a fragment of the right parietal may belong to this skull. The skull is little crushed, except for the lower rim of the left orbit where the jugal and ectopterygoid have been pushed upward. The second skull, the paratype, is completely disarticulated, and the elements were cataloged separately. Overall, the bones differ in shape from those of the articulated juvenile *Pinacosaurus* skull described by Maryańska (1971, 1977). Because the individual cranial elements have never been described before, these are described in detail, with reference made to the elements in the articulated holotype skull. Bones missing in the paratype are described from the holotype. A composite reconstruction of a skull is shown in Figure 11.4.

Premaxilla. The left premaxilla is almost complete, lacking only the nasal process and a small part of the maxillary process (Fig. 11.5). Orienting the midline suture vertically, as it would when articulating with the other premaxilla, the nasal process would be vertical and the body of the premaxilla would slope buccoventrally. The lateral margins are parallel, rather than diverging posteriorly as in all other known ankylosaurids. Anterior to the nasal process is a short rostrum, with a small subtriangular patch of cranial ornamentation at the anterolateral corners. The ornamentation is localized and mushrooms upward and outward. The maxillary process projects posteriorly and probably formed a wedge between the maxilla and nasal, as it does in *Pinacosaurus* (Maryańska 1977: fig. 2A$_2$; cf. Fig. 11.4A). The ventral surface of the process is flat and wide for its contact with the maxilla, whereas the dorsal surface is a thin ridge indicating a narrow, weak suture with the nasal.

The medial side of the premaxilla has a deep sutural surface for contact with the opposite premaxilla. This surface is flat and lacks any interdigitation. Posteriorly, the sutural surface has a groove to accommodate the vomer as a wedge between both premaxillae. Ventrally, the premaxillary scoop is twice as long as it is wide and has six alveoli along the lateral margins, just lingual to the cutting edge. The cutting edge is variably damaged along its length. The tip of a replacement tooth is visible within the fifth alveolus, but unfortunately, the matrix is too hard to expose the crown safely. The first alveolus is the largest, and is

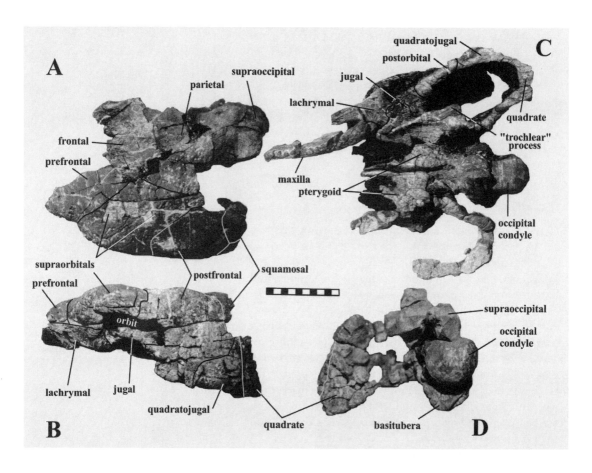

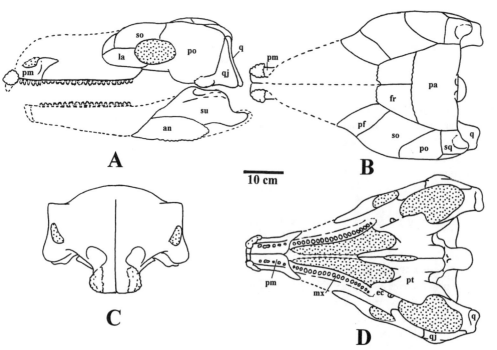

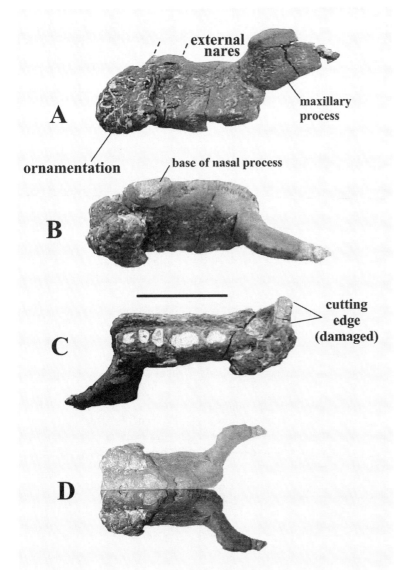

external nares

maxillary process

A

ornamentation

base of nasal process

B

cutting edge (damaged)

C

D

Figure 11.3. (opposite page top) Holotype skull of Cedarpelta bilbeyhallorum *(CEUM 12360) in (A) dorsal, (B) left lateral, (C) ventral, and (D) posterior views. Scale in centimeters.*

Figure 11.4. (opposite page bottom) Composite skull of Cedarpelta bilbeyhallorum *in (A) lateral, (B) dorsal, (C) anterior, and (D) ventral views. an = angular; ec = ectopterygoid; fr = frontal; la = lachrymal; mx = maxilla; pf = prefrontal; pm = premaxilla; po = postorbital; pt = pterygoid; q = quadrate; qj = quadratojugal; so = supraorbitals (fused); su = surangular. Scale bar = 10 cm.*

Figure 11.5. (this page) Left premaxilla of Cedarpelta bilbeyhallorum *(CEUM 10405) in (A) left lateral, (B) dorsal, (C) ventral, and (D) composite (mirrored) views. Scale bar = 5 cm.*

almost circular in cross section. The last four are somewhat oval and are wider than they are long. The cutting edge of the beak is confined to the anteriormost portion of the premaxilla. It is damaged, but it would have extended ventrally below the alveolar margins. A small gap separates the cutting edge from the first alveolus. Unlike many ankylosaurs, there is no premaxillary foramen.

Maxilla. A fragment of the right holotype maxilla has the crowns of all 18 teeth broken off (Fig. 11.6A). Replacement teeth are visible through the nutrient foramina on the lingual side and where the maxilla is damaged (Fig. 11.6B). No complete teeth are visible. However, the visible portions of the crown are reminiscent of nodosaurid teeth in that the teeth and their marginal denticles are relatively large as compared with the size of the maxilla. The cingulum on the lingual side is slightly developed as a ridge, whereas it is a thickened swelling on the buccal side.

Nasal. A small rectangular fragment may be part of the left nasal on the basis of the curvature of the surface. Unfortunately, the bone lacks almost all of the sutures, except possibly a short segment along the midline, but even that is damaged.

Prefrontal. The dorsal surface is quadrangular (Fig. 11.7), whereas it is rectangular in *Pinacosaurus* and polygonal in *Minmi*. Only a small part of the prefrontal is visible on the skull in lateral view, whereas proportionally more is visible in *Pinacosaurus* (Maryańska 1977: fig. 3A$_2$; cf. Fig. 11.4A). The surface of the paratype prefrontal is sculpted with a network of grooves (Fig. 11.7A). The posterior and posterolateral margins have interdigitating sutures for a supraorbital element, whereas the lachrymal suture, located anteroventrally, is wide and smooth. The lachrymal suture faces obliquely ventrolaterally. It is wide and has a broad contact with the prefrontal (Fig. 11.7C). In ventral view, the anterodorsal wall for the orbit slopes posterodorsally (Fig. 11.7B, D). In medial view (Fig. 11.7D), the preorbital wall forms a deep pocket, or fossa, medial to the lachrymal suture, here called the prefrontal fossa. In the holotype skull, the left prefrontal is preserved in situ (Fig. 11.3A, B). Medially, the prefrontal fossa is not as well developed as in the paratype. The underlapping medial suture to the nasal is slightly damaged in the paratype prefrontal.

Lachrymal. In lateral view, the right lachrymal is somewhat rectangular, being longer than deep (Fig. 11.8). In *Pinacosaurus*, the lachrymal is a parallelogram and is steeply angled anterodorsally, whereas in *Minmi*, if the lachrymal has been correctly identified, it is slender and vertical. The surface of the paratype lachrymal in *Cedarpelta* is ornamented with a network of grooves and tubercles. The posterolateral portion curves ventromedially and contributes to the maxillary emargination. At the posterodorsal corner is an interdigitated suture that is a continuation of the supraorbital suture on the prefrontal. Thus, the supraorbital overlaps the prefrontal–lachrymal suture. The prefrontal suture is wide, arcs anteroventrally, and posteriorly extends a short, thick process medially to contribute to the anterior wall of the orbit. Posteriorly, this wall is pierced by the lachrymal duct that extends the length of the lachrymal to exit just dorsal to the maxillary suture along the ventral margin; it is not known if this duct is connected to the maxillary sinuses. The ventral surface is triangular and bears the maxillary suture anteriorly, the jugal suture posterolaterally, and the ectopterygoid suture posteromedially; each sutural surface is grooved.

Supraorbitals. The supraorbitals are only known from the holotype, where they are located above the left orbit (Fig. 11.3A, B). The sutures between them are not visible, but presumably there were three, as in *Pinacosaurus* (one was identified as the postfrontal in Maryańska 1977). The anterolateral supraorbital is domed and thickened. Ventrally, the supraorbitals are concave to accommodate the eyeball.

Figure 11.6. (opposite page top) Right maxilla of Cedarpelta bilbeyhallorum (CEUM 12360) in (A) occlusal and (B) medial view. Scale bar = 10 cm.

Figure 11.7. (opposite page bottom) Right prefrontal of Cedarpelta bilbeyhallorum (CEUM 10421) in (A) dorsal, (B) ventral, (C) lateral, and (D) medial views. Scale bar = 10 cm.

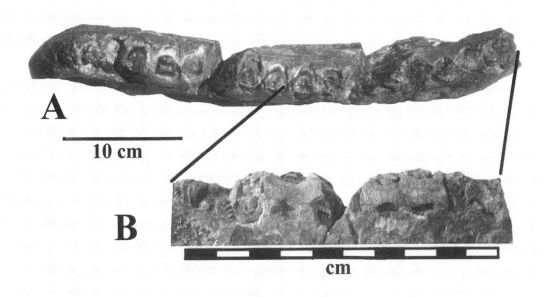

A

10 cm

B

cm

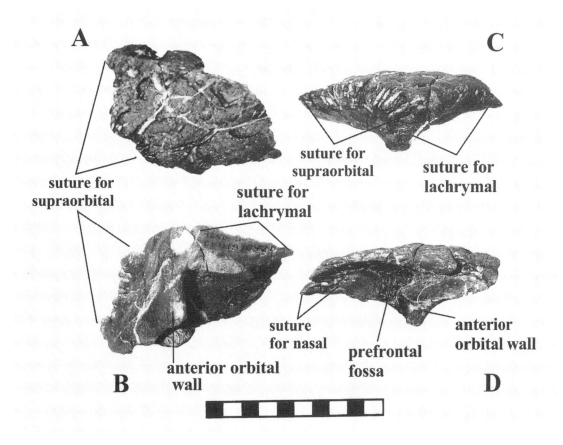

A

suture for
supraorbital

B

suture for
supraorbital

suture for
lachrymal

anterior orbital
wall

C

suture for
supraorbital

suture for
lachrymal

suture
for nasal

prefrontal
fossa

anterior
orbital wall

D

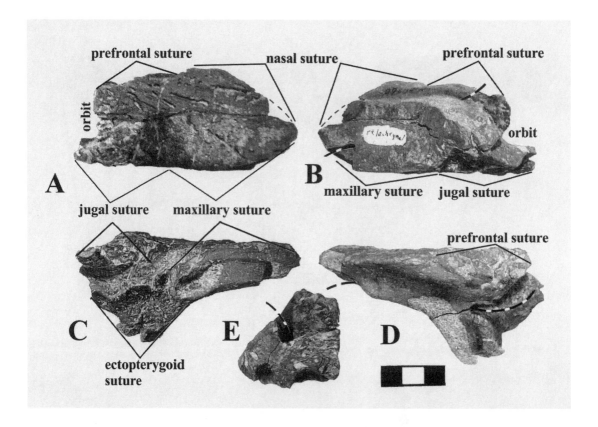

Figure 11.8. Right lachrymal of
Cedarpelta bilbeyhallorum
(CEUM 10560) in (A) lateral,
(B) medial, (C) ventral,
(D) dorsal, and (E) orbital views.
Heavy dashed line passes through
lachrymal foramen.
Scale bar = 3 cm.

Frontal. The right paratype frontal is almost square (Fig. 11.9A,
B), similar to that of *Pinacosaurus*, whereas it is more rectangular in the
holotype (Fig. 11.4A) and in an unnamed juvenile nodosaurid described
by Jacobs et al. (1994). Dorsally, the surface ornamentation of the
paratype is formed by a network of grooves, pits, and ridges. The
posterolateral margin of the paratype has coarse, vertical sutures for
interdigitating contact with the postorbital; all other edges are damaged. The anterior edge is about half as thick as in the holotype skull.
Ventrally, there is a damaged transverse ridge extending across the
paratype frontal. In the holotype, this ridge, which is also damaged,
forms part of the postocular wall. In the holotype, the suture for the
nasal is well defined; it is slightly concave posteriorly and its surface
denticulate for an interdigital contact with the nasals.

Parietal. Both parietals have fused along their midline and all traces
of the suture are gone (Fig. 11.9C, D). The combined parietals are
rectangular, being slightly wider than long. The dorsal surface of the
parietals are domed, a feature more commonly seen in nodosaurids
than in ankylosaurids. The surface of the paratype parietal is ornamented with grooves, ridges, pits, and nodes that radiate from the
midposterior margin. There is no trace of an anteroposteriorly narrow
nuchal ornamentation, as seen in nodosaurids. Interdigitating sutures
are present along the anterior and right lateral margins. Ventrally, the

dorsal cavity of the braincase is very wide and deep. The crushed remains of the supraoccipital are visible posteriorly. On either side of the lateral walls of the parietals are deep fossae for the adductors. In non-ankylosaurian thyreophorans, the adductors occupy the supratemporal fenestrae. *Cedarpelta* shows that the fenestrae in ankylosaurs were closed off dorsally by lateral expansion of the parietals and medial expansion of the squamosals. This interpretation is similar to that made by Tumanova (1987) from the cranial roof of the ankylosaurid *Talarurus*.

The parietal is damaged in the holotype and little can be added to the above, except to note that ornamentation is not as prominent. There is also a fragment (Fig. 11.10) that may be from the right posterior side of the holotype that provides information about the back of the skull. Neither this fragment nor the more complete paratype parietal show any trace of the "tabular" reported in *Pinacosaurus* (Maryańska 1977). The parietal fragment is shaped like an inverted L in cross section. The dorsal and posterodorsal surfaces are rugose, with grooves and pits (Fig. 11.10A, B). More ventrally along the posterior side, the bone is smooth. On the medial side, the bone is damaged, but it does show a transverse groove.

Postorbital. The right postorbital is slightly damaged, including many of the sutural areas; it forms half of the dorsal, all of the posterior, and a small portion of the orbital border (Fig. 11.11). The outer surface is rugose, with grooves and pits (Fig. 11.11A). On the dorsal surface is a low dome above the posterior edge of the orbit; this may be analogous to the larger postorbital boss seen in polacanthids (Kirkland 1998), as well as some nodosaurids. In the holotype, the horizontal dorsal ramus contacts the supraorbitals above the orbit, and the vertical ventral portion below the orbit contacts the jugal much like it does in *Pinacosaurus* (Maryańska 1971, 1977). On the medial surface, the postocular wall is not very extensive. However, extensive sutures indicate a large contribution by a descending process of the frontal (Fig. 11.3B). Unfortunately, this process in the holotype skull is damaged.

Quadratojugal. A pair of quadratojugals is present, one of which is fused to the left quadrate (Fig. 11.12A–C). In lateral profile, they are L-shaped. The quadratojugal is considerably larger and better developed than in *Pinacosaurus* (Maryańska 1971, 1977) and *Minmi* (Molnar 1996). The anteroventral process is not sutured to the jugal but to the postorbital. The postorbital and quadratojugal have a tongue-and-groove suture between them. The groove is developed on the quadratojugal along the anterior and dorsal margins of the horizontal process. The posteromedial part of the quadratojugal overlaps the lateral surface of the quadrate and extends to about half or more the quadrate height.

On the basis of the narrowness of the vertical process of the quadratojugal and large size of the postorbital, *Cedarpelta* lost the lateral temporal fenestra by rearward expansion of the postorbital, rather than forward expansion of the quadratojugal. The same apparently occurred in *Pinacosaurus* (see Maryańska 1977: fig. 2A$_2$), suggesting

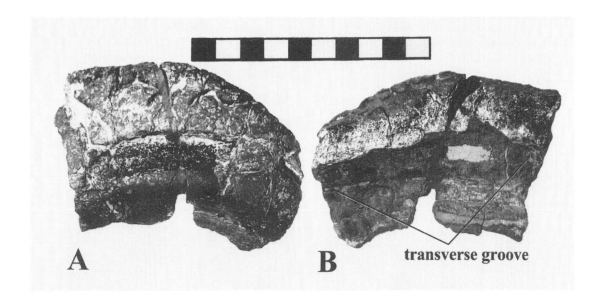

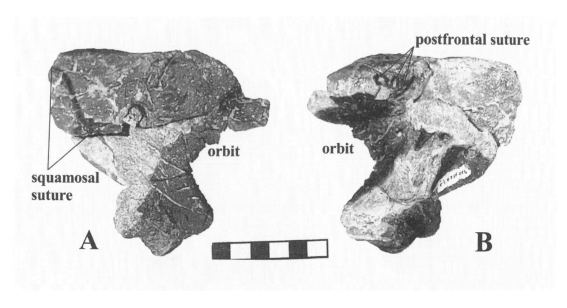

Figure 11.9. (opposite page) Right frontal in (A) dorsal and (B) ventral views.
Parietals in (C) dorsal and (D) ventral views. (E) Composite of frontals
(mirrored) and parietals. Scale for A–D = 10 cm.

Figure 11.10. (top) Fragment of the posterior portion of right parietal Cedarpelta
bilbeyhallorum (CEUM uncataloged), possibly to the holotype skull.
(A) Posterior (external) and (B) anterior (internal) views. Scale bar = 10 cm.

Figure 11.11. (bottom) Right postorbital of Cedarpelta bilbeyhallorum (CEUM
10352) in (A) lateral and (B) medial views. Scale bar = 6 cm.

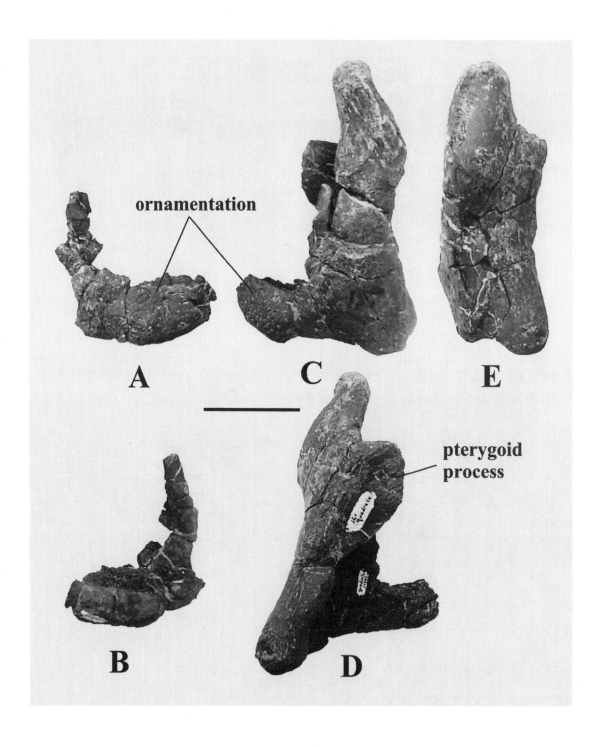

ornamentation

pterygoid
process

A   C   E

B   D

Figure 11.12. Right quadratojugal of Cedarpelta bilbeyhallorum (CEUM 10561) in (A) lateral and (B) medial views. Left quadratojugal and quadrate in (C) lateral, (D) medial, and (E) posterior views. Scale = 10 cm.

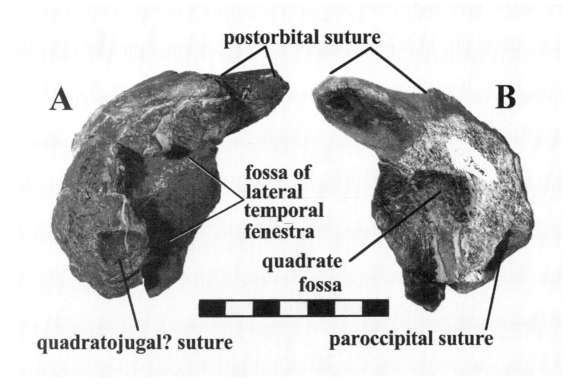

postorbital suture

A

B

fossa of lateral temporal fenestra

quadrate fossa

quadratojugal? suture

paroccipital suture

that this was how this fenestra was lost in ankylosaurs. The lateral surface of the quadratojugal is rugose, with pits and some grooves, and there is a boss along the ventrally located anteroventral process.

Quadrate. The disarticulated quadrate is sigmoidal in lateral view, rather than straight or arched as in most ankylosaurs. The quadrate head is well developed and rounded (Fig. 11.12C–E). It shows no evidence of having been coossified with the paroccipital process as in *Shamosaurus.* A depression along the anterior edge just above the vertical process of the quadratojugal is probably the contact for the ventral or the descending process of the squamosal (see below). There is no evidence for a foramen between the quadrate and quadratojugal as in *Pinacosaurus* (Maryańska 1977: fig. 2A$_1$). The pterygoid process is high on the quadrate shaft. A plane through the distal articular condyles is about 35° relative to the lateral–medial axis of the quadrate head. The condyles are asymmetrical, with the medial one more prominently developed, as in most dinosaurs.

Squamosal. In lateral view, the bone is C-shaped and has two vertical flanges separated by a deep oval fossa (Fig. 11.13). This fossa is the posterodorsal portion of the adductor fossa. The lateral flange is short and has a flat surface for overlap by the posterior extension of the postorbital (Fig. 11.13A). The more medial flange separates the adductor fossa from the cotyle for the quadrate head. Unfortunately, much of the medial side of this medial flange is damaged, and the full extent of the quadrate cotyle is unknown. Posterodorsally, the squamosal has a

Figure 11.13. Right squamosal of Cedarpelta bilbeyhallorum (CEUM 10345) in (A) lateral and (B) medial views. The quadrate fossa is all that remains of the more enveloping cotyle. Scale bar = 7 cm.

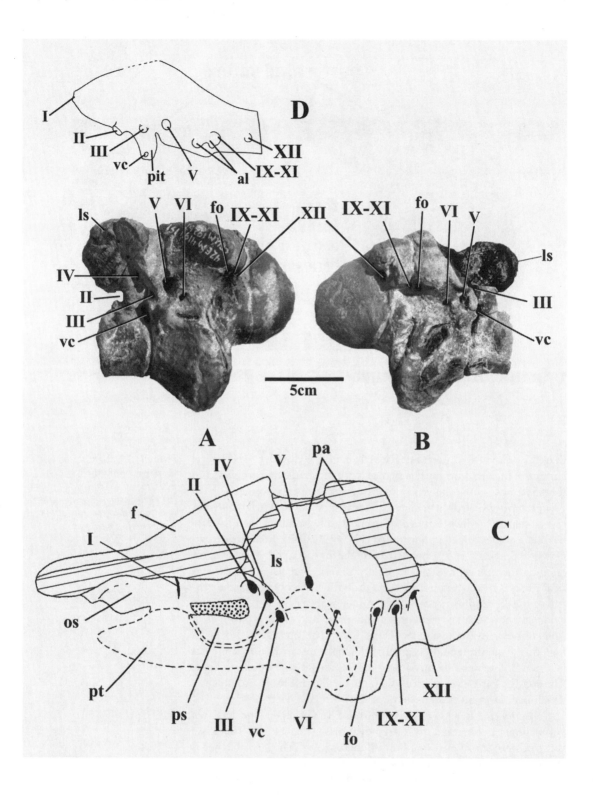

small patch of rugose ornamentation. The dorsal horizontal process has a flat suture for the posterodorsal extension of the postorbital. In posterior view, the lateral and medial flanges are separated by a U-shaped notch set within a shallow fossa; this fossa accommodated the dorsolateral part of the paroccipital process.

Pterygoid. The pterygoids are only preserved in the holotype skull; the right one is more complete. The pterygoids cover a large area ventral to the orbits. In contrast, the pterygoids in most ankylosaurids are narrow, triradiate structures. The largeness of the pterygoids in *Cedarpelta* is due to their being anteroposteriorly elongated, especially along the midline (Fig. 11.3C). The vomerine process is a thin, vertical sheet of bone on each side of the parasphenoid. The process is paired and extends forward, toward the vomers (a possible fragment is preserved anteriorly between them). Lateral to the vomerine processes, the pterygoid is vaulted toward the underside of the frontals; crushing of the skull has pushed the orbitosphenoid down between the vaults (see Fig. 11.14C). The vault is narrow and inclined posteroventrally in a manner somewhat similar to that seen in nodosaurids (e.g., *Silvisaurus*, Carpenter and Kirkland 1998). The full extent of the vault is unknown because the anterior edge is damaged on both sides. A moderate-size foramen, probably for the lateral head of the internal carotid, pierces the vomerine process at the base of the pterygoid vault. Anterolaterally, the pterygoid is sutured to the ectopterygoid. Laterally, the ectopterygoid is sutured to the ventroposterior part of the lachrymal, rather than to the maxilla as is typical of dinosaurs.

The posterolateral edge of the pterygoid is expanded anteroposteriorly into a thickened process analogous to the trochlear process of pleurodire turtles (see Gaffney 1979). This process in turtles is capped by the transiliens cartilage, which provides a gliding surface for the external tendons that insert the adductor mandibulae to the coronoid process (Schumacher 1973). A similar function may have occurred in *Cedarpelta* because the process is situated adjacent to the coronoid process when the lower jaw is closed (see below).

Posteriorly, the pterygoid has a complex shape that includes one process to the basisphenoid and another to the pterygoid process of the quadrate. These two processes are about the same length but are divergent; the pterygoid between the two processes is vaulted up alongside of the braincase adjacent to the internal carotid foramen. Posteromedially, the basisphenoid process laps against the anterior base of the basitubera, and the quadrate process of the pterygoid overlaps the pterygoid process of the quadrate.

Supraoccipital. The supraoccipital is present only in the holotype and is unfortunately damaged (Fig. 11.3A). It forms the dorsal rim of the circular foramen magnum. The sutures to the exoccipitals are not visible. The supraoccipital slopes anterodorsally from the foramen magnum toward the underside of an overhanging parietal.

Basicranium. The basicranium is complete in both specimens. Interestingly, there is variation in the relative lengths of the necks of the occipital condyles (Fig. 11.14A–C), suggesting that neck length may not be a reliable character for ankylosaur taxonomy. Similar variation

Figure 11.14. (opposite page) Braincase of Cedarpelta bilbeyhallorum: *paratype (CEUM 10267) in (A) left and (B) right lateral views. (C) Holotype (CEUM 12360) in left lateral view; horizontal lines = broken surface; dashed lines = portion of pterygoids adjacent to the basicranium. There is variation in the shape of occipital condyles, the length of condylar neck, and the shape of basitubera between A and B, and C. (D) Composite endocast made from the holotype and paratype skulls. Note lack of strong flexures. All images are to the same scale. al = auditory lacuna; f = frontal; fo = fenestra ovalis; ls = laterosphenoid; os = orbitosphenoid; pa = parietal; pit = pituitary; ps = parasphenoid; pt = pterygoid (denoted by dashed lines); vc = Vidian canal; I–XII nerve foramina.*

has also been seen in *Euoplocephalus* (Carpenter, personal observation). In addition, the neck for the occipital condyle does not face obliquely downward as in most ankylosaurs, but rather projects posteriorly, almost parallel with the floor of the braincase. Ventrally, a very large, prominent, wedge-shaped basitubera is developed along the basioccipital–basisphenoid suture. In all other ankylosaurs, the basitubera are paired and developed from the basisphenoid. Scars for the M. rectus capitis ventralis are present at the ventrolateral corners, just posterior to where the pterygoids articulate. Without the atlas, it is not possible to determine what role the straight occipital condyle and unusual basitubera had in limiting neck motion. On the lateral side of the basicranium, the opisthotic, prootic, and lateral sphenoid are coossified to each other and to the basioccipital and basisphenoid; all traces of the sutures are obliterated. The holotype skull still has the braincase attached to the frontals and parietals and has an orbitosphenoid.

An endocranial cast was reconstructed from two peels taken from the holotype skull and paratype braincase (Fig. 11.14D). The endocast is moderately flexed, more so than in *Euoplocephalus* (Coombs 1978), but less than that of cf. *Polacanthus* (Norman and Faiser 1996). Not all the foramina of the endocranial cavity or semicircular canals are distinct because of damage or matrix. The olfactory nerve (I) appears to have been smaller than in *Euoplocephalus*, although the anterior part of the braincase is not completely cleaned and is damaged. The cerebral hemisphere is very prominent and is proportionally larger than in *Euoplocephalus*.

The various cranial foramina are best seen on the outer surface of the paratype braincase, supplemented with the holotype braincase. The basic pattern is similar to that seen in other ankylosaurs (Carpenter et al., in press), although some foramina are larger and some are coalesced. Anterodorsally is the laterosphenoid, which is pierced by enlarged foramina II–IV. Posterior to the laterosphenoid is the prootic, which together with the opisthotic, forms most of the lateral wall of the braincase and the auditory chamber. The prootic is pierced by a large foramen for the trigeminal nerve (V) near the laterosphenoid suture. The auditory chamber along the prootic–opisthotic suture is pierced laterally by the fenestra ovalis ventral to the paroccipital process. Immediately posterior to the fenestra ovalis and separated from it by a thin bone wall (crista interfenestralis) is the fenestra vena jugularis internus. Foramina IX–XII are combined, as they are in *Pinacosaurus*, into a large foramen at the base of the exoccipital. The foramen for the internal carotid artery pierces the basisphenoid lateral to the pituitary fossa.

Surangular. The lower jaw is incomplete and consists only of the left surangular and angular of the paratype (Fig. 11.15). The surangular has the characteristic low coronoid process of ankylosaurids, but it differs in having a large, somewhat oval, medially placed process for the insertion of the external adductor tendon for the M. adductor mandibularis externus medialis. When the lower jaws were closed, this process would be opposite the trochlearlike process on the pterygoid;

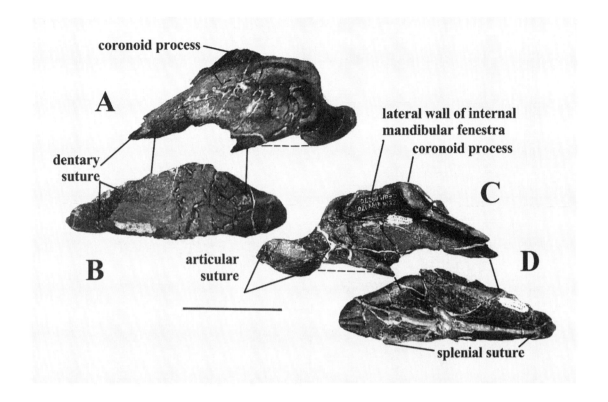

coronoid process

A

lateral wall of internal
mandibular fenestra
coronoid process

dentary
suture

C

B

articular
suture

D

splenial suture

both processes are unique to ankylosaurs. On the lateral surface, the ornamentation consists of a few large grooves, but mostly pits and nodes, on the main body; the bone is smooth on the posteriorly projecting articular process. Along the dorsal margin is an irregular surface for the M. adductor mandibularis externus superficialis. Ventrally, the suture with the angular is complicated. The posterior half is a tongue and groove, with the groove developed in the surangular. Anteriorly, the suture is flat for overlap by the angular. Medially, the lateral wall of the internal mandibular fenestra is concave and is pierced by the posterior surangular foramen. The articular process is spoon-shaped, and the sutural surface for the articular extends anteriorly to the posterior edge of the internal mandibular fossa. The articular is unfortunately missing.

Angular. The angular is triangular and has prominent ornamentation on the lateral surface that projects ventrolaterally. The ornamentation consists mostly of grooves and pits, with irregular rows of nodes along the ventrolateral margin. Anteriorly, the angular has a large, flat surface where it underlapped the posterior portion of the dentary. Medially, there is a large triangular suture for the prearticular.

Postcrania

Of the postcrania, only the ischium is distinctive enough to warrant description here (a description of the entire postcrania will be presented elsewhere). As is characteristic of ankylosaurids (Coombs and Maryańska 1990), the shaft is straight, lacking the flexion noted in

Figure 11.15. Cedarpelta bilbeyhallorum *(A) surangular (CEUM 10270) and (B) angular (CEUM 10529) in (C) lateral and (D) medial views. Scale bar = 10 cm.*

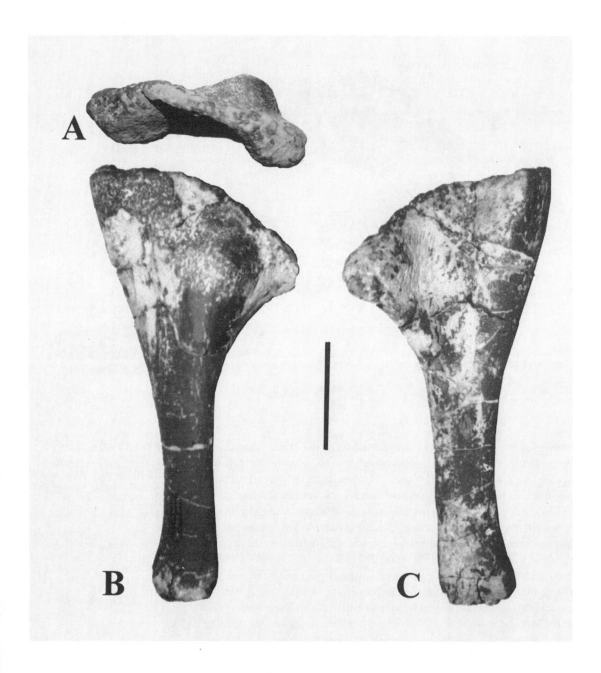

Figure 11.16. Left ischium of
Cedarpelta bilbeyhallorum
(CEUM 10266) in (A) dorsal,
(B) medial, and (C) lateral views.
Note the prominent dome on the
medial side of A and B.
Scale bar = 10 cm.

nodosaurids. On the medial side is a prominent dome opposite the
acetabulum (Fig. 11.16A). This dome is identical in both the left and
right ischium and does not appear to be pathologic.

Discussion

The skulls of *Cedarpelta* provide important information about the
origins of the cranial "armor" that characterizes ankylosaurs. Traditionally, the armor was thought to be thin, dermally derived scutes
coossified to the surface of the skull. Coombs (1971), however, sug-

gested that the "armor" might be formed from modification of the bone surface under the influence of the overlying epidermal scute (see review of both hypotheses in Vickaryous et al. 2001). In *Cedarpelta*, the "armor" is restricted to the individual cranial elements, indicating that the "armor" is not formed by fused dermal scutes but rather by reworking of the cortical surface by the periosteum ("dermatocranial elaboration" of Vickaryous et al. 2001). This reworking, however, is not the same as the differential resorption of the bone surface seen in the crocodile skull, where the ridges represent the original bone surface (Buffrénil 1982). In *Cedarpelta*, the cranial bones are fibrolamellar and show no evidence of growth lines or arrest lines; nor do they show evidence of the cementing lines at the bottom of pits as seen in the crocodile skull. Instead, the ornamentation is by modification of the skull surface below the periosteum.

Other ankylosaur specimens seem to support this interpretation of the cranial armor. Maryańska (1971, 1977) described a juvenile *Pinacosaurus* skull, noting that many of the skull bones were not hidden beneath coossified osteoderms. Thus exposed, the outlines for many of the individual cranial elements could be described. More recently, Godefroit et al. (1999) described another, slightly older subadult skull of *Pinacosaurus* and noted that although the sutures between individual cranial elements were mostly discernible, the surfaces of the elements were very rugose. They concluded that thin sheets of osteoderms had already coossified to the skull surface. However, because the rugosities are limited to individual elements as they are in *Cedarpelta*, there is ample cause to believe that the rugosities are not coossified osteoderms but rather ornamentation formed by periosteal osteogenesis. A similar condition is also apparently true of *Shamosaurus* as well. The near obliteration of the skull sutures and subdued surface ornamentation of the holotype skull of *Cedarpelta* appears nearly as featureless as the holotype skull of *Shamosaurus,* suggesting that the cranial "armor" of *Shamosaurus* is developed in the surface of the cortical bone.

Ornamentation has been reported by Jacobs et al. (1994; called "excrescence") on some of the individual disarticulated cranial elements of a juvenile nodosaurid (estimated body length 0.7 m), suggesting a similar periosteal modification of the skull surface in some nodosaurids. Such an interpretation is further supported by the holotype cranium of *Struthiosaurus australis* (Carpenter and Kirkland, personal observations) in which the nasal–frontal suture is distinct and indicates a loss of the nasals in a manner similar to the holotype skull of *Cedarpelta*. We do not, however, argue that all cranial "armor" in ankylosaurs is modified cortical bone, because the cheek plates of *Panoplosaurus* and *Edmontonia* (Carpenter 1990) are clearly epidermal in origin. Furthermore, as Vickaryous et al. (2001) have shown, some ankylosaurid skulls (notably of *Pinacosaurus* and possibly *Euoplocephalus*) apparently had both epidermal-derived armor fused to the skull surface as well as modified cortical bone. On the basis of Vickaryous et al. (2001), the cranial "armor" of *Minmi* is modified cortical bone, although this hypothesis has yet to be tested.

The two skulls of *Cedarpelta* are almost the same size, yet as noted above, the paratype is disarticulated (except for the braincase), and the holotype has lost the maxillae and frontals along their sutures. We conclude that neither individual was very old, although the holotype individual was mature enough so that its skull begin fusing the individual elements. The pattern of fusion apparently progressed anteriorly from the posterior region of the skull. This fusion was rapid enough that some sutures have been obliterated, despite the separation of the frontal and maxilla. A similar pattern of fusion may also be inferred for *Struthiosaurus* on the basis of the open nasal–frontal suture, although all sutures have been lost on the postorbital portion of the cranium.

The disarticulated paratype skull of *Cedarpelta* provides the first three-dimensional look at individual cranial elements of an ankylosaur skull. Some of the elements resemble their counterparts in a disarticulated *Stegosaurus* skull (Carpenter and Miles, in preparation), such as the lachrymal, but others are so unusual (e.g., prefrontal) that their identification was initially difficult. The outline for most of the elements of *Cedarpelta* only vaguely resemble those of the juvenile *Pinacosaurus* (Maryańska 1971, 1977) or *Minmi* (Molnar 1996), suggesting that there is no standard shape for the individual elements of the ankylosaurid skull. As for comparisons with nodosaurids, the only cranial elements of a nodosaurid known are those of a juvenile described by Jacobs et al. (1994). The occipital condyle is on a very short neck that is more horizontal than seen in adults, especially if the basioccipital–basisphenoid suture is oriented vertically. This more horizontal occipital condyle is reminiscent of the condition in *Cedarpelta*, suggesting that the horizontal occipital condyle in *Cedarpelta* may be pedomorphic, although this is far from certain.

The skull of *Cedarpelta* most closely resembles that of *Shamosaurus scutatus* Tumanova (1983, 1985, 1987) from the Lower Cretaceous Dzun Bayn Formation of southeastern Mongolia. Tumanova erected the subfamily Shamosaurinae because of the primitiveness of *Shamosaurus*. However, as shown elsewhere (Carpenter 2001), the subfamily Shamosaurinae cannot be supported phyletically. Nevertheless, *Cedarpelta* is more derived than *Shamosaurus* in the unusually straight occipital condyle (but see above), highly domed parietal, elongate, platelike pterygoids, and trochlearlike process on the pterygoid. On the other hand, the retention of premaxillary teeth is a primitive feature that is phyletically uninformative.

Another undescribed *Shamosaurus*-like ankylosaurid has been collected in China and includes a skull and postcrania (Currie, personal communication). The skull appears to be more similar to *Shamosaurus* than to *Cedarpelta*, and we agree with Currie and Vickaryous (personal communication) that it is a distinct genus (e.g., has very large, nodosaurid-style cheek teeth). Nevertheless, we do not yet know the autapomorphies of this genus to place it as the closest sister taxon to *Shamosaurus* or to *Cedarpelta*. Clearly, the diversity of *Shamosaurus*-like ankylosaurids is considerably greater than Tumanova (1983) originally realized and may yet demonstrate that the Shamosaurinae is a valid taxon.

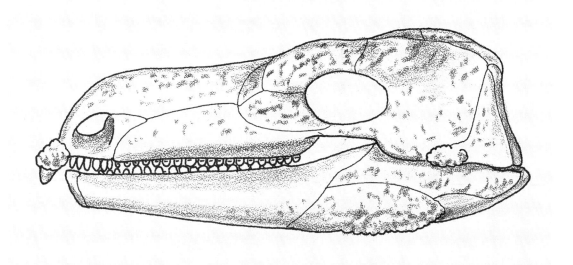

The size of the holotype *Cedarpelta* skull (estimated premaxilla–occipital condyle length 60 cm; Fig. 11.17) indicates a very large individual 7.5–8.5 m long. *Shamosaurus* (skull approximately 36 cm long) is estimated to have been 5 m long. This compares with 3.7 m for the primitive Late Jurassic polacanthid *Gargoyleosaurus* (skull length 29 cm) and 7 m for the Late Cretaceous nodosaurid *Edmontonia* (skull length 50 cm).

Acknowledgments. Thanks to all those from the Utah Friends of Paleontology who assisted in the field, and especially to Sue Ann Bilbey and Evan Hall for finding the type specimen. Matthew Vickaryous reviewed an earlier version of the manuscript, and his comments are appreciated. Technical reviews by Mike Lowe and Mike Hylland are appreciated. Bob Peyton assisted in making the measured section.

Figure 11.17. Reconstructed skull of Cedarpelta bilbeyhallorum. *Estimated premaxilla–occipital condyle length = 60 cm.*

References Cited

Blows, W. 1987. The armoured dinosaur *Polacanthus foxi* from the Lower Cretaceous of the Isle of Wight. *Palaeontology* 30: 557–580.
———. 1996. A new species of *Polacanthus* (Ornithischia: Ankylosauria) from the Lower Cretaceous of Sussex, England. *Geological Magazine* 133: 671–682.
Buffrénil, V. de. 1982. Morphogenesis of bone ornamentation in extant and extinct crocodilians. *Zoomorphology* 99: 155–166.
Carpenter, K. 1990. Ankylosaur systematics: Example using *Panoplosaurus* and *Edmontonia* (Ankylosauria, Nodosauridae). In K. Carpenter and P. J. Currie (eds.), *Dinosaur Systematics: Perspectives and Approaches,* pp. 282–298. New York: Cambridge University Press.
———. 2001. Phylogenetic analysis of the Ankylosauria. In K. Carpenter (ed.), *The Armored Dinosaurs.* Bloomington: Indiana University Press [this volume, Chapter 21].

Carpenter, K., and J. I. Kirkland. 1998. Review of Lower and Middle Cretaceous ankylosaurs from North America. In S. G. Lucas, J. I. Kirkland, and J. W. Estep (eds.), *Lower and Middle Cretaceous Terrestrial Ecosystems. New Mexico Museum of Natural History Science Bulletin* 14: 249–270.

Carpenter, K., J. I. Kirkland, D. Burge, and J. Bird. 1999. Ankylosaurs (Dinosauria: Ornithischia) of the Cedar Mountain Formation, Utah, and their stratigraphic distribution. In D. Gillette (ed.), *Vertebrate Paleontology in Utah. Utah Geological Survey Miscellaneous Publication* 99-1: 244–251.

Carpenter, K., T. Maryańska, and D. Weishampel. In press. Ankylosauria: In P. Dodon, D. Weishampel, and H. Osmólska (eds.), *The Dinosauria.* Berkeley: University of California Press.

Cifelli, R. L., J. I. Kirkland, A. Weil, A. L. Dino, and B. J. Kowallis. 1997. High-precision $^{40}Ar/^{39}Ar$ geochronology and the advent of North American's Late Cretaceous terrestrial fauna. *Proceedings of the National Academy of Sciences USA* 94: 11163–11167.

Coombs, W. P., Jr. 1971. *The Ankylosauria.* Ph.D. diss. Columbia University, New York.

———. 1978. An endocranial cast of *Euoplocephalus* (Reptilia, Ornithischia). *Palaeontographica* 161: 176–182.

———. 1995. A new nodosaurid ankylosaur (Dinosauria: Ornithischia) from the Lower Cretaceous of Texas. *Journal of Vertebrate Paleontology* 15: 298–312.

Coombs, W. P., Jr., and T. Maryańska. 1990. Ankylosauria. In D. B. Weishampel, P. Dodson, and H. Osmólska (eds.), *The Dinosauria*, pp. 456–483. Berkeley: University of California Press.

Eaton, T. H. 1960. A new armored dinosaur from the Cretaceous of Kansas. *University of Kansas Paleontological Contribution* 8: 1–24.

Gaffney, E. S. 1979. Comparative cranial morphology of recent and fossil turtles. *American Museum of Natural History Bulletin* 164: 69–376.

Godefroit, P., X. Pereda Suberbiola, H. Li, and Z. Dong. 1999. A new species of the ankylosaurid dinosaur *Pinacosaurus* from the Late Cretaceous of Inner Mongolia (P.R. China). *Bulletin de l'Institut des Sciences Naturelles de Belgique, Sciences de La Terra* 69(Suppl. B): 17–36.

Hulke, J. W. 1881. *Polacanthus foxii*, a large undescribed dinosaur from the Wealden Formation in the Isle of Wight. *Philosophical Transactions of the Royal Society, London* 178: 169–172.

Jacobs, L. L., D. A. Winkler, P. A. Murry, and J. M. Maurice. 1994. A nodosaurid scutling from the Texas shore of the Western Interior Seaway. In K. Carpenter, K. Hirsch, and J. H. Horner (eds.), *Dinosaur Eggs and Babies,* pp. 337–346. New York: Cambridge University Press.

Kirkland, J. I. 1998. A polacanthine ankylosaur (Ornithischia: Dinosauria) from the Early Cretaceous (Barremian) of Eastern Utah. In S. G. Lucas, J. I. Kirkland, and J. W. Estep (eds.), *Lower and Middle Cretaceous Terrestrial Ecosystems. New Mexico Museum of Natural History Science Bulletin* 14: 271–281.

Kirkland, J. I., B. Britt, D. L. Burge, K. Carpenter, R. Cifelli, F. DeCourten, J. Eaton, S. Hasiotis, and T. Lawton. 1997. Lower to Middle Cretaceous dinosaur faunas of the Central Colorado Plateau: A key to understanding 35 million years of tectonics, sedimentology, evolution, and biogeography. *Brigham Young University, Geological Studies* 12: 69–103.

Kirkland, J. I., S. G. Lucas, and J. W. Estep. 1998. Cretaceous dinosaurs of the Colorado Plateau. In S. G. Lucas, J. I. Kirkland, and J. W. Estep (eds.), *Lower and Middle Cretaceous Terrestrial Ecosystems. New Mexico Museum of Natural History Science Bulletin* 14: 79–89.

Lee, Y.-N. 1996. A new nodosaurid ankylosaur (Dinosauria: Ornithischia) from the Paw Paw Formation (late Albian) of Texas. *Journal of Vertebrate Paleontology* 16: 232–345.

Mantell, G. 1833. Memoir on the *Hylaeosaurus*, a newly discovered fossil reptile from the strata of Tilgate Forest. *Geology of the South East of England.* London: Longman, Rees, Orme, Brown, Green and Longman.

Maryańska, T. 1971. New data on the skull of *Pinacosaurus grangeri* (Ankylosauria). *Palaeontologica Polonica* 25: 45–53.

———. 1977. Ankylosauridae (Dinosauria) from Mongolia. *Palaeontologica Polonica* 37: 85–151.

May, M. T., L. C. Furer, E. P. Kvale, L. J. Suttner, G. D. Johnson, and J. H. Meyers. 1995. Chronostratigraphy and tectonic significance of Lower Cretaceous conglomerates in the foreland of central Wyoming. Stratigraphic Evolution of Foreland Basins. *Society of Economic Paleontology and Mineralogy, Special Publication* 52: 97–110.

Molnar, R. 1980. An ankylosaur (Ornithischia: Reptilia) from the Lower Cretaceous of Southern Queensland. *Memoir, Queensland Museum* 20: 77–87.

———. 1996. Preliminary report on a new ankylosaur from the Early Cretaceous of Queensland, Australia. *Memoir, Queensland Museum* 39: 653–668.

Moodie, R. 1911. An armoured dinosaur from the Upper Cretaceous of Wyoming. *University of Kansas Science Bulletin* 5: 257–273.

Norman, D. B., and T. Faiser. 1996. On the first partial skull of an ankylosaurian dinosaur from the Lower Cretaceous of the Isle of Wight, southern England. *Geological Magazine* 133: 299–310.

Ostrom, J. H. 1970. Stratigraphy and paleontology of the Cloverly Formation (Lower Cretaceous) of the Bighorn Basin area, Wyoming and Montana. *Peabody Museum of Natural History Bulletin* 35: 1–234.

Pereda Suberbiola, X. 1993. *Hylaeosaurus, Polacanthus*, and the systematics and stratigraphy of Wealden armoured dinosaurs. *Geological Magazine* 130: 767–781.

———. 1994. *Polacanthus* (Ornithischia, Ankylosauria) a transatlantic armoured dinosaur from the Early Cretaceous of Europe and North America. *Palaeontographica Abt. A* 232: 133–159.

Pereda Suberbiola, X., and P. M. Barrett. 1999. A systematic review of ankylosaurian dinosaur remains from Albian–Cenomanian. *Palaeontology* 60: 177–208.

Russell, D. A. 1993. The role of Central Asia in dinosaurian biogeography. *Canadian Journal of Earth Sciences* 20: 2002–2012.

Schumacher, G.-H. 1973. The head muscles and hyolaryngeal skeleton of turtles and crocodilians. In C. Gans and T. Parson (eds.), *Biology of the Reptilia*, Vol. 4, pp. 101–199. New York: Academic Press.

Tumanova, T. 1983. The first ankylosaurs from the Lower Cretaceous of Mongolia [in Russian]. *Transactions of the Joint Soviet–Mongolian Paleontological Expedition* 24: 110–120.

———. 1985. Skull morphology of the ankylosaur *Shamosaurus scutatus* from the Lower Cretaceous of Mongolia. In D. Baubis (ed.), *Les Dinosaures de La Chine a La France*, pp. 75–79. Toulouse, France: Muse D'History Naturelle.

————. 1987. Armored dinosaurs of Mongolia [in Russian]. *Transactions of the Joint Soviet–Mongolian Paleontological Expedition* 32: 1–77.

Vickaryous, M. K., A. P. Russell, and P. J. Currie. 2001. Cranial ornamentation of ankylosaurs (Ornithischia: Thyreophora): Reappraisal of developmental hypotheses. In K. Carpenter (ed.), *The Armored Dinosaurs*. Bloomington: Indiana University Press [this volume, Chapter 15].

12. Carlsbad Ankylosaur (Ornithischia, Ankylosauria): An Ankylosaurid and Not a Nodosaurid

Tracy L. Ford and
James I. Kirkland

Abstract

The Carlsbad, California, ankylosaur is described as a new genus and species. Originally it was identified as nodosaurid on the basis of its teeth, the interpretation of its shoulder blade, and the similarity of its pelvic armor to that of the nodosaurid *Stegopelta*. It is reidentified as an ankylosaurid on the basis of its hollow dorsal armor and the morphology of its limbs. It is distinguished from all other known ankylosaurids in its limb proportions and its tooth morphology, as well as in the morphology of its armor.

Introduction

Armor occurs in several different clades of dinosaurs, the two most famous being stegosaurs and ankylosaurs. Determining the exact position of osteoderms is difficult because so few articulated specimens are known, and even then their position is debatable—for example, the orientation of armor in the Polacanthidae (Blows 1987; Kirkland 1998; Pereda-Suberbiola 1994). Osteoderms change shape cranial–caudally and laterally in ankylosaurs. Some osteoderms do not change morphology much from genera to genera or family to family, but others do (Carpenter 1990). For this reason, specimens of ankylosaurs with in situ armor are significant. Such a specimen was discovered in 1987 (not 1985, as reported by Coombs and Demere 1996) in Carlsbad, California. It was originally believed to be a nodosaurid because the in situ pelvic armor looked similar to that of *Stegopelta* (Coombs and Demere 1996; Demere 1988); other armor is present on the limbs. This specimen is one of the most complete dinosaurs known from California.

The specimen was accidentally discovered by a backhoe operator during construction of College Boulevard on the northwest side of Palomar-McClellan Airport, Carlsbad, California (Coombs and Demere 1996). Unfortunately, parts of the specimen were broken and lost. It was excavated by the San Diego Natural History Museum, where it is now held. The details of the excavation are given by Demere (1988).

The specimen was recovered from marine strata that indicated that the carcass had apparently floated a short distance from land and sunk to the seafloor with its ventral side up. The carcass then became an artificial reef on the soft substrate and was subsequently colonized by encrusting and nestling epibionts (Coombs and Demere 1996). Invertebrates burrowed into the long bones so that they are now hollow.

Additional preparation of the specimen by one of us (T.L.F.) has revealed new information that was not available to Coombs and Demere (1996). We now believe that the specimen is an ankylosaurid and can be diagnosed, primarily by the armor.

Institutional Abbreviations. AMNH: American Museum of Natural History, New York. BMNH: Natural History Museum (formerly British Museum [Natural History]), London. FMNH: Field Museum of Natural History, Chicago, Illinois. NMC: Canadian Museum of Nature (formerly the National Museum of Canada), Ottawa, Ontario. ROM: Royal Ontario Museum, Toronto, Ontario. TMP: Royal Tyrrell Museum of Paleontology, Drumheller, Alberta, Canada. SDNHM: San Diego Natural History Museum, San Diego, California. USNM: National Museum of Natural History (formerly United States National Museum), Washington D.C. YPM: Peabody Museum of Natural History, Yale University, New Haven, Connecticut.

Systematic Paleontology
Order Ornithischia Seeley, 1888
Suborder Ankylosauria Osborn, 1923
Family Ankylosauridae Brown, 1908
Aletopelta n. g.

Etymology. Aletopelta, "wandering shield," from Greek *aletes,* "wanderer, wandering," and Greek *pelte,* "shield," a word commonly used for ankylosaurs in reference to their armor. *Aletes* was chosen because originally, the plate containing the Peninsular Ranges Terrane, where Carlsbad and San Diego, California, are today, was somewhere opposite the middle of Mexico (Ford et al. 1998; Lund and Bottjer 1992; Lund et al. 1991); this plate has thus been wandering northward, carrying the specimen with it.

Diagnosis. As for species.

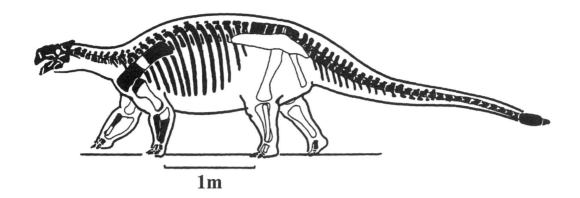

1m

A. *coombsi* n. sp.

Figure 12.1. Skeleton of Aletopelta coombsi *showing known elements in white from SDNHM 33909.*

Etymology. In honor of Walter P. Coombs, Jr., for his ground-breaking work on ankylosaurs and his years of research, which have inspired many an enthusiast as well as professional paleontologist.

Holotype. SDNHM 33909. Eight teeth, fragmentary scapulae, partial humerus, partial ulna, possible fragment of right ?radius, ulna, partial left and possibly right ischium, femora, tibiae, fibulae, four or five partial vertebrae, dorsal neural arch, neural arches of the sacrum, fragmentary ribs, partial armor over pelvic girdles in situ (Fig. 12.1), at least 60 detached armor plates, and numerous indeterminate fragments, most embedded in matrix and many of which are probably broken armor or ribs.

Type Locality. Point Loma Formation, Upper Campanian (Upper Cretaceous). College Boulevard between El Camino Real and Palomar Airport Road, northwest of the Palomar-McClellan Airport, City of Carlsbad, California (SDNHM Locality 3392), approximately longitude 117°15'W, latitude 33°9'N. Exact locality data are available through the San Diego Natural History Museum.

Diagnosis. A medium-size ankylosaurid, teeth wider than tall; femur greatly longer than tibia and fibula; three metatarsals; ilia covered in sutured, polygonal, low-peaked scutes set in a mosaic pattern; massive, short-pointed spike in shoulder region; hollow caplike scutes across back; hollow pup tent–like scutes over neck and shoulders; triangular, dorsally compressed caudal scutes that are highly asymmetrical top to bottom; nearly all osteoderms hollow and thin.

Description

Teeth. The eight teeth known from *Aletopelta* are all unworn (Coombs and Demere 1996) and were recovered from the region of the rib cage. Three nearly complete crowns with partial roots are still encased in matrix, and two fragmentary crowns and three complete crowns with partial roots are completely out of the matrix (Fig. 12.2). The crowns are compressed labiolingually and are rounded rather than triangular in outline in labial or lingual view. The base of the best

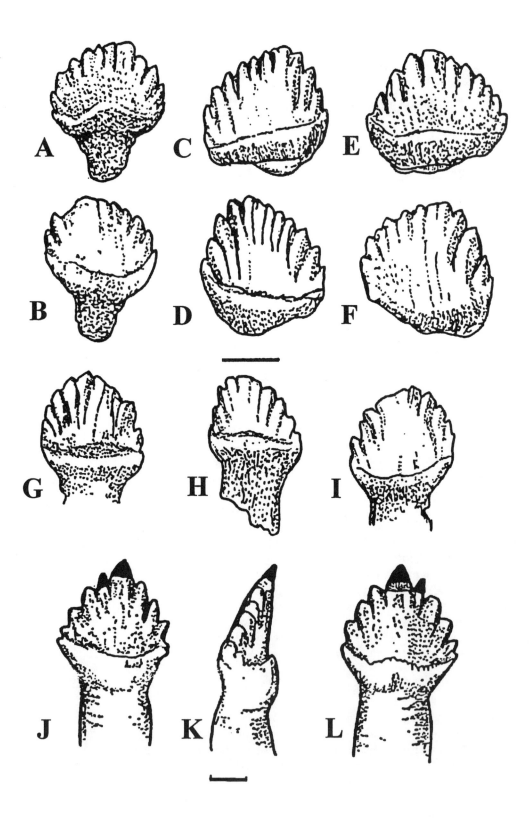

preserved tooth is longer than tall, although the smaller teeth are proportionally longer than the larger tooth. The posterior end of this tooth has three pointed denticles, with a short curve to the apex of the tooth. The apex denticle is squared off. From the apex to the anterior end, there are six pointed denticles. The cingulum differs from tooth to tooth, as well as from the labial and lingual sides. The cingulum in one tooth is irregularly wavy on one side, whereas on the other side, the cingulum arches downward and is covered by fine vertical grooves. Another tooth has a cingulum that arches toward the root and is irregularly lumpy, whereas on the other side, it is fairly smooth and obliquely angled. The apex of the crown has a small, blunt, pointed, bifurcated denticle on the posterior edge of the tooth. There are six small denticles from the lower edge of the cingulum to the apex and three large denticles on the posterior edge. The cingulum has small cusps on both ends. There are grooves that start at the denticles but do not extend to the cingulum.

Axial Skeleton. A few fragmentary vertebrae are known, including a possible fragmentary cervical. A distal cervical or anterior dorsal neural arch has recently been uncovered on one of the blocks that had the distal cervical or shoulder armor. The neural arch is split nearly down the middle, with parts in two blocks of the concretion. The neural spine is short, and the diapophysis arches upward and is about the same height as the neural spine (typical of ankylosaurs).

The sacral neural arches have fragmentary coossified sacral ribs, lack centra, and are exposed on the ventral side of the pelvic blocks. The sacral ribs are between the ilia and sacral centra but are not attached to the ilia (the ilia was slightly misplaced during deposition). The neural canal is larger anteriorly and tapers posteriorly; this condition is in opposition to other ankylosaurs, which typically have a shallower anterior and expanded posterior neural canal. This would seem to indicate that the sacral arches were flipped around, but the sacral ribs indicate that this is not the case. The sacral ribs are present on the last two dorsosacrals and the three sacrals.

Two caudal vertebrae are represented by their centra, with a possible third, fragmentary one as well. They are part of a block with three caudal plates, and judging from the size of the dorsal neural arch, they are from the medial to distal caudal region.

The ribs are still concealed in matrix, and it is not known if they are flat or oval, or if they had a typical ankylosaur T shape. A rib head was recently uncovered lying on top of the dorsal neural arch. The rib head is smaller than it should be at the position it was found, and it is probably from the distal cervical area. It was probably displaced by either ocean currents or animal scavenging. The rib head has a very shallow arch, indicating that *Aletopelta* has a wide torso.

Pectoral Girdle and Forelimb. The forelimb includes fragments of the scapula blade, humeri, ulna, and radius (Fig. 12.3). The fragments of the scapula blade show no evidence of distal expansion. Although Coombs and Demere (1996) report a nodosauridlike acromion process, we found no fragments preserving any of the acromion process. Both humeri have the very distal and proximal ends missing. The distal

Figure 12.2. (opposite page) Teeth of nodosaurs. (A–F) Labial and lingual sides of teeth of Aletopelta. (G, H) Teeth from the type specimen of Edmontonia longiceps (NMC 8531). (I) Tooth of a referred specimen of Edmontonia rugosidens (AMNH 5381). (J–L) Tooth of Niobrarasaurus coleii (MU 650 VP). Figures A–I after Coombs and Demere (1996); figures J–L after Carpenter et al. (1995). Scale bar = 5 mm.

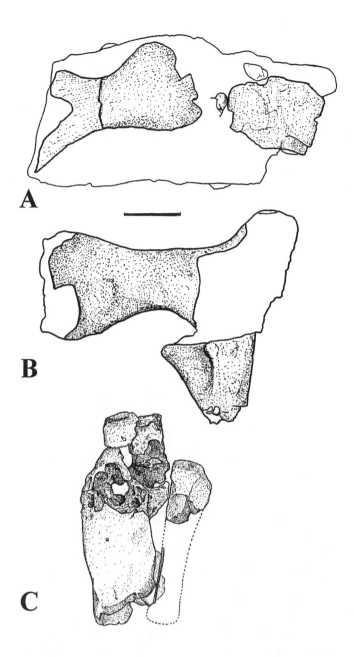

Figure 12.3. Forelimb skeletal elements of Aletopelta. (A) Block showing a fragmentary right humerus (right) and portion of scapula (left). (B) Block showing partial left humerus and portion of left ulna?. (C) Right ulna (pathologic?) and portion of radius. After Coombs and Demere (1996). Scale bar = 10 cm.

end is better preserved on one humerus and indicates a distal width of approximately 18 cm (Coombs and Demere 1996). Although incomplete, the broad rounded deltopectoral crest is better preserved on the left humerus (Fig. 12.4A). The shaft is of medium length and is estimated to be at least 40 cm.

The proximal end of the left ulna is preserved near the proximal end of the left humerus. It is massively constructed with a large olecranon and more closely resembles that of ankylosaurids than nodosaurids. In nodosaurids, the olecranon is thick and tapers (Coombs 1978).

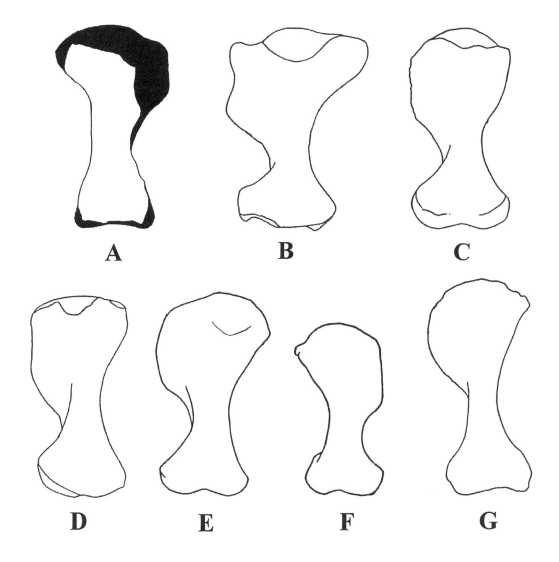

A B C

D E F G

A large, enigmatic bone (Figs. 12.3C, 12.5A–C) was originally thought to be a skull, but computed tomography scans have eliminated this possibility. It now seems more likely that the element is a heavily bioeroded ulna. The bone is large and rounded, with both ends bioeroded away. The anterior end is heavily bioeroded, with part of the upper section broken and slightly displaced; bioerosion has also made a trough through the bone surface along the shaft. A possible proximal end of the right radius was preserved next to the ulna.

Pelvic Girdle and Hind Limbs. The pelvic region is the best preserved portion of the skeleton. The ilia are about 90 cm long and 31 cm wide at the broad, shallow acetabulum. The preacetabular blade is elongate, compressed, diverges from the midline anteriorly, and is twisted to a near vertical orientation at the anterior end. Coombs and Demere (1996) suggest that this anterior twist is preservational distor-

Figure 12.4. Humeri of ankylosaurs. (A) Aletopelta. (B) Saichania. (C) Talarurus. (D) Pinacosaurus (after Maryańska 1977). (E) Ankylosaurus. (F) Euoplocephalus (after Coombs 1978). (G) Sauropelta (after Ostrom 1970).

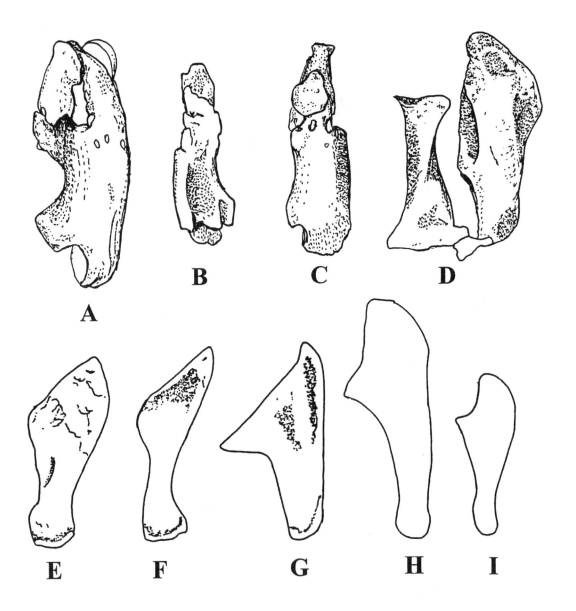

Figure 12.5. Ulna of ankylosaurs. Pathologic right ulna of Aletopelta, *in (A) lateral, (B) cranial, and (C) caudal views. (D)* Euoplocephalus *(= Scolosaurus). (E)* Talarurus. *(F)* Pinacosaurus *(Maryańska 1977). (G)* Saichania. *(H)* Sauropelta *(after Ostrom 1970). (I)* Mymoorapelta *(Kirkland and Carpenter 1994).*

tion because it departs from the nodosaurid pattern in which the ilium typically lies obliquely and flat against the ribs (Coombs 1979). However, because none of the other skeletal elements are warped, it seems doubtful that this part of the skeleton would be distorted. In fact, the twisted ilia of *Aletopelta* is twisted in a manner seen in other ankylosaurids.

Coombs and Demere (1996) described what they called the left ischium lying next to the left ilium. This bone is elongate and expands proximally. The distal end tapered to a slightly flared end that is roughly circular or oval in cross section. An alternative identification is that this bone may be the anterior portion of the synsacrum that broke away

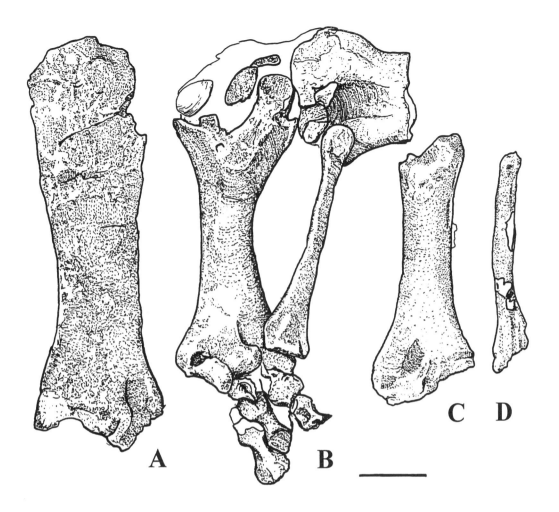

from the rest of the sacrum. Only further preparation will fully determine the identity of this bone. Clearly, there is not yet evidence that *Aletopelta* has a bent ischium, which is found in all known nodosaurids and polacanthids.

The femora are incomplete at the articular ends (Fig. 12.6A). The head of the left femur is slightly offset (though still in matrix). Femoral shafts are straight and anteroposteriorly compressed. The fourth trochanter is low and proximal to the femoral midlength as in the Ankylosauridae (Coombs 1978). The left femur is at least 51 cm long, and the right femur is at least 56 cm; if the femora were complete, the estimated length would be about 60 cm (Coombs and Demere 1996). *Niobrarasaurus coleii* MU 650VP has a femur length of 54 cm (Carpenter et al. 1995); *Hoplitosaurus marshi* USNM 4752 has a length of 50 cm (Gilmore 1914); *Polacanthus foxii* BMNH R175 has a length of 56 cm; *Nodosaurus textilis* YPM 1815 has a length of 60 cm (Lull 1921); *Edmontonia longiceps* NMC 8531 has a length of 66 cm (Sternberg 1928); *Sauropelta edwardsorum* YPM 5456 has a length of 77 cm (Ostrom 1970); *Gastonia burgei* has a length of 59 cm (Gaston et al.

Figure 12.6. Hindlimb skeletal elements of Aletopelta. *(A) Right femur. (B) Distal half of left femur, left tibia and fibula, and partial left pes. (C) Right tibia. (D) Right fibula. After Coombs and Demere (1996). Scale = 10 cm.*

2001); *Euoplocephalus tutus* BMNH R5161 has a length of 60 cm ("*Scolosaurus cutleri*" of Nopsca 1928); and *Euoplocephalus tutus* AMNH 5404 has a length of 57 cm (Coombs 1978).

The tibia is expanded on both ends, with a narrow midshaft (Fig. 12.6B, C). The left tibia is better preserved, with a strongly compressed distal end and a centrally positioned subcircular astragalus. There is a deep cleft that slants obliquely across the distal end of the tibia proximally and lateral to the astragalus where the fibula articulated. There is also a small prong or flange that is bent over the cleft on the medial side. The left tibia is about 36 cm and the right is about 39 cm. *Niobrarasaurus coleii* MU 650VP has a tibia length of 36.5 cm (Carpenter et al. 1995); *Polacanthus foxii* BMNH R175 has a right tibia length of 40.5 cm (Pereda-Suberbiola 1994); *Nodosaurus textilis* YPM 1815 has a length of 44 cm (Lull 1921); *Sauropelta edwardsorum* YPM 5456 has a length of 57 cm (Ostrom 1970); *Gastonia burgei* has a length of 30 cm (Gaston et al. 2001); *Euoplocephalus tutus* BMNH R5161 has a length of 41.5 cm ("*Scolosaurus cutleri*" of Nopsca 1928); and *Euoplocephalus tutus* AMNH 5404 has a length of 33 cm (Coombs 1978).

The left fibula is excellently preserved (Fig. 12.6B). It is 32 cm long and has a rounded, expanded proximal end, with a compressed shaft that is twisted distally, so that it is flattened at the distal end and is rotated about 75° to the proximal end. An oblique angle at the mediodistal corner probably fit into the cleft at the distal end of the tibia. The right fibula (Fig. 12.6D) is fragmentary and compressed, and it has a gently arched shaft that is 31 cm long (Coombs and Demere 1996).

The tibiae and fibulae are shorter in relation to the length of the femur than is known in nodosaurids, but this proportion is close to that observed in polacanthids and ankylosaurids. *Aletopelta* has a tibia/femur ratio of 0.67–0.63 (Fig. 12.7). This animal has a long upper leg and short lower leg with small feet. *Niobrarasaurus coleii* MU 650VP has a ratio of 0.75; *Polacanthus foxii* BMNH R175 has a ratio of 0.63; *Nodosaurus textilis* YPM 1815 has a ratio of 0.77; *Sauropelta edwardsorum* YPM 5456 has a ratio of 0.74; *Gastonia burgei* (Gaston et al. 2001) has a ratio of about 0.50 (the relative lengths of the femur and tibia in *Gastonia* are estimated because all the bones were found disarticulated in a bone bed; it is clear that the tibia is much shorter than the femur in *Gastonia*; Kirkland 1998); *Euoplocephalus tutus* BMNH R5161 has a ratio of 0.69; and *Euoplocephalus tutus* AMNH 5404 has a ratio of 0.57.

The astragalus is oval and obliquely positioned at the distal end of the left tibia. It is not possible to determine whether the astragalus was fused to the tibia or whether it was held in place by matrix. It is common for the astragalus to be fused to the tibia in adult ankylosaurs. In juvenile ankylosaurs, the astragalus is not fused to the tibia and is in general shape similar to *Aletopelta*. There is no firm evidence to indicate the ontogenetic age of *Aletopelta,* although the absence of fusion of the neural spines to the vertebral centra suggests that this specimen may not be an adult.

There are three left metatarsals that are slightly displaced at the end of the tibia and fibula and two isolated right metapodials that were

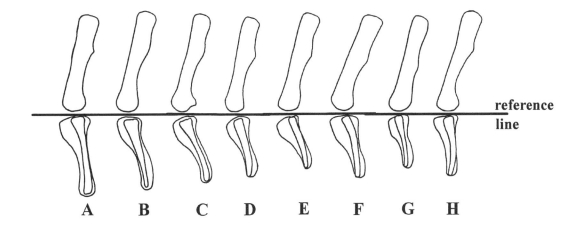

reference
line

A   B   C   D   E   F   G   H

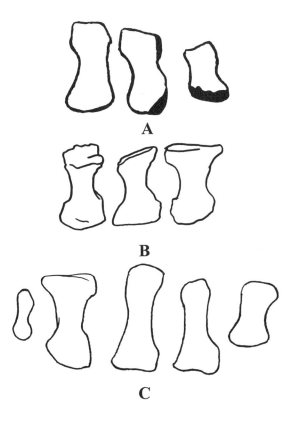

A

B

C

Figure 12.7. *(above) Illustration showing the hindlimb proportions of (A)* Niobrarasaurus coleii, *(B)* Nodosaurus textilis, *(C)* Sauropelta edwardsorum, *(D)* Polacanthus foxii, *(E)* Gastonia burgei, *(F)* Euoplocephalus tutus *(= Scolosaurus cutleri), (G)* Euoplocephalus tutus, *and (H)* Aletopelta coombsi.

Figure 12.8. *(left) Pes of ankylosaurids. (A) Aletopelta. (B) Dyoplosaurus (after Coombs 1986). (C) Sauropelta (after Ostrom 1970). Scale bar = 10 cm.*

identified as metatarsal II and metatarsal IV (Fig. 12.6B). These metatarsals are short and thick (Fig. 12.8).

Armor. The scutes in ankylosaurids and nodosaurids differ in structure. In nodosaurids, the scutes were solid or slightly concave just under the apex, whereas in ankylosaurids, they were thin and hollowed out ventrally (Coombs and Maryańska 1990; Kirkland 1998).

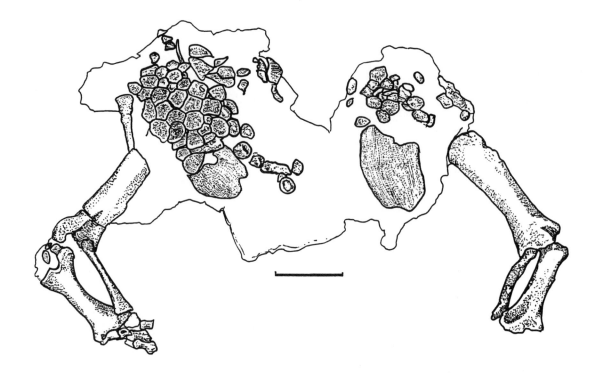

Figure 12.9. Osteoderms on block of pelvis and hind limbs of Aletopelta, *after Coombs and Demere (1996). Scale = 25 cm.*

Several different types of osteoderms are known for *Aletopelta*, with the majority of them being thin and hollow. The best preserved region of articulated armor is over the pelvic area (Fig. 12.9). The scutes over the ilia are irregular polygons in shape, with low or flat subcentral peaks. They are uniform in size, abut each other, and are sutured together in a mosaic pattern. Some scutes near the inner edge of the ilia have a more laterally offset central peak. The apexes of some of these scutes have small holes in them. These holes may have been caused by a parasite similar to that which causes shell pit disease in turtles, or these holes may have been formed by marine invertebrates. Similar pits are known from a number of ankylosaurs, including *Stegopelta* and *Gastonia* (Kirkland 1998), but they have yet to be studied. Some pelvic scutes (four?) are isolated and partly prepared, but it is not yet known if these scutes are hollow (Fig. 12.10E, F). The scutes are not fused to the ilia, and there are several millimeters of matrix between the overlying scutes and the ilia. The ilia are not in natural articulation but have been displaced several centimeters away from each other, as noted above. On the outer margin of the left ilia, some of the scutes appear to be fused to each other, although thin sectioning would be needed to determine this for certain.

The area between the ilia has several isolated scutes, some with small, distinctively round, cap shapes (approximately two-thirds height to width). These scutes may have covered several areas of the body. There are approximately seven isolated, cap-shaped scutes in various degrees of preservation. The height of the apex of these scutes is from 50%–70% of the length of the scutes, with the ventral side completely hollowed (Fig. 12.10G, H). A similar caplike scute (TMP 97.12.112) is

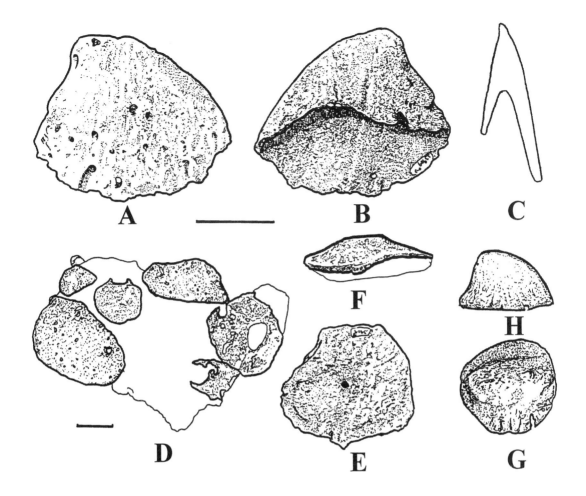

A   B   C

D   F   H

E   G

the only specimen not belonging to *Aletopelta*. It is cap shaped, hollow ventrally, and has a ridge on the ?caudal end extending from the apex to lateral edge.

The lateral caudal scutes of *Aletopelta* are dorsoventrally asymmetrical and have a triangular shape, with the dorsal surface being significantly more extensive than the ventral surface (Fig. 12.10A–C). These tall scutes have rounded posterior apices. This morphology suggests that these scutes covered the dorsal surface of the tail. There are approximately 12 caudal scutes in various degrees of preservation. As noted above, there is a block of three caudal scutes associated with caudal centra. These scutes are believed to be from the distal medial part of the tail because they are large in comparison to the caudal centra and would have covered several vertebrae. The scutes are more highly vascularized than other scute types and are also hollow ventrally.

The articulated caudal scutes of *Saichania* (Psihoyos 1994) and *Pinacosaurus* are similar in shape to those in *Aletopelta*, but they are more symmetrical dorsoventrally. They are evenly spaced and placed about one scute length apart along the lateral margins of the tail. There is little size decrease of these scutes between the hips and the tail club in

Figure 12.10. Osteoderms of Aletopelta. (A) Dorsal, (B) ventral, and (C) cross-section view of caudal scute. (D) Block of two caudal vertebrae and three caudal scutes. (E) Dorsal and (F) lateral view of median pelvic scute with black dot in the middle, perhaps indicating shell pit disease (as in turtles); cap-shaped scute in (G) dorsal and (H) lateral view. A, B, D, and G after Coombs and Demere (1996). Scale bar = 5 cm.

these Asian taxa. Polacanthid caudal plates are more posteriorly inclined and not as hollow ventrally. Additionally, they appear to have been much more numerous, and they become progressively smaller from the hips to the distal portion of the tail (Kirkland 1998; Kirkland and Carpenter 1994; Pereda-Suberbiola 1994). In the nodosaurid *Sauropelta*, the proximal caudal scutes have a more posteriorly placed apex (Ostrom 1970) than *Aletopelta*, and they decrease in size more rapidly (Carpenter 1984).

Two small oblong scutes (one larger than the other) have a long ridge and a short ridge separated by a concave center (most likely because of bioerosion). These scutes may have been between the caudal scutes on the dorsal surface of the tail. A single, solid, subrectangular, flat-ridged scute is present; it is similar to those of various nodosaurids.

Three articulated blocks contain scutes that are believed to be from the neck–shoulder region. These are the blocks with the split posterior neural arch, as mentioned above. One small block contains two hollowed out, pup tent–shaped scutes. These scutes are longer than tall, with a ridge that extends from the lower anterior end to the posterior apex. One side is vertical or nearly vertical, and the other side is angled upward; one end is lower than the other. One scute was complete; the other was about half complete. There are also three partial pup tent scutes. One pup tent scute is split at or near the apex, with one side of the scute in the small block and the other side in the adjacent middle block. The pup tent scutes are similar to scutes present in other nodosaurids and ankylosaurids. In nodosaurids, the pup tent scutes are on the side of the body. In *Euoplocephalus* (= *Scolosaurus*, Nopsca 1928) there are pup tent scutes near and on the front leg. Next to the block with the pup tent scutes is a large, somewhat flattened scute.

The larger block (Fig. 12.11) contains a small, oval, flat scute with an off-center keel, and next to this is a partial large scute that one of us (J.I.K.) identified as a shoulder spine. This block and its spine were unfortunately damaged by the backhoe. A small teardrop-shaped keel is present on the anterior end of the large spine. It is lying parallel to what is thought to be the posterior edge of the large shoulder spine. The dorsal side of the spine has a very rough, rugose texture. There is also a small, oval, convex scute lying on top of the large spine with a few millimeters of matrix between the two. The "back" of the scute is solid, whereas both "sides" are hollowed, suggesting that the scute is completely hollow. Whether this is due to bioerosion or its being naturally hollow is not known. The base of the spine is roughly triangular in cross section, with the large end slightly concave cranially and the caudal end having a small expansion or bulb. The base of the spine is not beveled, as are the spines on the lateral side of the body. The posterior edge of the spine is widest at the base and tapers to a point. The anterior edge is much thinner than the posterior edge. When complete, the spine would have had a round cranial edge and a flat caudal edge. This shoulder spine may either have been laterally located on the body, or, most probably because the base is level, been a medial spine situated on the back as in *Euoplocephalus* (Carpenter 1982).

There is a smaller scute in *Aletopelta* on the posterior edge of the

Figure 12.11. (opposite page) Pectoral and cervical scutes of Aletopelta. *Large pectoral spike in (A) dorsal, (B) lateral, (C) anterior, and (D) posterior view. Broken scute immediately posterior to large lateral scute in (E) dorsal, (F) posterior, and (G) lateral view. (H) Illustration showing these two scutes in natural articulation; cervical pup tent–shaped scute that was abutted against the cranial edge of the pectoral spike in (I) dorsal, (J) ventral (showing the partial bioerosion), (K, L) lateral (dashed line indicates how the scute may have fit in the neck region), (M) anterior, and (N) posterior view. Small pup tent scute found near the large shoulder scute in (O) dorsal, (P) lateral, and (Q) anterior view. Small, flat scute found next to the pup tent scute in (R) dorsal, (S) ventral, (T) lateral, and (U) anterior view.*

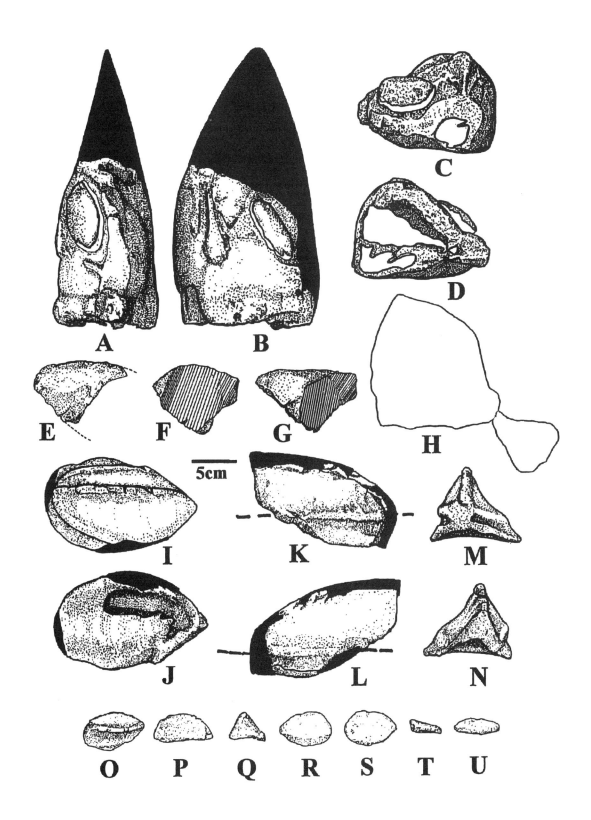

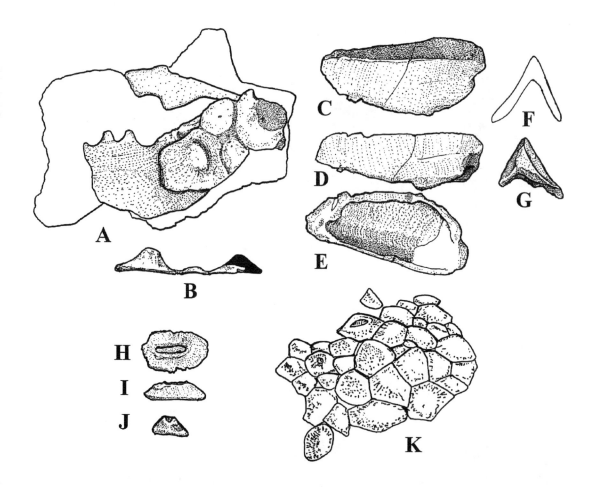

Figure 12.12. Osteoderms of Aletopelta. *Cervical ring? armor in (A) dorsal and (B) lateral view; pup tent scute in (C) dorsal, (D) lateral, (E) ventral, (F) cross section, and (G) anterior views. Oblong scute in (H) dorsal, (I) lateral, and (J) anterior view. (K) Pelvic osteoderms in oblique view.*

large spine and separated from it by matrix. This scute is about one fourth the size of the large scute and is similar to the lateral scutes of *Edmontonia longiceps* (USNM 11868). Near the posterior end of the large scute are two small scutes: one is small and pup tent–shaped, with an oblong base and a short ridge; a flat, oval scute is adjacent to it. The pup tent scute has been displaced and has its ventral side adjacent to the cranial edge of the large scute. Both ends of this scute were either broken away during deposition or when it was uncovered by the backhoe. A third of the anterior portion of the ventral side is convex and smooth. The other two thirds are split nearly down the middle, with one side having a convex, spongy nature and the other side being bioeroded. A small ossicle (type D of Coombs and Demere 1996) is ventral to the pup tent scute and is separated by a few millimeters of matrix from the scute. This scute is partially bioeroded and is similar in shape to the cervical ring scutes of *Euoplocephalus tutus* (NMC 0210, Lambe 1902). There is a ridge dorsally on the middle of the scute, indicating that it is a midline scute. The lower anterior end has a small ridge on both sides of the scute. The anterior, ventral ridge begins about one third the height of the scute, then angles upward to nearly the middle of the scute. The lower edge of the ridge is concave and flares out to the base of the scute. The ridge on the scute compares well with cervical scutes of other known nodosaurids and ankylosaurids (Fig. 12.12). This ridge would have been within the skin of the neck.

Half of the cervical band shows the bony ring, whereas the other half has several scutes overlying it (Fig. 12.12A, B). Extending along the band is a slightly offset low-domed polygonal scute; next to this are three smaller irregular flat scutes, two on the ventral side and the third one attached to both of these.

The anterior end of the pathologic ?ulna has a small oval scute with a low point. Two small scutes were found near where the knee would have been. Coombs and Demere (1996) believed that these scutes were misplaced there. Scutes are known on the legs of ankylosaurs *Niobrarasaurus coleii* (Carpenter et al. 1995), *Euoplocephalus* (Nopcsa 1928), and *Saichania* (in Psihoyos 1994). Several small ossicles (type D scute of Coombs and Demere 1996) are also present; they are believed to have been in between the larger scutes, much like in other ankylosaurs.

Discussion

Tooth size does not reflect the size of the animal. For example, the largest ankylosaurid, *Ankylosaurus*, has the smallest teeth relative to its size of any known ankylosaurid. *Aletopelta* has large teeth more similar to those of nodosaurids and some shamosaurine ankylosaurids than those of more derived ankylosaurids (J.I.K., personal observation). The teeth of *Edmontonia rugosidens* (AMNH 5381; Coombs 1990; Coombs and Demere 1996) are taller than long, have four denticles on the anterior edge and three on the posterior edge, have an apex that is either bifurcated or that is one large denticle, and have small denticles on the edge of the cingulum. The teeth of *Euoplocephalus tutus* (holotype of *Palaeoscincus asper,* NMC 1349; Coombs 1990; Coombs and Demere 1996) have five small denticles along the anterior and posterior edges, a large central denticle, and small cusps on the cingulum. The tooth is taller than long and has several grooves extending down both sides of the crown. The teeth of the nodosaurid *Niobrarasaurus coleii* (MU 650 VP; Carpenter et al. 1995) have a well-developed, swollen cingulum on the ?buccal side. The cingulum is rimmed by 28 small denticles. There are three denticles on the anterior edge, a broken apical denticle, and four denticles on the posterior side. Some of the teeth of *Aletopelta* have a rough, pebbly surface similar to the texturing that is on the single tooth of *Niobrarasaurus coleii* (Coombs, in Coombs and Demere 1996). This texture may be due to weathering in the marine environment.

The humeri of *Aletopelta* have a large deltopectoral crest that expands sharply. The crests in *Sauropelta edwardsorum* and *Animantarx ramaljonesi* have a slight curve that expands to the distal end, as is the case generally among nodosaurids (Carpenter et al. 1999; Coombs 1978; Ostrom 1970; Russell 1940). *Saichania* has a massive deltopectoral crest and a short shaft, whereas *Talarurus* and *Pinacosaurus* have a large crest and a short shaft (Maryańska 1977). *Ankylosaurus* and *Euoplocephalus* have a slightly longer shaft and a large, rounded crest (Coombs and Maryańska 1990). Among the polacanthids, the humerus is short, with a large, anteriorly deflected deltopectoral process (Kirkland 1998; Pereda-Suberbiola 1994). *Aletopelta* humeri are more like those of ankylosaurids or polacanthids than those of nodosaurids.

The right ulna of *Aletopelta* appears to be pathologic, as in "*Scolosaurus*" (Nopsca 1928). On the interior edge of the ulna of *Aletopelta*, there is a concave groove (keel and groove) that the radius crossed over. This groove would have inhibited rotation of the lower foreleg. This type of keel-and-groove arrangement with the ulna and radius was discussed by Johnson and Ostrom (1995) for *Torosaurus*. This condition also occurs in several other quadrupedal dinosaurs (Carpenter et al. 1994).

Euoplocephalus tutus (AMNH 5266 and ROM 784) have three metatarsals (Carpenter 1982; Coombs 1986), and *Pinacosaurus* and *Talarurus* have four (Maryańska 1977). Among the nodosaurids, *Sauropelta* has four metatarsals and *Nodosaurus* has four or five (Carpenter and Kirkland 1998; Coombs 1990). The primitive polacanthid *Mymoorapelta* has at least four metatarsals (Kirkland et al. 1998). *Euoplocephalus* (ROM 784; Coombs 1986) has short, massive metatarsals, whereas those of *Sauropelta* (AMNH 3032) are much longer and larger. If the left pes of *Aletopelta* is complete, then it is more like ankylosaurids than nodosaurids or polacanthids in the number of metatarsals, as well as in their relative lengths.

Demere (1988) and Coombs and Demere (1996) compared the pelvic osteoderms of *Aletopelta* to those of *Stegopelta landerensis* (FMNH UR88; Moody 1910). Superficially, the scutes covering the ilia are similar. The pelvic scutes of *Stegopelta* are irregularly hexagonal, pentagonal, or quadrilateral in outline (type B of Coombs and Demere 1996). The scutes are either flat or have a very slight peak. These scutes abut each other, and there are no type D scutes of Coombs and Demere (1996) as in *Aletopelta*. The scutes of *Stegopelta* are much larger in size relative to the size of ilia than is the case in *Aletopelta*. The scutes of *Aletopelta* have taller peaks and have a different mosaic pattern than those over the ilia of *Stegopelta* (Moody 1910). Also, although the sacral armor is superficially similar to that in *Stegopelta*, this is probably due to both having pelvic regions covered by closely packed scutes of equal size; it is probably not a shared character between the two.

In contrast, the pelvic region of polacanthids is covered by a large sacral shield. The shield consists of fused oval scutes and small ossicles (Blows 1987; Hulke 1888; Kirkland 1998; Pereda-Suberbiola 1994). The pelvic osteoderms of *Aletopelta* did not fuse into a sacral shield, as did those of polacanthids. In situ pelvic armor is known for the ankylosaurid *Euoplocephalus* (= *Scolosaurus*; Nopsca 1928). It has a medial and primary row of three oval scutes surrounded by large, irregular ossicles. Between the ossicles are smaller ossicles, unlike *Aletopelta*.

The large, spined cervical–pectoral scute of *Aletopelta* does have some similarities to those of nodosaurids. For example, *Sauropelta* has massive, wide-based cervical scutes and a long pectoral scute situated high on the shoulder (Carpenter 1984; Carpenter and Kirkland 1998). These scutes in *Sauropelta* have an angled base, whereas that of *Aletopelta* has a flat base.

Conclusions

Aletopelta has more ankylosaurid characteristics than nodosaurid ones and is believed to represent a new genus of ankylosaurid. It is possible that the holotype of *Aletopelta* is an immature individual on the basis of the unfused astragalus, partly fused scutes, and unfused neural spines. *Aletopelta* has broad teeth, a large deltopectoral crest, femur longer than the tibia and fibula, and horizontal ilium, as well as unique armor. Although Coombs and Demere (1996) identified the specimen as being an undiagnostic nodosaurid, the majority of skeletal elements indicate that *Aletopelta* belongs to a distinctive ankylosaurid genus.

Coombs and Demere (1996) did not believe that there were enough different types of armor in *Aletopelta* to be of diagnostic value. Further preparation has revealed shoulder–cervical armor. The large spine differs from lateral spines of nodosaurids, and its erect base indicates it is more probably a medial dorsal spine. Pup tent–shaped scutes show similarities to the scutes of the medial cervical ring of *Euoplocephalus*. Also, at the shoulder region of *Aletopelta*, there are large pup tent scutes, smaller pup tent scutes, a flat, oval scute, and hollow, cap-shaped scutes. Such scutes have not been described from nodosaurids, polacanthids, or other ankylosaurids.

The pelvic armor does show superficial similarities to that of *Stegopelta*. The pelvic scutes of *Aletopelta* are uniform in shape, although of different sizes. The scutes of *Stegopelta* are more irregularly shaped and are larger in relative size to the ilia than are those of *Aletopelta*. Like *Stegopelta, Aletopelta* lacks small ossicles separating the scutes. The lateral caudal armor is similar to that of *Sauropelta* and polacanthids, but also to *Saichania* and *Pinacosaurus*. Because of the beveled inner edge, it is possible that the caudal scutes were situated higher on the tail. A hypothetical reconstruction of *Aletopelta* showing the interpreted organization of the armor is shown in Figure 12.13.

Aletopelta was found in marine sediments, as have nodosaurids. Some, such as *Nodosaurus, Stegopelta, Pawpawsaurus,* and *Niobrarasaurus,* are only known from marine strata (Carpenter and Kirkland 1998; Carpenter et al. 1995). *Aletopelta* is the first ankylosaurid to be recovered from marine strata. Because the site is considered to be close to shore and abutting a rugged terrain, finding a taxon more typical of inland environments is not completely unexpected.

The position of the Carlsbad, California, area during the Late Campanian is believed to be just off the coast of middle Mexico or just north of middle Mexico, making this the southernmost ankylosaurid known from North America (Ford et al. 1998).

Acknowledgments. We thank Ben Creisler, who suggested the name *Aletopelta*; Ralph Molnar, who encouraged our selecting this name; and Tom Demere and the staff at the San Diego Natural History Museum for letting us prepare and describe the specimen. Reviews by Kenneth Carpenter, Xabier Pereda Suberbiola, Mike Lowe, and Mike Hylland are gratefully acknowledged. We also thank Kenneth Carpen-

Figure 12.13. Lateral view of Aletopelta coombsi showing where the known armor may have been placed (black).

ter for scanning and reformatting the figures. Finally, T.L.F. would like to thank the junior author, J.I.K., for strongly suggesting that he undertake redescription of the Carlsbad ankylosaur. T.L.F. also thanks Miners' Gem in Old Town, San Diego, California, for allowing him to do additional preparation of the specimen behind the store.

References Cited

Blows, W. T. 1987. The armored dinosaur *Polacanthus foxi* from the Lower Cretaceous of the Isle of Wight. *Palaeontology* 30: 557–580.

Carpenter, K. 1982. Skeletal and dermal armor restoration of *Euoplocephalus tutus* (Ornithischia: Ankylosauridae) from the Late Cretaceous Oldman Formation of Alberta. *Canadian Journal of Earth Science* 21: 1491–1498.

———. 1984. Skeletal reconstruction and life restoration of *Sauropelta* (Ankylosauria: Nodosauridae) from the Cretaceous of North America. *Canadian Journal of Earth Science* 21: 1491–1498.

———. 1990. Ankylosaur systematics: Example using *Panoplosaurus* and *Edmontonia* (Ankylosauria: Nodosauridae) from the Cretaceous of North America. In K. Carpenter and P. J. Currie, (eds.), *Dinosaur Systematics: Approaches and Perspectives*, pp. 281–299. Cambridge: Cambridge University Press.

Carpenter, K., D. Dilkes, and D. B. Weishampel. 1995. The dinosaurs of the Niobrara Chalk Formation (Upper Cretaceous, Kansas). *Journal of Vertebrate Paleontology* 15: 275–297.

Carpenter, K., and J. I. Kirkland. 1998. Review of Lower and Middle Cretaceous Ankylosaurs from North America. In S. G. Lucas, J. I. Kirkland, and J. W. Estep (eds.), *Lower and Middle Cretaceous Terrestrial Ecosystems. New Mexico Museum of Natural History and Science Bulletin* 14: 249–270.

Carpenter, K., J. I. Kirkland, D. Burge, and J. Bird. 1999. Ankylosaurs (Dinosauria: Ornithischia) of the Cedar Mountain Formation, Utah, and their stratigraphic distribution. In D. D. Gillette (ed.), *Vertebrate Paleontology in Utah. Utah Geological Survey, Miscellaneous Publication* 99-1: 243–251.

Carpenter, K., J. H. Madsen, and A. Lewis. 1994. Mounting of fossil vertebrate skeletons. In P. Leiggi and P. May (eds.), *Vertebrate Paleontological Techniques,* Vol. 1, pp. 285–322. Cambridge:, Cambridge University Press.

Coombs, W. P., Jr. 1978. The families of the ornithischian dinosaur order Ankylosauria. *Palaeontology* 21: 143–170.

————. 1979. The osteology and myology of the hindlimb in the Anky-losauria (Reptilia, Ornithischia). *Journal of Palaeontology* 53: 666–684.

————. 1986. A juvenile ankylosaur referable to the genus *Euoplocephalus* (Reptilia, Ornithischia). *Journal of Vertebrate Paleontology* 6: 162–173.

————. 1990. Teeth and taxonomy in ankylosaurs. In K. Carpenter and P. J. Currie (eds.), *Dinosaur Systematics: Approaches and Perspectives,* pp. 269–279. Cambridge: Cambridge University Press.

Coombs, W. P., Jr., and T. A. Demere. 1996. A Late Cretaceous nodosaur-id ankylosaur (Dinosauria: Ornithischia) from marine sediments of coastal California. *Journal of Paleontology* 70: 311–326.

Coombs, W. P., Jr., and T. Maryańska. 1990. Ankylosauria. In D. B. Weis-hampel, P. Dodson, and H. Osmólska (eds.), *The Dinosauria,* pp. 456–483. Berkeley: University of California Press.

Demere, T. A. 1988. An armored dinosaur from Carlsbad. *Environment Southwest* 523: 12–15.

Ford, T. L., J. I. Kirkland, and W. P. Elder. 1998. The surfing nodosaur or riding up the Pacific Plate. In D. L. Wolberg, K. Gittis, S. Miller, L. Carey, and A. Raynor (eds.), *Dinofest Symposium,* pp. 16–17. Phila-delphia: Academy of Natural Sciences.

Gaston, R. W., J. Schellenbach, and J. I. Kirkland. 2001. Mounted skeleton of the polacanthine ankylosaur *Gastonia burgei.* In K. Carpenter (ed.), *The Armored Dinosaurs.* Bloomington: Indiana University Press [this volume, Chapter 18].

Gilmore, C. W. 1914. Osteology of the armored Dinosauria in the United States National Museum, with special reference to the genus *Stegosau-rus. Memoirs of the United States National Museum* 89: 1–316.

Hulke, J. W. 1888. Supplemental note on *Polacanthus foxii,* describing the dorsal shield and some parts of the endoskeleton imperfectly known in 1881. *Philosophical Transactions of the Royal Society* 178: 169–172.

Johnson, R. E., and J. H. Ostrom. 1995. The forelimb of *Torosaurus* and an analysis of the posture and gait of ceratopsian dinosaurs. In J. J. Thomason (ed.), *Functional Morphology in Vertebrate Paleontology,* pp. 205–218. Cambridge: Cambridge University Press.

Kirkland, J. I. 1998. A polacanthinae ankylosaur (Ornithischia: Dino-sauria) from the Early Cretaceous (Barremian) of Eastern Utah. In S. G. Lucas, J. I. Kirkland, and J. W. Estep (eds.), *Lower and Middle Cretaceous Terrestrial Ecosystems. New Mexico Museum of Natural History and Science Bulletin* 14: 271–281.

Kirkland, J. I., and K. Carpenter. 1994. North America's first pre-Creta-ceous ankylosaur (Dinosauria) from the Upper Jurassic Morrison For-mation of western Colorado. *Brigham Young University Geology Studies* 40: 25–41.

Kirkland, J. I., K. Carpenter, A. Hunt, and R. Scheetz. 1998. Ankylosaur (Dinosauria) from the Upper Jurassic Morrison Formation. In K. Car-penter, D. Chure, and J. I. Kirkland (eds.) *The Upper Jurassic Morrison Formation: An Interdisciplinary Study. Modern Geology* 23: 145–177.

Lambe, L. 1902. New genera and species from the Belly River Series (mid-Cretaceous). *Geological Survey of Canada, Contributions to Cana-dian Paleontology* 3: 25–81.

Lund, S. P., and D. J. Bottjer 1992. Paleomagnetic evidence for microplate tectonic development of southern and Baja California. In J. P. Dauphin and B. R. T. Simoneit (eds.), *The Gulf and Peninsular province of the*

Californias. American Association of Petroleum Geologists Memoir 47: 231–248.

Lund, S. P., D. J. Bottjer, K. J. Whidden, J. E. Powers, and M. C. Steele. 1991. Paleomagnetic evidence for Paleogene terrane displacements and accretion in southern California. In P. L. Abbott and J. A. May (eds.), *Eocene Geologic History of San Diego Region. Pacific Section Society of Economic Paleontology and Mineralogy* 68: 99–106.

Lull, R. S. 1921. The Cretaceous armored dinosaur, *Nodosaurus textilis* Marsh. *American Journal of Science*, Series 5, 1: 97–126.

Maryańska, T. 1977. Ankylosauridae (Dinosauria) from Mongolia. *Palaeontologica Polonica* 37: 85–151.

Moody, R. L. 1910. An armored dinosaur from the Cretaceous of Wyoming. *Kansas University Science Bulletin* 5(14): 257–273.

Nopcsa, F. 1928. Palaeontological notes on reptiles. *Geologica Hungarica, Series Palaeontologica* 1: 1–84.

Ostrom, J. H. 1970. Stratigraphy and paleontology of the Cloverly Formation (Lower Cretaceous) of the Bighorn Basin Area, Wyoming and Montana. *Peabody Museum of Natural History, Yale University, Bulletin* 35: 1–234.

Pereda-Suberbiola, J. 1994. *Polacanthus* (Ornithischia, Ankylosauria): A transatlantic armored dinosaur from the Early Cretaceous of Europe and North America. *Palaeontographica Abt. A* 232: 133–159.

Psihoyos, L. 1994. *Hunting Dinosaurs*. New York: Random House.

Russell, L. S. 1940. *Edmontonia rugosidens* (Gilmore), an armored dinosaur from the Belly River Series of Alberta. *University of Toronto Studies, Geological Series* 43: 3–28.

Sternberg, C. M. 1928. A new armored dinosaur from the Edmonton Formation of Alberta. *Transactions of the Royal Society Canada* 4: 93–103.

13. Variation in Specimens Referred to *Euoplocephalus tutus*

PAUL PENKALSKI

Abstract

A systematic study of ankylosaurid specimens from the Upper Campanian of Alberta and Montana indicates greater diversity than is generally recognized. A multivariate analysis of skull measurements revealed several clusters. More than one taxon is probably represented. A survey of skeletal characters and armor morphology supports a hypothesis of multiple taxa. Taxonomically useful characters may be found in the vertebral column, humerus, pelvis, pes, and armor. Certain characters are attributed to sexual dimorphism.

Introduction

In 1902, Lawrence Lambe described *Euoplocephalus tutus* (originally as *Stereocephalus tutus*; see Lambe 1910) on the basis of a skull fragment and cervical armor from the Dinosaur Park Formation (formerly the Judith River Formation; Eberth 1997) of southern Alberta. Before the naming of *E. tutus*, the only well-known armored dinosaur was *Stegosaurus* from the Jurassic Morrison Formation of Colorado and Wyoming. In 1908, Barnum Brown erected the family Ankylosauridae on the basis of another new genus, *Ankylosaurus,* from the Hell Creek Formation of Montana. To this day, only three specimens of *Ankylosaurus* are known, and they are the only upper Maastrichtian ankylosaurid specimens from North America. Coombs (1990) concluded from tooth morphology that there is a single valid species, *Ankylosaurus magniventris.* The recently described genus *Nodocephalo-*

saurus (Sullivan 1999), from the Upper Cretaceous Kirtland Formation of New Mexico, is substantially different from all other North American forms and may be allied with some Asian taxa.

At least 40 specimens of *E. tutus* are now known from the Upper Campanian and Lower Maastrichtian of Alberta and Montana, including 17 complete or partial skulls (see Coombs 1978; Coombs and Maryańska 1990). Some of these specimens became the holotypes of other taxa. In 1923, Charles Gilmore described a fine skull from the Dinosaur Park Formation of Alberta. It was the first well-preserved ankylosaurid skull to be described, and he tentatively referred it to *Euoplocephalus tutus*, fully aware that other well-preserved skulls awaited preparation at the American Museum of Natural History in New York. By 1929, three more North American genera had been named and described. *Dyoplosaurus acutosquameus* Parks (1924) is a fine specimen that includes the articulated caudal half of the axial skeleton, the pelvis, one hindlimb, armor, skin impressions, and several skull fragments. *Scolosaurus cutleri* Nopsca (1928) is a rare specimen consisting of most of the postcranial skeleton with armor in situ, but which is missing the skull and distal half of the tail. *Anodontosaurus lambei* Sternberg (1929) includes an almost complete, but badly crushed, skull, a mandible, skeletal fragments, and armor. In 1930, Gilmore described another skull (USNM 11892), referring this one to *Dyoplosaurus* on the basis of similarities in the teeth.

Over the next 40 years, ankylosaurs received relatively little attention in Canada and the United States until Walter Coombs studied them for his doctoral work. Coombs (1978), in the first detailed survey of the Ankylosauria, referred all Campanian forms—including the types of *Dyoplosaurus acutosquameus, Scolosaurus cutleri,* and *Anodontosaurus lambei*—to *Euoplocephalus tutus,* concluding that there was but a single valid species of ankylosaurid from the Campanian of North America. Coombs and Maryańska (1990) maintain this conclusion with reservations. In retrospect, Coombs' (1971) dissertation focused on anatomy and familial characteristics and was not a rigorous systematic analysis. Indeed, he expressed discomfort in synonymizing several of the specimens, particularly the holotype of *Scolosaurus*. Blows (personal communication) has pointed out several differences between *Scolosaurus* and other specimens referred to *Euoplocephalus* and believes that *Scolosaurus cutleri* is a valid taxon. In a reconstruction of *Euoplocephalus,* Carpenter (1982) suggested that differences in the armor could not be resolved and based the postcranial part of his reconstruction on the type of *Scolosaurus*. However, he later (Carpenter 1990, 1997) refers to his reconstruction as *Dyoplosaurus,* having determined that the differences were probably significant. More recently, Carpenter (1998) reexamined the holotype of *Scolosaurus* and concluded that it was not distinct from *Euoplocephalus,* as he previously thought.

Evaluation of the holotypes is difficult because of the lack of complementary material. For instance, the holotypes of *Euoplocephalus* and *Dyoplosaurus* include no common elements. The resulting taxonomic quandary is not surprising. Another complicating factor is the likely presence of sexual dimorphism and ontogenetic stages. Some of the differences among specimens that have been referred to *Euoplo-cephalus* are discussed below and provide a framework for future taxonomic studies.

Institutional Abbreviations. AMNH: American Museum of Natural History, New York. BMNH: Natural History Museum (formerly British Museum [Natural History]), London. MOR: Museum of the Rockies, Bozeman, Montana; NMC: Canadian Museum of Nature, Ottawa, Canada; ROM: Royal Ontario Museum, Toronto, Canada; TMP: Royal Tyrrell Museum of Palaeontology, Drumheller, Alberta, Canada; UA: University of Alberta, Edmonton, Canada. USNM: National Museum of Natural History (formerly United States National Museum), Washington, D.C. ZPAL: Zaklad Paleobiologii, Polish Academy of Sciences, Warsaw, Poland.

TABLE 13.1.

Important Campanian–Early Maastrichtian ankylosaurid specimens from Alberta (AB) and Montana (MT). Postcranial means part of the postcranial endoskeleton; half-ring indicates cervical armor; armor indicates loose osteoderms (scutes).

Specimen	Material	Horizon
AMNH 5238	Skull, teeth	Dinosaur Park Formation, AB
AMNH 5266	Postcranial	Upper Horseshoe Canyon Formation, AB
AMNH 5337	Skull, teeth, postcranial, half-rings, armor	Dinosaur Park Formation, AB
AMNH 5403	Skull, teeth, postcranial, half-rings, armor	Dinosaur Park Formation, AB
AMNH 5404	Skull, teeth, postcranial, half-rings, armor	Dinosaur Park Formation, AB
AMNH 5405	Skull, teeth, postcranial, half-rings, armor	Dinosaur Park Formation, AB
AMNH 5406	Postcranial, half-rings, armor	Dinosaur Park Formation, AB
AMNH 5409	Pelvis	Dinosaur Park Formation, AB
BMNH R4947	Skull	Dinosaur Park Formation, AB
BMNH R5161	Postcranial skeleton, half-rings, armor in situ (Type of *Scolosaurus cutleri*)	Dinosaur Park Formation, AB
MOR 433	Skull, postcranial, half-rings, armor	Upper Two Medicine Formation, MT
NMC 0210	Fragmentary skull, half-ring (Type of *Euoplocephalus tutus*)	Dinosaur Park Formation, AB
NMC 8530	Skull, half-ring, postcranial, armor (Type of *Anodontosaurus lambei*)	Lower Horseshoe Canyon Formation, AB
ROM 784	Skull fragments, teeth, postcranial, armor (Type of *Dyoplosaurus acutosquameus*)	Dinosaur Park Formation, AB
ROM 813	Partial postcranial skeleton with some armor in situ	Dinosaur Park Formation, AB
ROM 1930	Skull, teeth, postcranial, armor	Dinosaur Park Formation, AB
TMP 91.127.1	Skull	Dinosaur Park Formation, AB
UA 31	Skull (postcranial unprepared)	Dinosaur Park Formation, AB
USNM 7943	Half-ring	Upper Two Medicine Formation, MT
USNM 11892	Skull, teeth	Upper Two Medicine Formation, MT

Material

At least 17 specimens are complete enough to be relevant to this study (see Table 13.1). Many of these are described in the literature. Some of the published articles include detailed descriptions, whereas others cover only a small portion of the specimen. Features pertinent to this study are summarized below.

NMC 0210 (see Lambe 1902). This specimen is the holotype of *Euoplocephalus tutus*, and unfortunately, it is poor. The skull (Fig. 13.1A, B) fragment consists only of the central portion between the orbits and external nares and is not badly crushed. The dorsal surface is well preserved; the palatal side is not. There is a large (although damaged), centrally situated nasal plate located just caudal to the external nares on the skull roof. This plate is present, though not usually as prominently, in almost all ankylosaurid specimens, including most Asian forms. Exceptions are MOR 433, perhaps USNM 11892, and possibly the Asian genus *Talarurus*.

The postcranial material consists of a single cervical armor half-ring, a bony arch that protected the dorsal surface of the neck in ankylosaurs. The half-ring (Fig. 13.1C, D; Lambe 1902: pl. 12) has five keeled scutes, but it is broken, and as noted by Lambe, one or two scutes may be missing. Because all known complete cervical half-rings of ankylosaurids have an even number of scutes, there were probably six in *Euoplocephalus*. Judging by the small girth, even after inclusion of a sixth scute, this is undoubtedly the first half-ring (all ankylosaurids have two). Lambe identified the "concave side" (in dorsal view; see Fig. 13.1C) as cranial because this edge was damaged, and he suggested that the half-ring might have been an ornamental crest on the back of the skull (Lambe 1902: 56). In fact, the edge is probably caudal, but without careful examination of a well-preserved series of cervical vertebrae, the orientation is unclear. Coombs and Maryańska (1990) note differences in the cervical vertebrae between North American and Asian ankylosaurids, and the half-ring morphology probably correlates as well.

The morphology of the half-ring is distinctive. The scutes are closely set, oval, and sharply keeled, with the apices of the keels situated ?caudally on the outer two pairs of scutes and centrally on the medial pair. Their surfaces are finely vascularized and relatively smooth compared with most other ankylosaurid scutes. All the scutes are solidly fused to the bony half-ring underneath. The half-ring is narrower than the scutes are long, so that the scutes overhang the edges of the half-ring both cranially and caudally (see Fig. 13.1C).

AMNH 5406. This specimen lacks the skull, but it has a substantial amount of postcranial material, all presacral, including dorsal vertebrae, ribs, scapula, humerus, ulna, radius, metacarpals, one phalanx, complete first half-ring and partial second half-ring, about 80 scutes, and a small patch of skin impression. This specimen is minimally crushed.

The humerus is smaller than in other specimens, but the cervical half-ring is not. The dorsal neural arches are fused to their centra, so AMNH 5406 might be mature, or at least is close to adult size. The half-

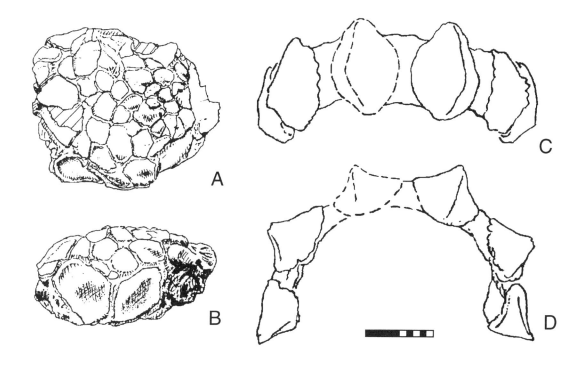

A

B

C

D

Figure 13.1. NMC 0210,
holotype of Euoplocephalus
tutus. Skull in (A) dorsal and
(B) left lateral view, and first
cervical half-ring in (C) dorsal
and (D) ?caudal views.
Scale bar = 10 cm.

ring is identical in all observable characters to that of NMC 0210, but it is about 15% larger. Given this unique similarity, AMNH 5406 is unquestionably referable to *Euoplocephalus tutus*. The postcervical scutes exhibit a variety of shapes and sizes, ranging from small (3–4 cm) and circular with an abbreviated conical point, to large (32 cm), high pitched, and laterally compressed. All of the scutes are thin walled, and all of the moderate to large ones are sharply keeled. Some have a peculiar L shape to the keel when viewed dorsally.

NMC 8530 (see Sternberg 1929). Originally described by Sternberg (1929) as *Anodontosaurus*, meaning "toothless reptile," this specimen consists of a badly crushed skull (Fig. 13.2G), mandible, one vertebra, ischium, a cervical half-ring, and scutes. Many of the skull plates have incipient nodes as in NMC 0210, although they are difficult to see because of crushing. It is likewise difficult to tell whether the nasal plate is large as in most other specimens. The edge of the premaxillary beak is missing.

Both Sternberg (1929) and Coombs (1971) alluded to a possibly unusual palatal structure, but the palate is badly weathered. Much of the palatal surface is eroded away, revealing the internal sinuses in many places. The palate probably did not differ significantly from that of other specimens. Although no teeth remain, the badly eroded alveolar border of the left maxilla is visible, so the name "*Anodontosaurus*" was not appropriate.

In his original description, Sternberg (1929) stated that there were about 100 scutes preserved with the specimen, but during my visit to the collections of the Canadian Museum of Nature, only about 10 labeled as NMC 8530 could be found. According to Kieran Shepherd (personal communication), some additional preparation was done on many of the scutes about 15 years ago, and some might have been mislabeled as NMC 8531, a nodosaur. Indeed, several scutes labeled 8531 appear ankylosaurid and probably belong to 8530. Among these is the distal extremity of an ankylosaurid cervical half-ring (Fig. 13.9K). The fragment has a fused scute, much like that of NMC 0210 and AMNH 5406. There is a partial half-ring labeled 8530, and although it is poorly preserved, it appears to have oval, keeled scutes, and one distal extremity scute is missing. It is unclear whether the loose fragment is the missing end from this half-ring. The scutes of NMC 8530 are similar to those of AMNH 5406, being smooth and sharply keeled, with some scutes having an L shape to the keel in dorsal view. NMC 8530 appears to be referable to *Euoplocephalus tutus*. The minor differences between NMC 8530 and the other moderate-size skulls may be attributable to the fact that NMC 8530 is from the Horseshoe Canyon Formation and is at least 2 million years younger than any of the other specimens.

UA 31 (see Gilmore 1923). This specimen, the first good North American skull to be described, was referred to *Euoplocephalus tutus* by Gilmore (1923), probably for lack of any other choice. In fact, his assignment appears to have been correct because the skull shares several characters with NMC 0210, NMC 8530, BMNH R4947, and TMP 91.127.1. Gilmore gave a detailed description, so here I summarize only the features relevant to this study.

Overall, the skull (Figs. 13.2B, 13.12A) is striking in having a trapezoidal or wedge shape in dorsal view, with virtually no constriction in front of the orbits. The surface is finely vascular to somewhat rugose. The external nares are more rectangular than in most of the other skulls; this condition is not due entirely to crushing because the morphology of the rugosities above the nares also differs. The quadratojugal "horn" (a hornlike secondary dermal ossification fused to the quadratojugal region) is broad, with a centrally situated apex.

According to Gilmore (1923), there is substantial postcranial material, including two cervical half-rings. As of yet, only the skull has been prepared for study (Fox, personal communication). When it is finally prepared, it will be interesting to see whether half-ring morphology confirms this specimen as *Euoplocephalus tutus*.

TMP 91.127.1. A fine skull was collected in 1992 and will be described in detail by M. T. Vickaryous (personal communication). It is the smallest North American ankylosaurid skull in this study (Fig. 13.2J), with a basal length of about 320 mm. It is slightly crushed, but the skull is generally well preserved. It is similar to UA 31 in many characters, and the two cluster together in the morphometric analysis (Appendix 13.1).

BMNH R4947 (see Nopsca 1928). This isolated skull (Fig. 13.2D) is fairly well preserved, although the left side is incomplete and the right quadratojugal horn is missing. Nopsca (1928) stated that the arrangement of plates on the skull was exactly the same as in UA 31, but

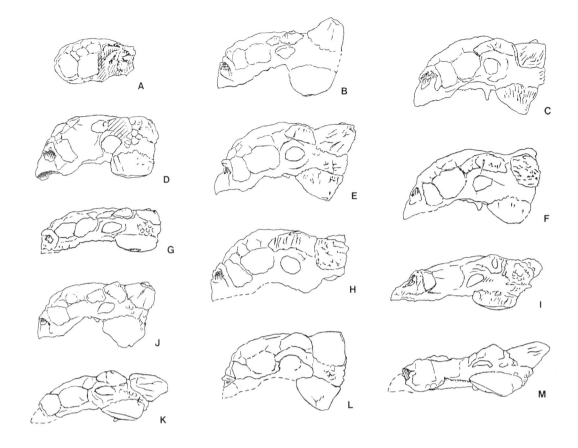

this is not true. On the other hand, no two skulls have exactly the same plate arrangement, and the pattern appears to be of limited taxonomic utility, as discussed below.

USNM 11892 (see Gilmore 1930). This skull (Fig. 13.2K) was referred to *Dyoplosaurus acutosquameous* by Gilmore (1930), mainly on the basis of similarities in the teeth between this specimen and that of ROM 784. The five preserved teeth of USNM 11892 all have a sharp lingual-side cingulum and a bulbous, shelflike labia-side cingulum that is more prominent than in any other North American ankylosaurid teeth. Whether these characters are taxonomically significant is a subject for future study.

The skull of USNM 11892 is crushed, the nasal area fragmented, the premaxillary beak missing, and the palate almost entirely removed by erosion. However, as with most ankylosaurid skulls, the occipital region is well preserved. In separating USNM 11892 from *Euoplocephalus,* Gilmore placed weight on the fact that the snout is constricted rostral to the orbits, unlike in *Euoplocephalus* (i.e., UA 31). The constriction is not entirely an artifact of crushing. He also believed that the presence of four nuchal scutes separated this specimen from *Euoplocephalus* (i.e., UA 31), which appears to have only two; in fact, it also has four, but the medial two are tiny. USNM 11892 most strongly

Figure 13.2. Skulls used in the morphometric analysis.
(A) NMC 0210.
(B) UA 31.
(C) AMNH 5405.
(D) BMNH R4947.
(E) ROM 1930.
(F) AMNH 5337.
(G) NMC 8530.
(H) AMNH 5404.
(I) AMNH 5403.
(J) TMP 91.127.1.
(K) USNM 11892.
(L) AMNH 5238.
(M) MOR 433.

resembles MOR 433. Perhaps not coincidentally, these are the only two skulls from the Two Medicine Formation of Montana.

ROM 784 (see Parks 1924). This specimen is the holotype of *Dyoplosaurus acutosquameus*, most of which Parks (1924) described in detail. I add only a few comments that are based on what is known today that was not known in 1924. Little can be said about the preserved skull fragments, except that the skull was of about average size.

Three tooth crowns were preserved in the maxillary fragments, but Parks (1924) only illustrated the best one and only from one side. Unfortunately, Parks' drawing is not good; Gilmore (1930) listed several mistakes, errors that Parks (1924: 10) had acknowledged. Therefore, the published drawing is unreliable for comparisons with other teeth. To make matters worse, the best tooth now appears to be lost. Only two teeth remain; one lacks enamel and the other is partial. The teeth lack the sharp, lingual-side cingulum present in the preserved teeth of USNM 11892.

Parks (1924) did not illustrate the hindlimb or pes because they were not completely prepared at the time. The left femur is only moderately well preserved, but it appears to be typically ankylosaurid. The femoral head is slightly offset, unlike the bluntly rounded head of other ankylosaurid femora, but the offset could be due to postmortem deformation. The tibia/femur ratio is typical of ankylosaurids, being a bit greater than that of AMNH 5404. The pedal unguals are unusual, as discussed below.

Much of the pelvic and caudal armor is preserved. Many of the scutes are tall, laterally compressed, and rugose. ROM 784 has a very small tail club. This could be a primitive or juvenile character.

ROM 813. This specimen is only partially prepared in three large blocks, and it is best left that way. Some of the material is preserved in situ and includes vertebrae, ribs, the right ilium and femur, left ischium, tibia, ?fibula, metatarsals, an almost complete left pes, numerous scutes, and skin. The pes was labeled as ROM 833 for many years, but field notes show that it is part of ROM 813 (Seymour, personal communication). The pedal unguals differ significantly from those of ROM 784.

Examination of the two largest blocks, both from the dorsal or pelvic region, reveal that the ilium slid caudally after death by about 40 cm relative to the underlying femur, which is still encased in matrix. The armor and skin also moved caudally as well as laterally. Therefore, few conclusions are possible regarding the armor pattern in this area. At least eight large (25 cm), relatively flat, keeled plates are present. They are rugose and perforate. No other Campanian specimen has so many platelike scutes preserved, although the Late Maastrichtian *Ankylosaurus* does. Some of the large plates have pointed, beaklike tips on the keel that overhang the perimeter of the scute, a morphology also present in AMNH 5337 (cf. Fig. 13.11C, F). Some of the large plates lie in close proximity to one another, but because of the displacement, it is hard to determine whether this is their natural position.

AMNH 5238 (see Coombs 1972). This skull (Fig. 13.2L) is well preserved but incomplete ventrally and rostrally. The surface is relatively smooth as in some other specimens, including the holotype. However, in the morphometric analysis, this specimen appears to cluster with the other skulls in the American Museum of Natural History. Perhaps the smoother texture, narrower nuchal shelf, and moderate size are juvenile features. Several teeth are preserved, and these are most like those of AMNH 5337, 5403, and 5405.

AMNH 5266 (Coombs 1986). This is the most complete juvenile ankylosaurid known from North America. The specimen includes vertebrae, a rib, a partial pelvis, complete right hindlimb with pes, and partial left pes. As discussed below, the morphology of the pes differs from that of ROM 784. AMNH 5266 is from the Upper Horseshoe Canyon Formation.

AMNH 5337. This is perhaps the best North American ankylosaurid specimen. The material includes a well-preserved skull with teeth, both lower jaws, both scapulocoracoids, humeri, radii, and ulnae, numerous vertebrae and ribs, pelvis with sacrum, two finely preserved cervical half-rings, and several armor plates. The hindlimbs and tail are missing. The specimen is largely undescribed, although Haas (1969) illustrated parts of the skull and mandible for a study of jaw musculature, and Coombs (1971) discussed portions of the specimen.

The skull (Fig. 13.2F) is striking in several features. Its surface is rugose and unlike that in any of the above specimens. The nuchal shelf is very wide, and the snout is constricted rostral to the orbits, giving the skull a bell-shaped outline in dorsal view. AMNH 5337 has scutes that are very similar to those of AMNH 5403, ROM 813, and ROM 1930.

AMNH 5337 appears to be a mature adult. The dorsal ribs are fused to their vertebrae, the scapulae and coracoids are solidly fused, the long bones have rugose, well-developed articular surfaces, and the skull and armor are rugose, a condition associated with advanced age.

AMNH 5403 (see Coombs 1971). This specimen includes a skull with teeth, lower jaws, vertebrae, ribs, sternum, scapulae, forelimbs, both half-rings, and several scutes. Much of the material, including the skull, is badly crushed. The skull (Fig. 13.2I) is the same in all observable characters as that of AMNH 5337, and the two cluster together in the morphometric analysis despite the severe crushing in AMNH 5403. These two specimens also appear to have identical armor, and the two are probably the same sex of the same species.

AMNH 5404 (see Coombs 1971). This specimen includes the skull (Figs. 13.2H, 13.12B), teeth, vertebrae, ribs, forelimbs, hindlimbs, the first half-ring, and a few scutes. The skull is the largest known from the Campanian–Early Maastrichtian of North America (basal length about 402 mm). It is generally well preserved, but the edge of the premaxillary beak and the quadratojugal horns are missing. It appears to be most similar to the skull of ROM 1930. The first half-ring is similar to that of AMNH 5405.

AMNH 5405 (see Coombs 1978). This specimen includes the best preserved North American skull (Fig. 13.2C) with an almost complete set of maxillary teeth, both lower jaws, the predentary, a humerus and ulna, a partial tail club, partial cervical half-rings, and just two scutes.

In dorsal view, the skull is strikingly similar to AMNH 5337, with an extremely wide nuchal shelf and a marked constriction of the snout rostral to the orbits. The arrangement of plates in the nasal area is more similar to that in AMNH 5404 and ROM 1930.

BMNH R5161 (see Nopsca 1928). The holotype of *Scolosaurus cutleri,* this specimen was described in detail by Nopsca in 1928. The remains are substantial, consisting of the postcranial skeleton with armor in situ, and are virtually complete, except for the skull and distal half of the tail.

Nopsca (1928) assigned the specimen a new genus rather than referring it to *Euoplocephalus tutus* or *Dyoplosaurus acutosquameus,* or defining it as a new species of either of these genera. He suggested that the tails of BMNH R5161 and *Dyoplosaurus* were very different, not realizing that the distal half is missing in R5161. He made no comparisons whatsoever with the holotype of *Euoplocephalus tutus,* even though both specimens have the first half-ring preserved. These are quite different, and as discussed below, BMNH R5161 does not appear to be referable to *Euoplocephalus.*

ROM 1930. This specimen includes the skull (Fig. 13.2E), teeth, postcranial material, and armor. The preserved armor is very much like that of BMNH R5161. In particular, several conical scutes are radially ribbed, perforate, and open dorsally, exactly the same as in BMNH R5161. No other specimen's preserved scutes show this morphology. In fact, every preserved scute of ROM 1930 has a close morphologic equivalent in BMNH R5161, and these two specimens are probably conspecific.

MOR 433. This is an undescribed specimen from the Upper Two Medicine Formation of northern Montana. It includes a partial skull, several vertebrae and ribs, sacrum, a partial scapula, two partial humeri, sections of the cervical half-rings, and numerous scutes. Some of the material is well preserved, but the skull is badly crushed. According to John Horner (personal communication), all of the material was found associated in a small area, with no extraneous elements.

The skull (Fig. 13.2M) is fairly well preserved on its dorsal surface, but the palate is almost completely eroded away and the premaxillary beak is missing. No teeth are present. The skull is large, measuring 415 mm long as preserved. It lacks the large, central nasal plate found on other ankylosaurid skulls. The squamosal and quadratojugal horns are prominent.

Portions of six cervical scutes with fused bone underneath are preserved (see Figs. 13.9L, 13.10A). The most unusual item is a nodosaurlike cervical spike. The spike measures about 40 cm long and was situated laterally at the distal extremity of a cervical half-ring, probably the second. The end opposite the spike is a rugose sutural surface for union with the neighboring segment—that is, medial to the spike. Another piece has the same type of sutural surface, although it does not appear to be the piece that adjoined the spike; it may be a fragment of the spike from the animal's other side. The cervical armor of MOR 433 is unlike that in any other North American ankylosaurid, but it is similar in some ways to nodosaur cervical armor. MOR 433 probably represents a new taxon and will be described in detail elsewhere (Penkalski, in preparation).

Morphometric Analysis of the Skulls

To date, the taxon Ankylosauridae has never been evaluated quantitatively. Previous systematic studies include a cladistic analysis of Ankylosauria (Coombs and Maryańska 1990), a qualitative examination of the teeth of *Ankylosaurus* and several nodosaurid genera (Coombs 1990), and a systematic comparison of the two nodosaurids *Edmontonia* and *Panoplosaurus* (Carpenter 1990). Maryańska (1977) and Tumanova (1987) discussed ankylosaurid phylogeny with an emphasis on Asian forms.

Several biometric methodologies have been employed in taxonomic studies of other dinosaur groups. Dodson (1975) used bivariate plots and principal coordinates analysis to assess the genera and species of lambeosaurine hadrosaurs from Alberta. At the time, there were three accepted genera and 12 named species. Dodson's study showed that most of the variation was attributable to juvenile stages and sexual dimorphism, and in fact a taxonomy with three species in two genera was much more realistic. Interestingly, all of the discriminant characters were confined to the cranial crest, which was evidently for purposes of sexual display. Dodson (1976) also used principal coordinates analysis and bivariate plots to establish the presence of sexual dimorphism in *Protoceratops.*

Chapman et al. (1981) analyzed pachycephalosaurid crania of the genus *Stegoceras* for variability by use of principal components analysis (PCA) and bivariate plots. Two phenons were discovered and interpreted as elements of sexual dimorphism rather than as separate species because of the overall similarity between them. PCA is a method of evaluating correlation between variables that are typically measurements taken from skulls or skeletal elements. Weishampel and Chapman (1990) used PCA to investigate the taxonomy of the genus *Plateosaurus,* distinguishing two morphs on the basis of femoral dimensions. Dodson (1976) employed a landmark-based method of shape analysis called resistant-fit theta-rho analysis to assess ceratopsid taxonomy from a morphometric standpoint. Forster (1996) examined the taxonomic structure of the genus *Triceratops,* using PCA to test her cladistic results. Smith (1998) used PCA to analyze variation in *Allosaurus* specimens from the Morrison Formation and concluded that there was a single, sexually dimorphic species (*Allosaurus fragilis*).

It is significant that in many studies, including PCA of Chapman et al. (1981) and Forster's (1996) analysis, multiple taxa overlap (i.e., they do not separate along any axis). The utility of any biometric study is dependent on the quality of the specimens and the appropriateness of characters (measurements) selected, or on the characters available when using incomplete specimens. The fact that distinct taxa sometimes cluster together in part reflects the similarities between them. However, the inability to discriminate taxa in some cases can be related to variance among the specimens and reflects a fundamental limitation of morphometrics. Dodson pointed out that the majority of bivariate allometric plots (43 of 48) in his 1975 study did nothing to help discriminate among taxa. Thus, regardless of the methods used in a study, significant interpretation is usually necessary to glean meaningful information from it. As suggested by Chapman (1990), more than one morphometric methodology should be used, if possible.

Because ankylosaurids are relatively scarce, at least in North America, and because many specimens are moderately crushed, they are perhaps not an ideal group for a morphometric analysis. Nor do they show any obvious cranial displays that might be sexually dimorphic and that are present in some other dinosaur groups (e.g., hadrosaurs and ceratopsians). A cladistic analysis with discrete character data would be desirable. However, because of extensive dermal sculpturing, sutural contacts on the skull roof are never visible in adult ankylosaurs (Coombs and Maryańska 1990). Furthermore, about half of the North American skulls are so poorly preserved on the palatal side that a reliable character analysis is probably not possible. In other skulls, many ventral sutures are obliterated by fusion. Hence, a morphometric analysis is appropriate.

Coombs and Maryańska (1990) note that more skulls have been referred to *Euoplocephalus* than to any other ankylosaur. This fact

Figure 13.3. Measurements used in the morphometric analysis. (A) Basal length. (B) Occipital condyle diameter. (C) Nuchal crest width. (D) Squamosal horn "height" = (x + y)/2. (E) Width at dorsal edge of orbits. (F) Width of muzzle. (G) Width of beak caudally. (H) Beak "height." (I) Orbit to naris distance. (J) Orbit to back of skull. (K) Orbit to tip of beak. (L) Length of quadratojugal horn base. (M) Paroccipital span. (N) Minimum distance between tooth rows. (O) Width of skull postorbitally. Skull is AMNH 5405.

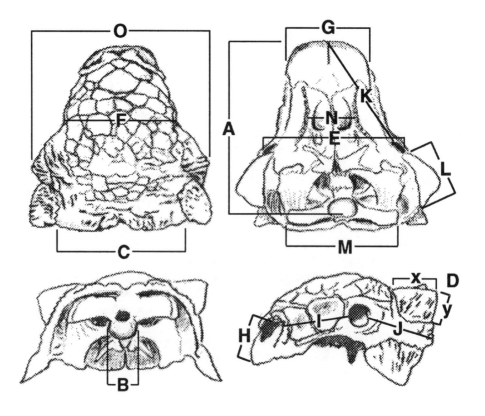

alone warrants a careful, systematic look at the specimens. Because of the poor condition of the type skull of *Euoplocephalus tutus*, it could not be included in PCA. Therefore, a morphometric analysis cannot show which skulls actually pertain to *Euoplocephalus tutus* but can only serve to elucidate potential groupings. Discrete characters are needed to diagnose any taxa. One of the goals of this study is to use morphometrics to draw attention to possible qualitative differences among the skulls.

Materials and Methods

Various skull measurements (see Fig. 13.3) were taken on all well-preserved skulls (excluding NMC 0210, ROM 784, ROM 832, and AMNH 5223, which are poorly preserved, and NMC 8876, which is not completely prepared). Measurements were taken with calipers to the nearest millimeter. Vertical measurements were avoided because many of the skulls are dorsoventrally crushed. There is a risk of obtaining groupings on the basis of the degree of crushing, but as explained below, this eventuality was avoided by careful choice of measurements. Where a skull was well enough preserved to take bilateral measurements, the left and right values were averaged. Measurements are given in Table 13.2.

The data were analyzed by NTSYS (Numerical Taxonomy System; Rohlf 1990). Because of the poor condition of several skulls, there is considerable missing data in the matrix. NTSYS accounts for missing elements by calculating correlations only on the basis of the variables that are present for any two specimens. For PCA, the values were log transformed and then standardized and a correlation matrix was calculated, and from this, the principal components (PCs) were computed. Log transformation minimizes the effects of absolute skull size on the analysis and provides a good quantitative characterization of shape (Rohlf 1990). The relative importance of each principal component is indicated by the size of its eigenvalue, which is proportional to the percentage of the total variation captured by that principal component (see Table 13.3). Chapman et al. (1981) provides a more detailed explanation of the use of PCA in dinosaur studies.

Results

Variable loadings on the first three principal components are given in Table 13.3. Loadings represent the correlation between each variable and each principal component axis. PC I accounted for 61.8% of the total variance, PC II for 14%, and PC III for 11.2%, for a total of 87% across the first three principal components. The high loadings on PC I for most variables indicate that this axis contains mostly variation due to absolute skull size. However, one variable, D, has a negative value on PC I, showing that PC I includes some shape information as well. The negative value for variable D indicates that there is a general decrease in squamosal horn height with increasing skull size, a relationship noted by Coombs (1971).

Three size classes are evident in Fig. 13.4A: an isolated small skull (T, Fig. 13.4A), a group of small- to moderate-size skulls (A, B, N, U),

TABLE 13.2.

Skull measurements (variables) used in the morphometric analysis, in millimeters, for each specimen.

	USNM 11892 (U)	AMNH 5238 (2)	AMNH 5337 (7)	AMNH 5404 (4)	AMNH 5405 (5)	AMNH 5403 (3)	NMC 8530 (N)	TMP 911271 (T)	MOR 433 (M)	UA 31 (A)	ROM 1930 (R)	BMNH R4947 (B)
A. Basal length	345e	360e	386	402	377	390e	350e	320	400e	363	371	362
B. OC diameter	54	76	80	67	65	80	56	56	61	999	69	61
C. Nuchal width	175	209	250	263	256	264	219	172	237	160	243	220
D. SQ horn height	102	56	54	64	66	60	65	74	88	75	80	63
E. Width at orbits	299	334	337	364	334	350	308	289	355	310	351	280e
F. Width at muzzle	215e	265	240	276	238	250	239e	235	285e	263	270	250
G. Width at beak	175	180e	186	207	193	210e	200	163	999	181	205e	177
H. Beak "height"	999	999	106	95	116	103	999	63	999	76	96	85
I. Orbit–naris	147	150	152	167	146	147	141	148	165	145	164	152
J. Orbit–back of skull	114	112	115	118	122	117	115	98	120	115	113	103
K. Orbit–tip of beak	999	999	262	287	270	270e	220	227	999	242	275	254
L. QJ horn base	110	123	120	999	118	126	117	117	128	129	125	129
M. Paroccipital span	210	999	266	268	253	281	212	205e	999	999	244	220e
N. Tooth row spacing	999	999	113	130	109	130e	115e	99	999	112	121	999
O. Postorbital width	332	346	365	395	368	363	325	320	372e	328	388	313

NOTE. A value of 999 indicates missing data due to poor preservation. A suffix of "e" means an estimate. OC = occipital condyle; QJ = quadratojugal; SQ = squamosal.

TABLE 13.3.
Variable (VAR) loadings and variance on first four principal components
(see Appendix 13.1).

	Component			
	1	2	3	4
VAR A	0.944	0.055	−0.068	−0.071
VAR B	0.748	−0.230	0.506	0.159
VAR C	0.859	−0.253	0.044	0.146
VAR D	−0.373	0.417	−0.771	0.004
VAR E	0.923	0.051	−0.221	−0.028
VAR F	0.639	0.704	0.177	−0.112
VAR G	0.836	−0.109	−0.175	−0.378
VAR H	0.824	−0.608	−0.086	0.093
VAR I	0.588	0.619	−0.239	0.407
VAR J	0.698	−0.268	−0.452	−0.361
VAR K	0.889	0.075	−0.060	0.373
VAR L	0.467	0.640	0.538	−0.209
VAR M	0.966	−0.154	0.213	0.073
VAR N	0.863	0.182	0.015	−0.382
VAR O	0.882	0.043	−0.332	0.224
Eigenvalue	9.263	2.100	1.685	0.895
% Variance	61.8	14.0	11.2	6.0
Cumulative %	61.8	75.8	87.0	93.0

NOTE. Variables are as follows: A = basal length; B = occipital condyle diameter; C = nuchal crest width; D = squamosal horn "height" = $(x + y)/2$; E = width at dorsal edge of orbits; F = width of muzzle; G = width of beak caudally; H = beak "height." I = orbit to naris distance; J = orbit to back of skull; K = orbit to tip of beak; L = length of quadratojugal horn base; M = paroccipital span; N = minimum distance between tooth rows; O = width of skull postorbitally.

and one of larger skulls. The smallest skull is about four-fifths the length of the largest (see Fig. 13.2), so no young juveniles are present. The skulls arrange roughly in order of increasing basal length along PC I, verifying that PC I is primarily a size axis. However, two skulls are out of place along PC I as compared with where they would plot if PC I were strictly a size axis: AMNH 5238 and MOR 433 (2, M, Fig. 13.4A). MOR 433, despite being one of the two largest skulls on the basis of all measurements, is situated away from the right end of the plot (maximum size) toward the left side of the large skull cluster, confirming that PC I contains some shape information. AMNH 5238 (2, Fig. 13.4) is

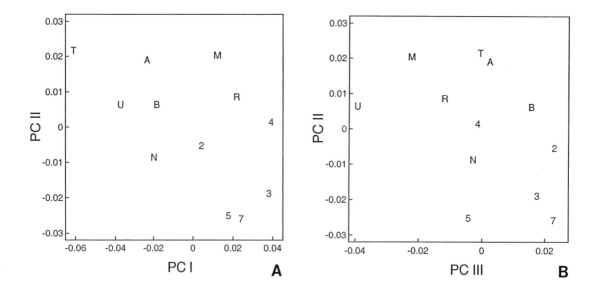

A   B

Figure 13.4. Orthogonal principal component biplots. (A) PC II versus PC I; (B) PC II versus PC III. 2 = AMNH 5238; 3 = AMNH 5403; 4 = AMNH 5404; 5 = AMNH 5405; 7 = AMNH 5337; A = UA 31; B = BMNH R4947; M = MOR 433; N = NMC 8530; R = ROM 1930; T = TMP 91.127.1; U = USNM 11892.

moderate in size (basal length 360 mm) but plots with the larger skulls along PC I. The anomalous positions of MOR 433 and AMNH 5238 may be due in part to missing data. However, when PCA was rerun without variables H, K, M, and N (the ones having the greatest number of missing values; see Table 13.2), the positions of MOR 433 and AMNH 5238 along PC I did not change significantly, although AMNH 5405 moved to the left of MOR 433. In Figure 13.4B, AMNH 5238 clusters with AMNH 5337 and 5403. These three skulls all have relatively large occipital condyles and very small squamosal horns (see Fig. 13.5).

PC II and PC III each account for roughly the same amount of variation (14% and 11.2%, respectively). Neither of these values is particularly high, but separation among specimens along these axes highlights some subtle differences among the skulls. PC II contains both positive and negative values. Variables with the highest positive loadings are quadratojugal horn width, squamosal horn height, muzzle width, and orbit–naris distance (i.e., muzzle length). Two of these variables involve measurements of the snout. Three of six negative variables are associated with breadth of the skull caudally, although beak height has the highest negative loading at −0.61. PC II separates specimens having wider or longer snouts from those with more constricted snouts, broader occipital regions, and narrower quadratojugal horns.

PC III also contains contrasting positive and negative values. Variables B and L (condyle diameter and width of quadratojugal horn) have the highest positive loadings. Variable D (squamosal horn height) has a particularly high negative loading at −0.77. Variable J also has a high negative loading, and it is surprising that there is so little correlation between variables J (distance from orbit to back of skull) and L (breadth of quadratojugal horn). Evidently these structures were unrelated. Specimens to the right along PC III tend to have larger occipital condyles and smaller squamosal horns. Although there is a strong negative cor-

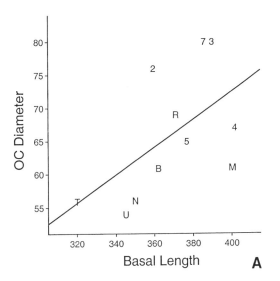

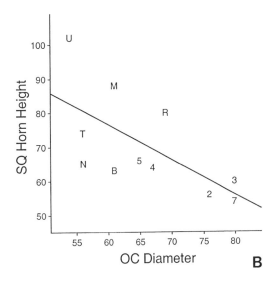

relation between quadratojugal horn width and squamosal horn height, position along PC III is only weakly related to a specimen's quadratojugal horn width.

Figure 13.4B shows a plot of PC II versus PC III (orthogonal to the plot in Fig. 13.4A) and may provide the best summary of shape similarities and disparities among the specimens. Several interesting associations are visible in Figure 13.4B. AMNH 5337 and AMNH 5403 (7, 3, Fig. 13.4B) form a tight pair and may be part of a larger cluster that includes AMNH 5238 and 5405 (2, 5, Fig. 13.4B). AMNH 5404, ROM 1930, and NMC 8530 (4, R, N, Fig. 13.4B) cluster together. MOR 433 (M, Fig. 13.4B) falls at the edge of this group. BMNH R4947, TMP 91.127.1, and UA 31 (B, T, A, Fig. 13.4B) all fall outside (although nearby) the above group. USNM 11892 (U, Fig. 13.4B) clusters with the latter three specimens in Figure 13.4A because of its small size, but in Figure 13.4B has MOR 433 as its nearest neighbor. As discussed below, these two specimens probably belong to a separate taxon.

The geometry of the clusters agrees in most ways with the qualitative characters observed in each of the skulls. AMNH 5337 and AMNH 5403, two of the largest skulls, are virtually identical in all details, including proportions, texture, ornamentation, nasal shape, and nuchal morphology. The fact that these two skulls cluster together despite severe crushing in the latter specimen would appear to validate the choice of measurements (e.g., nonvertical) used in the analysis. AMNH 5337 and AMNH 5403 are probably the same sex and species. The fact that they do not plot exactly the same in three-dimensional morphospace reveals the approximate range of individual variation. It may also indicate the potential margin of error due to crushing.

UA 31 and TMP 91.127.1 are likewise similar and undoubtedly belong to the same species. They share numerous characters, including a relatively smooth texture; tall, sharply keeled squamosal horns; a

Figure 13.5. Bivariate plots of selected skull measurements to show the relationship between occipital condyle diameter and other variables. (A) Condyle diameter versus basal skull length. (B) Squamosal horn height versus condyle diameter. Measurements are in millimeters. 2 = AMNH 5238; 3 = AMNH 5403; 4 = AMNII 5404; 5 = AMNH 5405; 7 = AMNH 5337; A = UA 31; B = BMNH R4947; M = MOR 433; N = NMC 8530; R = ROM 1930; T = TMP 91.127.1; U = USNM 11892.

narrow nuchal crest; distinct preorbital and postorbital hornlets (small bosses above the orbit that are confluent on most other specimens); broad, centrally pointed quadratojugal horns; low, rectangular nares; and a broad snout that is trapezoidal in dorsal view. Indeed, there are few notable differences between the two skulls. The squamosal horns of UA 31 are more erect than those of TMP 91.127.1, which might be a feature of preservation, although neither skull is badly crushed. The nasal plate of TMP 91.127.1 is much larger than that of UA 31, and the surrounding plates consequently have different shapes. Yet the similarities between these two skulls far outweigh any differences, so evidently the details of nasal plate arrangement are not taxonomically significant. It is noteworthy that in only two specimens do the rostrolateral plates on the side of the skull rise to contact the large nasal plate: TMP 91.127.1 and NMC 0210 (holotype of *E. tutus*). Further study of the plate arrangement may be worthwhile.

UA 31 is about 10% larger than TMP 91.127.1 and so is presumably closer to maturity, or perhaps it is mature. AMNH 5238 has a basal length about equal to that of UA 31, but it is larger in most other measurements although the squamosal and quadratojugal horns are smaller. AMNH 5238 has a very smooth surface and well-developed osteoderms on its nuchal crest and is probably immature, but it separates from UA 31 morphometrically. These two specimens may represent distinct taxa.

NMC 8530 is a puzzling skull. It is badly crushed, but it appears to share several characters with UA 31, TMP 91.127.1, and BMNH R4947, characters that are not present in AMNH 5404 or ROM 1930. Yet in the graph of PC II versus PC III (Fig. 13.4B), NMC 8530 clusters with AMNH 5404 and ROM 1930 as well as BMNH R4947. There are two possible reasons for this. First, all the specimens in the study (except perhaps USNM 11892 and MOR 433) may belong to *Euoplocephalus tutus*. If this is the case, then the characters listed above that unite UA 31 and TMP 91.127.1 must be juvenile characters because neither AMNH 5238 nor the larger skulls show these characters. Alternatively, if BMNH R4947, NMC 8530, TMP 91.127.1, and UA 31 are *Euoplocephalus tutus* and the others are not, then the variance must be such that PCA cannot separate the skulls into distinct clusters on all axes. The probability of this being the case increases with the number of taxa present. Perhaps the subtle differences between BMNH R4947 and NMC 8530 on the one hand, and TMP 91.127.1 and UA 31 on the other, represent sexual dimorphism.

Unfortunately, both skulls from the Two Medicine Formation, MOR 433 and USNM 11892, are badly weathered. Nevertheless, PCA provides evidence that the superficial similarity between them is real. In Figure 13.4B, USNM 11892 is an outlier, yet MOR 433 is its nearest neighbor. These two skulls agree in many characters, and the differences that exist may be due to poor preservation. For example, the squamosal horns of MOR 433 are more erect than those of USNM, but MOR 433 is more severely crushed, and crushing could have affected horn orientation. MOR 433 has several unusual features and, along with USNM 11892, probably represents a separate taxon.

Variation in Postcranial Skeleton

Axial Skeleton

There are undoubtedly many subtle taxonomic differences in the vertebrae, but the fragmentary nature of most specimens makes elucidation impractical. Much of the variation—for example, in shape of the neural canal, angle of transverse processes, and orientation of prezygapophyses and postzygapophyses—can be attributed to crushing. Nevertheless, differences are evident in the pelvis and caudal vertebrae, including the tail club.

Pelvis. Few good ankylosaurid pelves are preserved. Fortunately, the holotypes of *Dyoplosaurus acutosquameus* and *Scolosaurus cutleri* both have partial pelves (see Fig. 13.6). In addition, AMNH 5337 has

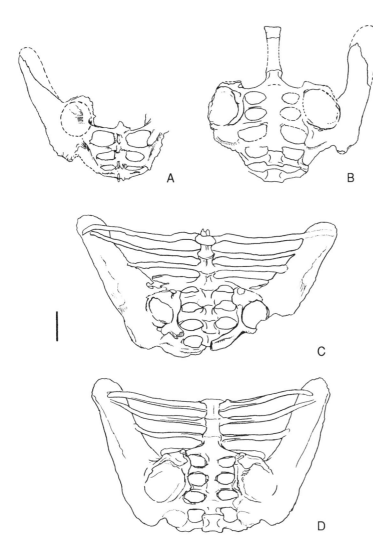

Figure 13.6. Pelves of (A) ROM 784, (B) BMNH R5161 (after Nopsca 1928), (C) AMNH 5409 (after Coombs 1978), and (D) AMNH 5337. Scale bar = 20 cm.

a well-preserved pelvis, and there are at least two good isolated pelves: AMNH 5409 and TMP 82.9.3. Ankylosaur pelves are large, complex, three-dimensional objects, and it is not always possible to observe them from the desired perspective. The most apparent differences among them are synsacrum vertebral count, acetabulum shape, and angle of divergence between the ilia.

In *Dyoplosaurus,* only three poorly preserved sacral vertebrae are present. A free caudal (sacrocaudal, according to Parks 1924) vertebra posterior to these has its transverse processes fused to the ilia. The dashed line in Figure 13.6A shows the approximate boundary of the large acetabulum on the underside of the ilium. It is unclear whether the anterior iliac blade was pulled outward after death, although the skin clearly was. In his reconstruction of the pelvis of ROM 784, Parks (1924) straightened the ilia, making them less divergent, but offered no rationale for this. There is some distortion of the left ilium, which makes it difficult to ascertain the true angle.

The pelvis of *Scolosaurus* is similar to that of *Dyoplosaurus,* but the ilia do not appear to be as widely divergent. Nopsca (1928) described a presacral rod of four "lumbaroid" (dorsal) vertebrae, which, along with five sacrals, makes a total synsacrum count of nine. There is a similar presacral rod of four fused vertebrae preserved with ROM 1930. As in *Dyoplosaurus,* the first caudal of BMNH R5161 has its transverse processes fused to the ilia (see Fig. 13.6B).

AMNH 5337 has a virtually complete and finely preserved pelvis (Fig. 13.6D). There are eight vertebrae in the synsacrum, which measures 85 cm long. The acetabulae are large, and the preacetabular processes of the ilia are not strongly divergent, conditions consistent with those seen in *Scolosaurus.* However, in AMNH 5337, the postacetabular processes are relatively short compared with those of *Scolosaurus* and *Dyoplosaurus.* The length appears to be correlated with the number of vertebrae in the synsacrum, which has one fewer than in *Scolosaurus.*

AMNH 5409 (Fig. 13.6C) is an almost complete and very well preserved isolated pelvis. The synsacrum has seven vertebrae, but an eighth may have been present. The length without an eighth vertebra (one has been added in plaster) is about 73 cm. The acetabulae are relatively smaller than in any of the above specimens, and the ilia are more divergent. TMP 82.9.3, from the Horseshoe Canyon Formation, also has a synsacrum of seven vertebrae, and its ilia are strongly divergent cranially. The synsacrum of TMP 82.9.3 measures 62 cm in length, so this pelvis is slightly smaller than the others. The fact that the smallest pelvis has the fewest vertebrae may be evidence of an ontogenetic increase in the number of vertebrae in the synsacrum.

The significance of the angle of iliac divergence is unclear. There is a tendency to anthropomorphize and wonder whether the difference might be attributable to the animal's sex. There are not enough preserved pelves to test this hypothesis. Some of the variation might be individual, but some is probably taxonomic.

Tail Clubs. There is considerable variation among tail clubs. The largest is around 50 times as massive as the smallest, yet the smallest is almost 20 cm long (see Coombs 1995 for a comprehensive survey of tail clubs). Coombs concluded that a display function for the tail club was improbable and that most of the variation is individual, ontogenetic, or specific, not sexual. He also classified the clubs into three categories on the basis of shape, which could denote separate taxa. Unfortunately, only three Campanian North American tail clubs have any significant associated material: AMNH 5245, AMNH 5405, and ROM 784.

Coombs (1995) noticed differences in the free caudal vertebrae among specimens referred to *Euoplocephalus*. In ROM 784, the zygapophyseal surfaces of the proximal free caudals are more horizontal than in AMNH 5404, a condition that would allow greater horizontal flexibility. The tail club of AMNH 5404 is not preserved, but that of ROM 784 is unusually small. Perhaps the highly inclined zygapophyseal surfaces in AMNH 5404 prevented excessive lateral movement of the tail that, with a larger mass at the tip, might have resulted in injury were the tail swung too strongly from side to side. This hypothesis could be verified by examination of free caudals associated with large clubs. Unfortunately, few are preserved. One of the largest known tail clubs, ZPAL MgD-I/43 (Maryańska 1969), has the second to last free caudal preserved. The prezygapophyses and postzygapophyses are sticklike, horizontal, almost parallel protrusions, but the articular surfaces are strongly inclined. This would have severely limited lateral movement of the middle part of the tail, which is to be expected near the base of the tail club. The condition in ROM 784 is slightly different; here the second to last free caudal vertebra has divergent prezygapophyses and reduced, nubbinlike postzygapophyses.

MOR 433 has at least three caudal vertebrae preserved. One of these has a fused chevron and is probably the fourth or fifth caudal, judging by the presence of the chevron and the length of the transverse processes. The zygapophyseal surfaces are strongly inclined, making an acute angle between them of about 60° or 70°. This contrasts again with the condition in *Dyoplosaurus*, where the angle between the zygapophyseal surfaces reaches almost 180° by the third caudal. The small tail club in ROM 784 is probably a taxonomic, not a juvenile, feature.

Appendicular Skeleton

The two most remarkable differences are found in the forelimb and pes.

Forelimb. Differences are present in the scapulocoracoid and humerus. Most of the preserved humeri are similar, with the exception of AMNH 5406 and MOR 433 (see Fig. 13.7). The humerus of AMNH 5406 is smaller than in other specimens, and the deltopectoral crest does not appear to extend as far distally along the shaft. The coracoid of AMNH 5406 is missing, but its sutural surface on the scapula is undamaged, so clearly these bones had not yet fused. This lack of fusion may be evidence that AMNH 5406 was immature, although Carpenter (1990) considered fusion of scapula and coracoid to be of taxonomic value for nodosaurs.

The forelimb of MOR 433 appears to be ankylosaurid in nature, but the humerus is unusual. In absolute terms, it is larger than in any

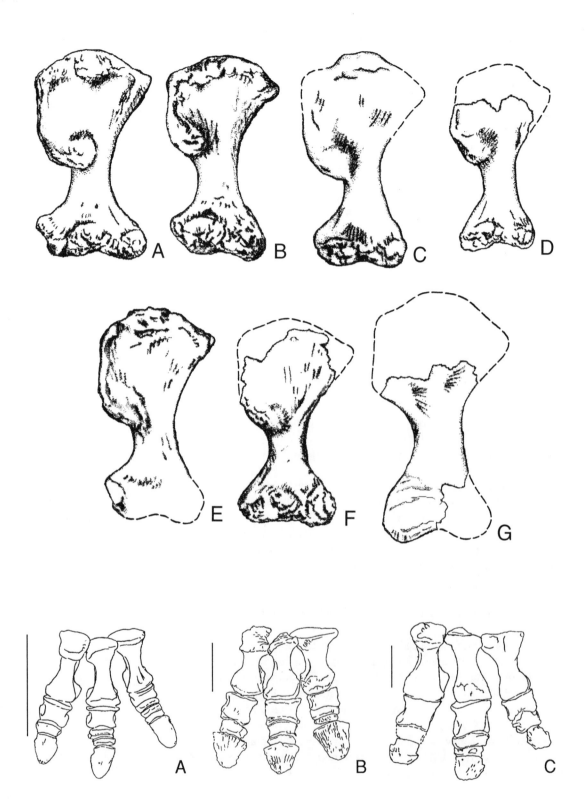

other contemporaneous specimen from North America. The deltopectoral crest is not rotated cranially around the shaft, as it is in nodosaurids. Yet the crest, although pronounced, does not appear to extend as far distally as in other ankylosaurids, so unless it was very short, the humerus must have been relatively longer than in other ankylosaurids. Either way, it is significantly different.

Pes. There are evidently at least two distinct types of pes. Coombs (1986) noted differences in the unguals of AMNH 5266 and ROM 784. In ROM 784, the unguals are more triangular and are widest at their proximal end (see Fig. 13.8B). In AMNH 5266 (Fig. 13.8A), the unguals are widest about one third of the way from their proximal end and are narrower overall. Coombs assumed that the difference was due to an ontogenetic change in ungual shape, following Galton's (1982) conclusions regarding *Stegosaurus*. However, this would not explain the condition seen in the very large pes of ROM 813 (Fig. 13.8C), which has unguals like those of AMNH 5266 (although relatively more robust). In ceratopsians, ungual shape has proved to be a useful taxonomic character (Penkalski and Dodson 1999). The pes of AMNH 5266 has a phalangeal formula of 0-3-4-5-0, whereas the pes of ROM 784 (as preserved) has a formula of 0-3-4-4-0. Although it is always possible that one or more unguals are missing, both phalangeal count (Coombs and Maryańska 1990) and ungual shape may be taxonomically useful.

Variation in Armor

Armor has proved useful in the systematics of glyptodonts (Gillette and Ray 1981) and nodosaurids (Carpenter 1990). Long and Ballew (1985) classify aetosaurs on the basis of scute morphology and overall armor pattern, but they did not discuss the possibility of sexual dimorphism. Heckert and Lucas (1999) diagnose aetosaurs largely on the basis of their armor. Brazaitis (1973) successfully use armor for identification of extant crocodilians even though the differences among genera are subtle. Ross and Mayer (1983) report in a comprehensive study of crocodilians that there is little ontogenetic change in the armor pattern of an individual.

Ankylosaurs had an enormous physiologic potential for producing dermal armor. The skin itself was practically ossified, containing tens of thousands of tiny (3–8 mm) ossicles. In addition, there were numerous bony, keeled scutes covering the neck, back, and tail. Even the cheeks and eyelids of some species contained osteoderms.

In all ankylosaurs, there is a range of scute shapes, with considerable overlap between species. Hence, it may not be possible to assign an isolated scute (or even several scutes) to a particular species. Because most specimens are fragmentary, the temptation to use negative evidence must be resisted. For example, large plates are not present with some specimens, but this does not mean they did not have any. For any given specimen, all scutes tend to have a similar texture (e.g., smooth or rugose). Despite this consistency, it is still unclear whether or not texture is a useful taxonomic character. The presence of identical scutes

Figure 13.7. (opposite page top) Humeri in ventral view. (A) AMNH 5337. (B) AMNH 5404. (C) AMNH 5405. (D) AMNH 5406. (E) BMNH R5161 (after Nopsca 1928). (F) ROM 1930. (G) MOR 433. Scale bar = 10 cm.

Figure 13.8. (opposite page bottom) Pes of (A) AMNH 5266 (after Coombs 1986), (B) ROM 784, and (C) ROM 813. Scale bars = 10 cm.

(shape, texture, and sculpturing) in two specimens may be evidence of conspecificity. If intraspecific variability in the armor can be quantified, armor has the potential to be highly useful for ankylosaur taxonomy.

Cervical Armor. An important characteristic uniting all ankylosaurids is the presence of two cervical armor bands protecting the dorsal surface of the neck (in nodosaurids, there are usually three). These bands, or half-rings, consist of pairs of scutes fused to an underlying band of bone. This bone is presumed to be dermal in origin because it does not articulate with any part of the skeleton. The bony layer appears to originate from discrete elements, one beneath each scute. The scutes are separate from one another, and bone is normally visible in between. The bone may show the impressions of the former polygonal scale epidermis. The spacing between scutes is variable; in some specimens there is little separation between scutes, whereas in others the scutes are widely spaced and the sutures between half-ring segments are visible. In contrast, the scutes in nodosaurids are usually in contact with one another, and the half-rings may consist of unfused sections. Except for MOR 433, no adult ankylosaurid specimen shows unfused sutures between half-ring segments, although baby *Pinacosaurus* from Asia do. Evidently the half-ring segments fused during ontogeny, at least in some taxa. In specimens with two well-preserved half-rings (AMNH 5337, AMNH 5403, BMNH R5161, *Pinacosaurus, Saichania,* and *Shamosaurus*) the second is always larger by a ratio of about 3:2.

First Half-Ring. There appear to be two general types of first half-ring. One consists of six smooth, closely set, suboval, sharply keeled scutes, all similar in size and shape. This morph is seen in NMC 0210 (Fig. 13.1C, D), AMNH 5406 (Fig. 13.9E), probably in MOR 433 (Fig. 13.9L), and possibly in NMC 8530 (Fig. 13.9K). MOR 433 differs from the others in that the scutes are packed more closely and have lower keels with depressions on either side of the keel. Thus, half-rings with six scutes come in two different varieties. Six scutes are also present in the first half-ring of *Shamosaurus,* the most primitive known ankylosaurid (Coombs and Maryańska 1990; Sereno 1997), in *Saichania* (see Maryańska 1977), and may be present in *Tianchisaurus* (Dong 1993), an early ankylosaur from Asia. The three pairs of scutes are hereafter termed medial, lateral, and distal. Six close-set scutes in the first half-ring probably represent the plesiomorphic condition for ankylosaurids.

The second general type consists of two or four scutes, more widely spaced and variably rugose, with the center pair roughly oval to rectangular and weakly keeled or having a conelike prominence. The lateral pair, when present, are more strongly keeled. This morph is seen in AMNH 5337 (Fig. 13.9A–C), AMNH 5403 (Fig. 13.9D), AMNH 5404, and AMNH 5405 (Fig. 13.9F), and USNM 7943 (Fig. 13.9I, J). In BMNH R5161, the medial scutes are lacking, but the lateral scutes are present and are fused to the half-ring. In AMNH 5337, AMNH 5403, AMNH 5405, and BMNH R5161, the scutes are incredibly pitted and rugose. In AMNH 5337, the distal extremities of the half-ring are vascular and perforate, and sutures are visible about 8 cm from

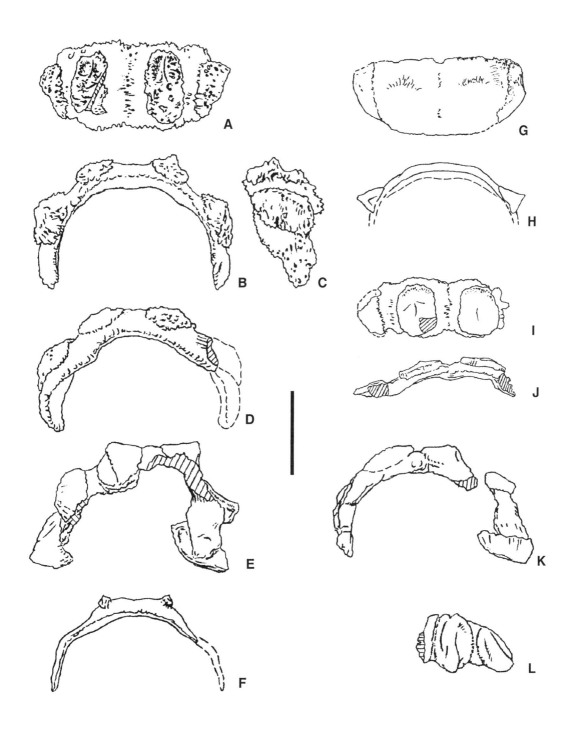

Figure 13.9. *First cervical half-rings of AMNH 5337 in (A) dorsal, (B) ?cranial, and (C) lateral views; AMNH 5403 in (D) ?caudal view; AMNH 5406 in (E) ?caudal view; (F) AMNH 5405; (G, H) BMNH R5161 in dorsal and cranial views (cranial modified from Nopsca 1928); (I, J) partial half-ring of USNM 7943 in dorsal and ?cranial views; (K) poorly preserved NMC 8530 (fragment with scute is NMC ?8531); (L) distal part of MOR 433 in dorsal view. Scale bar = 20 cm.*

either end. It is unclear whether the distal scutes no longer developed or are simply missing. In AMNH 5404 and AMNH 5405, the medial scutes are tiny (see Fig. 13.9F), and the lateral scutes are missing. Perhaps the two varieties (large medial scutes and tiny or missing medial scutes) represent a sexual difference. Some half-rings are flatter—that is, less strongly arched than others (cf. Fig. 13.9B, H, J). This is probably due simply to crushing, although taxonomic differences cannot be ruled out.

Figure 13.10. Second cervical half-rings. (A) MOR 433 fragments in dorsal view. (B) BMNH R5161 in dorsal view. (C, D) AMNH 5337 in dorsal and caudal views. Scale bar = 20 cm.

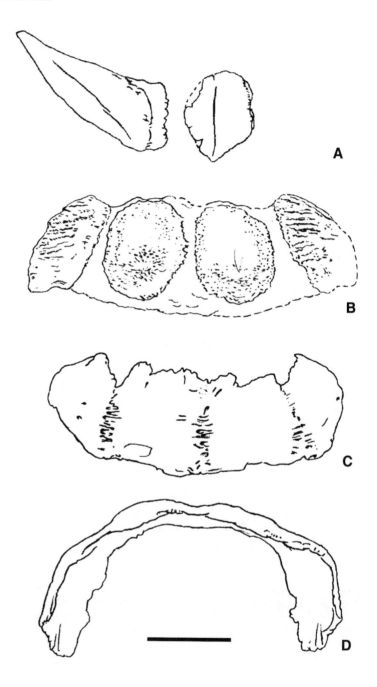

Second Half-Ring. Several specimens have material from the second half-ring preserved (Fig. 13.10). Unfortunately, the second half-ring is missing from NMC 0210, but AMNH 5406 has a fragment preserved. The piece is one distal extremity with no scute attached. AMNH 5406 has a long (30 cm), high-pitched scute (Fig. 13.11B) that appears to fit against the preserved end of the second half-ring. Evidently, in some cases, scutes are present but do not fuse to the underlying bone. In AMNH 5337, from a mature adult, no scutes are fused to the second half-ring (Fig. 13.10C, D). However, several loose scutes are preserved that might have occupied the second half-ring. In BMNH R5161, four scutes are present and all appear to be fused to the bony half-ring layer. The medial two are flat, oval shaped, and have a conelike prominence in lieu of a keel (see Fig. 13.10B). The lateral scutes are elliptical and strongly keeled, with parallel ridges or "ribbing" on the cranial half.

Dorsal Armor. In all ankylosaurids, the dorsal armor consists of scutes arranged in transverse rows, parasagittal rows, or both. Scutes typically do not touch one another, and the area in between is covered with smaller, polygonal scales or ossicles. Some specimens have smooth, finely vascular scutes, whereas others have extremely rugose, even perforate scutes (cf. Fig. 13.11B and Fig. 13.11C–F). The texture is generally consistent within an individual and could be ontogenetic in nature. However, some textural differences may be specific because other differences, including scute shape and half-ring morphology, are correlated.

Some scutes have an unusual double-edge, a sort of escarpment around the perimeter. It is most pronounced on the medium to large oval scutes (visible in Fig. 13.11B). The nature of this morphology is unknown, but it is present in several ankylosaurid specimens, including AMNH 5406, MOR 433, NMC 0210, NMC 8530, TMP 82.9.3, and UA 31, as well as in *Gargoyleosaurus,* a basal ankylosaurid from the Jurassic of Wyoming (Carpenter et al. 1998). A similar morphology is found in nodosaur cervical spines, and the general assumption has been that the lower edge anchored the spikes in the skin. This makes sense for a long spike having much of its mass situated distally, but it does not follow for a small, relatively flat scute. Moreover, it does not explain the total lack of a double edge in the majority of scutes. Consequently, the double-edge described here may be of taxonomic importance.

Only three Campanian specimens have large (>25 cm), platelike scutes preserved: ROM 813, AMNH 5337, and AMNH 5403. As noted above, this does not necessarily mean that other specimens did not have large plates. The latest Cretaceous *Ankylosaurus* has numerous large plates. Nevertheless, it can be stated with reasonable certainty that BMNH R5161 (holotype of *Scolosaurus*) lacked them because most of its armor is preserved.

The preserved scutes of AMNH 5337 are all ovalform and incredibly rugose, even perforate. Two are large and sharply keeled. Another is more conical, with a circular hole where the tip of the cone would be (see Fig. 13.11C). The largest plate (32 cm long) has a ribbed structure near its cranial end and resembles the lateral scutes on the second half-

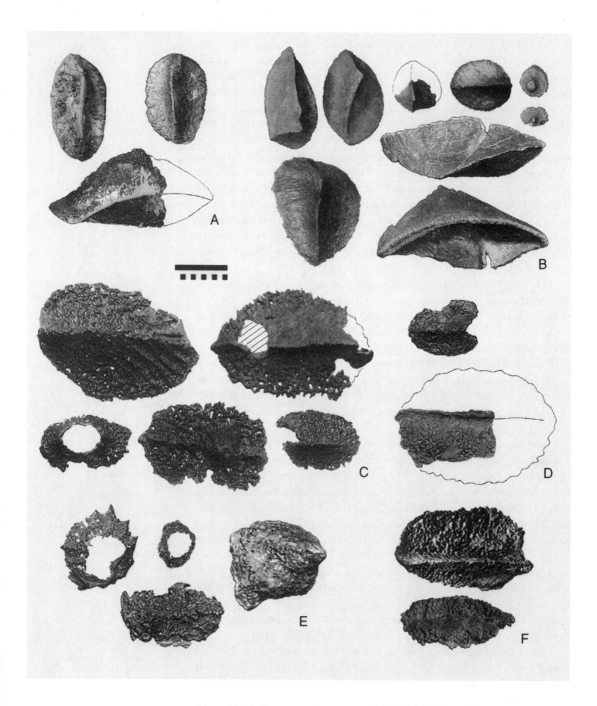

Figure 13.11. Representative armor of (A) NMC 8530, (B) AMNH 5406, (C) AMNH 5337, (D) AMNH 5403, (E) ROM 1930, and (F) ROM 813. The largest scute in B is shown in two views. The largest scute in E is conical, with the cone pointing out of the page and toward the right. Scale bar = 10 cm.

ring of BMNH R5161. A similar morphology is seen in the Asian ankylosaurid *Talarurus*, in ROM 1930, in BMNH R5161, and probably in AMNH 5403 and AMNH 5404. In fact, most specimens have some scutes that show an incipient corrugation or ribbing that is apparently unrelated to vascularization, but the phenomenon is far more prominent in the above specimens. ROM 1930 also has some scutes that are open dorsally (see Fig. 13.11E). The scutes are undamaged, so the openings appear to be natural, although they could be the result of a pathology.

Discussion

Sexual dimorphism and ontogenetic stages are potential complicating factors in any systematic study of fossil vertebrates. Chapman et al. (1997) discuss sexual dimorphism in various dinosaur groups and note that in thyreophorans, because of their relatively small sample size, the presence of sexual dimorphism probably cannot be evaluated statistically. Galton (1982) and Coombs (1986) studied juvenile thyreophorans (stegosaurs and ankylosaurs, respectively) and compiled a list of apparent ontogenetic changes. These include increasing rugosity of bone surfaces, relative thickening of long bones and diapophyses, fusion of dorsal arches and centra, fusion of dorsal ribs to their vertebrae, a relative widening of the ilia (in ankylosaurs), fusion of astragalus and tibia, and expansion of the proximal end of the ungual phalanges. Only one ontogenetic change in armor was listed: the possible lack of dorsal plates in juvenile *Stegosaurus*, but at least one dorsal plate is now known from a juvenile specimen of *Stegosaurus* (Carpenter, personal communication). Several young juvenile *Pinacosaurus* specimens lack tail clubs, and in fact the only secondary dermal ossifications are a few skull elements and the bony layers of the cervical half-rings (Currie 1991). The bony half-rings lack any scutes, bringing into question the developmental origin of these structures. A lack of secondary dermal ossifications in young juveniles was probably the general condition for thyreophorans. In extant crocodilians, the scutes are fibrous pads at birth and ossify early during ontogeny (Brochu, personal communication).

Variability in the Skull

A comparison between large and small skulls revealed some possible ontogenetic differences. In general, the larger the skull, the greater the rugosity of its dorsal surface, including the squamosal horns. In all skulls with a basal length of less than 365 mm, the nuchal crest has four well-defined osteoderms, whereas in all but one of the skulls over 365 mm basal length, the osteoderms of the crest are poorly defined; the lone exception is MOR 433, in which the crest is well defined despite the specimen's large size, but this might be a taxonomic distinction. In the other large skulls, AMNH 5337, 5403, and 5404, the nuchal crest is rugose and pitted, and the four osteoderms are difficult to make out. In AMNH 5403, they are indistinguishable. The nuchal crest also tends to be relatively wider in the larger specimens. However, TMP 91.127.1

and UA 31, which as discussed above are similar, both have extremely narrow nuchal crests, yet the crest is relatively wider in the smaller of these two specimens, TMP 91.127.1. Overall rugosity and perhaps relative width of the nuchal crest are ontogenetic indicators.

Another character varying between the large and small skulls is quadratojugal horn shape. In all of the large skulls (2, 3, 5, 7, M, R, Fig. 13.4), the horn curves caudally, whereas in four of five smaller skulls, it is more symmetrical and has a central point. The only small skull with a caudal curvature is USNM 11892. In *Nodocephalosaurus* (Sullivan 1999), the horn is directed rostroventrally, so the morphology may be of taxonomic significance.

Two other highly variable characters are occipital condyle diameter and squamosal horn height. Three specimens (AMNH 5238, AMNH 5337, and AMNH 5403) have relatively large occipital condyles (see Fig. 13.5A). AMNH 5238 is a moderate-size skull with a smooth surface that lacks the extreme rugosity of the larger skulls and has well-defined osteoderms on its nuchal crest, so it was probably immature. Yet its condyle is relatively as large as those of AMNH 5337 and AMNH 5403. Hence, it is unlikely that relative condyle size is an ontogenetic trait. Further study of juvenile Asian ankylosaurids may yield additional ontogenetic characters.

Could the relatively large occipital condyles of these three specimens be due to sexual dimorphism? Perhaps a larger condyle was associated with intraspecific combat—for example, head butting (see Coombs 1995)—in which case these specimens would be male. This hypothesis is highly speculative. Additional study on the structure of the skull and cervical vertebrae would be desirable. Interestingly, the three specimens with the largest occipital condyles have the smallest squamosal horns (see Fig. 13.5B). Perhaps these characters are related. AMNH 5405, which clusters with AMNH 5238, AMNH 5337, and AMNH 5403 in the morphometric analysis (5, 2, 7, 3, respectively, Fig. 13.4), has a relatively smaller condyle and larger squamosal horns. The morphologies of the first cervical half-rings of these specimens are correlated. AMNH 5337 and AMNH 5403 each have four large scutes on their first half-ring, whereas AMNH 5405 has only the tiny medial pair, although unfused lateral scutes were probably present (in AMNH 5238, the half-ring is not preserved). If the half-ring scutes served a display function, this would seem to be consistent with the interpretation of AMNH 5337 and AMNH 5403 as males and AMNH 5405 as a female. AMNH 5404, too, has a small condyle, moderate squamosal horns, and tiny medial scutes on the first half-ring. ROM 1930 is consistent in having a moderately small condyle and large squamosal horns, but its half-ring is not preserved. If the hypothesis that the specimens with larger condyles are male is correct, then males apparently had smaller squamosal horns than females.

A taxonomic significance to condyle size cannot be ruled out. It is interesting that the specimen with the smallest occipital condyle, USNM 11892, has the largest squamosal horns of all. The relationship between condyle and squamosal horn size does not appear to have an allometric basis. The much larger MOR 433 has the second largest squamosal horns, and along with USNM 11892, these two specimens have, proportionately, the smallest condyles of any of the skulls examined (see Fig. 13.5A). This difference is probably taxonomic.

All of the skulls appear to be more similar to one another than any is to *Nodocephalosaurus* or *Ankylosaurus*. As suggested by Carpenter (1982), cranial convergence may explain this similarity. It might also be the result of conservative evolution within the North American clade. Even MOR 433, by far the most distinctive of the specimens analyzed for this study, has a skull that is similar to the others in overall proportions and ornamentation. At the same time, there is substantial variation in the postcranial armor. Perhaps armor, as proposed by Carpenter (1997), was the primary means of display for ankylosaurids.

Cervical Half-Ring Variability

The principal question concerning the cervical half-rings is to what degree taxonomy, ontogeny, sexual dimorphism, and simple individual variation contributed to the differences in morphology. There appear to be two general classes of first half-ring: six scutes or two/four scutes, and four or five distinct subtypes. The lack of intermediate forms implies that most of the differences are not ontogenetic or individual, although some details evidently varied among individuals—for example, the spacing between the scutes (cf. Fig. 13.9B, D). As suggested above, sexual dimorphism could explain some of the differences.

The question arises as to whether the tiny medial scutes on the half-rings of AMNH 5404 and AMNH 5405 might be a juvenile character. The large size of both specimens argues against this interpretation. Nor can fusion of scutes to the underlying half-ring be an ontogenetic character because the second half-ring of AMNH 5337 has no scutes, even though AMNH 5337, for reasons listed above, is probably an adult. Hence, the absence of scutes on some half-rings cannot be interpreted as a juvenile character. Ross and Roberts (1979), in attempting to quantify variability within populations of the American alligator, noted variation in the number of scutes in the transverse rows of the cervical and dorsal regions. There are not enough specimens to evaluate such variability in ankylosaurids, if it is present at all. Six scutes is the typical number for the first half-ring of Asian ankylosaurids (Maryańska 1977; Tumanova 1987). However, in North American specimens, a majority of half-rings have only two or four scutes attached. The half-rings of NMC 0210 and AMNH 5406, were their distal scutes removed, would resemble the half-rings of AMNH 5337 and AMNH 5403. Yet the smallest of these, NMC 0210, is one with six scutes.

USNM 7943 has a half-ring resembling that of AMNH 5337 in most characters (see Fig. 13.9), but the half-ring of USNM 7943 is slightly smaller and its scutes are relatively smooth. Hence, there is some evidence that armor texture changed during ontogeny. Against this interpretation is MOR 433, which, despite its very large size, has a smooth skull and armor. In general, for all specimens with both a skull and armor preserved, the rugosity of the armor is similar to that of the skull. Also, in both *Ankylosaurus* specimens with armor preserved, the scutes are all relatively smooth, although highly vascularized, and none exhibit the extreme rugosity seen in many of the Campanian specimens.

Fusion might be due to individual variation, but it seems odd that the scutes would always fuse in pairs. It is unlikely that scutes would become unfused (i.e., separated from the bony half-ring) during ontogeny. Moreover, scute morphology differs consistently between half-rings having six scutes and those having two or four. Interestingly, no known complete half-ring has an odd number of scutes, implying a genetic rather than environmental cause for missing scutes. Other processes such as resorption may be involved, and this is an area for future study. The discrete half-ring morphologies are probably due to a combination of sexual dimorphism and distinct taxa.

Taxonomic Implications of Postcranial Variation

Given the striking differences between the first half-rings of NMC 0210 (*Euoplocephalus tutus*) and BMNH R5161 (?*Scolosaurus cutleri*), it seems unlikely that these two specimens are conspecific. However, BMNH R5161 shares several features with ROM 784 (holotype of *Dyoplosaurus acutosquameus*). The preserved ilia are similar in having large acetabulae and long postacetabular processes. The ilia of ROM 784 are more widely divergent than those of BMNH R5161, but perhaps this is an individual or sexual difference; as noted above, there are not enough specimens to test this hypothesis. Unfortunately, the zygapophyses on the caudal vertebrae of BMNH R5161 are hidden by armor, so they cannot be compared with the caudals of ROM 784. The armor appears to differ. In ROM 784, many pelvic and caudal scutes are preserved, some of which are tall, laterally compressed, and sharply keeled. No such scutes are present in the pelvic region of BMNH R5161, although there might have been lateral scutes that were not preserved. Scutes from the pelvic region of BMNH R5161 are circular and low, and some have radial ribbing and are truncated dorsally, unlike any of the preserved scutes of ROM 784. The texture is highly rugose in both specimens. The most intriguing piece of evidence for distinct armor is the presence of a pair of tall, laterally compressed, conical scutes at the midpoint of the tail in both holotypes. Could this be a generic trait? Alone, it certainly is not sufficient evidence to synonymize *Scolosaurus cutleri* with *Dyoplosaurus acutosquameus*. These two specimens may represent sexual dimorphs or distinct species of *Dyoplosaurus* (the name has precedence over *Scolosaurus*). Further study is needed.

ROM 813 is unique in having numerous large plates, although other specimens, such as AMNH 5337 and AMNH 5403, might have had more plates than are preserved. Could ROM 813 be referable to *Euoplocephalus*? AMNH 5406, which is clearly referable to *Euoplocephalus*, has approximately 75 scutes preserved. The shapes and textures between the two specimens are quite different, so ROM 813 does not appear to be referable to *Euoplocephalus tutus*. However, all of the material from AMNH 5406 appears to be from the front part of the animal, whereas ROM 813 consists of the pelvic and caudal regions. It is thus conceivable that ROM 813 is in fact *Euoplocephalus*. The large number of platelike scutes preserved with ROM 813 plainly cannot be accommodated by the holotype of *Scolosaurus cutleri*. If it is assumed that ROM 784 and BMNH R5161 both belong to the same taxon, and if ROM 813 is then included, this would imply major sexual differences in the morphology of the postcranial armor. ROM 813 may represent an undescribed taxon that had numerous large plates. If future work shows that ROM 813 is in fact referable to *Euoplocephalus tutus*, then armor texture would not be a useful taxonomic character.

As discussed above, BMNH R4947, NMC 8530, TMP 91.127.1, and UA 31 share several cranial characters. The preserved armor of NMC 8530 is most similar to that of AMNH 5406, a specimen that is certainly referable to *Euoplocephalus tutus*, and NMC 8530 appears to be referable to *Euoplocephalus* too. The armor of ROM 1930 is quite different, resembling that of *Scolosaurus*. If ROM 1930 and AMNH 5404 are adult specimens of *Euoplocephalus* and NMC 8530 is a sub-adult, then there were substantial ontogenetic changes in armor texture and shape, a phenomenon that is undocumented in other archosaurs. Armor morphology supports the presence of at least three taxa, including the one represented by MOR 433.

Status of MOR 433

As discussed above, the skull of MOR 433 is similar to that of USNM 11892, which Gilmore (1930) referred to *Dyoplosaurus* (e.g., ROM 784) primarily on the basis of the teeth. However, Gilmore's evaluation of the similarity between the teeth must be taken in the context in which it was made—that is, the poor knowledge of anky-losaur teeth in 1930. Coombs and Demere (1996) concluded that an-kylosaur teeth are of highly questionable taxonomic value below the familial level, so the status of both MOR 433 and USNM 11892 needs to be established with other characters. A careful comparison between MOR 433, ROM 784, and USNM 11892 is warranted.

As discussed above, MOR 433 and USNM 11892 cluster together in the skull morphometric analysis. MOR 433 shows several primitive characters, including the nodosaurlike cervical armor and six closely set scutes in the first half-ring. ROM 784 also appears to have several primitive characters, including a relatively long postacetabular process of the ilium and a small tail club. Unfortunately, MOR 433 consists primarily of the cranial half of the skeleton and ROM 784 the caudal half (skull fragment notwithstanding). On the basis of the material preserved, the only characters separating MOR 433 from ROM 784 (i.e., *Dyoplosaurus*) are armor texture, relatively greater length and robustness of the preserved sacral vertebrae in *Dyoplosaurus,* and dif-fering zygapophyses on the proximal caudals. Yet MOR 433 is substan-tially different than any of the other specimens and might even belong to a separate subfamily. Other unusual, morphologically intermediate dinosaur taxa have been described from the Upper Two Medicine For-mation of Montana (Horner et al. 1992).

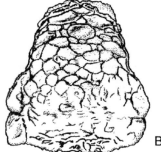
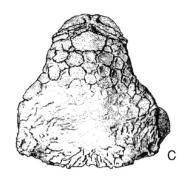

Figure 13.12. Individual variation or distinct taxa? (A) UA 31. (B) AMNH 5404. (C) AMNH 5337. Scale bar = 10 cm.

Conclusions

Few unambiguous statements can be made about *Euoplocephalus tutus* because of the incomplete nature of NMC 0210, the holotype. Given the bimodal nature of cervical half-ring morphologies, AMNH 5406 is certainly referable to *Euoplocephalus tutus*. AMNH 5406 is important because it significantly extends the definition of the genus to include the forelimb and numerous scutes. The half-ring of AMNH 5406 is as large as any, yet the humerus is relatively small. Perhaps this is a real difference. Few specimens have the bulk of their armor preserved, but many have a substantial number of isolated scutes, albeit from different parts of the body. Although these cannot demonstrate the overall pattern, they can give vital information about the range of variation in the scutes of an individual. The scutes of *Euoplocephalus* (AMNH 5406) are all thin walled and sharply keeled. Further study is needed to establish the range of intraspecific variation. A morphometric study of postcranial elements might also be revealing.

Many of the striking differences among the skulls are illustrated in Figure 13.12. These three skulls all have been referred to *Euoplocephalus tutus*. None of them is badly crushed, so the dorsal outlines are approximately correct. Given the associated differences in the postcranial skeleton and armor, it seems likely that more than one taxon is represented.

Acknowledgments. I am grateful to the following for granting me access to material under their care: Charlotte Holton and Mark Norell (American Museum of Natural History), John R. Horner and Pat Leiggi (Museum of the Rockies), Kevin Seymour and Janet Waddington (Royal Ontario Museum), Philip Currie and Tim Schowalter (Royal Tyrrell Museum of Palaeontology), Sandra Chapman (Natural History Museum), Kieran Shepherd (Canadian Museum of Nature), Richard Fox (University of Alberta), and especially the late Nicholas Hotton III and Mike Brett-Surman (National Museum of Natural History). A very special thanks goes to John Horner for allowing me to study undescribed material at the Museum of the Rockies. I am also indebted to Philip Currie and M. T. Vickaryous for permission to examine an undescribed skull at the Royal Tyrrell Museum of Palaeontology (Drumheller). Comments from Ralph Chapman helped to considerably improve the morphometrics section. I thank Sandra Chapman for providing me with excellent high-quality photographs of the Natural History Museum (London) specimens, and I thank William Blows for supplying additional data on them. Exchanges with Walter Coombs, Kenneth Carpenter, Robert Sullivan, and M. T. Vickaryous have been valuable. My mentor John Kirsch provided more help and encouragement than he will likely admit.

References Cited

Brazaitis, P. 1973. The identification of living crocodilians. *Zoologica* 58: 59–101.

Brown, B. 1908. The Ankylosauridae, a new family of armored dinosaurs from the Upper Cretaceous. *Bulletin of the American Museum of Natural History* 24: 187–201.

Carpenter, K. 1982. Skeletal and dermal armor reconstruction of *Euoplocephalus tutus* (Ornithischia: Ankylosauridae) from the Late Cretaceous Oldman Formation of Alberta. *Canadian Journal of Earth Sciences* 19: 689–697.

———. 1990. Ankylosaur systematics: Example using *Panoplosaurus* and *Edmontonia* (Ankylosauria: Nodosauridae). In K. Carpenter and P. J. Currie (eds.), *Dinosaur Systematics: Approaches and Perspectives,* pp. 281–298. Cambridge: Cambridge University Press.

———. 1997. Ankylosaurs. In J. Farlow and M. Brett-Surman (eds.), *The Complete Dinosaur,* pp. 307–316. Bloomington: Indiana University Press.

———. 1998. Ankylosaur odds and ends [abstract]. *Journal of Vertebrate Paleontology* 18(3, Suppl.): 31A.

Carpenter, K., C. Miles, and K. Cloward. 1998. Skull of a Jurassic ankylosaur. *Nature* 393: 782–783.

Chapman, R. E. 1990. Shape analysis in the study of dinosaur morphology. In K. Carpenter and P. J. Currie (eds.), *Dinosaur Systematics: Approaches and Perspectives,* pp. 21–42. Cambridge: Cambridge University Press.

Chapman, R. E., P. M. Galton, J. J. Sepkoski, Jr., and W. P. Wall. 1981. A morphometric study of the cranium of the pachycephalosaurid dinosaur *Stegoceras. Journal of Paleontology* 55: 608–618.

Chapman, R. E., D. B. Weishampel, G. Hunt, and D. Rasskin-Guutman. 1997. Sexual dimorphism in dinosaurs. In D. Wolberg, E. Stump, and G. D. Rosenberg (eds.), *Dinofest International,* pp. 83–93. Philadelphia: Academy of Natural Sciences.

Coombs, W. P., Jr. 1971. The Ankylosauria. Ph.D. diss. Columbia University, New York.

———. 1972. The bony eyelid of *Euoplocephalus* (Reptilia, Ornithischia). *Journal of Paleontology* 46: 637–650.

———. 1978. The families of the ornithischian dinosaur order Ankylosauria. *Palaeontology* 21: 143–170.

———. 1986. A juvenile ankylosaur referable to the genus *Euoplocephalus* (Reptilia, Ornithischia). *Journal of Vertebrate Paleontology* 6: 162–173.

———. 1990. Teeth and taxonomy in ankylosaurs. In K. Carpenter and P. J. Currie (eds.), *Dinosaur Systematics: Approaches and Perspectives,* pp. 269–279. Cambridge: Cambridge University Press.

———. 1995. Ankylosaurian tail clubs of middle Campanian to early

Maastrichtian age from western North America, with description of a tiny tail club from Alberta and discussion of tail orientation and tail club function. *Canadian Journal of Earth Sciences* 32: 902–912.

Coombs, W. P., Jr., and T. A. Demere. 1996. A Late Cretaceous nodosaurid ankylosaur (Dinosauria: Ornithischia) from marine sediments of coastal California. *Journal of Paleontology* 70: 311–326.

Coombs, W. P., Jr., and T. Maryańska. 1990. Ankylosauria. In D. Weishampel, P. Dodson, and H. Osmólska (eds.), *The Dinosauria,* pp. 456–483. Berkeley: University of California Press.

Currie, P. J. 1991. The Sino/Canadian dinosaur expeditions 1986–1990. *Geotimes* 36(4): 18–21.

Dodson, P. 1975. Taxonomic implications of relative growth in lambeosaurine hadrosaurs. *Systematic Zoology* 24: 37–54.

———. 1976. Quantitative aspects of relative growth and sexual dimorphism in Protoceratops. *Journal of Paleontology* 50: 929–940.

Dong, Z. 1993. An ankylosaur (ornithischian dinosaur) from the middle Jurassic of the Junggar Basin, China. *Vertebrata Palasiatica* 31: 258–265.

Eberth, D. 1997. Judith River Wedge. In P. J. Currie and K. Padian (eds.), *Encyclopedia of Dinosaurs,* pp. 379–385. San Diego: Academic Press.

Forster, C. A. 1996. Species resolution in *Triceratops:* Cladistic and morphometric approaches. *Journal of Vertebrate Paleontology* 16: 259–270.

Galton, P. 1982. Juveniles of the stegosaurian dinosaur *Stegosaurus* from the Upper Jurassic of North America. *Journal of Vertebrate Paleontology* 2: 47–62.

Gillette, D. D., and C. E. Ray. 1981. Glyptodonts of North America. *Smithsonian Contributions to Paleobiology* 40: 1–255.

Gilmore, C. W. 1923. A new species of *Corythosaurus* with notes on other Belly River Dinosauria. *Canadian Field-Naturalist* 37: 1–9.

———. 1930. On dinosaurian reptiles from the Two Medicine Formation of Montana. *Proceedings U.S. National Museum* 77: 1–39.

Haas, G. 1969. On the jaw musculature of ankylosaurs. *American Museum Novitates* 2399: 1–11.

Heckert, A. B., and S. G. Lucas. 1999. A new aetosaur (Reptilia: Archosauria) from the Upper Triassic of Texas and the phylogeny of aetosaurs. *Journal of Vertebrate Paleontology* 19: 50–68.

Horner, J. R., D. Varricchio, and M. Goodwin. 1992. Marine transgressions and the evolution of Cretaceous dinosaurs. *Nature* 358: 59–61.

Lambe, L. M. 1902. New genera and species from the Belly River Series (mid-Cretaceous). *Contributions to Canadian Palaeontology, Geological Survey of Canada* 3: 25–81.

———. 1910. Note on the parietal crest of *Centrosaurus apertus,* and a proposed new generic name for *Stereocephalus tutus. Ottawa Naturalist* 14: 149–151.

Long, R. A., and K. L. Ballew. 1985. Aetosaur dermal armor from the Late Triassic of southwestern North America, with special reference to material from the Chinle Formation of Petrified Forest National Park. *Museum of Northern Arizona Bulletin* 54: 45–68.

Maryańska, T. 1969. Remains of armored dinosaurs from the uppermost Cretaceous in Nemegt Basin, Gobi Desert. *Palaeontologia Polonica* 21: 22–34.

———. 1977. Ankylosauridae (Dinosauria) from Mongolia. *Palaeontologia Polonica* 37: 85–151.

Nopsca, F. 1928. Palaeontological notes on reptiles. *Geologica Hungarica, Series Palaeontologica* 1: 1–84.

Parks, W. A. 1924. *Dyoplosaurus acutosquameus*, a new genus and species of armored dinosaur; with notes on a skeleton of *Prosaurolophus maximus. University of Toronto Studies, Geological Series* 18: 1–35.

Penkalski, P., and P. Dodson. 1999. The morphology and systematics of *Avaceratops,* a primitive horned dinosaur from the Judith River Formation (Late Campanian) of Montana, with the description of a second skull. *Journal of Vertebrate Paleontology* 19: 692–711.

Rohlf, F. J. 1990. *NTSYS-pc: Numerical Taxonomy and Multivariate Analysis System. Applied Biostatistics.* Setauket, N.Y.: Exeter Software.

Ross, C. A., and C. D. Roberts. 1979. Scalation of the American alligator. *Special Scientific Report—Wildlife No. 225.* Washington, D.C.: U.S. Department of the Interior, Fish and Wildlife Service.

Ross, F. D. and G. C. Mayer. 1983. On the dorsal armor of the Crocodilia. In A. Rhodin and K. Miyata (eds.), *Advances in Herpetology and Evolutionary Biology: Essays in Honor of Ernest E. Williams,* pp. 305–331. Cambridge, Mass.: Museum of Comparative Zoology.

Sereno, P. C. 1997. The origin and evolution of dinosaurs. *Annual Review of Earth and Planetary Sciences* 25: 435–489.

Smith, D. 1998. A morphometric analysis of *Allosaurus. Journal of Vertebrate Paleontology* 18: 126–142.

Sternberg, C. M. 1929. A toothless armored dinosaur from the Upper Cretaceous of Alberta. *National Museum of Canada Bulletin* 54: 28–33.

Sullivan, R. M. 1999. *Nodocephalosaurus kirtlandensis,* gen. et sp. nov., a new ankylosaurid dinosaur (Ornithischia: Ankylosauria) from the Upper Cretaceous Kirtland Formation (Upper Campanian), San Juan Basin, New Mexico. *Journal of Vertebrate Paleontology* 19: 126–139.

Tumanova, T. A. 1987. Armored dinosaurs of Mongolia [in Russian]. *Transactions of the Joint Soviet–Mongolian Paleontological Expedition* 32: 1–77.

Weishampel, D. B., and R. E. Chapman. 1990. Morphometric study of *Plateosaurus* from Trossingen (Baden-Wurttemberg, Federal Republic of Germany). In K. Carpenter and P. J. Currie (eds.), *Dinosaur Systematics: Approaches and Perspectives,* pp. 43–51. Cambridge: Cambridge University Press.

APPENDIX 13.1.
Scores for each specimen in the morphometric analysis on the first four principal components.

Symbol	Component 1	2	3	4	Specimen
2	0.0048	−0.0053	0.0233	0.0027	AMNH 5238
3	0.0383	−0.0189	0.0177	−0.0014	AMNH 5403
4	0.0396	0.0015	−0.0011	0.0044	AMNH 5404
5	0.0182	−0.0250	−0.0042	0.0073	AMNH 5405
7	0.0247	−0.0259	0.0228	0.0123	AMNH 5337
A	−0.0233	0.0190	0.0024	−0.0129	UA 31
B	−0.0183	0.0062	0.0159	0.0003	BMNH R5161
M	0.0123	0.0203	−0.0222	−0.0005	MOR 433
N	−0.0199	−0.0087	−0.0029	−0.0169	NMC 8530
R	0.0217	0.0087	−0.0119	0.0058	ROM 1930
T	−0.0609	0.0217	−0.0005	0.0019	TMP 91.127.1
U	−0.0372	0.0063	−0.0393	−0.0029	USNM 11892

14. Evidence of Complex Jaw Movement in the Late Cretaceous Ankylosaurid *Euoplocephalus tutus* (Dinosauria: Thyreophora)

N. Rybczynski and M. K. Vickaryous

Abstract

Previous work has described the chewing mechanics of thyreophorans, pachycephalosaurs, and stegosaurs as "orthal pulping." Such orthal pulpers generally possess a loosely packed dentition of small, cuspate teeth and a rigidly constructed skull. Their power stroke is presumed to have been strictly orthal. New evidence indicates that the jaw movement of the Late Cretaceous ankylosaurid *Euoplocephalus tutus* was not restricted to simple orthal adduction. Examination of in situ upper dentition reveals that tooth wear facets are contiguous between adjacent teeth. In addition, scanning electron microscopy of these surfaces shows numerous microscopic striations that are roughly parallel with the line of the tooth row. Tooth wear evidence is used in conjunction with morphologic cranial data to support a new model of jaw action that requires more complex movements. The model describes a bilateral power stroke in which the lower teeth move retractively against the upper teeth. The mandibular rami also pivot slightly at the jaw joint, resulting in a symmetrical medial displacement of the lower tooth rows. The latter movement is accommodated by mobile dentary–dentary and dentary–predentary contacts. *Euoplocephalus tutus* is the first thyreophoran to be diagnosed with a complex jaw action. Similar studies of other ornithischian orthal pulpers may elucidate a greater diversity of comminutory systems than has previously been suspected.

Introduction

Dinosaurian paleobiology, as in all vertebrate paleobiology, relies on our ability to infer unpreserved attributes, such as behavior, physiology, and soft tissue morphology, from skeletal evidence. Two approaches are generally used. One is ahistorical and makes extrapolatory, inference-based established biological generalizations, as might be identified by comparative methods or by engineering principles (see Bryant and Russell 1992: 406; also see Weishampel 1995; the nonhistorical method of Bock 1989; and the paradigm method of Lauder 1995; Kay et al., in press). The other approach is historical and is represented by the recently systematized method of crown group phylogenetic bracketing (Bryant and Russell 1992; Weishampel 1995; Witmer 1995, 1997; the historical method of Bock 1989). Ahistorical extrapolatory modeling has frequently been used to evaluate the trophic paleoecologic role of dinosaurs through the analysis of postcranial and especially craniodental evidence (Fiorillo 1998; Galton 1986; Weishampel 1984). We reevaluate the functional morphology of the craniodental system apparatus of the Late Cretaceous ankylosaur *Euoplocephalus tutus* by use of ahistorical extrapolatory modeling. Our study differs from previous works on this taxon, in part because it focuses on one particularly reliable predictor of function: tooth wear (see Lauder 1995).

Once the feeding biomechanics of a fossil organism have been reconstructed, it becomes possible to circumscribe aspects of its paleoecologic identity. The present study considers only a part of the craniodental system (in particular, it does not consider the function of the beak region), and so it does not represent a complete assessment of the feeding apparatus. Even so, the results presented here suggest that current paleoecologic interpretations of the clade Ankylosauria require reappraisal (also see Barrett 1988, 2001).

The jaw mechanics of ankylosaurs have been the subject of a variety of interpretations. For instance, it has been postulated that ankylosaurs were insectivorous (Nopsca 1928) and that the mandible may have adducted lateral to the maxillary tooth row (Haas 1969; Russell 1940). Although subsequent, more rigorous work has cast these hypotheses in doubt (Coombs 1971), the functional morphology of the ankylosaur jaw apparatus has remained elusive. Most recently, the question of ankylosaur feeding biomechanics was deemphasized when ankylosaurs were grouped with various other ornithischians in a functional wastebasket (e.g., *Lesothosaurus;* Galton 1986; see also Weishampel and Norman 1989).

The most detailed attempts at analyzing ankylosaur jaw mechanics have focused largely on the interpretation of mandibular adductor musculature, beginning with the contribution of Haas (1969). He examined two well-preserved, relatively uncrushed specimens of *Euoplo-*

cephalus (AMNH 5337 and AMNH 5405) and compared the size and position of muscle scars with those of modern structural grade reptiles. On the basis of his analysis, he concluded that ankylosaurs had a relatively weak jaw apparatus and likely consumed soft plant material. Although he assumed that the mandible adducted lateral to the maxilla, he did not suggest a model for the jaw action.

The work of Haas was followed up by Coombs (1971), who studied the skull of *Euoplocephalus* as well as the nodosaurid taxa *Panoplosaurus mirus* and *Edmontonia rugosidens*. Coombs's reassessment of the evidence resulted in a more comprehensive review of adductor musculature. The orientation of the adductor musculature was of particular interest because it implied that a strictly orthal adduction of the mandible was unlikely. In contrast, examination of worn ankylosaur teeth appeared to reveal only vertically oriented wear striations on the wear facets. The vertical nature of the striae supports an orthal power stroke. This contradictory evidence was further confounded by the mandibular articulation that readily permitted anteroposterior mobility of the mandibles. To resolve this paradox, Coombs (1971) proposed that the propalinal movement functioned to bring the rostral tomium into occlusion. This propalinal movement would be followed by a strictly orthal adduction of the mandible to bring the teeth into occlusion (i.e., the power stroke). However, because of the unusual morphology of the tooth rows on both the maxillae and the dentaries, he further posited that during orthal adduction, the mandibles would be "brought from a slightly lateral position to a more medial position at full occlusion" (Coombs 1971: 347). Coombs (1971, 1978) also proposed that more complex masticatory movements were facilitated in some taxa by metakinesis or limited streptostyly.

More recent reviews of ornithischian jaw mechanics have considered the ankylosaur cranium to be akinetic (Galton 1986; Weishampel and Norman 1989), and most authors have accepted that members of the clade used an orthal power stroke (Galton 1986; King 1996; Weishampel 1984; Weishampel and Norman 1989). Weishampel and Norman (1989) qualitatively assessed the feeding mechanics of ankylosaurs to be typical of orthal pulpers. However, contiguous wear facets between adjacent teeth (Coombs and Maryańska 1990) appear to contradict this interpretation. The presence of contiguous wear facets has been used as a means of predicting a propalinal power stroke in extant lepidosaurs (e.g., *Uromastyx*: Throckmorton 1976; *Sphenodon*: Gorniak et al. 1982)

On the basis of a more intensive review of available evidence, the mechanics of the ankylosaur masticatory apparatus is reevaluated. Analysis of the data indicates that the power stroke was predominantly retractive. Furthermore, experimental modeling of the skull suggests that a purely retractive power stroke would be only feasible if the tooth rows could displace mediolaterally. The kinetic cranium and orthal pulping hypotheses are not supported.

Institutional Abbreviations. AMNH: American Museum of Natural History, New York. TMP: Royal Tyrrell Museum of Palaeontology, Drumheller, Alberta, Canada.

Materials and Methods

Skull

We chose *Euoplocephalus* as a model for the examination of anky-losaur feeding apparatus because of the relative abundance and avail-ability of specimens and the considerable amount of previous work conducted on this taxon. Among the specimens incorporated in the present study are many of the skulls examined by Haas (1969) and Coombs (1971, 1978) in their treatments of jaw mechanics. Additional material assigned to other ankylosaur taxa was examined whenever possible. The principal *Euoplocephalus* specimen used in this study was AMNH 5405 because it is well preserved (although slightly dis-torted) and because it possesses an in situ maxillary dentition (Figs. 14.1, 14.2).

The taphonomic distortion of AMNH 5405 is seen in the asym-metrical disarticulation of the pterygoid–quadrate contact, parasagittal displacement of the pterygoid complex, and slightly sinusoidal vomeral keel; these suggest that the specimen was subjected to an anterodorsal crushing force. This force also displaced the quadrates so that the po-sition of the jaw joint within AMNH 5405 is more anterior to where it would have been in life. Therefore, previous functional analyses based

Figure 14.1. Lateral view of (A) left side of skull, modified from Coombs and Maryańska (1990), and (B) right mandible of Euoplocephalus tutus *(AMNH 5405). Note dorsolateral curvature of the tooth row. Scale bar = 10 cm.*

Figure 14.2. (opposite page top) (A) Ventral view of cranium, modified from Coombs and Maryańska (1990). (B) Dorsal view of right mandible of Euoplocephalus tutus *(AMNH 5405). Scale bar = 10 cm.*

Figure 14.3. (opposite page bottom) Schematic, parasagittal section through the cranium of Euoplocephalus tutus *(AMNH 5405), showing preserved (solid line) and reconstructed positions (dotted line) of quadrate. Adapted from Coombs (1971).*

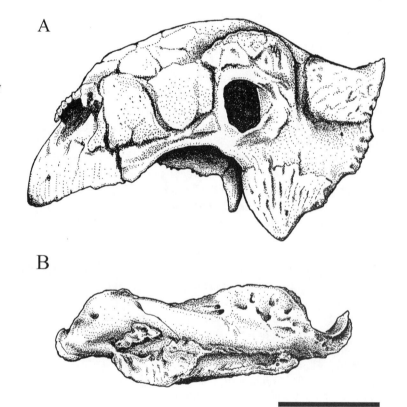

A

B

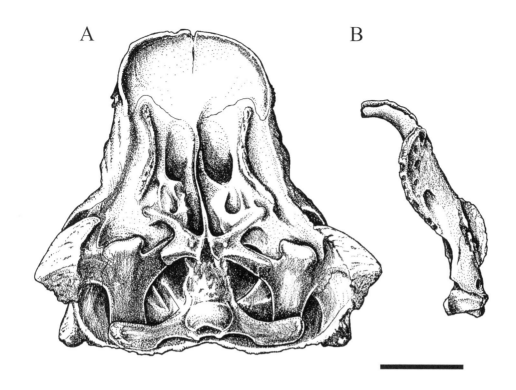

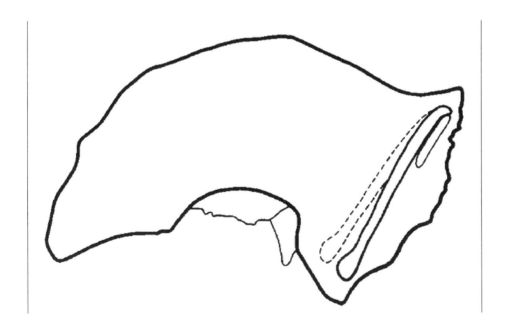

on this specimen (Coombs 1971; Haas 1969) must be viewed with caution. Before modeling jaw mechanics, the position of the quadrate was reconstructed on a cast of AMNH 5405 (TMP 82.1.1) (Fig. 14.3). This modification is significant from a functional perspective because it results in a significant improvement to the occlusal fit between the opposing tooth rows.

Dentition

Both macrowear features (e.g., angle of the wear facet) and microwear features (e.g., striae) of various worn ankylosaur teeth were examined. Because of a dearth of in situ dentition, isolated teeth were also inspected. Detailed examinations of the microwear attributes were conducted with the use of a scanning electron microscope.

Macrowear. Ankylosaur teeth have commonly been described as possessing irregular wear patterns (Coombs 1971, 1978; Coombs and Maryańska 1990; Sander 1997). This description encompasses reports of wear patterns that imply conflicting jaw actions. It has been claimed, for instance, that ankylosaur teeth with double wear surfaces are not unusual (Russell 1940, cited in Galton 1986; Weishampel and Norman 1989). Double wear facets on a single tooth crown occur if two adjacent teeth occlude in an orthal motion against a single tooth on the opposing jaw. This wear pattern results from orthal adduction and is characteristic of basal ornithischians, such as *Lesothosaurus* (Sereno 1991). On the other hand, as noted above, some ankylosaur specimens demonstrate wear surfaces aligned across adjacent teeth (i.e., contiguous; Coombs and Maryańska 1990), implying propalinal movement.

To discern the pattern of tooth macrowear in *Euoplocephalus,* the in situ dentition of AMNH 5405, as well as published descriptions and illustrations of tooth macrowear in other ankylosaurs, was considered. Particular attention was given to double-wear facets, so numerous isolated teeth referable to both *Euoplocephalus* and at least two nodosaurid taxa were also examined.

Microwear. Striations on tooth wear facets form as a result of tooth–tooth occlusion or when mineralized particles are caught between shearing teeth (Weishampel 1984). Therefore, striations reflect the line of jaw motion during the power stroke. Striations might indicate, for example, that the movement of the lower jaw was parallel to the tooth row. However, the striations would not indicate whether the power stroke is from back to front (proal) or front to back (palinal) (see Krause 1982). The direction of the power stroke can be determined by examining the enamel–dentine interface at the periphery of a wear facet (Costa and Greaves 1981; Greaves 1973; Teaford and Byrd 1989). The first edge contacted during occlusion (the leading edge) bears a flush enamel–dentine interface. This occurs because the hard enamel protects the adjacent, softer dentine from abrasion. At the opposite edge of the wear facet (the trailing edge), the softer dentine is worn away more quickly than the harder enamel. The result is that at the trailing edge, the surface of the enamel is raised relative to the surface of the dentine.

Following the methodology of Teaford (1991), both isolated teeth

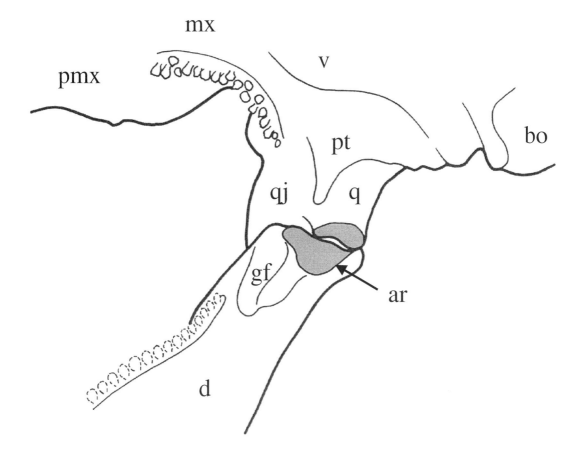

mx

pmx

v

bo

pt

qj

q

gf

ar

d

(recovered in close proximity to a *Euoplocephalus* skull, TMP 96.75.1) and in situ teeth (AMNH 5405) were molded in polyvinylsiloxane (3M President Jet regular body). Epoxy casts were sputter-coated with gold and imaged by scanning electron microscopy. Dental microwear features and the morphology of the enamel–dentine interface were then examined.

Figure 14.4. Anteromedial view of right jaw joint of Euoplocephalus tutus (AMNH 5405). ar = articular; bo = basioccipital; d = dentary; gf = glenoid fossa; mx = maxilla; pmx = premaxilla; pt = pterygoid; q = quadrate = qj =quadratojugal; v = vomer.

Description and Results

Skull

The skull of *Euoplocephalus* is considered here in terms of mechanical units, beginning with the cranium. In ankylosaurids, the contacts between the basipterygoid processes and pterygoid complex, and the quadrates, squamosals, and paroccipital processes do not fuse (Coombs 1978). However, cranial kinetism is prohibited by fusion of other elements in the head skeleton. For example, the quadrate is fused to the dermatocranium via a nonarticulating contact with the quadratojugal (AMNH 5337, AMNH 5403, AMNH 5405, TMP 91.127.1, TMP 97.132.1; Coombs 1971).

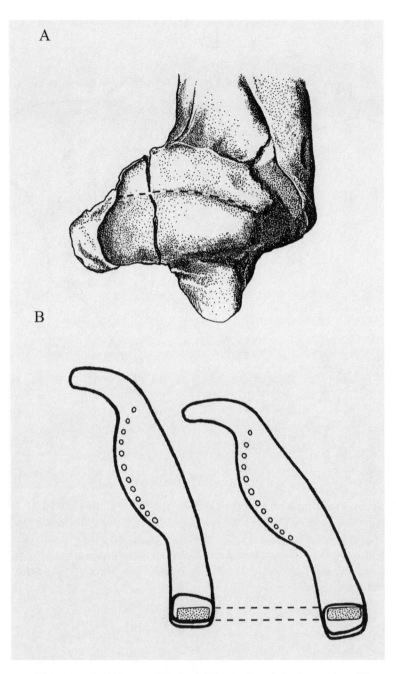

A

B

Figure 14.5. (A) Dorsal view of right articular–mandibular joint of Euoplocephalus tutus *(AMNH 5405). The anterior portion of the mandible is toward the top of the page. The dotted line roughly delineates division between two positions of stability (see text). (B) Schematic diagram showing the position of jaw joint in relation to quadrate with mandible in two positions of stability, protracted (P) and retracted (R).*

The two remaining mechanical units make up the lower jaw. They include the predentary, a neomorphic element of the Ornithischia, and the mandible exclusive of the predentary. The predentary is considered a separate mechanical unit because it is never found fused to the mandible proper, suggesting that it may have been mobile on the latter. Between these three mechanical units there are three joint surfaces: quadrate–articular joints, dentary–dentary joints, and dentary–predentary joints.

It is impossible to precisely reconstruct the movement that would

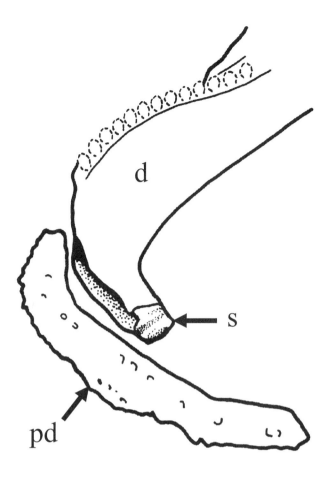

Figure 14.6. Anteromedial exploded view of predentary and right dentary of Euoplocephalus tutus *(adapted from AMNH 5405). Note the relatively small size of the symphysis.*
d = dentary; pd = predentary; s = dentary symphysis.

have been have permitted at the quadrate–articular joint. The osteologic morphology, however, suggests that the movement of the jaw was relatively unconstrained in the anteroposterior direction (Fig. 14.4). The joint surface of the articular (the cotyle) is relatively flat and slightly anteroposteriorly elongate relative to the quadrate. We suggest that the cotyle provided two distinct regions of stability for the quadrate, a slightly elevated anterolateral region and a transversely expanded posteromedial region. In addition, the two positions are at an angle relative to one another, with the posteromedial region at a more oblique angle relative to the long axis of the tooth row compared with the anterolateral region (Fig. 14.5A). The implications of this cotyle structure are twofold. First, the jaw is capable of fore-and-aft translation (propalinal motion). Propalinal displacement, relative to the medial condyle, is approximately 17 mm. Second, during retraction, each mandible pivots, displacing the tooth rows medially (Fig. 14.5B). As mentioned above, the possibility of medial displacement of the tooth rows during occlusion and propaliny were previously noted by Coombs (1971).

The dentary–dentary joint of *Euoplocephalus* is gracile, unfused, and small. In AMNH 5405, the length of the mandible is about 290 mm, whereas the length of the symphysis is only 20 mm. The surface of the symphysis is subcircular and possesses an anteroposteriorly elongated ridge on its ventral margin (Fig. 14.6). Unfortunately, only the right mandible of AMNH 5405 is preserved, so it is not known how the two dentaries articulated. The horizontal symphyseal ridge could potentially provide some degree of stabilization against dorsoventral shearing. Most significantly, the dentary–dentary joint was unfused and would not have precluded symmetrical, mediolateral pivoting of the mandibles.

The predentary–dentary joint of ankylosaurs does not appear to have formed a tight synarthrosis, a finding made in part on the basis of the frequent absence of the predentary in otherwise well-preserved specimens. The predentary is transversely elongate and recurved distally to articulate with the anterior margin of the mandible. The entire length of the posterior surface of the predentary is wedge-shaped and fits into a deeply pitted groove at the anteriormost edge of the dentaries

Figure 14.7. Unworn, isolated tooth of Euoplocephalus tutus *(TMP 92.36.1226). Scale bar = 500 μm.*

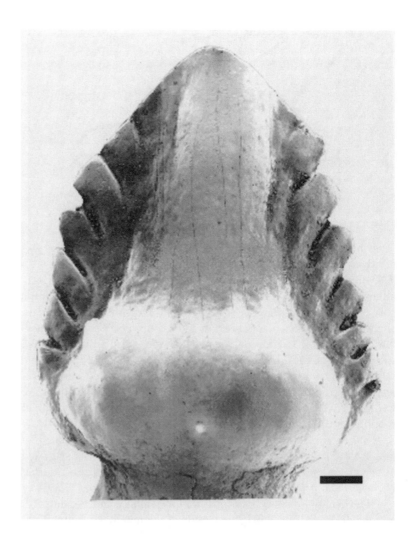

(Fig. 14.6). This joint would have permitted only a slight transverse movement of the dentaries. Rotation around the long axis of the mandibles, as seen in *Lesothosaurus* (Sereno 1991) and heterodontosaurids (Weishampel 1984), was precluded.

Dentition

As in most derived ankylosaurids (e.g., *Ankylosaurus*, *Saichania*, and *Tarchia*), the predentary and paired premaxillae in *Euoplocephalus* are edentulous and presumably served to support a rhamphotheca (see Coombs 1971; Coombs and Maryańska 1990). The upper dentition consists of relatively short tooth rows inset from the lateral margin of the maxillae. Both the upper and lower tooth rows arch dorsally (Fig. 14.1) and medially (Fig. 14.2), although the dorsal curvature of the upper tooth row is less than that of the lower tooth row. The teeth are relatively small (mesial–distal length 3–5 mm) and single rooted. Unworn, the teeth are buccolingually compressed, with coarse denticles extending along the mesial–distal carina (Fig. 14.7). This tooth morphology is considered plesiomorphic for Ornithischia (Coombs and Maryańska 1990).

The right upper tooth row of AMNH 5404 possesses at least 11 worn teeth. Five additional teeth, preserved with broken crowns, may also have formed part of the worn surface. The upper left tooth row has nine worn teeth, with another eight lost to breakage. None of the teeth in the lower jaw was preserved.

Macrowear. We have been unable to substantiate the claim by Weishampel and Norman (1989: 91) that Russell (1940) stated that ankylosaurs exhibit "obliquely inclined double-wear facets." Furthermore, to our knowledge, such double-wear facets do not appear in figures or descriptions of any known ankylosaur teeth (Baszio 1997; Carpenter 1990; Coombs 1990; Coombs and Maryańska 1990), nor were we able to locate them on any isolated or in situ teeth examined. A single possible example of a double-wear facet was found, but, as will be discussed below, this particular facet is considered to be the result of propalinal motion. It is suggested, therefore, that double-wear facets do not appear to occur as regular features of ankylosaur dental macrowear.

In ankylosaurs, the occlusal surfaces commonly change angles along the tooth row. Coombs (1971) indicates that wear facets located near the center of the tooth row have occlusal surface angles approximating 20° to 25° from the vertical and that this angle decreases towards either extremity. However, there appears to be variation within this pattern. For example, a specimen of *Edmontonia rugosidens* has near horizontal wear facets across the most posterior teeth only (Russell 1940). Because of poor preservation, the posteriormost teeth of *Euoplocephalus* AMNH 5405 cannot be assessed. The remainder of the dentition conforms to the pattern described by Coombs (1971). Teeth near the middle of the tooth row in AMNH 5405 possess wear facets that are, for the most part, contiguous between adjacent teeth. This is consistent with a proal or palinal power stroke.

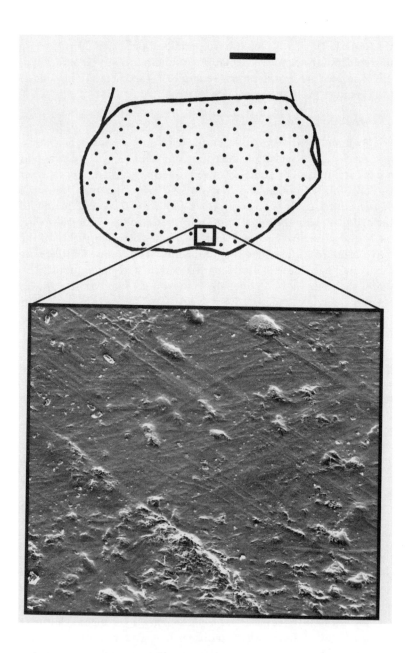

Figure 14.8. Scanning electron
microscopy image of (microwear)
striations on upper right tooth
(tooth 4; Euoplocephalus tutus;
AMNH 5405). Wear facet is
indicated by stippling. Scale bar
of tooth = 1 mm. Width of
imaged surface = 800 μm.

Previous works have described ankylosaur teeth as exhibiting irregular wear indicative of imprecise occlusion (Coombs 1971, 1978;
Coombs and Maryańska 1990; Sander 1997). As described above, tooth
facets vary in their angle predictably. Furthermore, the wear facets
themselves are distinct, with sharp edges indicating that they are largely the result of tooth–tooth contact, rather than food–tooth contact
(Rensberger 1973; also Popowics and Fortelius 1997). Thus, occlusion
must also have been precise.

Microwear. Microwear features on the in situ dentition (AMNH
5405) were sometimes concealed by oxidized precipitate, glue, or pre-

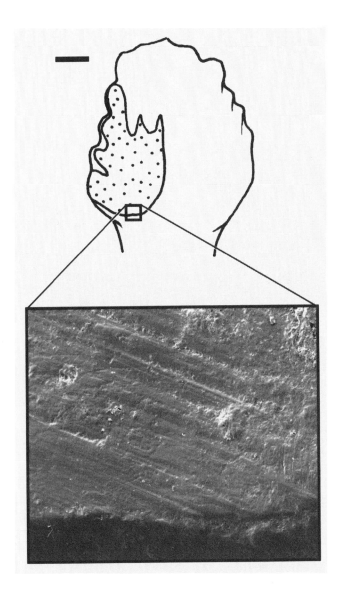

paratory artifacts. Those teeth that could be examined were found to exhibit low-angled striations that are nearly parallel with the tooth row (Fig. 14.8). None of the in situ dentition possesses vertically oriented striations. However, vertical preparatory artifacts were observed (see Teaford 1989), and we presume that these are the vertical striations observed by Coombs (1971: 350).

The eight isolated (but associated) *Euoplocephalus* teeth (TMP 96.75.1) that we examined also exhibited low-angled striations (Fig. 14.9). In addition, a limited number of vertical striations (parallel to the tooth axis) were also found on two teeth, near the tooth apex. One of the isolated teeth appears to possess a double facet, but both occlusal surfaces are crossed by nearly horizontal striations.

Figure 14.9. Scanning electron micrograph of microwear striations on an isolated tooth (Euoplocephalus tutus; TMP 96.75.1). Wear facet is indicated by stippling. Scale bar of tooth = 1 mm. Width of imaged surface = 400 μm.

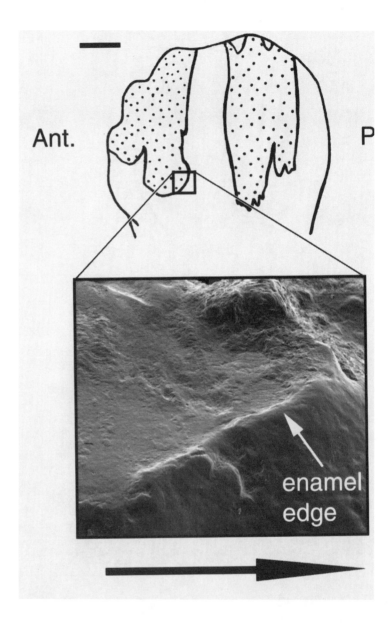

Ant.   P

enamel
edge

Figure 14.10. Scanning electron micrograph of the trailing edge of the upper left tooth (tooth 4; Euoplocephalus tutus; AMNH 5405). The raised enamel edge on the wear facet (white arrow) of this tooth appears only on the distal margins of the tooth, indicating that the power stroke was retractive. Black arrow indicates relative motion of the opposing tooth row. Wear facet is indicated by stippling. Scale bar of tooth = 1 mm. Width of imaged surface = 800 µm.

Examination of enamel–dentine interfaces in AMNH 5405 revealed that the mesial edges bore flush interfaces, whereas the distal edges possessed a stepped interface (Fig. 14.10).

Interpretation of Jaw Action

Tooth wear evidence clearly indicates that the power stroke in *Euoplocephalus* was primarily palinal. These findings are contrary to previous interpretations (Coombs 1971; Galton 1986; Weishampel 1984) that postulated that the power stroke was orthal. This interpretation is in agreement with the structure of the jaw joint and the orientation of the musculature (Coombs 1971). The occurrence of sparse

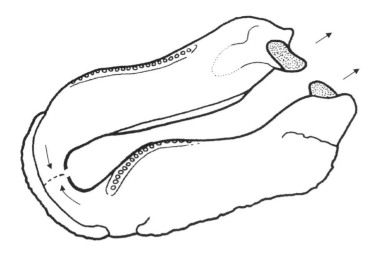

vertical striations near the apex of isolated teeth provides evidence of only limited orthal shearing. However, at least for AMNH 5405, the power stroke was unquestionably dominated by shearing within the plane of the tooth row.

A purely retractive power stroke was virtually impossible in AMNH 5405 because the tooth rows are curved in the horizontal plane (Fig. 14.2). We suggest that in order to accommodate the shape of the tooth row, the power stroke must have included a slight medial displacement of the tooth rows. This would be accomplished by a pivoting movement at the quadrate–articular joints. The structure of the quadrate–articular, dentary–dentary, and dentary–predentary joints could all accommodate this symmetrical medial displacement. Furthermore, medial displacement of the mandible is in some ways congruent with the model proposed by Coombs (1971). Pivoting of the mandible may account for the oblique inclination of the dental wear facets (relative to the vertical plane) at the mesial and distal ends of the tooth rows. Similar mandibular movements have been proposed for some nimravid mammals (Bryant and Russell 1995).

To conclude, tooth wear evidence indicates that the primary movement of the mandible during the power stroke was retractive, and therefore trituration in AMNH 5405 was accomplished principally by bilateral, palinal shearing. To accommodate the unusual tooth row morphology, the power stroke also involved pivoting the mandibles (Fig. 14.11).

Conclusion

This study and that of Barrett (2001) are among the first to use a detailed analysis of dental microwear features in the interpretation of thyreophoran jaw mechanics. Microwear analysis demonstrates that, contrary to previous assessments, *Euoplocephalus* did not employ or-

Figure 14.11. Proposed model of power stroke in Euoplocephalus tutus. *Arrows near the jaw joint indicate that the largest component of mandibular movement is retractive. Arrows near the symphysis indicate slight medial pivoting of left and right dentary. Arrows reflect direction of motion only.*

thal pulping as a primary means of food trituration. Rather, the power stroke was predominantly retractive, and comminution was dominated by shearing in a direction roughly parallel to the tooth row. The power stroke also involved a small, bilateral, medial pivoting of the mandibles. This movement was accomplished, in part, by a mobile dentary–predentary system. Evidence of dentary–predentary mobility has also been found for the basal ornithischian *Lesothosaurus* (Sereno 1991) and for heterodontosaurids (Weishampel 1984). These two forms differ from *Euoplocephalus* in that the dentary–predentary joint would have permitted rotation about the long axis of their mandibles.

The results of this study also have paleoecologic implications. Members of Ankylosauria are generally considered to have been predominantly or exclusively herbivorous (type A herbivores of Hotton 1955; *contra* Nopcsa 1928). Because of such factors as their small teeth, short tooth rows, unusual cranial architecture, and presumed orthal power stroke, they are usually considered to have been capable of only limited oral processing (Coombs and Maryańska 1990; Haas 1969; Russell 1940). It has been suggested that they must have eaten plant parts that were succulent (Russell 1940), soft (Galton 1986; Haas 1969), or nonabrasive (Coombs 1971). Further digestive considerations have ranged from the probable (use of gut fermentation, chemical digestion, or both; Coombs 1971; Coombs and Maryańska 1990) to the unlikely (suction feeding of plant material; Haas 1969). Even though the dentition is reduced, the findings presented here indicate that the dental apparatus was more effective at comminuting tough foodstuffs than has been previously suggested (e.g., Molnar and Clifford 2001). Thus, their herbivorous diet may not have been as limited as has been presumed. Future work will look at other thyreophorans to explore the evolution of this unusual jaw action.

Acknowledgments. We thank Kenneth Carpenter for inviting us to contribute a chapter to this book. Many thanks go to the staff of the Royal Tyrell Museum of Paleontology, Alberta. Phil Currie allowed us to borrow his laboratory space and also authorized the loan of material to Duke University. We are also thankful to Don Brinkman and others for valuable discussion and to Kevin Aulenback for preparation advice and assistance. Eugene Gaffney and Mark Norell permitted the loan of material from the American Museum of Natural History and also allowed one of us (M.K.V.) access to specimens in their care. Mary Maas and Leslie Eibest (Duke University) provided excellent training in scanning electron microscopy. The cost of the many hours of scanning electron microscopy work was covered by the Department of Biological Anthropology and Anatomy at Duke University. We also thank Lisa McGregor (University of Calgary) for her scanning electron microscopy technical expertise. We also thank those who kindly reviewed earlier drafts of this article: Kathleen Smith (Duke University), Anthony Russell (University of Calgary), Paul Barrett (University of Oxford), and David Weishampel (Johns Hopkins University). Cost of supplies and travel was offset by a Sigma Xi grant in aid of research.

References Cited

Barrett, P. M. 1998. Feeding in thyreophoran dinosaurs [abstract]. *Journal of Vertebrate Paleontology* 18(3, Suppl.): 26A.

———. 2001. Tooth wear and possible jaw action of *Scelidosaurus harrisonii* Owen and a review of feeding mechanisms in other thyreophoran dinosaurs. In K. Carpenter (ed.), *The Armored Dinosaurs*. Bloomington: Indiana University Press [this volume, Chapter 2].

Baszio, S. 1997. Systematic palaeontology of isolated dinosaur teeth from the Latest Cretaceous of South Alberta, Canada. *Courier Forschungsinstitut Senckenberg* 196: 33–77.

Bock, W. J. 1989. Principles of biological comparison. *Acta Morphologica Neerlando-Scandinavica* 27: 17–32.

Bryant, H. N., and A. P. Russell. 1992. The role of phylogenetic analysis in the inference of unpreserved attributes of extinct taxa. *Philosophical Transactions of the Royal Society of London, B* 337: 405–418.

———. 1995. Carnassial functioning in nimravid and felid sabertooths: Theoretical basis and robustness of inferences. In J. J. Thomason (ed.), *Functional Morphology in Vertebrate Paleontology*, pp. 116–135. Cambridge: Cambridge University Press.

Carpenter, K. 1990. Ankylosaur systematics: Example using *Panoplosaurus* and *Edmontonia* (Ankylosauria: Nodosauridae). In K. Carpenter and P. J. Currie (eds.), *Dinosaur Systematics: Perspectives and Approaches*, pp. 281–298. Cambridge: Cambridge University Press.

Coombs, W. P., Jr. 1971. The Ankylosauria. Ph.D. diss. Columbia University, New York.

———. 1978. The families of the ornithischian dinosaur order Ankylosauria. *Palaeontology* 21: 143–170.

———. 1990. Teeth and taxonomy in ankylosaurs. In K. Carpenter and P. J. Currie (eds.), *Dinosaur Systematics: Approaches and Perspectives*, pp. 269–279. Cambridge: Cambridge University Press.

Coombs, W. P., Jr., and T. Maryańska. 1990. Ankylosauria. In D. B. Weishampel, P. Dodson, and H. Osmólska (eds.), *The Dinosauria*, pp. 456–483. Berkeley: University of California Press.

Costa, R. L., and W. S. Greaves. 1981. Experimentally produced tooth wear facets and the direction of jaw motion. *Journal of Paleontology* 55: 635–638.

Fiorillo, A. R. 1998. Dental microwear patterns from the sauropod dinosaurs *Camarasaurus* and *Diplodocus*: Evidence for resource partitioning in the Late Jurassic of North America. *Historical Biology* 13: 1–16.

Galton, P. M. 1986. Herbivorous adaptations of Late Triassic and Early Jurassic dinosaurs. In K. Padian (ed.), *The Beginning of the Age of Dinosaurs*, pp. 203–221. Cambridge: Cambridge University Press.

Gorniak, G. C., H. I. Rosenberg, and C. Gans. 1982. Mastication in the tuatara, *Sphenodon punctatus* (Reptilia: Rhynchocephalia): Structure and activity of the motor system. *Journal of Morphology* 171: 321–353.

Greaves, W. S. 1973. The inference of jaw motion from tooth wear facets. *Journal of Paleontology* 47: 1000–1001.

Haas, G. 1969. On the jaw muscles of ankylosaurs. *American Museum Novitates* 2399:1–11.

Hotton, N., III. 1955. A survey of adaptive relationships of dentition to diet in North American Iguanidae. *American Midland Naturalist* 53: 88–114.

Kay, R. F., B. A. Williams, and F. Anaya. In press. The adaptations of *Branisella boliviana*, the earliest South American monkey. In J. M. Plavcan, R. F. Kay, W. Jungers, and C. Van Schaik (eds.), *Reconstructing Behavior in the Primate Fossil Record*. Plenum Publishing.

King, G. 1996. *Reptiles and Herbivory*. New York: Chapman and Hall.

Krause, D. W. 1982. Jaw movement, dental function, and diet in the Paleocene multituberculate *Ptilodus*. *Paleobiology* 8: 265–281.

Lauder, G. V. 1995. On the inference of function from structure. In J. J. Thomason (ed.), *Functional Morphology in Vertebrate Paleontology*, pp. 1–18. Cambridge: Cambridge University Press.

Molnar, R. E., and H. T. Clifford. 2001. An ankylosaurian cololite from the Lower Cretaceous of Queensland, Australia. In K. Carpenter (ed.), *The Armored Dinosaurs*. Bloomington: Indiana University Press [this volume, Chapter 19].

Nopcsa, F. 1928. Paleontological notes on reptiles. *Geologica Hungarica, Series Palaeontologica* 1: 1–84.

Popowics, T. E., and M. Fortelius. 1997. On the cutting edge: Tooth blade sharpness in herbivorous and faunivorous mammals. *Annales Zoologici Fennici* 34:73–88.

Rensberger, J. M. 1973. An occlusion model for mastication and dental wear in herbivorous mammals. *Journal of Paleontology* 47: 515–528.

Russell, L. S. 1940. *Edmontonia rugosidens* (Gilmore), an armored dinosaur from the Belly River Series of Alberta. *University of Toronto Studies, Geology Series* 43: 1–28.

Sander, P. M. 1997. Teeth and jaws. In P. J. Currie and K. Padian (eds.), *Encyclopedia of Dinosaurs,* pp. 717–725. New York: Academic Press.

Sereno, P. C. 1991. *Lesothosaurus*, "Fabrosaurids," and the early evolution of the Ornithischia. *Journal of Vertebrate Paleontology* 11: 168–197.

Teaford, M. F. 1989. Scanning electron microscope diagnosis of wear patterns versus artifacts on fossil teeth. *Scanning Microscopy* 2: 1167–1175.

———. 1991. Dental microwear: What can it tell us about diet and dental functions? In M. A. Kelley and C. S. Larsen (eds.), *Advances in Dental Anthropology*, pp. 341–356. New York: Liss.

Teaford, M. F., and K. E. Byrd. 1989. Differences in tooth wear as an indicator of changes in jaw movement in the guinea pig *Cavia porcellus*. *Archives of Oral Biology* 34: 929–936.

Throckmorton, G. S. 1976. Oral food processing in two herbivorous lizards, *Iguana iguana* (Iguanidae) and *Uromastyx aegyptius* (Agamidae). *Journal of Morphology* 148: 363–390.

Weishampel, D. B. 1984. Evolution of jaw mechanisms in ornithopod dinosaurs. *Advances in Anatomy, Embryology and Cell Biology* 87: 1–110.

———. 1995. Fossils, function, and phylogeny. In J. J. Thomason (ed.), *Functional Morphology in Vertebrate Paleontology*, pp. 34–54. Cambridge: Cambridge University Press.

Weishampel, D. B., and D. B. Norman. 1989. Vertebrate herbivory in the Mesozoic; jaws, plants, and evolutionary metrics. *Geological Society of America Special Paper* 238: 87–100.

Witmer, L. M. 1995. The extant phylogenetic bracket and the importance of reconstructing soft tissues in fossils. In J. J. Thomason (ed.), *Functional Morphology in Vertebrate Paleontology,* pp. 19–33. Cambridge: Cambridge University Press.

———. 1997. The evolution of the antorbital cavity in archosaurs: A study in soft-tissue reconstruction in the fossil record with an analysis of the function of pneumaticity. *Journal of Vertebrate Paleontology Memoir* 4: 1–73.

15. Cranial Ornamentation of Ankylosaurs (Ornithischia: Thyreophora): Reappraisal of Developmental Hypotheses

M. K. VICKARYOUS,
A. P. RUSSELL, AND
P. J. CURRIE

Abstract

The presence of cranial ornamentation on the dorsal and lateral surfaces of the skull has long been considered diagnostic for the Ankylosauria. Ornamentation and the highly fused nature of the ankylosaur head skeleton have constrained our understanding of cranial anatomy. Two alternative hypotheses have been proposed to explain the origin of cranial ornamentation. The most widely accepted hypothesis suggests that superficial osteoderms coossified with the external surface of the skull, whereas the alternative hypothesis proposes that ornamentation was derived from the elaboration of the dermatocranium. Evidence from the Late Cretaceous ankylosaurids *Euoplocephalus tutus* and *Pinacosaurus grangeri*, and comparative data from extant squamates demonstrating similar conditions support both hypotheses. The developmental model we propose suggests that the dermis of ankylosaurs demonstrated a propensity for the formation of osteoderms under normal ontogenetic conditions. This propensity accounts for the postcranial "armor," the buccal ossifications, and some of the cranial ornamentation. However, evidence from subadult *Euoplocephalus* and *Pinacosaurus* specimens suggests that a second mechanism, dermatocranial elaboration, also plays a role in the development of ornamentation. The demonstrable association between developmental processes and osteologic correlates illustrated here provides a means of inferring ontogenetic mechanisms in extinct taxa.

Introduction

The ankylosaur head skeleton is arguably the most enigmatic and poorly known area of osteology within the Dinosauria. In contrast to what is known about most other extinct vertebrates, the paucity of knowledge of ankylosaur skull anatomy appears to be more the result of peculiar morphology than a limited fossil record. In particular, members of the Ankylosauria demonstrate three major architectural novelties: secondary closure of the supratemporal and antorbital fenestrae, nearly complete obliteration of cranial sutures in adult-size specimens, and extensive development of ornamentation across the dermatocranium (Fig. 15.1; Sereno 1986). Whereas all three architectural modifications obscure the topography and morphology of individual dermatocranial elements (and otherwise impede research dealing with homology and phylogeny), the development of cranial ornamentation is the most conspicuous synapomorphy of the taxon. Despite its conspicuous nature and near ubiquitous presence within the clade Ankylosauria, the nature of this ornamentation has received scant attention. Through a reappraisal of the fossil evidence and an analysis of extant nondinosaurian vertebrates with comparable morphology, we attempt to elucidate the developmental mechanisms intrinsic to the formation of cranial ornamentation.

Ornamentation, defined as novel anatomic attributes or modifications (i.e., elaborations) of a preexisting structure not appearing to have a primary role in food acquisition or locomotion (Vickaryous and Rybczynski, in press), is prevalent throughout the Dinosauria, and as such, it is by no means unique or restricted to the Ankylosauria. Although such features are generally osseous in a paleontological context (e.g., horn cores, bony crests and frills, bosses, nodes, and foveae or pitting), modern avian dinosaurs demonstrate a suite of soft tissue derivatives (e.g., wattles, combs, feather crests, and colored loreal regions) that are unlikely to withstand the rigors of preservation. In ankylosaurs, the osteologic manifestation of cranial ornamentation has relatively little surface relief, with few elongate projections or protuberances. Elements of the dermatocranium (those parts of the skull derived from intramembranous ossification—e.g., the nasals, frontals, and parietals) appear to have become externally embossed with continuous, amorphous, or rugose bone, or some combination of these. Additionally, the skull may also exhibit a series of interconnected superficial furrows that subdivide the external surface into a mosaic of polygons and hornlike bosses, resulting in the manifestation of cranial sculptur-

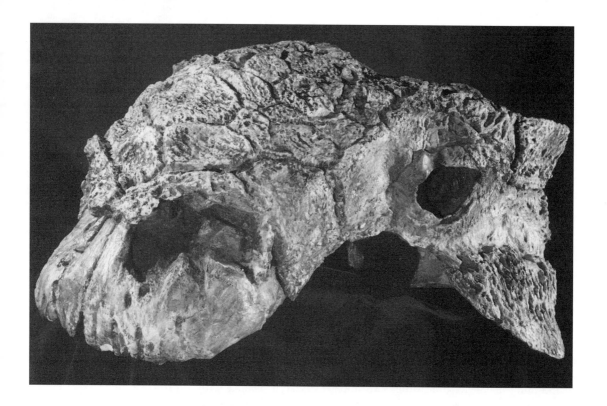

Figure 15.1. Euoplocephalus tutus *adult cranium (AMNH 5405), oblique rostrodorsolateral view.*

ing (Fig. 15.1; Coombs 1971). Variants of this phenomenon have often been cited as a means by which the Ankylosauria may be subdivided (e.g., nodosaurids versus ankylosaurids; see Carpenter 1990; Coombs 1971; Coombs and Maryańska 1990; Sereno 1986). Cranial sculpturing is most prominent across the antorbital region, with many taxa demonstrating a gradation of the mosaic pattern into a more random, amorphous texture posterior to the orbits.

Previous work has resulted in the erection of two competing hypotheses that attempt to explain the development of cranial ornamentation (including cranial sculpturing) in ankylosaurs: overlying bony plates (osteoderms) coossifying with the dermatocranium, or osteologic elaboration of the dermatocranial elements. Within the limited confines of the fossil record, problems of development are difficult to examine. Therefore, it is necessary to draw on ontogenetic processes that occur within extant taxa to further elucidate the applicability of these two competing, but not necessarily mutually exclusive, hypotheses. The issue of cranial ornamentation development in ankylosaurs was investigated by reappraising fossil evidence and reviewing extant taxa that demonstrate similar conditions of osseous ornamentation. In this instance, application of the extant phylogenetic bracket (Witmer 1995, 1997) results in a decisively negative assessment (i.e., a level 3 inference); neither modern birds nor modern crocodylians possess osteologic correlates of the cranial ornamentation of ankylosaurs. However, examination of more remote outgroups identifies a common pattern of development present within modern amniotes. This common pattern of development is herein considered to represent a biological generalization (Bryant and Russell 1992), and as such, this study relies heavily on the ahistorical extrapolatory approach (Bryant and Russell 1992; also see Weishampel 1995; Witmer 1995). Our observations lead us to conclude that the cranial ornamentation of ankylosaurs developed as a result of two independent developmental processes: the co-ossification of overlying intramembranous bone to the skull and the elaboration of existing cranial elements. The primacy of each of these mechanisms is apparently quite region specific.

Because of the erratic and inconsistent nature of terminology used to describe cranial ornamentation, some clarification is required at the outset. The term "osteoderm" ("bone-skin") refers to any superficial dermal ossification that supports overlying epidermal scales (e.g., Hildebrand 1988; Romer 1956; Zylberberg and Castanet 1985; cf. osteoscute: de Buffrénil 1982; Meszoely 1970; cf. dermal scute: Camp 1923). Although it is generally accepted that at least some members of most tetrapod clades possess or possessed osteoderms (e.g., mammals, squamates, and crown-group archosaurs; Fig. 15.2), the accuracy of the term has been disputed. Moss (1969) found that although the dermal skeletal tissue of the various tetrapod lineages is homologous, the histologic structure of the tissue is not homogeneous, ranging from dense calcified tendon to true bone. Often these tissues grade seamlessly into one another, preventing segregation of distinct types. Consequently, Moss (1969: 510) advocated use of the term "sclerification" to emphasize the inconsistent tissue histology. However, with the excep-

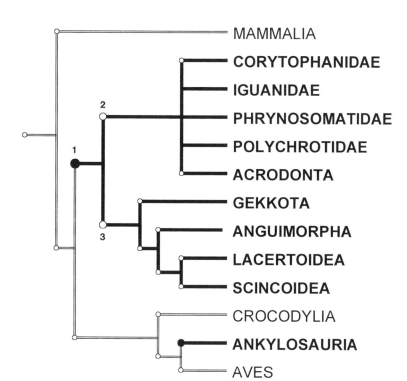

Figure 15.2. Modified phylogeny of the Amniota, providing a phylogenetic framework for the taxa discussed in this article (after Estes et al. 1988; Gauthier et al. 1988; Witmer 1997). Taxa in bold typeface are of primary importance to this study. Node 1 = Squamata; node 2 = Iguania; node 3 = Scleroglossa.

tion of teeth and perhaps eggshell, preservation of nonosseous tissues in the fossil record is rare, and the term "sclerification" (although technically more accurate) is not appropriate for the present study. The term "coossification" ("ossified with"; Trueb 1966) refers to the fusion of two or more structures through the deposition of bone, reorganization of bone, or both.

Materials

The development of cranial ornamentation is nearly ubiquitous within the Amniota, and thus, a comprehensive systematic examination is beyond the scope of this article. For reasons of practicality (relative size and abundance, or ease of transport), the comparative (homoplasious) portion of this study focuses on the development of cranial ornamentation in extant squamates. All amniote lineages appear to maintain the capacity to develop true osteoderms (both under normal ontogenetic conditions and pathologically) by similar processes (Moss 1972), although this is variably expressed within a given clade. The Squamata (see Fig. 15.2) is a monophyletic clade encompassing the majority of taxa commonly considered as structural grade "reptiles." Squamates may be subdivided into two major lineages: the Iguania and the Scleroglossa (Fig. 15.2). Members of both lineages exhibit a wide array of cranial ornamentation, including osseous and soft tissue structures. A survey of alcohol-preserved and skeletal specimens representing most of the major squamate clades (see Appendix 15.1), supplemented with data from the literature (see reviews by Camp 1923; Estes et al. 1988; and Gadow 1901), suggests that particular patterns of osteology are strongly correlated with phylogeny. The presence of both cephalic and postcranial osteoderms in squamates appears to be restricted to members of the Scleroglossa, in particular scincoids and anguimorphs (exclusive of the Ophidia), as well as some members of the Gekkota and Lacertoidea. In contrast, members of the Iguania (with the exception of *Amblyrhynchus cristatus*) rarely, if ever, develop osteoderms (Camp 1923; Gadow 1901). However, many iguanian taxa exhibit the development of osseous horn cores, bosses (e.g., phrynosomatids and some acrodonts), and exostoses (polychrotids).

Extant iguanians serve as the basis for modeling the dermatocranial elaboration hypothesis. Of primary concern for this study were several embryo and adult combinations of *Phrynosoma* (including *P. cornutum*, *P. hernandesi*, and *P. modestum*), embryonic *Chamaeleo pumilus*, neonate *C. calyptratus*, and adult *C. jacksonii*, *C. montium*, and *C. parsoni* (see Appendix 15.1). Extant scleroglossans were similarly used as models for testing the osteodermal coossification hypothesis. The principal specimens examined for this study were subadult and adult *Heloderma suspectum*, adult *Gerrhosaurus major*, and adult *Tarentola mauritanica* (see Appendix 15.1).

The developmental models were generated through observations made from a combination of cleared and double-stained, alcohol-preserved, fresh-frozen, and dried skeletal specimens. Alcohol-preserved and fresh-frozen specimens were subjected to radiographic imaging.

Despite demonstrating prominent cranial ornamentation and a relatively close phylogenetic relationship to dinosaurs, members of the Crocodylia were not considered for use as extant developmental models. The ornamentation of crocodylians, a collection of shallow pits and grooves, is morphologically divergent from that of known ankylosaurs. Although the developmental mechanism giving rise to this morphology (differential resorption of the periosteum; de Buffrénil 1982) is not precluded from occurring within the Ankylosauria, herein, we do not consider it to play a major role. Postcranial osteoderms of crocodylians are presumed to share a homologous developmental pathway with those of scleroglossans (*sensu* Moss 1972).

A superficial consideration of osseous cranial ornamentation in mammals suggests that the developmental pathways noted in squamates are representative of a more inclusive clade. Among extant mammals, only dasypodids (armadillos) commonly develop osteoderms (DeBlase and Martin 1974). Morphologically, the condition appears to parallel that of scleroglossans. The elaboration of the frontals in the Bovidae is likewise similar to the developmental process noted in iguanians.

A review of original ankylosaur material was based largely on two taxa; *Euoplocephalus tutus* (Fig. 15.1) and *Pinacosaurus grangeri* (Fig. 15.3). Both taxa were medium-size ankylosaurid ankylosaurs, from the Late Cretaceous of North America and Asia, respectively. We selected these taxa because both are relatively common and skull material is

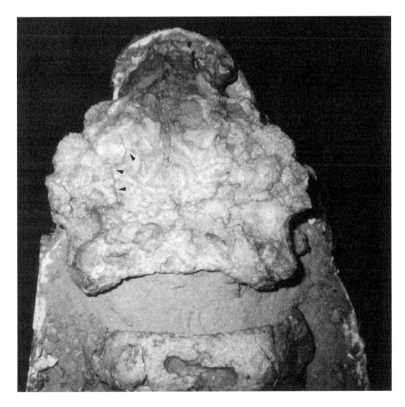

Figure 15.3. Pinacosaurus grangeri *subadult cranium and first cervical half-ring (TMP 90.301.1), dorsal view. Small arrowheads indicate the lateral sutural border of the left frontal.*

abundantly available. Although the majority of *Euoplocephalus* specimens appear to represent adult-size individuals, a number of undescribed elements referable to subadult individuals have recently been identified (Vickaryous, unpublished data). Nearly all the material referable to *Pinacosaurus* represents subadult individuals. Wherever possible, additional material assigned to other ankylosaur taxa was examined.

Institutional Abbreviations. AMNH: American Museum of Natural History, New York. APRC: Personal reference collection of A. P. Russell, University of Calgary, Calgary, Canada. LACM: Los Angeles County Museum, Los Angeles, California. NMC: Canadian Museum of Nature, Ottawa, Canada. PJBC: Personal reference collection of P. J. Bergmann, University of Calgary, Calgary, Canada. TMP: Royal Tyrrell Museum of Palaeontology, Drumheller, Alberta, Canada. UCMZ: University of Calgary Museum of Zoology, Calgary, Canada.

Hypothesis 1: Coossification of Osteoderms

Historical Development

Overwhelmingly, the majority of dinosaur researchers have advocated (usually incidentally) the hypothesis of osteoderm coossification with the dermatocranium as the developmental mechanism responsible for ankylosaur cranial ornamentation (e.g., Brown 1908; Lambe 1902, 1919; Russell 1940; Sereno 1997; Sternberg 1929). Often this interpretation has been invoked circumstantially (and inappropriately) during discussion of morphology, but it has yet to be critically tested or substantiated.

Development of Cranial Ornamentation in Extant Scleroglossans

A review of the literature suggests that the development of osteoderms (both cranial and postcranial) in scleroglossans is a character that undergoes repetitive reversal within the clade (Arnold 1973; Camp 1923; Estes et al. 1988; Gadow 1901; Zylberberg and Castanet 1985). Morphologically, there is a high degree of variability among members demonstrating the condition, ranging from the imbricated "shingle" morphotype of *Gerrhosaurus* (Scincoidea) and *Anguis* (Anguimorpha) (the so-called primitive condition; Otto 1909), to the polygonal "pavement" morphotype of *Heloderma* (Anguimorpha; see Fig. 15.4) and *Tarentola* (Gekkota) (the so-called advanced condition; Otto 1909). Developmental mechanisms investigated to date have been largely restricted to the polygonal pavement morphotype, although all osteoderms are presumed to be homologous (Camp 1923; Moss 1969).

Moss (1969) investigated the ontogenetic development of osteoderms in the anguimorphan *Heloderma horridium*. Presumptive osteoderms develop as domed regions of thick collagen within the dermis, first appearing over the ossified head skeleton, then spreading caudally (Fig. 15.4). Topographic distribution of these collagen "domes" over the cranium bears no direct relationship to the underlying, and well-

established, dermatocranial elements. As the individual matures, the collagen domes increase in size before finally becoming ossified. With later development, the osteoderms may fuse directly to the dermatocranium (e.g., *Heloderma*) or remain suspended within the overlying dermis (e.g., some gekkotans such as *Tarentola*). An ontogenetic study of the anguimorphan *Anguis fragilis* by Zylberberg and Castanet (1985) indicates that the shingle morphotype undergoes a similar process.

Figure 15.4. Heloderma suspectum *subadult specimen (UCMZ [R] 2000.001; snout–vent length = 192 mm), radiographic image illustrating the pervasive development of osteoderms in the taxon. The osteoderms located in the head region are pronounced as compared with the more posterior areas of the body. Scale bar = 10 mm.

Osteologic Correlates

Because of the nature of osteoderm development, a number of biological generalizations (*sensu* Bryant and Russell 1992) can be made with regard to the resulting osteology. Osteoderms develop within the dermis, superficial to the head and postcranial skeletons, and as such, they are not confined to the topographic limitations of individual elements. Thus, osteoderms may originate in positions that overlap several elements, sutural boundaries, or both. Additionally, osteoderms may form in regions where underlying skeletal elements are absent (e.g., between adjacent ribs, superficial to the temporal fenestrae and orbits; Fig. 15.4). Consequently, the position and morphology of the cranial openings and the sutural arrangement of dermatocranial elements may become obscured in mature individuals. Before fusion with the cranium, however, it is possible to remove the osteoderms by removing the integument.

Although the gross morphology of postcranial osteoderms may be highly variable, those that develop over the skull generally form polygons (Fig. 15.4). These polygons appear to develop as the incipient osteoderms (centers of ossification within individual collagen domes) expand radially and begin to encroach on one another. Continued growth is constrained by adjacent (incipient) osteoderms, thereby giving rise to the polygonal morphology. Abutment of adjacent polygonal osteoderms may cover the cranial openings with extradermatocranial dermal bone (e.g., the supraorbital ossifications of the gekkotan *Tarentola*; Bauer and Russell 1989). The polygonal configuration of the cranial osteoderms closely resembles the morphologic condition noted in the carapace of placodonts (Westphal 1976) and some mammals (e.g., dasypodids). A brief analysis of dasypodid material suggests that the presence of osteoderms in mammals is derived from developmental processes similar to those of scleroglossans. Dried adult *Dasypus novemcinctus* skeletal material and alcohol-preserved adult *Chaetophractus* sp. demonstrates the presence of both the polygonal pavement morphotype and the imbricated shingle morphotype osteoderms, covering the entire dorsal surface of the head and body. However, no osteoderms were identified through the radiographic imaging of alcohol-preserved fetal *Chaetophractus vellerosus*. Because of the presence of the osteologic correlates identified in scleroglossans, dasypodids are presumed to undergo a similar developmental process.

Hypothesis

Despite their phylogenetically distant relationship to ankylosaurs, many scleroglossans appear to present a pattern of cranial ornamentation that is morphologically congruent with them. On the basis of this gross similarity, it may be hypothesized that ankylosaur cranial ornamentation developed in an analogous manner—that is, osteoderms overlying the dermatocranium fuse with the skull, resulting in a polygonal pattern of ornamentation that obscures sutural contacts and cranial openings. Before maturation (and subsequent fusion), the dermatocranium of ankylosaurs should not exhibit cranial ornamentation in any areas associated with this mode of formation.

Testing the Hypothesis

To test the coossification hypothesis, osteologic correlates within the Ankylosauria must be identified. Requirements for satisfying the hypothesis of osteoderm coossification include the following: (1) the complete absence of ornamentation in immature specimens, (2) the presence of ornamentation that obscures (overlaps) the sutural arrangement in mature specimens, and (3) the presence of osseous ornamentation in regions of the skull without underlying dermatocranial elements.

The ubiquitous presence of cranial ornamentation has been purported to obscure the sutural arrangement of the head skeleton of ankylosaurs (Coombs and Maryańska 1990). Evidence to support this assertion is derived from the examination of material referred to *Pinacosaurus grangeri*. Several specimens of *Pinacosaurus* (TMP 90.301.1; Fig. 15.3; Maryańska 1971, 1977) illustrate the morphology of individual dermatocranial elements. On the basis of, in part, the lack of cranial ornamentation and the relatively small and unfused nature of the skeletal elements, these specimens are considered to represent subadult or immature individuals. Examination of the holotype (AMNH 6523), a larger specimen with ornamentation, suggests that the state of development of cranial sculpturing is ontogenetically regulated. Assuming that skull morphology is conservative (although see Carpenter et al. 2001 for a new ankylosaur taxon with a divergent element morphology), comparison of subadult *Pinacosaurus* crania with those of other, related ankylosaurids suggests that there is no relationship between the individual dermatocranial elements and any overlying polygonal ornamentation. Subadult cranial material is not yet known for all taxa, thus thwarting a more comprehensive systematic review. (Cranial material referred to an undescribed species of *Minmi* also appears to demonstrate the sutural arrangement of the dermatocranium, although further study is required; Molnar 1996.)

The majority of ankylosaur taxa are known, at least in part, from cranial material considered to represent adult or mature individuals. In every case, the specimens demonstrate extensive development of cranial ornamentation (most often cranial sculpturing). With the exception of some specific topographic regions (e.g., the premaxillary beak), the sutural arrangement across the skull roof in these specimens is unknown and presumed to be obscured by the presence of cranial ornamentation.

In addition to the widespread development of postcranial osteoderms, a number of ankylosaur taxa have been found preserved with in situ osseous eyelids (*Euoplocephalus tutus*, AMNH 5238, AMNH 5404; Coombs 1972) and buccal ossifications (osseous cheek plates; *Edmontonia rugosidens*, AMNH 5381; *Panoplosaurus mirus*, NMC 2759). All these dermal elements develop in the apparent absence of underlying skeletal tissue.

The hypothesis of osteodermal coossification giving rise to cranial ornamentation was originally dismissed by Coombs (1971) in his comprehensive review of the Ankylosauria. Coombs cited two major flaws with the argument, one based on gross morphology and the other on bone histology. Before Coombs's research, no subadult ankylosaur skull had been described, and thus all the cranial material he examined demonstrated extensive development of cranial ornamentation. He reasoned that if the cranial ornamentation of ankylosaurs was the result of osteoderms coossifying with the skull surface, then a specimen should exist where the dermatocranium proper was visible. Despite referring to an unseen subadult specimen of *Pinacosaurus* (via communication with Teresa Maryańska), he submitted that "no such specimen exists" (Coombs 1971: 157).

Coombs found further evidence to refute the coossification hypothesis by examining the bone histology of an ankylosaur skull. He suggested that if overlying osteoderms did contact and fuse with the dermatocranium, then an appreciable thickening of the skull roof should be noted when compared with that of other nonpachycephalosaurian ornithischians. A single cranium of *Euoplocephalus tutus* (AMNH 5403) was transversely sectioned across the antorbital region. Examination of the cross sections indicated that the thickness of the *Euoplocephalus* skull roof was "modest" (Coombs 1971: 156) and generally comparable to that of most other ornithischians. He also noted that there was no indication of a juncture between the dermatocranium and the ornamentation. This led Coombs to surmise that *Euoplocephalus*, as a representative of the Ankylosauria, did not have osteoderms fused to the skull roof proper.

A review of the present evidence, supplemented by new cranial thin sections, accounts for the misgivings of Coombs (1971). Detailed examination of subadult *Pinacosaurus* specimens confirms the absence of cranial ornamentation in immature ankylosaurs (see above). Additionally, an undescribed bony plate (TMP 89.36.183; Fig. 15.5) may represent unfused cranial ornament. The element has a rugose and pitted texture, similar to that of postcranial osteoderms (to which it was originally referred). However, the specimen is thin (<10 mm thick), flat with a slightly concave upper surface, and hexagonal in dorsal view. A comparison of this specimen with other known ankylosaur postcranial osteoderms suggests that the morphology is unique. It most closely resembles the polygons associated with the rostral cranial sculpturing of *Euoplocephalus* and is herein referred to as an unfused cranial osteoderm.

To review ankylosaur cranial bone histology, the original thin sec-

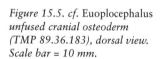

Figure 15.5. cf. Euoplocephalus *unfused cranial osteoderm (TMP 89.36.183), dorsal view. Scale bar = 10 mm.*

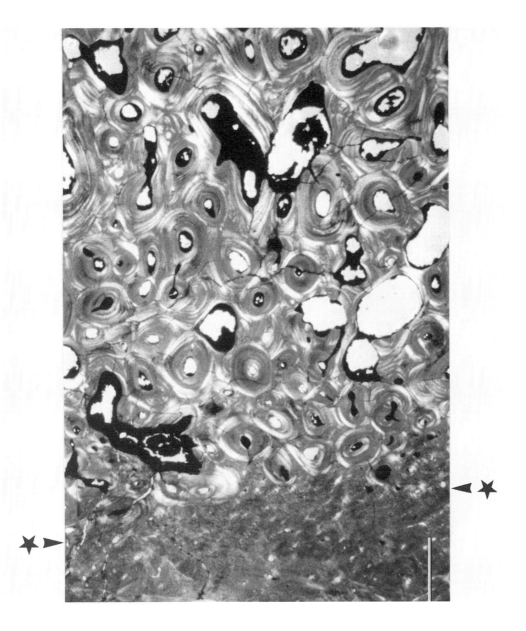

tions of Coombs (from AMNH 5403) have been augmented with new material (TMP 67.20.20 and TMP 98.115.2). Two main histologic layers may be differentiated (Fig. 15.6); a thin (<2.5 mm) layer of isolated, unorganized primary osteons superficial to deeper, extensively remodeled Haversian bone (Coombs 1971). Thorough examination of all thin sections failed to identify any sutural junctions, either between overlying osteoderms and the dermatocranium or between individual dermatocranial elements. The highly reorganized nature of the bone histology suggests that the cranium underwent continual remodeling throughout ontogeny and that any sutural contacts have long since been obliterated. The functional implications of remodeled bone in the ankylosaur cranium are not presently understood. However, a super-

Figure 15.6. cf. Euoplocephalus thin section (TMP 98.115.2) from an isolated fragment of skull representing the frontal–nasal region. Section was taken in the transverse plane, with the periosteum toward the top of the page. Position of star symbols indicates the junction between the two main histological layers: a superficial layer of woven bone and a deeper layer of highly remodeled Haversian bone. Scale bar = 1 mm.

ficial layer of bone lacking a strict organizational pattern (i.e., woven bone), coupled with the presence ornamentation, has been suggested as a structural mechanism for stress diffusion (Coldiron 1974). Thus, in a teleologic sense, the peripheral layer of an ankylosaur skull may have arisen in response to dispersing and dissipating any stresses incurred.

Conclusions

All the fundamental requirements for advocating the hypothesis of osteoderm coossification are fulfilled by ankylosaur cranial osteology. The expression of cranial sculpturing in ankylosaurs is governed by ontogeny, and the entire clade demonstrates a propensity for the production of osseous, nonpathologic dermal tissue. We consider the superficial furrows that subdivide the cranium (giving rise to the cranial sculpturing) to represent the areas of coossification between adjacent cephalic osteoderms. The unusual distribution of cranial sculpturing (i.e., concentrated around the rostrum) may reflect the degree of interaction between the epidermis and the dermis. However, the role of the epidermis in mediating osteodermal growth is not presently understood. In the anuran *Hyla septentrionalis,* dense connective tissue of the dermis is effectively replaced by bone over areas of the cranium (Trueb 1966). This integumentary derivative then coossifies with the dermatocranium, creating a casque. The lack of epidermal scutes may partially explain the absence of polygonal organization in this secondary dermal bone.

Whether the production and subsequent coossification of dermal bone to the skull is responsible for all the cranial ornamentation noted in ankylosaurs has yet to be addressed. The subadult crania of *Pinacosaurus grangeri,* although devoid of cranial sculpturing, do demonstrate small hornlike bosses over the orbits and at the posterior corners of the skull. In addition, Jacobs et al. (1994: 338) noted the presence of what they termed "excrescences" on some disarticulated elements of a very small, subadult nodosaurid skull. The development of these structures cannot be accounted for by osteoderm coossification.

Hypothesis 2: Elaboration of the Dermatocranium

Historical Development

As noted above, Coombs (1971) rejected the notion of osteoderm coossification as the predominant developmental mechanism of cranial ornamentation because of a number of perceived morphologic and histologic inconsistencies. Alternatively, he proposed that the osseous ornamentation of the dermatocranium was the result of individual cranial elements becoming elaborated under the influence of epidermal structures. Coombs noted that elaborate modifications of the cranium were not without precedent among the Ornithischia (e.g., the premaxillary–nasal crests of lambeosaurines and the squamosal–parietal frills of neoceratopsians). Therefore, although the possibility of some extradermatocranial contributions to the ankylosaur skull was not entirely precluded, it was thought to play a minor role.

A more recent review of the Ankylosauria by Coombs and Maryańska (1990) puts forward a less polarized view by stating that the cranial ornamentation might be the result of either elaboration of the dermatocranium or the coossification of osteoderms.

Development of Cranial Ornamentation in Extant Iguanians

Analysis of an ontogenetic sequence of cleared and double-stained *Phrynosoma* specimens provide the basis for our discussion of the development of cranial ornamentation in iguanians.

Members of the taxon *Phrynosoma* are characterized, in part, by the presence of laterally and posteriorly directed cranial horns (Montanucci 1987). These horns can initially be distinguished as protuberances on the squamosals and parietal of neonates (e.g., *Phrynosoma modestum*; Fig. 15.7). Throughout the process of ontogeny, the protuberances may increase in size and become modified in appearance to form a variety of horns and bosses that are always confined to individual dermatocranial elements. Cranial sutures are generally not obscured by the development of osseous ornamentation, although the degree of elaboration is taxonomically variable. Examination of cleared and double-stained, alcohol-preserved, and dried skeletal material of various specimens of *Chamaeleo* corroborates these findings. The role of epidermal structures in the development of dermatocranial protuberances is not presently understood, although the horns of phrynosomatids and *Chamaeleo jacksonii* are sheathed by a single, much enlarged conical scale in life.

Osteologic Correlates

Early ontogenetic development is characterized by osseous outgrowths from dermatocranial elements, resulting in the development of horns, bosses, and protuberances in iguanians. Cranial ornamentation derived from elaborations of the head skeleton proper cannot be removed at any time during ontogeny without causing physical injury to the skeletal elements of origin. In addition, the cranial ornamentation of iguanians is restricted to individual elements and thus does not generally transgress sutural boundaries except sometimes during the late growth stages (e.g., *Chamaeleo jacksonii*). This condition appears to be paralleled by bovids in the development of frontal horn cores.

Hypothesis

The morphology of the posterolaterally directed horns and superficial protuberances on iguanian crania appears to parallel the condition of cranial ornamentation demonstrated by ankylosaurs. On the basis of this osteologic resemblance, we hypothesize that the cranial ornamentation of ankylosaurs developed in an analogous manner— that is, by elaboration of individual dermatocranial elements.

Testing the Hypothesis

Requirements for satisfying the hypothesis of dermatocranial elaboration include the presence of ornamentation in immature specimens and the restriction of ornamentation to individual elements.

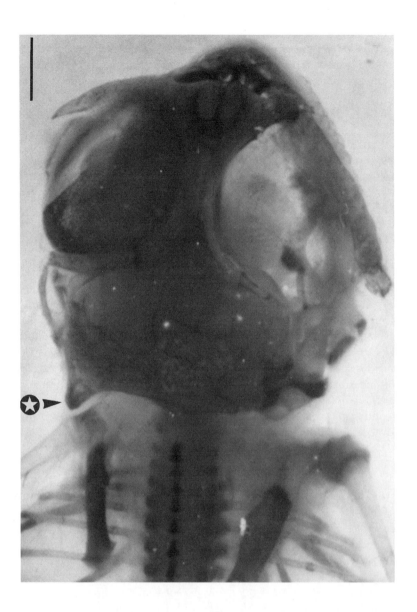

Figure 15.7. Phrynosoma modestum *neonate, cleared and double stained (LACM 123344; snout–vent length = 30 mm), dorsal view (right eye removed). Circle-star symbol indicates the position of an occipital (parietal) horncore. Scale bar = 1 mm.*

The developmental nature of ankylosaur skulls is such that the majority of cranial sutures are rarely visible. Examination of material referred to subadult *Pinacosaurus grangeri* (TMP 90.301.1; Fig. 15.3; Maryańska 1971, 1977) suggests that this feature is related to ontogeny. Although the cranial sutures are readily apparent (the result of the specimens lacking the coossification of dermally derived cranial sculpturing), they are not without osseous ornamentation. The squamosals, quadratojugals, and supraorbitals all demonstrate the incipient development of triangular or pyramidal bosses. These bosses correspond to individual elements on a one-to-one basis and appear as direct outgrowths of the dermatocranium. Additional supporting evidence includes several undescribed subadult cranial elements. These unrelated elements (none are considered to represent the same individual) include a left supraorbital (TMP 88.106.5; Fig. 15.8), two left squamosals

Figure 15.8. cf. Euoplocephalus left supraorbital (TMP 88.106.5), dorsal view. Arrowheads mark the position of the sutural contacts. po = postorbital; pr = prefrontal; so = additional supraorbitals. Scale bar = 10 mm.

(TMP 67.19.4 and TMP 93.36.79; Fig. 15.9), and a right and left quadratojugal (TMP 67.20.20 and TMP 79.14.164, respectively; Fig. 15.10). On the basis of morphology, all of these elements may be referred to the Ankylosauridae cf. *Euoplocephalus*. Of particular significance are the sutural boundaries on each element and the triangular to pyramidlike outgrowths or bosses.

Currently, little is known about the osteology of subadult nodosaurids. Several disarticulated cranial elements from the Albian of Texas suggest that at least some subadult nodosaurids developed rugose, encrusting elaborations ("excrescences"; Jacobs et al. 1994: 338). A number of nodosaurid taxa are also known to develop incipient bosses above the orbits and at the posterior corners of the cranium (e.g., *Pawpawsaurus campbelli*; Lee 1996).

Conclusions

Osteologic evidence suggests that the cranial ornamentation of ankylosaurs is partly the result of elaboration of outgrowths from dermatocranial elements. Such expression is consistent throughout known ontogenetic stages, at least among members of the Ankylosauridae. There is no evidence, however, to support the hypothesis that all of the cranial ornamentation of ankylosaurs is derived in this way.

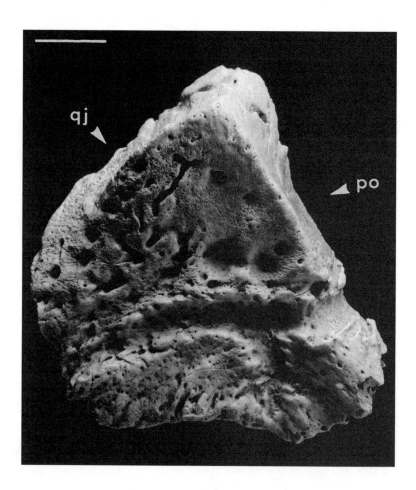

Figure 15.9. cf. Euoplocephalus
left squamosal (TMP 93.36.79),
occipital view. Arrowheads mark
the position of the sutural
contacts. po *= postorbital;* qj *=*
quadratojugal.
Scale bar = 10 mm.

A Synthetic Approach

A comparison of osteologic correlates associated with extant developmental processes with those demonstrated by fossils permits the reappraisal of ontogenetic mechanisms in extinct taxa. Previous work had suggested two alternative and competing hypotheses to explain the development of cranial ornamentation in ankylosaurs. A review of modern squamates demonstrates the independent occurrence of each process within selected taxa and presents the opportunity to determine the resulting osteologic expression of each mechanism. Comparisons of these skeletal correlates with ankylosaur cranial material suggest that ankylosaurs use both developmental processes. The degree of expression of each mechanism is highly variable among, and even within, taxa. Thus, the expressed pattern may be used to diagnose a taxon (e.g., ankylosaurids versus nodosaurids) or a particular ontogenetic stage. Carpenter (1990) noted the variability of cranial ornamentation expressed within the taxon *Edmontonia rugosidens*, and suggested that it may reflect a state of maturation. Contention over the taxonomic validity of the many synonyms of *Euoplocephalus* is based in part on subtle characters associated with cranial ornamentation morphology (Pen-

Figure 15.10. cf. Euoplocephalus
left quadratojugal (TMP
79.14.164), lateral view. Arrow
marks the position of the sutural
contact. sq = squamosal.
Scale bar = 10 mm.

kalski 1998, 2001). Resolution of these issues is likely embedded within a greater overall understanding of developmental mechanics.

Summary

The developmental biology of dinosaurs has rarely received extensive consideration. Because of the inherent limitations of the fossil record, these investigations require alternative (e.g., extrapolatory) techniques to test the validity of interpretation. Despite being well established, the use of modern taxa as functional models for fossil organisms is often ineffectively applied. Efforts to minimize speculation rely upon the stringent analysis of available evidence and the identification of relevant osteologic correlates. Use of modern ontogenetic mechanisms as a research protocol for the induction of fossil structures has broad applications paleobiologically. A review of the ankylosaur head skeleton in this context suggests that the development of cranial ornamentation is the result of two independent and generally segregated processes involving the dermatocranium proper and extradermatocranial ossifications.

Acknowledgments. We thank L. M. Witmer (Ohio University), L. McGregor (University of Calgary), and K. Carpenter (Denver Museum of Natural History) for critically reviewing and improving this article. A debt of gratitude is owed to R. Etheridge (San Diego State University) for an extremely valuable discussion on squamate cranial ornamentation and J. Resultay (University of Calgary) for skillful sectioning of the ankylosaur material. A. L. Vickaryous photographed figures 1, 5, 8, 9, and 10; all others are by the senior author. E. Gaffney and M. Norell (American Museum of Natural History), A. Newman, J. Wilke, and L. Cook (Royal Tyrrell Museum of Palaeontology), W. Fitch (University of Calgary Museum of Zoology), and P. J. Bergmann provided access to, and facilitated loans of, specimens in their care. This research was supported by a Duerksen memorial scholarship and a Heaton student support grant awarded to M.K.V. and an National Science and Engineering Council operating grant awarded to A.P.R.

References Cited

Arnold, E. N. 1973. Relationships of the Palearctic lizards assigned to the genera *Lacerta, Algyroides* and *Psammodromus* (Reptilia, Lacertidae). *Bulletin of the British Museum (Natural History)* 25: 291–366.

Bauer, A. M., and A. P. Russell. 1989. Supraorbital ossifications in geckos (Reptilia: Gekkonidae). *Canadian Journal of Zoology* 67: 678–684.

Brown, B. 1908. The Ankylosauridae, a new family of armored dinosaurs from the Upper Cretaceous. *Bulletin of the American Museum of Natural History* 24: 187–201.

Bryant, H. N., and A. P. Russell. 1992. The role of phylogenetic analysis in the inference of unpreserved attributes of extinct taxa. *Philosophical Transactions of the Royal Society of London, B* 337: 405–418.

Camp, C. L. 1923. Classification of the lizards. *Bulletin of the American Museum of Natural History* 48: 289–481.

Carpenter, K. 1990. Ankylosaur systematics: Examples using *Panoplosaurus* and *Edmontonia* (Ankylosauria: Nodosauridae). In K. Carpenter and P. J. Currie (eds.), *Dinosaur Systematics: Approaches and Perspectives,* pp. 281–298. Cambridge: Cambridge University Press.

Coldiron, R. W. 1974. Possible functions of ornament in labyrinthodont amphibians. *Occasional Papers of the Museum of Natural History, The University of Kansas* 35: 1–19.

Coombs, W. P., Jr. 1971. The Ankylosauria. Ph.D. diss. Columbia University, New York.

———. 1972. The bony eyelid of *Euoplocephalus* (Reptilia, Ornithischia). *Journal of Paleontology* 46: 637–650.

Coombs, W. P., Jr., and T. Maryańska. 1990. Ankylosauria. In D. B. Weishampel, P. Dodson, and H. Osmólska (eds.), *The Dinosauria,* pp. 456–483. Berkeley: University of California Press.

DeBlase, A. F., and R. E. Martin. 1974. *A Manual of Mammalogy with Keys to the Families of the World*. Dubuque, Iowa: William C. Brown.

de Buffrénil, V. 1982. Morphogenesis of bone ornamentation in extant and extinct crocodilians. *Zoomorphology* 99: 155–166.

Estes, R., K. de Queiroz, and J. Gauthier. 1988. Phylogenetic relationships within Squamata. In R. Estes and G. Pregill (eds.), *Phylogenetic Relationships of the Lizard Families; Essays Commemorating Charles L. Camp*, pp. 119–270. Stanford, Calif.: Stanford University Press.

Gadow, H. 1901. *The Cambridge Natural History*. Vol. 8, *Amphibia and Reptiles*. London: Macmillan.

Gauthier, J., A. G. Kluge, and T. Rowe. 1988. Amniote phylogeny and the importance of fossils. *Cladistics* 4: 105–209.

Hildebrand, M. 1988. *Analysis of Vertebrate Structure*. New York: Wiley.

Jacobs, L. L., D. A. Winkler, P. A. Murry, and J. M. Maurice. 1994. A nodosaurid scuteling from the Texas shore of the Western Interior Seaway. In K. Carpenter, K. F. Hirsch, and J. R. Horner (eds.), *Dinosaur Eggs and Babies*, pp. 337–346. Cambridge: Cambridge University Press.

Lambe, L. M. 1902. New genera and species from the Belly River Series (mid-Cretaceous). *Contributions to Canadian Palaeontology, Geological Survey of Canada* 3: 25–81.

———. 1919. Description of a new genus and species (*Panoplosaurus mirus*) of an armored dinosaur from the Belly River Beds of Alberta. *Proceedings and Transactions of the Royal Society of Canada* 13: 39–51.

Lee, Y.-N. 1996. A new nodosaurid ankylosaur (Dinosauria: Ornithischia) from the Paw Paw Formation (late Albian) of Texas. *Journal of Vertebrate Paleontology* 16: 232–245.

Maryańska, T. 1971. New data on the skull of *Pinacosaurus grangeri* (Ankylosauria). *Palaeontologia Polonica* 25: 45–53.

———. 1977. Ankylosauria (Dinosauria) from Mongolia. *Palaeontologia Polonica* 37: 85–151.

Meszoely, C. 1970. North American anguid lizards. *Bulletin of the Museum of Comparative Zoology* 139: 87–149.

Molnar, R. E. 1996. Preliminary report on a new ankylosaur from the early Cretaceous of Queensland, Australia. *Memoirs of the Queensland Museum* 39: 653–668.

Montanucci, R. R. 1987. A phylogenetic study of the horned lizards, genus *Phrynosoma*, based on skeletal and external morphology. *Natural History Museum of Los Angeles County, Contributions in Science* 390: 1–36.

Moss, M. L. 1969. Comparative histology of dermal sclerifications in reptiles. *Acta Anatomica* 73: 510–533.

———. 1972. The vertebrate dermis and the integumental skeleton. *American Zoologist* 12: 27–34.

Otto, H. 1909. Die Beschuppung der Brevilinguier und Ascalaboten. *Jenaische Zeitschrift* 44: 247–285.

Penkalski, P. 1998. A preliminary systematic analysis of Ankylosauridae (Ornithischia: Thyreophora) from the Late Campanian of North America [abstract]. *Journal of Vertebrate Paleontology* 18(3, Suppl.): 69A–70A.

———. 2001. Variation in specimens referred to *Euoplocephalus tutus*. In K. Carpenter (ed.), *The Armored Dinosaurs*. Bloomington: Indiana University Press [this volume, Chapter 13].

Romer, A. S. 1956. *Vertebrate Paleontology.* Chicago: University of Chicago Press.

Russell, L. S. 1940. *Edmontonia rugosidens* (Gilmore), an armored dinosaur from the Belly River Series of Alberta. *University of Toronto Studies, Geological Series* 43: 3–27.

Sereno, P. C. 1986. Phylogeny of the bird-hipped dinosaurs (order Ornithischia). *National Geographic Research* 2: 234–256.

———. 1997. The origin and evolution of dinosaurs. *Annual Review of Earth and Planetary Sciences* 25: 435–489.

Sternberg, C. M. 1929. A toothless armored dinosaur from the Upper Cretaceous of Alberta. *Bulletin of the National Museum of Canada* 54: 28–33.

Trueb, L. 1966. Morphology and development of the skull in the frog *Hyla septentrionalis. Copeia* 1966: 562–573.

Vickaryous, M. K., and N. Rybczynski. In press. Ornamentation. In P. J. Currie and K. Padian (eds.), *Dinosaur Biology.* New York: Academic Press.

Weishampel, D. B. 1995. Fossils, function, and phylogeny. In J. J. Thomason (ed.), *Functional Morphology in Vertebrate Paleontology,* pp. 34–54. Cambridge: Cambridge University Press.

Westphal, F. 1976. The dermal armor of some Triassic placodont reptiles. In A. d'A. Bellairs and C. B. Cox (eds.), *Morphology and Biology of Reptiles. Linnean Society Symposium Series* 3: 31–41.

Witmer, L. M. 1995. The extant phylogenetic bracket and the importance of reconstructing soft-tissue in fossils. In J. J. Thomason (ed.), *Functional Morphology in Vertebrate Paleontology,* pp. 19–33. Cambridge: Cambridge University Press.

———. 1997. The evolution of the antorbital cavity in archosaurs: A study in soft-tissue reconstruction in the fossil record with an analysis of the function of pneumaticity. *Journal of Vertebrate Paleontology Memoir* 4: 1–73.

Zylberberg, L., and J. Castanet. 1985. New data on the structure and the growth of the osteoderms in the reptile *Anguis fragilis* L. (Anguidae, Squamata). *Journal of Morphology* 186: 327–342.

Species; *n*; age	Specimen type	Collection
Iguania		
Corytophanidae		
Corytophanes cristatus; 1; ad	FF	PJBC*
Iguanidae		
Iguana iguana; 2; subad	C&S	APRC*
I. iguana; 1; ad	SD	UCMZ/R/1978–11
Dipsosaurus sp.; 1; ad	PR	UCMZ/R/1975–67
Polychrotidae		
Anolis equestris; 1; ad	PR	UCMZ/R/1986–21
A. equestris; 1; ad	SD	PJBC*
Phrynosomatidae		
Phrynosoma cornutum; 1; neo	C&S	LACM 19897(12185)
P. cornutum; 1; subad	C&S	LACM 4307
P. cornutum; 2; ad	SD	UCMZ/R/1979–6,7
P. hernandesi; 3; neo	C&S	APRC*
P. hernandesi; 1; ad	FF	APRC*
P. modestum; 2; neo	C&S	LACM 19692(5536), 19694(5538)
P. modestum; 1; ad	SD	TMP 90.7.162
P. mcalli; 2; neo	C&S	LACM 123343, 123344
Acrodonta		
Chamaeleo calyptratus; 2; neo	C&S	APRC*
C. jacksonii; 1; ad	SD	TMP 90.7.350
C. montium; 1; ad	FF	PJBC*
C. parsoni; 2; ad	PR	APRC*
C. pumilus; 3; neo	C&S	APRC*
Draco volans; 1; ad	C&S	APRC*
Moloch horridus; 1; ad	PR	UCMZ/R/1975–106
Uromastyx hardwickii; 1; ad	PR	UCMZ/R/1975–98
Scleroglossa		
Gekkota		
Gecko gekko; 1; ad	C&S	APRC*
Tarentola annularis; 1; ad	PR	APRC 10
T. mauritanica; 2; ad	C&S	APRC*
Lialis burtonis; 1; ad	C&S	APRC*
Lacertoidea		
Xantusia vigilis; 1; ad	PR	UCMZ/R/1975–136
Cnemidophorus soki; 2; ad	PR	UCMZ/R/1975–134, 135
Scincoidea		
Eumeces sp.; 2; ad	PR	UCMZ/R/1975–115, 116
Tiliqua scincoides; 2; ad	PR	UCMZ/R/1975–119, 120
Gerrhosaurus major; 1; ad	SD	PJBC*
G. nigrolineatus; 1; ad	PR	PJBC*
Anguimorpha		
Gerrhonotus multicarinatus; 1; ad	PR	UCMZ/R/1975–137
Heloderma suspectum; 1; ad	SD	APRC*
H. suspectum; 1; subad	PR	UCMZ (R) 2000.001
Heloderma sp.; 1; subad	SD	TMP 90.7.26
Varanus bengalensis; 1; subad	SD	UCMZ/R/1993–1
V. nebulosus; 1; ad	PR	UCMZ/R/1976–32

Species; *n*; age	Specimen type	Collection
Mammalia		
Dasypodidae		
Chaetophractus vellerosus; 1; neo	PR	UCMZ/M/1975–67
Chaetophractus sp.; 1; ad	PR	UCMZ/M/1984–7
Dasypus novemcinctus; 2; ad	SD	UCMZ/M/1975–65, 1977–119

NOTE. C&S = cleared and double stained for bone and cartilage; SD = dried skeletal material; FF = fresh frozen; PR = alcohol-preserved whole specimen. All specimens were assigned to one of three different ontogenetic stages, as follows: neo = neonate/embryo; subad = subadult; ad = adult. An asterisk denotes an unnumbered specimen.

16. Armor of the Small Ankylosaur *Minmi*

RALPH E. MOLNAR

Abstract

An articulated specimen of the Australian Early Cretaceous ankylosaur *Minmi* preserves much of the dermal in its original disposition. The armor of the trunk consists of elliptical keeled scutes arranged in longitudinal rows. Larger scutes are found in the shoulder region and form a single partial nuchal band. Large scutes, each bearing a posteriorly directed spike, were situated posterolateral to the ilia. Scutes are also found on the forelimbs and hindlimbs. The dorsum and proximal limbs and tail were covered with a sheet of ossicles intervening between the scutes, and this sheet probably extended across the belly as well. The tail may have borne a row of horizontal triangular plates on each side distally, but the evidence for this is weaker than for the other features.

Introduction

Few ankylosaurian specimens retain extensive portions of armor in situ, preserved as it was arranged on the living animal. These include *Scolosaurus cutleri* (BMNH R5161, Nopcsa 1928; now referred to the genus *Euoplocephalus*), *Edmontonia rugosidens* (AMNH 5665), and *Sauropelta edwardsorum* (AMNH 3036; the two latter figured in Coombs and Maryańska 1990; Glut 1997). The armor of all three was described by Carpenter (1982, 1984, 1990). Extensive armor is also preserved in place with the holotype of *Saichania chulsanensis* (Maryańska 1977), but has yet to be described in detail. A specimen of the plesiomorphic thyreophore *Scelidosaurus harrisonii* (BMNH R1111) was found with substantial portions of the dermal armor only slightly disturbed (figured in Glut 1997).

These specimens were supplemented in 1990, when the Queensland Museum collected a skeleton of *Minmi* (QM F18101) from just south of the Flinders River, north-central Queensland. Although it was not the first ankylosaur found in Queensland, it was largely articulated, and much of armor was in place (Molnar 1996). Color photographs of the specimen, clearly showing the disposition of the armor, have been published in popular books (Lambert 1993; Lambert and Bunting 1995; Tomida and Sato 1995). Like the other Queensland ankylosaurian specimens, it was found in Early Cretaceous marine rocks, the Albian Allaru Mudstone.

Two other specimens of Australian ankylosaurs also contribute substantially to our knowledge of the disposition of the armor in this genus. The holotype specimen of *Minmi paravertebra* (QM F10329), from the Aptian Bungil Formation of southeastern Queensland, shows ventral belly armor (Molnar 1980). An undescribed, partly disarticulated specimen (QM F33286), also from the Allaru of north-central Queensland, shows patches of ossicles of what is apparently belly armor that remain undisturbed.

The state of QM F18101 when discovered was described by Molnar (1996), together with a preliminary taphonomic interpretation. Some of this material is repeated here to provide background for understanding whether or not the armor suffered disturbance. The specimen, largely enclosed within a carbonate nodule, was found upside down so that the dorsal armor was preserved on the bottom. The dorsal armor (and dorsal surfaces of the ribs and ilia) were exposed, either due to chemical weathering in the ground or having never been enclosed in the nodule. Most of the ribs of the left side have suffered plastic distortion with the result that, except for the proximal 70 mm, they are now almost straight; it is on this side that most of the dorsal armor is preserved. Those of the right side have rotated about 90° so that they lie flat without losing their original curvature (cf. Molnar 1996: fig. 1). Many of the exposed ossicles suffered some erosion so that their exposed surfaces have been lost. However, the elements of the dorsal armor seem to have largely retained the disposition and orientation they had in life.

QM F10329 was found as a group of nodules containing bone (Molnar 1980). The ossicles were present as a single sheet, one ossicle thick, much of which remained embedded in carbonate, protecting the enclosed bone from erosion. The specimen was apparently preserved in articulation, but most of the skeleton had been destroyed by erosion before discovery. Thus, the description here is based largely on QM F18101, with some comments about QM F10329 and QM F33286. Interpretation of the arrangement of the armor in life follow description of the elements of the armor and their preservation.

Institutional Abbreviations. AMNH: American Museum of Natural History, New York. BMNH: Natural History Museum (formerly British Museum [Natural History]), London. QM: Queensland Museum, Brisbane, Australia.

Figure 16.1. (opposite page top) Dorsal view of the reassembled skeleton of Minmi *(QM F18101). Portions of the sheet of ossicles may be clearly seen just behind the skull (at left), in the shoulder region, along the left margin of the trunk (at bottom), and over the distal femur. Ossicles may also be seen between the ribs to the left of center. This photograph was taken before the pelvic region was prepared. The armor of that region is shown in Figure 16.9. Scale bar = 300 mm.*

Figure 16.2. (opposite page bottom) Disposition of the dorsal armor of Minmi *(QM F18101); cf. Figure 16.1. Elements of the armor are shown in heavy outline. Three adjacent pieces of the tail and two blocks preserving the tarsal regions (one at the distal end of the left limb) shown in Figure 16.1 are omitted here. Anterior is to the left. Scale = 100 mm. b = sheet of bone roofing anterior sacral ribs; lf = left femur; lh = left humerus; li = left ilium; lu = left ulna; n = neural spines; ri = right ilium; s = skull; sr = sacral rib; t = transverse process.*

Description

Elements of the Armor

The armor consists of several kinds of elements: ossicles, small elliptical keeled scutes (trunk), large elliptical scutes lacking a keel (snout), large elliptical keeled scutes (neck, shoulders, and possibly tail), large keeled scutes bearing a sharp apex or spike (hips), rhomboid keeled scutes, elongate ridged scutes, and (probably) triangular plates (all three

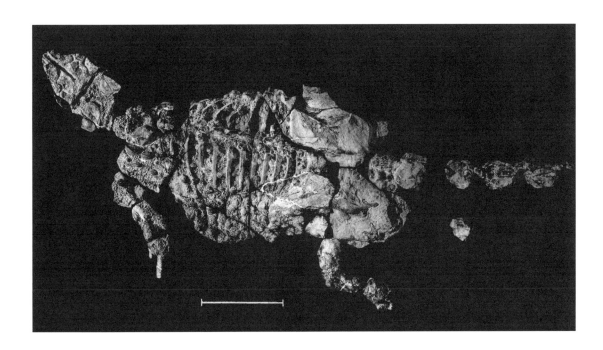

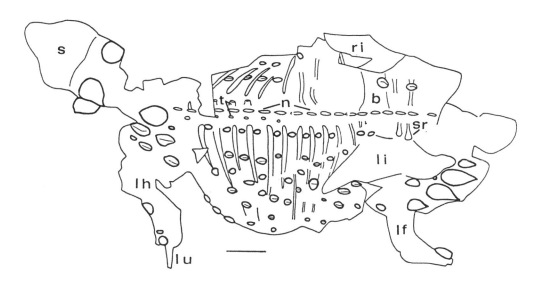

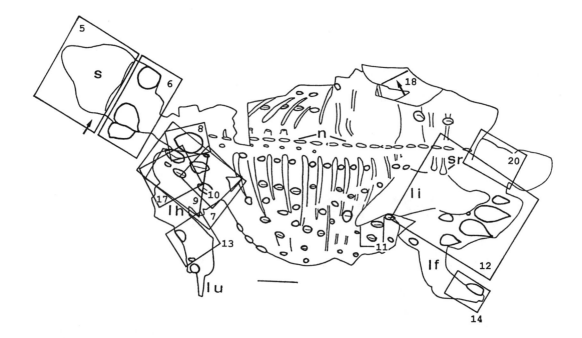

Figure 16.3. Key to figures of specific regions and dermal elements. Figure numbers given for each. The figures show isolated pieces rather than those in the reassembled skeleton. Arrows indicate view for Figure 16.5 (center) and Figure 16.15. b = sheet of bone roofing anterior sacral ribs; lf = left femur; lh = left humerus; li = left ilium = lu = left ulna; n = neural spines; ri = right ilium; s = skull; sr = sacral rib = t = transverse process.

on the tail). The general disposition of the elements of the armor of *Minmi* (QM F18101), as preserved, is shown in Figure 16.1. Figure 16.2 illustrates the arrangement of the scutes of the neck, trunk and (left) limbs, and Figure 16.3 indicates the locations of the material illustrated in the other figures of this specimen.

Ossicles. The individual ossicles are approximately equidimensional, 4–5 mm in diameter, and pillow-shaped (i.e., oblate spheroids). The better preserved of the isolated ossicles are often rectangular, sometimes almost square, and show a orthogonal pattern of ridges and grooves on what is presumably the internal surface. This pattern seems to be formed by the edges of a series of bony sheets of successively smaller size. The internal surface is often penetrated by one or more foramina. The external surface is nearly flat and covered with shallow pits separated by narrow ridges, somewhat resembling the sculpture of crocodilian dermal cranial elements, but on a much finer scale. The ossicles of QM F10329 are similar in form (Fig. 16.4) but noticeably larger (about 6 by 8 mm) than most of those of QM F18101, and those of QM F33286 are approximately square (5 by 5 mm) and worn.

As far as can be determined, the entire dorsum of the trunk is covered with a pavement of small ossicles (Fig. 16.1). This sheet certainly also extends onto the proximal segments of the forelimbs and hindlimbs and the base of the tail. As preserved, in some places the ossicles are closely spaced, with about 1 mm between adjacent ossicles, but in others, they are 2–3 mm apart. The ossicles form a layer varying from one to four ossicles thick. No ossicles are preserved on the dorsal surfaces of the ilia, but they are preserved between the sacrum and the ilium posteriorly on the right side, with a few isolated ossicles just in front.

In QM F10329, the sheet of ossicles was preserved in the nodules 75–80 mm below the ventral surfaces of the (articulated) ribs. Because the flat sheet shows no evidence of substantial disruption and the external surfaces of the ossicles were generally directed ventrally (although a few individual ossicles have been inverted), the ossicles were taken to be a ventral sheet, armoring the belly of the animal (Molnar 1980).

Cranial Armor. The skull of *Minmi* (QM F18101) lacks the dermal armor plates associated with the cranial roof in other ankylosaur skulls (Fig. 16.5). Small ossifications are found in the orbital and cheek regions and a large one in the region of the nares. The sheet of cervical ossicles is continuous to the quadrate region of the skull, and ossifications overlying the jugal and squamosal have the appearance of fused ossicles.

A flat, shallowly crescentic ossification overlies the posterior part of the "squamosal" (see Molnar 1996 for comments on the composition of this element) on the cranial roof on both sides (Fig. 16.5). On the left, this ossification is matched by a triangular ossification on the lateral face of the squamosal, with a few small, isolated ossicles in front of it ventrally (Fig. 16.5). Below this is another shallowly crescentic ossification that has the appearance of being composed of parallel sheets of bone, and along the ventral edge (here broken) is a triangular ossification (Fig. 16.5). All of these are located just anterior to the occipital face of the skull.

Small ossifications overlie the supraorbital–squamosal contact posteromedially, some of which have been removed during preparation. A large (50 by 35 mm), flat, triangular ossification remains adjacent to the narial rim of the nasals on the left; a thin sheet of bone about 3 mm

Figure 16.4. Ossicles of the ventral armor of Minmi *paravertebra (QM F10329). The internal surface of the ossicle in the center of the image, the others show the external surface. Orientation unknown; scale in millimeters.*

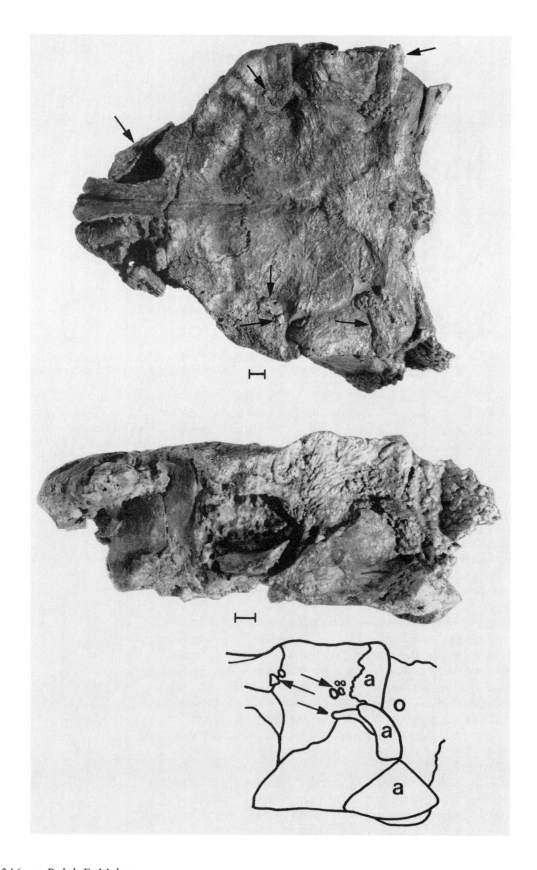

thick, it is covered with coarse foramina on its external face. What seem to be ossicles are also found in the orbital region, but this area remains incompletely prepared and poorly understood, so these will not be discussed further.

Armor of the Neck and Trunk. The individual scutes of the trunk are oval or elliptical in form, approximately symmetrical, and where obviously preserved in situ, their long axes are oriented longitudinally (Figs. 16.1, 16.2). The scutes are prominently ridged or keeled, generally being triangular in transverse section. The ridges run along the midlines of the scutes and are about one third as high as the scutes are wide transversely. The scutes have a rough surface, with small pits (or foramina) on well-preserved specimens. A break reveals a dense superficial laminar cortex around a more open, cancellous interior. The internal surfaces of the scutes are not exposed, but a scute broken marginally shows that the internal face was concave.

These scutes are all of approximately the same size, but they slightly increase in size with increased lateral distance from the vertebral column. Concomitant with this, the longitudinal ridge of the scutes also increases in height (relative to the length of the scute) with distance from the column. The more medial scutes are clearly arranged in longitudinal rows that were either not present or have been severely disrupted laterally. The more lateral scutes become nearly circular opposite dorsals 6–7, and there are also nearly circular scutes between the sacrum and ilium. The most lateral of these nearly circular scutes lacks the keel; instead, it has a slightly overhanging posterior apex.

The cervical scutes form a rudimentary transverse band, with the lateral pairs in contact (one of which is missing on the left side), but the medial pair are well separated (Fig. 16.6). On both medial and lateral scutes, the external surface medial to the keel faces dorsally and the lateral faces laterally, inclined at about 90° to the medial portion (Fig. 16.6). The keel is laterally placed. Large foramina penetrate the external surfaces, and a pattern of thin, shallow grooves on both surfaces radiates from approximately the center of the keel. The line of the keel is slightly convex over its anterior two thirds and mildly concave over the rest. Both medial and lateral portions of the external surface are also weakly convex anteriorly and concave posteriorly. The lateral scute is slightly smaller than the medial and tapered posteriorly, making the scute about half as broad at the posterior as anteriorly. Its keel curves medially at the back, so that the nearly vertical, lateral surface faces posteriorly. The keel is elevated posteriorly, rendering the dorsally facing surface concave posteriorly but mildly convex anteriorly. This face also retains a set of thin, shallow, radiating grooves. The internal faces of these scutes are not exposed, and the scutes lie over and adjacent to a layer of ossicles three to four ossicles thick.

In the shoulder region, the scutes are larger than on the trunk, but smaller than those of the neck. They are generally similar in form to the more posterior trunk scutes, but unlike those, these scutes are asymmetrical (Fig. 16.7). On the largest scute of the shoulder, the keel is placed far laterally and thus does not extend the entire length of the scute (Fig. 16.8). It is basically similar in form to the cervical scutes.

Figure 16.5. (opposite page) The skull of Minmi (QM F18101) in dorsal (top) and left lateral (center) views. Anterior is to the left. In the dorsal view, elements of the armor are indicated with arrows, with the nasal scute at left, three ossicles in the middle, and both ossifications overlying the "squamosal" at right. For the lateral view armor is indicated on the diagram (bottom) by the letter a or an arrow. The irregular mass (o) behind the occipital face is part of the sheet of cervical ossicles.

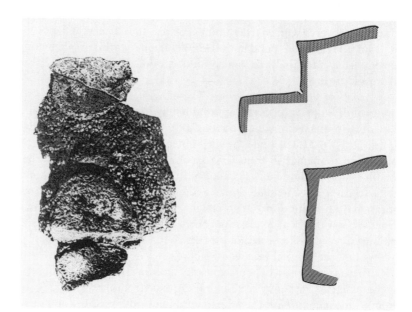

Figure 16.6. Dorsal view of the cervical region of Minmi (QM F18101). The scutes are paired, but the lateral is present only on the left side. Anterior is to the left, and that surface was adjacent to the occipital face of the skull. Transverse section through scutes at right, with suggested section below correcting for possible displacement of the lower scute.

Figure 16.7. The scutes of the scapular region of Minmi (QM F18101). Anterior is to the left; a small portion of the posterior end of the scapular blade may be seen at the far right.

This scute is located about 80 mm medial to the glenoid region of the scapula and less than 10 mm from the neural spines. As preserved, the dorsal surface medial to the keel forms a horizontal plane, which slopes upward to the keel at only a few degrees to the horizontal. Then it slopes laterally for about 15 mm at an angle of just less than 90° to curve laterally and form another almost horizontal surface along the lateral edge (Fig. 16.8). The surface medial to the keel bears several foramina and a few thin, shallow grooves that may be erosional features. What seem to be two scutes 30 and 50 mm long are lodged between the layer of ossicles and the glenoid end of the scapula. They are insufficiently exposed for description.

Laterally, two smaller scutes are asymmetrical, but not so mark-

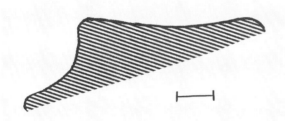

edly as the larger scute. These scutes have irregular grooves over their external faces, with several foramina. The anterior of them is just medial to the scapula, the posterior on its medial edge. Further lateral are two scutes similar in form to those of the trunk. One of these shows what appear to be ossicles arrested in the process of merging with the medial margin of the scute (Fig. 16.9). Anterior to this is what appears to be a very small, second scute, also in the process of merging with ossicles, and there are also three clumps of ossicles, one of which has what seems to be ossicles 9 and 12 mm long (Fig. 16.10), about twice the size of the usual ones. A similar apparent fusion may be seen with one of the scutes of the femoral region of the trunk in which three ossicles seem to be preserved in varying degrees of merging with the medial margin (Fig. 16.11).

Two round scutes remain between the right ilium and the sacrum. The posterior of these is 40 mm long by 35 mm broad, with a 17-mm-high keel.

The sacral region, like the neck, bears large elements of armor (Fig. 16.12). These are preserved only on the left. The elements range in size from 55 to 120 mm in length. In form, they vary from almost round or elliptical to (roughly) teardrop-shaped, rounded with an elongate, spikelike apex posteriorly. The keels of the smaller scutes are about 15 mm high. One small scute lies laterally alongside the edge of the ilium just behind the acetabulum, the other is just behind but touching the posterior extremity of the ilium. Between these is a much smaller ele-

Figure 16.8. The largest scute of the shoulder region of Minmi *(QM F18101). Anterior is to the left; the peglike structures along the anteromedial and posteromedial margins are the tips of transverse processes. Below, transverse section through middle of scute. Scale and scale bar in millimeters.*

Figure 16.9. One of the scutes with which ossicles appear to be merging, scapular region of Minmi *(QM F18101). This may be seen along the upper margin of the small scute near the center of the image. Anterior is to the left. Scale in millimeters.*

Figure 16.10. A clump of ossicles in the scapular region of Minmi *(QM F18101). Anterior is to the left. Scale in millimeters.*

350 • Ralph E. Molnar

Figure 16.11. Another of the scutes with which ossicles appear to be merging, in the femoral region (immediately lateral to the ilium) of Minmi (QM F18101). Anterior is to the left. Scale in millimeters.

Figure 16.12. The scutes of the pelvic region of Minmi (QM F18101). The three large, spike-bearing scutes are aligned diagonally along the lower right edge of the block; the sharp tip is present only on the central scute. Anterior is to the left. Scale in millimeters.

ment, 13 mm long. Lateral is a row of three elements arranged along an anterolateral–posteromedial line. The middle and posterior of these are about the same size (120 mm long), and the anterior is smaller (75 mm long as preserved). These are similar to the cervical scutes in having a large, flat dorsal face medial to the keel. But laterally, the keel strongly overhangs the other face, by at least 25 mm. The surfaces of all are weakly rugose, with shallow grooves and large foramina. The lateral and central of the large elements both have a peripheral row of pits (or large foramina) preserved along part of their margins. Breaks near the apices of two elements show that the bone structure is cancellous. The surface intervening between the elements is covered by a sheet of ossicles.

Armor of the Forelimb. Much of the left forelimb, the humerus and the proximal three quarters of the ulna, and one quarter of the radius are preserved. These are intimately associated with a sheet of ossicles adhering to the proximal part of the extensor surface of the humerus that extends distally to about one quarter the length of the ulna. This point marked the edge of the nodule, and the more distal parts of the ulna and radius were found as isolated pieces, so this does not indicate a limit to the extent of the ossicles. A small patch of ossicles also adheres to the flexor surface of the humerus just proximal to the elbow joint. The adherent patches of ossicles form layers a single ossicle thick, but along the distal extensor surface of the humerus, the layer is three ossicles thick. The ossicles are roughly square to hexagonal and about 4 by 6 mm, but two are 8 by 8 mm.

A large subcircular scute lays at the midlength of the humerus on its lateral side (Fig. 16.13). Its blunt keel rises to a distally directed sharp apex, and its internal face is obscured by a layer of ossicles. The scute is 40 mm long by 36 mm broad, and the keel is about 17 mm high at the apex. Its external surface has large foramina and shallow pits, although grooves are also present. A small plate of bone 15 mm medial

Figure 16.13. A brachial scute of Minmi (QM F18101). The underlying sheet of ossicles can be seen between the scute and the humeral shaft (the proximal piece of the humerus has been removed). Anterior is to the left. Scale in millimeters.

to the medial surface of the humerus, also at midlength, probably represents part of another scute.

Two scutes are associated with the proximal antebrachial elements. Both are located between the ulna and radius on what is presumed to be the extensor surface. The proximal of these, at the proximal articular surface of the ulna, is a rectangle 22 by 15 mm, but broken along one edge so was possibly slightly longer. It forms a thin, flat sheet, with a slightly convex external face. This face seems to lack sculpture but bears coarse foramina. The more distal scute is a regular ellipse of 40 by 32 mm. The external face is also mildly convex and bears large foramina, but it is weathered so that although some sculpture was apparently present, its form is no longer clear. The internal faces of both scutes are not accessible.

Armor of the Hindlimb. A layer of two ossicles covers the external side of the left knee region, apparently breaking up at the level of the proximal articular surfaces of the tibia and fibula, beyond which there are scattered ossicles. A strongly keeled scute of moderate size is found on the posterior surface of the proximal end of the fibula (Fig. 16.14). This scute is 40 mm long by 21 mm broad and is 7 mm high at the highest point, although the ridge line of the keel is worn, indicating that it was higher than this. Forty millimeters distal to this, also on the posterior surface of the fibula, is a larger scute, 45 by 38 mm by 15 mm high. It is roughly elliptical with a high, sharp keel that medially falls away vertically, but is more shallowly inclined laterally. The surface is not well preserved, but seems to have had a sculpture of small, shallow

Figure 16.14. The proximal fibular scute of Minmi (QM F18101). The distal portions of the tibia and fibula were not recovered. Anterior is to the left. Scale in millimeters.

pits as well as the usual large foramina. A scattering of small ossicles is present in the matrix between these scutes and against the internal face of the larger scute.

The right ankle region also preserves a sheet of ossicles and a roughly rectangular scute, 40 by 38 mm. This scute, the internal surface of which is exposed, was flat but with a keel along one diagonal. It lies just distal to and behind the astragalus.

Caudal Armor. The base of the tail to about 250 mm behind the sacrum is associated with large numbers of ossicles, mostly isolated or in scattered clumps. Proximally, a sheet of ossicles remains ventral to the centra and another to the left side of the neural arches. The distal of the two pieces from this region of the tail also has elongate scutes that are 40–45 mm long and a small, oval scute 20 mm long. One of the elongate scutes is exposed only internally; the other is a roughly rectangular element with a 12-mm-high keel along one of the long edges. The small scute is roughened by large foramina and shallow ridges and grooves.

The three more distal pieces of the tail, which may represent adjacent regions (and presumably are not from near the tip of the tail), have only a single small ossicle among them. They show two other types of dermal elements, scutes, and plates. Two of the plates are large, approximately triangular elements, about 110 mm long (Figs. 16.15, 16.16). These two plates were preserved, along with the caudals, in two blocks that fit together closely, indicating that they were consecutive. They are preserved along only one side of the tail, with no evidence remaining of what might have been on the other side. Both plates are crushed and distorted but seem to have been deeply excavated internally for at least 60 mm, with each limb 5–10 mm thick. Both plates are broken just beyond the junction of the limbs, but the bases of narrow

Figure 16.15. One of the triangular plates of the tail of Minmi (QM F18101), viewed from below. An elongate scute may be seen overlying the caudals on the right, and a second, immediately above that, projects from under the edge of the triangular plate. Anterior is to the left. Scale in millimeters.

keels, about 10 mm thick but of unknown height, remain. The broken surfaces reveal cancellous bone beneath only about 1 mm of lamellar bone.

The third piece also bears a broken and worn platelike element of about the same size as those just described, but different in shape. This consists of two sheets of bone that join at an approximately right angle (Fig. 16.17). This form is reminiscent of that of the cervical scutes described above. The large sheet is 7–10 mm thick; the smaller one is 4 mm thick near the edge but 11 mm thick near its junction with the large sheet. Judging from the widths of the caudal centra associated with each piece, this is the most proximal of the three pieces, and hence, it is the most proximal of the caudal plates. The surfaces of these plates generally seem to lack both the mild sculpture and the large foramina found on the scutes.

This piece also bears a scute (Fig. 16.17) 41 mm long by 31 mm broad, with a keel 15 mm high along one diagonal, that may originally have been rhomboidal in form but is now broken along two edges. The scute has the usual large foramina as well as a sculpture of shallow pits and grooves. The other two pieces have three low, elongate scutes, with lower, blunter keels. One of these is 38 by 19 mm (broken at one end), and other two are about 58 by at least 28 mm each; all bear very large foramina. The internal surface of one of the larger elongate scutes is visible and is slightly roughened and penetrated by large foramina, but the rectangular pattern seen on the internal faces of the ossicles is completely absent.

Figure 16.16. (above left) The second of the triangular plates of the tail of Minmi (QM F18101), viewed from above. Anterior is to the left. Scale in millimeters.

Figure 16.17. (above right) The proximal plate or scute of the tail of Minmi (QM F18101), viewed from behind to show the perpendicular meeting of the two lamina that make up the element. Ossified tendons are visible in the matrix between the two laminae, and a rhomboid scute may be seen against the caudal centrum at the top. Scale in millimeters.

Interpretation

Taphonomic Comments. The description permits some general observations about the armor of *Minmi*. Before this is done, how the carcass came to be preserved should be considered in order to explain some anomalies of the specimen and to assess whether the armor is completely undisturbed or if some small disruption has occurred. The general configuration of the elements of the dorsal armor and the fact that most of the endoskeletal elements remain articulated indicate that there was no major disturbance of the trunk armor. The ossicles are not found on the dorsal surfaces of the ilia or—generally—of the ribs. However, distally, ossicles do occur on the upper faces of the ribs, and in some regions, that sheet of ossicles becomes continuous over the ribs distally. This distribution suggests that originally, the sheet of ossicles extended over the ribs, and in those places where the ribs are exposed, the loss of ossicles is due to erosion. A covering sheet of ossicles is also seen at the anterior extremity of the preacetabular process of the left ilium. Although the dorsal neural spines and paravertebrae are largely free of ossicles, some ossicles are found among them, and in some places, the sheet of ossicles encroaches to the tips of the neural spines. This pattern suggests that the entire dorsal surface of the animal was covered by ossicles and that the regions free of ossicles on the specimen are the result of erosion. There may have been little, if any, disturbance of the armor, but apparently there was some loss.

There is no pelvic buckler, as seen in *Polacanthus*. Instead, some individual scutes are found in this region, and the ossicles have sunk down between the sacral ribs, showing that a rigid sheet of coalesced ossicles and scutes was absent. This individual is interpreted as approaching or having just reached maturity (Molnar 1996); possibly a buckler might have developed in older individuals, but there is no evidence to support this speculation.

Much of the armor of the right side of the specimen is missing, as may be seen by comparing the two sides in Figure 16.1. The region of the neck and shoulders is also incomplete and significantly better preserved on the left side than the right side. The disposition of scutes around the left scapula gives some notion of the arrangement of armor in this region, but the scutes lodged between the scapula and sheet of ossicles indicate that there were more scutes, possibly originally between the nuchal scutes and the scapula. A broad fold in the sheet of ossicles, about 20 mm wide, lies transversely just behind the dorsal edge of the occiput, indicating that the sheet was not rigid but permitted some flexibility.

The occurrence of the sheet of ossicles internal to some scutes and on the internal sides of the ribs requires explanation, because surely the coelomic cavity was not armored. Furthermore, on the right side (but not the left), the ilia lay over a sheet of ossicles as many as six ossicles thick, the thickest seen (Fig. 16.18). This sheet does not extend to the right side of the carcass. In view of the evidence for ventral armor in QM F10329 (Molnar 1980) and the inverted position of QM F18101,

the dorsal sheet of ossicles is interpreted as have been augmented by ossicles from the belly armor that sank onto the dorsal layer of ossicles. If the belly hide remained intact, it would be expected to have draped over the vertebrae. But there are only a few isolated ossicles adhering to the ventral surfaces of a few centra, and such draping is found only in one patch in the sacral region. Presumably after the carcass came to rest on the seafloor, the hide of the belly collapsed downward (dorsally) and deteriorated and the ossicles, mostly singly, sank to rest on the dorsal hide. This part of the integument either remained intact longer or the ossicles remained in place, supported by the seafloor. Beneath the left ilium, when the specimen is viewed right side up, a single sheet of ossicles lies 60 mm from the ventral iliac surface. These are presumably ossicles from the ventral hide that came to rest on sediment that had already accumulated in that region of the carcass. Some of these presumably slid or rolled over to the right side to add to the thick accumulation against the ilium there.

The armor of the forelimb and sacral regions was displaced because in several places, scutes or ossicles lie directly against endoskeletal elements, such as the scapula and humerus, which in life had muscular tissue between them and the dermal elements. The ossicles are often grouped together to form a sheet and perhaps represent pieces of hide.

General Pattern of the Armor. The dorsal sheet of ossicles forms a continuous layer, ranging from one to four ossicles thick, over the dorsum of the neck and trunk of the specimen. Over the glenoid region of the scapula, the layer of ossicles a single ossicle thick forms shallow longitudinal wrinkles (Fig. 16.19), possibly reflecting deformation of the hide in this region. There is also a single layer of ossicles over the external surface of the humerus. These occurrences suggest that the ossicle layer was generally a single ossicle thick and that the thicker layers of the specimen result from the displacement of ossicles from

Figure 16.18. Section through the left ilium seen from behind, showing the thick layer of ossicles beneath that element. Because the specimen was found upside down, these are interpreted to be ossicles from the belly armor that have sunk down to the ilium and accumulated there. Scale in millimeters.

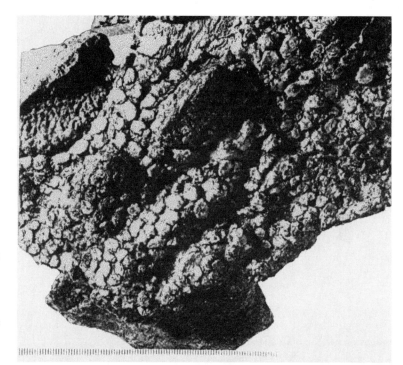

Figure 16.19. *The wrinkles in the sheet of ossicles overlying the scapula of* Minmi *(QM F18101). Anterior is to the lower left; a small portion of the glenoid region of the scapula may be seen at bottom. Scale in millimeters.*

Figure 16.20. *A block from the basal part of the tail of* Minmi *(QM F18101). Ossicles can be seen above the neural spines, intermingled with the ossified tendons. One ossicle lies between two adjacent neural spines. Anterior is to the left. Scale in millimeters.*

other parts of the carcass. In addition, QM F10329 retains portions of the ventral layer of ossicles that are one layer thick.

In the thoracic and proximal caudal regions, ossicles lay adjacent to, around, and between the ossified tendons (Fig. 16.20). Presumably,

at some time the dorsal hide disintegrated and allowed the vertebrae and their retinue of paravertebrae and other ossified tendons to sink into the ossicle layer.

QM F33286 is a specimen, presumably *Minmi,* from north-central Queensland consisting chiefly of two large blocks of closely associated, but mostly partly disarticulated, thoracic and abdominal remains. Large expanses of ossicles are preserved that, from their relationships to the endoskeletal elements, are thought to be from the ventral or lateroventral hide. Among these are patches of square or rhomboid ossicles that still fit closely, less than 1 mm apart, like tiles (Fig. 16.21). Because taphonomic processes can disturb fossils but would be unlikely to bring about this kind of pattern, I interpret this arrangement as representing the pattern in life. This, together with the squarish form of the ossicles in the other two specimens, suggests that the ossicles were tiled and so formed a complete "pavement" in the integument.

This pavement may well have covered the entire neck and trunk, dorsally and ventrally, except where interrupted by scutes. The ossicle layer extended at least onto the proximal segments of the limbs, and the ossicles associated with the animal's right ankle might suggest that they extended as far as the pes (and manus). However, the right crural region came to rest very close to the tail, and so both ossicles and scute may actually pertain to that structure; the scute resembles one associated with the proximal caudals. The ossicles also extended onto the proximal part of the tail, but apparently not as far distal as the region bearing the triangular plates.

Figure 16.21. Ossicles in situ in the ankylosaur QM F33286, showing their squared form and close, tilelike fit. Anterior is to the left. Scale in millimeters.

On the left side, a broad expanse of the dorsal armor is retained. The scutes of the trunk are arranged in parallel, parasagittal rows, at least medially (Figs. 16.1, 16.2). Laterally, the rows, if they exist, are difficult to discern. This may be due to distortion of the pattern associated with the deformation of the ribs mentioned above, but if so, the sheet of ossicles remained unbroken. Most scutes of the trunk occupy positions between the ribs, although anteriorly, two scutes overlie ribs and one in the sacral region is over a sacral rib. This intercostal distribution is maintained for the lateral scutes, where the sheet of ossicles remains continuous over the ribs, indicating that this pattern reflects that of the living animal. In one exception, the rib (possibly the first) appears to have been displaced, but the scute over the sacral rib gives no indication of being out of place.

The neck, in addition to ossicles, bears one incomplete transverse band of large scutes. The scutes seem to have formed a steplike structure on each side. However, the lateral scute may have been displaced by rotation, so that in life the structure was more ringlike in form (Fig. 16.6). If so, the posterior taper of the lateral scute suggests a constriction of the neck at this region, a feature not seen in any other tetrapods known to me. Large scutes also occupied the shoulder region, presumably between the scapulae. There is no sign of the kind of shoulder spines found in *Edmontonia*. However, the form of the sacral spine plates indicates that the keels, and hence the posteriorly directed spikes, projected from the flanks. In *Minmi*, the hip region, rather than the shoulders, bore projecting spines. The limbs bore scutes of moderate size on at least the lateral surface of the fore and posterior surface of the hind limb, as well as ossicles. These may have formed a row along the long axis of the limb.

The disposition and orientation of the caudal elements of armor is unclear. The triangular plates might represent a lateral row, as in *Gastonia* (Kirkland 1998). The elongate scutes might then represent a dorsal row, because they are about half the length of the triangular plates and were possibly arranged with two scutes per plate. This interpretation provides a plausible reconstruction for the tail, but it also implies the existence of a second row of triangular plates for which no evidence remains. Furthermore, as preserved, the plates are directed ventrally at angles of 30° and 45°. One of the ossified tendons is now inclined at about 25° to the anteroposterior axis, and an elongate scute lies internal to the base of one of the triangular plates (Fig. 16.17). Thus, because considerable displacement has occurred in this region, other arrangements will be considered. A row of ventrally directed triangular plates occurred on the tail of *Scelidosaurus*, in addition to a dorsal row and a lateral row on each side (Coombs et al. 1990). *Minmi* might have had medial ventral and dorsal rows. The interpretation of two lateral rows of plates (Fig. 16.22) provides a plausible reconstruction similar to that known to exist in other ankylosaurs but involves displacement of the caudal plates preserved and loss of those of one row; the interpretation of median rows provides a reconstruction not unlike that of *Scelidosaurus*, but such an interpretation assumes a greater displacement of the caudal plates and loss of those of the dorsal row. There may,

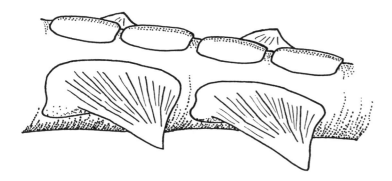

of course, have been only a single median (ventral or dorsal) row, assuming equally great displacement of the plates but no loss of elements.
Presumably large parts of the tail are missing, so it is reasonable to
expect that some plates may also have been lost. If the minimum amount
of displacement of the preserved plates is assumed, the existence of
lateral rows is most plausible. No evidence remains for the existence of
a tail club.

Fusion of Elements. Several scutes were apparently arrested in the
process of fusing with adjacent ossicles. The pelvic bucklers of *Polacanthus, Gastonia,* and *Gargoyleosaurus* indicate that fusion of elements of the armor did occur (cf. Kirkland 1998). However, interpret-
ing the existence and nature of such a process from fossil material,
which does not permit the process to be directly observed, is risky.
Examining microscopic sections of the margins of these scutes might
support or refute the interpretation of fusion if different degrees of
continuity of the histologic structure occur in different clusters of ossicles and scutes. The nature of the margins of the scutes of other ankylosaurs should reveal if this apparent process was of general occurrence.

*Figure 16.22. Sketch of a possible
arrangement of dermal elements
near base of tail of* Minmi, *in
oblique dorsolateral view. The
triangular plates are arranged
laterally and the smaller elongate
scutes dorsally.*

Summary

The armor of the trunk of *Minmi* consisted of small scutes arranged in parasagittal rows, at least medially, with the intervening regions occupied by a pavement of small ossicles. These ossicles probably
fitted closely together, like tiles. Larger scutes are found on the neck
behind the skull, where they possibly formed an incomplete ring, and
at the base of the neck and shoulders. Strongly keeled scutes, each bearing a posteriorly directed spike, formed a short row posterolateral to
the back portion of the ilium. The large cranial armor plates that coa-
lesce with the cranial bones in other ankylosaurs are absent. The limbs
were armored with ossicles, and scutes of moderate size are found on
both proximal (stylopodium) and middle (zeugopodium) segments. The
arrangement of the caudal armor remains unclear, and there is no evidence that a tail club was present.

Acknowledgments. Ian Ievers discovered this material and notified the Queensland Museum, and it was kindly donated by Margaret and the late Frank Ievers. Susan Walker and Mary Wade obtained the specimen that is now QM F33286 for the museum. Angela M. Hatch and Joanne Wilkinson prepared these specimens, and Laurie Beirne, Paul Stumkat, and Peter Trusler contributed to interpretation of the armor. William Blows and Kenneth Carpenter provided helpful comments on the text.

References Cited

Carpenter, K. 1982. Skeletal and dermal armor reconstruction of *Euoplocephalus tutus* (Ornithischia: Ankylosauridae) from the Late Cretaceous Oldman Formation of Alberta. *Canadian Journal of Earth Sciences* 19: 689–697.

———. 1982. Skeletal reconstruction and life restoration of *Sauropelta* (Ankylosauria: Nodosauridae) from the Cretaceous of North America. *Canadian Journal of Earth Sciences* 21: 1491–1498.

———. 1990. Ankylosaur systematics: Example using *Panoplosaurus* and *Edmontonia* (Ankylosauria: Nodosauridae). In K. Carpenter and P. J. Currie (eds.), *Dinosaur Systematics: Approaches and Perspectives*, pp. 281–298. Cambridge: Cambridge University Press.

Coombs, W. P., Jr., and T. Maryańska. 1990. Ankylosauria. In D. B. Weishampel, P. Dodson, and H. Osmólska (eds.), *The Dinosauria*, pp. 456–483. Berkeley: University of California Press.

Coombs, W. P., Jr., D. B. Weishampel, and L. M. Witmer. 1990. Basal Thyreophora. In D. B. Weishampel, P. Dodson, and H. Osmólska (eds.), *The Dinosauria*, pp. 427–434. Berkeley: University of California Press.

Glut, D. F. 1997. *Dinosaurs, the Encyclopedia*. Jefferson, N.C.: McFarland.

Kirkland, J. I. 1998. A polacanthine ankylosaur (Ornithischia: Dinosauria) from the Early Cretaceous (Barremian) of eastern Utah. In S. G. Lucas, J. I. Kirkland, and J. W. Estep (eds.), *Lower and Middle Cretaceous Terrestrial Ecosystems. Bulletin of the New Mexico Museum of Natural History and Science* 14: 249–270.

Lambert, D. 1993. *The Ultimate Dinosaur Book*. London: Dorling Kindersley.

Lambert, D., and E. Bunting. 1995. *The Visual Dictionary of Prehistoric Life*. London: Dorling Kindersley.

Maryańska, T. 1977. Ankylosauridae (Dinosauria) from Mongolia. *Palaeontologica Polonica* 37: 85–151.

Molnar, R. E. 1980. An ankylosaur (Ornithischia: Reptilia) from the lower Cretaceous of southern Queensland. *Memoirs of the Queensland Museum* 20: 77–87.

———. 1996. Preliminary report on a new ankylosaur from the Early Cretaceous of Queensland, Australia. In F. E. Novas and R. E. Molnar (eds.), *Proceedings of the Gondwanan Dinosaur Symposium. Memoirs of the Queensland Museum* 39: 653–668.

Nopcsa, F. 1928. *Scolosaurus cutleri*, a new dinosaur. *Geologica Hungarica, Series Paleontologica* 1: 54–74.

Tomida Y., and T. Sato. 1995. *Stegosaurus*. Tokyo: Kaisei-sha Publishing.

17. Dermal Armor of the Polacanthine Dinosaurs

WILLIAM T. BLOWS

Abstract

The polacanthines are a group of armored dinosaurs distinct from other nodosaurs by characteristic dermal bones, including skull coronuces, separate cervical ring elements, and a sacral shield. Dermal bone terminology is here clarified and new terms proposed. Polacanthines appear to have had a mixed dermal bone function, with some elements suited to either defense or thermoregulation; other elements were possibly for both functions. A change in thyreophoran armor morphology from the Jurassic stegosaurs to the Late Cretaceous ankylosaurs is noted; this change may have been necessary to accommodate a greater emphasis on defense.

Introduction

The distinct division of the Ankylosauria into the families Nodosauridae and Ankylosauridae by Coombs (1978) is complicated by new dermal skeleton fossils. The term "polacanthine" follows both Lapparent and Lavocat (1955) and Kirkland (1998) as the basal ankylosaurid subfamily Polacanthinae. The full analysis of this group, including endoskeletal features, in relationship to other armored dinosaurs is in progress elsewhere (Kirkland, personal communication). The polacanthines had both upright (i.e., a narrow base with a tall keel) and flat-mounted (i.e., a broad base and a very low keel height, sometimes flat) armor, which included a mix of primitive and advanced skull armor (Fig. 17.1), thin, upright plates and spines (Figs. 17.2, 17.3), and flat-mounted ossicles and sacral shield (Figs. 17.4–17.6). The ankylosaurines had a continuous compact armor, including well-developed skull

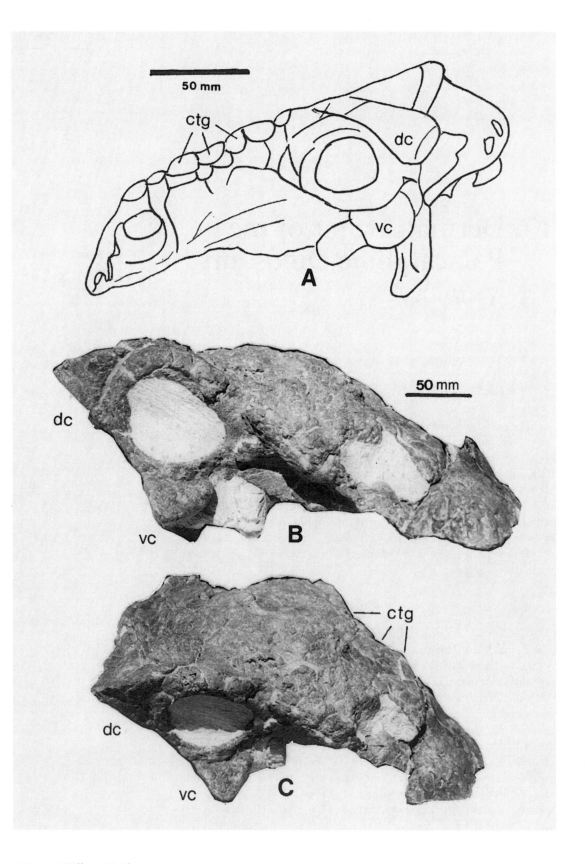

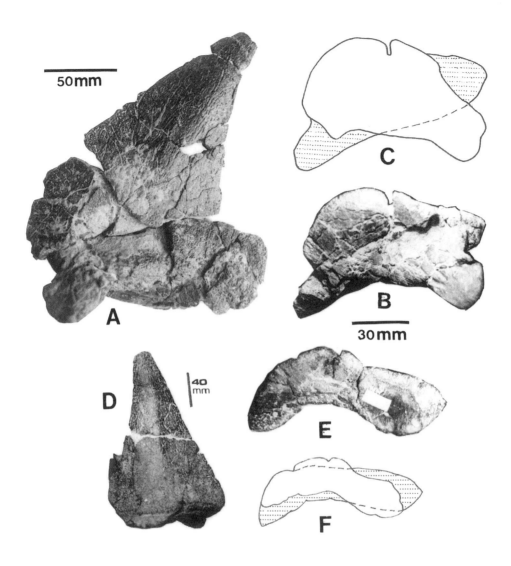

Figure 17.1. (opposite page) Major dermal bones of Lower Cretaceous polacanthine skulls from the United States. (A) Pawpawsaurus campbelli (after Lee 1996) and (B, C) Gastonia burgei (courtesy of Jim Kirkland); (A, B) Lateral view. (C) Dorsolateral view. ctg = capitegulae; dc = dorsal coronux; vc = ventral coronux.

Figure 17.2. (above) BMNH R9293 Polacanthus foxii cervicopectoral armor. Large spine in (A) ?lateral and (B) ventral (basal) views; (C) redrawn basal view to show basal bone (in white) and spine (stippled). Small spine in (D) ?lateral and (E) basal views; (F) redrawn basal view to show basal bone (in white) and spine (stippled). Lower Cretaceous, Wealden beds, Isle of Wight, UK.

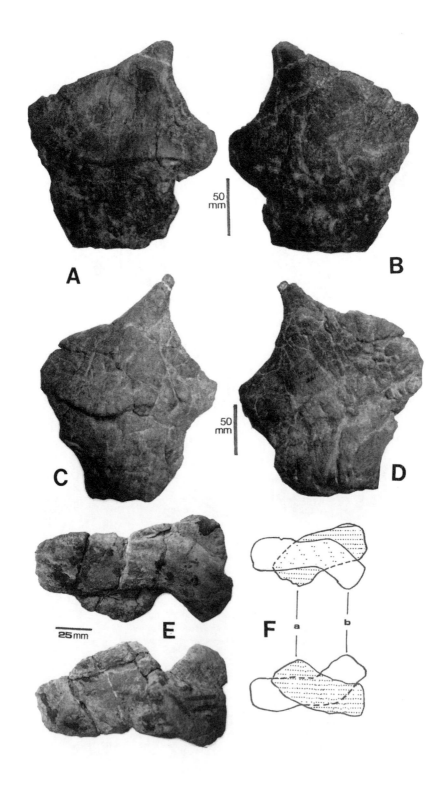

366 • William T. Blows

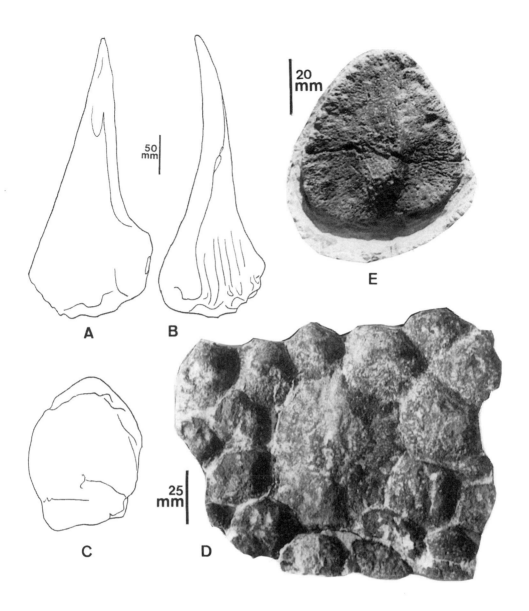

Figure 17.3. CEUM 1307 (opposite page) Gastonia burgei cervicopectoral armor. (A, B) Both sides of one spine. (C, D) Both sides of a second spine. (E) Ventral (top) and dorsal (bottom) of a third, flatter specimen. (F) Redrawn views of E to show the dorsal element (a, stippled) and the basal bone (b, white); Lower Cretaceous, Cedar Mountain Formation, Utah.

Figure 17.4. CEUM 1307 (above) Gastonia burgei presacral dorsal armor spine in (A) ?lateral, (B) ?anterior, and (C) ventral (basal) views. (D) Sacral shield fragment in dorsal view. Lower Cretaceous, Cedar Mountain Formation, Utah (courtesy of Jim Kirkland). (E) BMNH R13002 Polacanthus foxii ossicle; dorsal view, the first evidence of the species P. foxii from the English mainland. Lower Cretaceous, Wealden beds of Freshfields Quarry, Sussex, UK (found and donated by David Cooper, Isle of Wight).

Dermal Armor of the Polacanthine Dinosaurs • 367

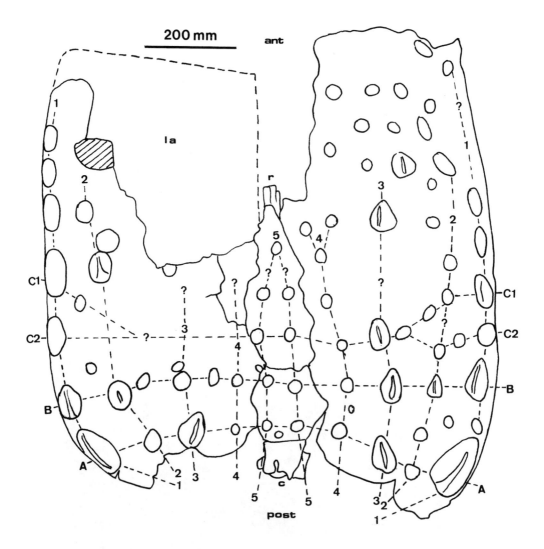

Figure 17.5. BMNH R175 *Polacanthus foxii* sacral shield in dorsal view from the Lower Cretaceous of the Isle of Wight, UK. Rows of bosses can be traced as follows. Transverse rows: A = most posterior row; B = next anterior row; C1 and C2 = most anterior row preserved. Longitudinal rows: 1 = most lateral rows, left and right, skirting periphery of the shield; 2 = next row medially; 3 = approximately middistance between periphery and shield center; 4 = just lateral to sacrum; 5 = most medial rows overlying sacrum. ant = anterior; la = lost area; post = posterior.

armor, transverse bands of scutes on the trunk, and clubs at the end of the tail.

It is difficult to accurately establish the position of the dermal armor on the animal, especially in nodosaurids and polacanthine ankylosaurs. The dermis is never preserved, and the armor is usually disarticulated and scattered.

Armor must have had a function (Kirkland 1998), and although some armor may have been used for protection against predator attack, it was a biological tissue, perfused with blood, and therefore prone to bleeding and infection if damaged. No dermal armor piece has been reported to show any evidence of fracture healing sustained during life.

Ankylosaurs are relatively rare as fossils (compared with hadrosaurs and ceratopsians), and their remains usually occur as isolated

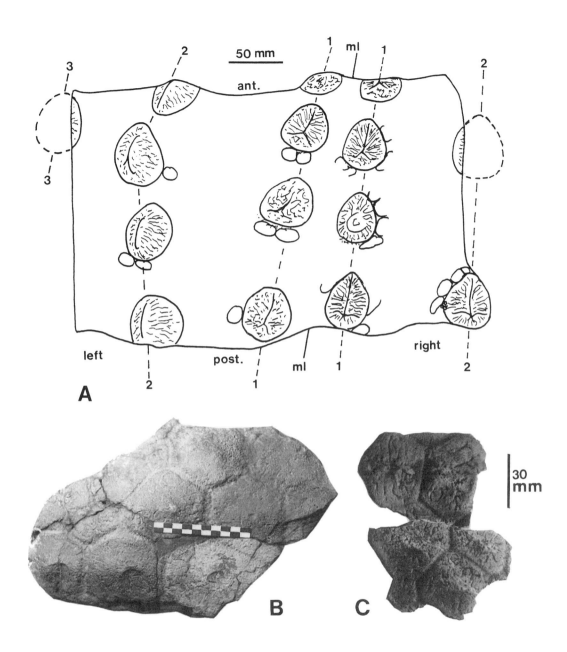

Figure 17.6. (A) Portion of the sacral shield of Mymoorapelta maysi *from the Upper Jurassic of Utah (courtesy of Jim Kirkland). Rows of bosses can be traced as follows: 1 = most medial rows on each side of the midline (ml); 2 = central rows; 3 = a trace of a lateral row. ant = anterior; post = posterior. (B, C) Nonpolacanthine shields that show variation from the boss and tubercle type of polacanthine shields: (B) FMNH UR88* Stegopelta landerensis, *part of the sacral shield in dorsal view (courtesy of A. Milner, Birkbeck College); (C) USNM 8610, a fragment of shield in dorsal view similar in pattern to FMNH UR88.*

and fragmentary specimens. Dermal armor usually constitutes a major proportion of any ankylosaur specimen, but descriptions have been brief. Older descriptions gave the impression of a lack of familiarity with this material; these also implied that the armor was considered to be the same across different species. There was difficulty in description or interpretation of dermal remains. Most descriptions of ankylosaurs have been made on the basis of endoskeletal fossils. Even illustrations of dermal pieces have been inadequate, with usually only a few scutes or spines figured. For example, Sternberg (1929) described the skull of *Anodontosaurus lambei* in detail over four pages, whereas the over 100 pieces of dermal armor found with it occupy just one page of description, accompanied by one text plate showing just five dermal pieces. There remains an accumulation of many thousands of dermal bones in museums that are not described, illustrated, or displayed. They remain a vast, underutilized source of valuable information about this dinosaur group.

The paucity of published data on dermal armor has resulted in downplaying these bones in phyletic analyses. If detailed descriptions, multiple illustrations, and a classification of dermal armor were available, a good overall picture might emerge that might contribute toward a better understanding of ankylosaur evolution and ecology.

Institutional Abbreviations. AMNH: American Museum of Natural History, New York. CEUM: College of Eastern Utah Prehistoric Museum, Price, Utah. DMNH: Denver Museum of Natural History, Denver, Colorado. FMNH: Field Museum of Natural History, Chicago. GM: Gosport Museum, Gosport, UK. HORSM: Horsham Museum, Horsham, UK. BMNH: Natural History Museum (formerly British Museum [Natural History]), London. MWC: Museum of Western Colorado, Grand Junction, Colorado. SMU: Shuler Museum of Paleontology, Southern Methodist University, Dallas, Texas. YPM: Peabody Museum of Natural History, Yale University, New Haven, Connecticut.

Terminology

The terminology used to describe dermal armor has been inconsistent. For example, the terms "scute" and "plate" are not clearly defined, and both have been used to represent the same type, and often multiple types, of dermal elements. "Osteoderm" has also been used to categorize similar elements, but it is a broad, generic term for all dermal elements that gives no indication as to the nature of the specimen. It would be useful to define the terms in a manner that clearly describes the form of the element and restrict the use of each term to the appropriate dermal piece. This limitation is becoming more important because particular dermal elements appear to have specific functions.

A synopsis of various terms, their author (in parentheses), an interpretation of their definition, and new terms are presented below.

Boss (Lydekker 1891) (Figs. 17.4–17.6), an elongate or rounded ridged keeled element incorporated into a pelvic shield. They range from 50 to 200 mm in diameter. They often have shallow grooves around them and are separated in shields by tubercles (example seen in *Polacanthus*).

Caputegulum (plural *caputegulae*) (new term, from the Latin words *caput* = skull and *tegula* = tile; i.e., skull tile) (Fig. 17.1), the flat bones that cover the skull roof and sides in ankylosaurian dinosaurs.

Caudorbitos (plural *caudorbitosa*) (new term, from the Latin words *cauda* = tail, *orbis* = circle, and *os* = bone; i.e., bony surround of the tail), the dermal bones of the tail club ball found in ankylosaurids (e.g., *Euoplocephalus*). The unit consists of a left lateral caudorbitos, a right lateral caudorbitos, and a terminal caudorbitos at the tail tip. Illustrations of caudorbitosa can be seen in Coombs (1978).

Coronux (plural *coronuces*) (new term, from the Latin word *corona* = crown) (Fig. 17.1), formally called caudolateral horns. These are usually conelike and extend laterally from the posterior skull margins of ankylosaurids (e.g., *Ankylosaurus* and *Euoplocephalus*). Those that occupy the upper posterior corners of the skull roof (squamosal) are dorsal coronuces; those that occupy a lower position on the skull sides (jugal) are ventral coronuces.

Ossicle (Owen 1863) (Fig. 17.4E), a small (5–70 mm) dermal element, often round, oval, or subtriangular. The larger ones have low ridges or peaks dorsally and have flat or slightly convex solid bases. They are often found loose, sometimes in large numbers, with armored dinosaur bones. They were possibly set as a mosaic between the major elements or as transverse rows along the trunk in some taxa.

Plate (Lee 1843) (Fig. 17.7), dermal bones comprising a tall dorsal keel with rounded or sharp points. The keel is flattened laterally but is long anteroposteriorly, with sharp edges. The long, narrow base shows rough areas for dermal insertion and hollows extending into the bone. They are probably placed dorsally in two rows toward the midline in stegosaurs and dorsolaterally or laterally on the tails of some ankylosaurs (e.g., *Polacanthus*). Lee (1843) used the term to describe parts of what has since then been consistently described as the shield.

Ring or *cervical ring* (Sternberg 1928) (Figs. 17.2, 17.3), curved group of two to six elements that surrounds the dorsal and lateral portions of the cervical region. Some are mounted dorsally on a flat band of bone (e.g., advanced ankylosaurids); others are separate elements, each with its own base. The dorsal keel varies from low, rounded mounds to tall points in different taxa, with some having rooflike ridged keels. Some taxa may have more than one ring (e.g., *Euoplocephalus*).

Scute (Owen 1863), low ridged, keeled elements of oval shape. The ridge extends along the longest axis and may be central or displaced laterally or overlying the posterior border of the base. The ridge has a rooflike appearance, usually longer and taller than the ridges of ossicles. Scutes were probably placed in transverse rows along the trunk (e.g., *Edmontonia*). Although the base of some scutes can be convex, most scute bases are slightly or extensively hollowed, which is not the case with ossicles. They are also generally larger than ossicles (50 mm to over 100 mm long). Some scutes form the cervical ring elements (e.g., *Panoplosaurus* and *Euoplocephalus*). Illustrations of scutes can be seen in Russell (1940).

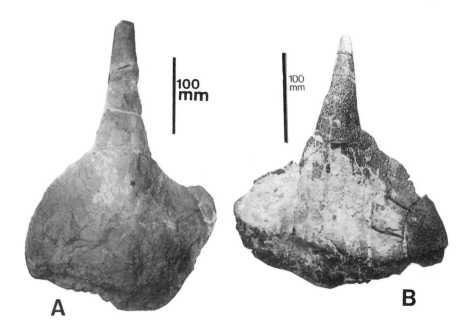

A

B

Figure 17.7. (A) USNM 4752
Hoplitosaurus marshi *sacral*
splate in lateral view;
(B) BMNH R9293 Polacanthus
foxii *sacral splate in medial view,*
from the Lower Cretaceous of
(A) South Dakota and (B) the
Isle of Wight, UK.

Shield (Hulke 1881) (Figs. 17.5, 17.6), a flat, broad layer of bone covering the pelvis and sacrum (including presacral rod) in some armored dinosaurs, such as *Polacanthus*. The thickness is typically 10 mm, and it is separated from the underlying endoskeleton by a further 10 mm or so of connective tissue (not preserved except sometimes ossified tendons). In the Polacanthinae, it is constructed from a mosaic of bosses and tubercles. The ventral surface often shows cross-patterning of ossified collagen bundles.

Spine (Mantell 1841) or *spike* (Figs. 17.2, 17.4), tall pointed elements with solid, rounded bases, whose diameter is less than the total height. The base is often offset relative to the spine. The tall, dorsal peak is round or oval in cross section. Some shorter spines have a lower peak that is displaced posteriorly on the base. Spines occur in bilateral rows close to the midline in some stegosaurs and are found within the remains of ankylosaurs such as *Polacanthus*. Specialized spines occur over the pectoral region in some taxa (e.g., *Edmontonia*) and some stegosaurs (e.g., the parascapular spines of *Kentrosaurus*), and these have very long, pointed dorsal keels with rounded cross sections and expanded bases set at extreme angles to the spine.

Splate (Olshevsky, personal communication) (Fig. 17.8), plates with a spinelike anterior leading edge that bear a point taller than the upper margin of the plate and that tend to incline posteriorly. Splates are intermediate between a plate and a spine. They are found in bilateral rows close to the midline in some stegosaurs (e.g., *Kentrosaurus*). A modified version of a splate is found in some polacanthines, notably *Polacanthus foxii* (Blows 1987: figs. 6A, 8) and *Hoplitosaurus marshi*

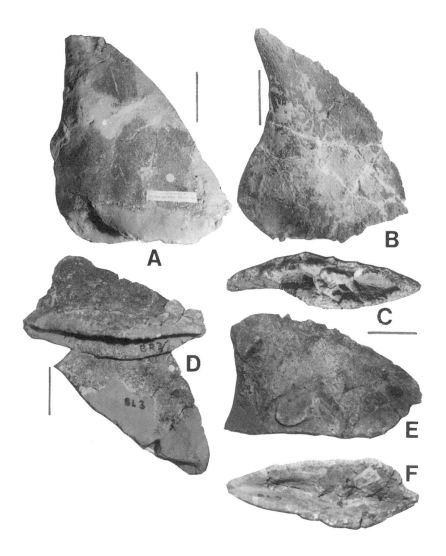

(Gilmore 1914: pl. 28). The superior point is developed into a cylindrical upright rod that occurs midway along the length of the element, emerging from both the anterior and posterior sloping margins of the plate equally. This creates a clear distinction between the stegosaurian and polacanthine splates (Fig. 17.8). This type of splate probably emerges from the lateral border of the shield (Kirkland, personal communication).

Tricorn (Norman 1985), a dermal element consisting of three small spines mounted on a single base, found in *Scelidosaurus* (BMNH R1111 and BMNH 42068) and possibly *Mymoorapelta* (MWC 1825, a single spine very similar to the tricorn spines of *Scelidosaurus*). In *Scelidosaurus,* they are positioned bilaterally, immediately behind the skull. An illustration of a tricorn can been seen in a life restoration of *Scelidosaurus* in Norman (1985: 159).

Figure 17.8. Polacanthus foxii *caudal plates. (A, D–F) BMNH R175 and (B, C) BMNH R9293 from the Lower Cretaceous of the Isle of Wight, UK. (A, B–E) ?Lateral views; (C, E) ventral views of B and E, respectively.*

Tubercle (Seeley 1881) (Fig. 17.4), the raised knobs of bone that are clustered and packed between the bosses of polacanthine sacral shields (e.g., *Polacanthus*). The base is the thickness of the shield, about 10 mm, and often bears bundles of ossified collagen. Tubercles are notably smaller (7–25 mm across) than the bosses they are fused to. They are usually not found separately, unless they come from a juvenile, unfused shield or have broken off an adult shield. Tubercles from an adult shield usually have edges created by fracture surfaces.

History of the Family Polacanthidae and Subfamily Polacanthinae

Wieland (1911) first published the name Polacanthidae without any qualification or description, probably because he considered the Nodosauridae (Marsh 1890) to be synonymous with the Polacanthidae. In his paper, it appears among a list of other family names, some of which are now regarded as synonyms:

> the startlingly strange, complex, and ornate aspect at once in-dicated by even lesser portions of the mail-clad Dinosaurs, has naturally led to so called "new families." Not to mention a long series of genera of convenience, we thus have the Scelidosaur-idae, Polacanthidae, Stegosauridae, Nodosauridae, Ankylosaur-idae, etc. (Wieland 1911: 118)

Wieland's concept of the Polacanthidae as being synonymous with the Nodosauridae becomes clearer in the same article as he discusses the family status of *Stegopelta* and *Hierosaurus*: "The family attribution of both genera along with *Polacanthus* Hulke, *Nodosaurus* Marsh, *Palaeoscincus* Leidy, *Stereocephalus* Lambe, and *Ankylosaurus* Brown, we think surely lies within the Nodosauridae of Marsh" (Wieland 1911: 117). It should be noted that *Palaeoscincus* is now a junior synonym of *Edmontonia,* and *Stereocephalus* has been renamed as *Euoplocephalus*.

Kirkland (1998) referred the name Polacanthinae, a subfamily of the Ankylosauridae, to Wieland (1911), but Polacanthinae does not appear in Wieland's text. The subfamily name Polacanthinae should be referred to Kirkland (1998), although Lapparent and Lavocat (1955) used the term "polacanthines," suggesting the existence of the term Polacanthinae as early as 1955.

Polacanthine Dermal Armor

The dermal armor related characteristics of the subfamily Pola-canthinae are:

1. Skull (Fig. 17.1) with paired dorsal and ventral coronuces extending laterally and slightly posteriorly from a position behind the orbit, giving the rear of the skull a box shape when viewed posteriorly. The coronuces are similar to, but less well developed than, those seen in the ankylosaurinae, such as *Euoplocephalus* and *Ankylosaurus*. The ankylosaurine coronux is apomorphic in relation to

the polacanthines, whereas in other nodosaurs, the coronuces are absent (e.g., *Edmontonia*).

2. Single, fused, continuous sheet of dermal bone covering the ilia and synsacrum, often called the "sacral shield" (Figs. 17.5, 17.6). This shield is a large but thin layer that is not fused to the underlying ilia and is usually ornamented on the dorsal surface by fused bosses and surrounding tubercles.

3. A bundle of ossified tendons occur ventral to the distal caudal verte-brae, with oval, flat dermal plates on each side (Blows 1987: fig. 3G, F) (Fig. 17.9). Similar ossification of distal caudal tendons is seen in ankylosaurines to support a caudal end club made from three dermal caudorbitos. Such tendons may therefore be indirect evidence of a similar primitive structure in polacanthines. This hypothesis awaits discovery of direct evidence in the form of a polacanthine preserved tail end.

4. Primitive cervical rings composed of separate ridged, laterally flat-

Figure 17.9. BMNH R175
Polacanthus foxii caudal mass in (A) left lateral and (B) right lateral views, from the Lower Cretaceous of the Isle of Wight, UK. cv = caudal vertebra; da = dermal armor; ot = ossified tendons.

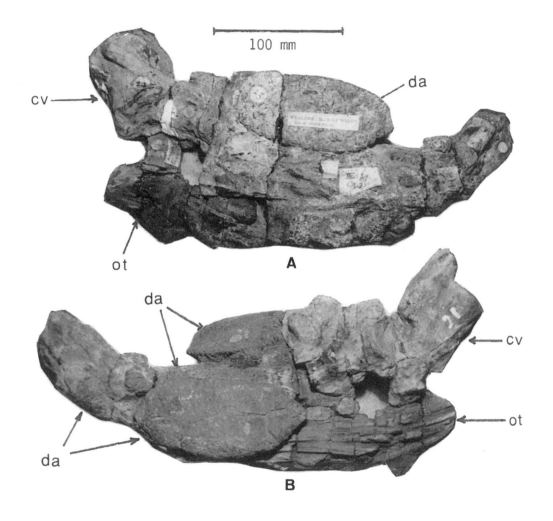

tened spines (Figs. 17.2D–F, 17.3), some mounted on basal narrow bands of bone.

5. Pectoral spines (Fig. 17.2A–C) of modest length, from small to large on solid bases, some with basal bones that extended into the dermis.

6. Presacral spines (Figs. 17.4A–C) that are laterally compressed and have flat, broad, solid bases.

7. Specialized splates extending laterally from the borders of the sacral shield (Kirkland, personal communication) (Fig. 17.8).

8. Tall, asymmetrical, hollow based caudal plates of varying heights (Fig. 17.7).

9. Small to large solid-based and dorsally ridged dermal ossicles (Fig. 17.4E).

The ankylosaur taxa included here within the Polacanthinae are:

1. *Polacanthus foxii* Hulke 1881 (BMNH R175) has characters 2, 3, 4, 5, 6, 7, 8, and 9 (Blows 1987). An entire skull from a specimen of this genus has never been found, and the partial skull described by Norman and Faiers (1996) does not preserve evidence of character 1. *Polacanthus rudgwickensis* Blows, 1996 (HORSM 1988.1546), has character 6 and may have had others, but these elements are missing.

2. *Hoplitosaurus marshi* (Lucas) 1901 (USNM 4752) has characters 5, 6, 7, 8, and 9; the other characters are unknown because those parts are missing (Gilmore 1914). The suggested synonymy of this taxon with *Polacanthus* has not been accepted by Carpenter and Kirkland (1998).

3. *Gastonia burgei* (Kirkland 1998) from the Lower Cretaceous of Utah (CEUM 1307) has characters 1, 2, 4, 5, 6, 8, and 9. Character 3 cannot be confirmed yet.

4. *Mymoorapelta maysi* Kirkland and Carpenter, 1994, has characters 1, 2, 4, 5, 8, and 9. The skull is largely unknown except for a jugal bone (MWC 2843) with a coronux, confirming this character in this taxon (Kirkland et al. 1998: fig. 1). The same article also illustrates (Fig.7F) a small, solid-based, sharp-pointed spine (MWC 1825), which was found with caudal plates. MWC 1825 very closely resembles spines from the tricorn element present in the cervical region of the Lower Jurassic *Scelidosaurus harrisonii* (BMNH R1111 holotype and BMNH 42068) and is probably also cervical in origin, suggesting the presence of character 4 in *Mymoorapelta*. Characters 3, 6, and 7 have not been found or reported.

5. *Gargoyleosaurus parkpinorum* Carpenter et al., 1998, is an Upper Jurassic ankylosaur known from a skull and postcranial skeleton (DMNH 27726). The features used in a cladistic study by Carpenter et al. (1998) indicated that this taxon was a member of the Ankylosauridae. The authors gave a description of the skull with only a brief outline of the postcranial remains, which were still in preparation. The skull shows that all four coronuces are well developed (character 1), and a random symmetrical pattern of caputegulae was present. The postcranial armor is said to be similar

to *Mymoorapelta* by Kirkland et al. (1998), who places this taxon within the Polacanthidae.

Tentative assignment to the Polacanthinae:

Pawpawsaurus campbelli Lee, 1996. SMU 73203 is a skull from the Lower Cretaceous of the United States that has character 1 present. On the grounds that the Edmontoniinae did not have any coronuces and that the Ankylosauridae had advanced developed coronuces, *Pawpawsaurus* had an intermediate state, similar to those taxa assigned to the Polacanthinae. However, because character 3 and all the postcranial armor is missing, this assignment remains tentative. Kirkland (personal communication) has now recovered sufficient material to indicate that this taxon is unlikely to be included within the Polacanthinae.

Taxa that appear to be excluded from the polacanthine nodosaurs:

1. *Hylaeosaurus armatus* (Mantell 1833) has only a small fragment of skull incorporated in matrix in the holotype (BMNH R3775), and this is inadequate to identify any coronuces. The cervical and pectoral armor in the holotype remains partly buried in the matrix, and its description awaits further preparation. The pectoral spines do not have broad bases, and from what can be seen in situ, there is no evidence that they were mounted on narrow bone bands, so they do not match character 5. There are no presacral spines of the type described in character 6, despite good anterior armor preserved in the holotype. Although sacral shield remains are a dominant feature of each *Polacanthus* specimen found, no evidence of a sacral shield is known from three major discoveries of *Hylaeosaurus* (Blows 1998; Mantell 1841). This could be a preservational or collecting anomaly, but it hints strongly at the absence of this feature in this taxon. The caudal sequence collected from one of the three specimens (BMNH 3789) shows no evidence of character 3. The ossicles are also small and simple, being round with a slight central peak, unlike character 6. This taxon is both primitive to *Polacanthus* (Pereda-Suberbiola 1994) and is geologically older (Blows 1987, 1998). Kirkland (1998) only tentatively included this taxon within the Polacanthinae, but further evidence is needed, from full preparation of the holotype or new discoveries, to positively include this taxon.

2. *Nodosaurus textilis* (Marsh 1889) was found without a skull, and only an incomplete, narrow sacral shield is preserved that covers the synsacrum and the medial margins of the ilia. The presence of a shield may mean that this taxon should be included within the polacanthines, but it appears not to be of the boss and tubercle type. Also, the elements involved in characters 3, 4, and 5 are basically unknown, and the presence of character 6 is doubtful. This taxon forms the foundation of the Nodosauridae, yet it is difficult to characterize because of its incomplete state and therefore calls into question the very basis of the Nodosauridae as a unified group.

3. *Stegopelta landerensis* Williston, 1905, the holotype (FMNH UR88), has no skull preserved, and although the specimen does have a sacral shield, it is constructed differently than the boss and tubercle arrangements seen in the polacanthines; that is, the shield is a series of regular fused elements (Fig. 17.6B–C). It has at least one cervical ring of fused scutes (Carpenter and Kirkland 1998: fig. 25). It does have ossicles, character 6, similar to *Polacanthus,* and this suggests that *Stegopelta* may be placed in the polacanthine group when more evidence is available.

4. *Silvisaurus condrayi* Eaton, 1960, has a skull with no coronuces. The postcranial armor is very limited, and Eaton's (1960) description of this armor is poor. Carpenter and Kirkland (1998) have redescribed the few pieces of postcranial armor preserved. The three cervical rings are preserved on the left side only and are apparently fused, not separate spines or scutes. The ?shoulder spine does not have a basal bone extending into the dermis, and the ?sacral shield fragment is identified from its similarity to *Nodosaurus* rather than *Polacanthus.* The armor, as currently known, does not include any of the stated characters.

5. *Sauropelta edwardsorum* Ostrom, 1970, has a wide arrangement of dermal armor attributed to the various specimens. Small coronuces exist on the skull (AMNH 3035; Carpenter and Kirkland 1998: fig. 9), but the animal had fused cervical rings (YPM 5178; Carpenter and Kirkland 1998: fig 11), contrary to character 4, and long pectoral spines (AMNH 3032), contrary to character 5. The sacral armor is of the boss and tubercle type (AMNH 3036), but it is not clear if this is fused as a solid sheet across the pelvis. Carpenter (1984) states that the sacral armor is tightly interlocking. Character 3 is absent (Coombs 1978), and characters 6, 7, and 8 are not reported in this taxon.

6. The Edmontoniinae (Russell 1940), a subfamily of nodosaurs occupied by the Upper Cretaceous, forms *Edmontonia* and *Panoplosaurus.* They have skulls with no cranial coronuces, broad cervical rings bearing large flat scutes, rows of scutes on the trunk, and a pelvis with no sacral shield or ossicles.

Several other taxa—for example, *Priconodon* and *Texasetes*—have either no preserved armor or inadequate dermal armor features to be sure of their status, and they must await further discoveries and analyses of their other features.

Discussion

The presence of primitive dorsal and ventral coronuces on the skull, the presence of ossified tendons in the distal tail, and tall, ridged cervical ring elements on narrow basal bones all place the polacanthines closer to the ankylosaurids than any other nodosaurids. Kirkland (personal communication), for example, has indicated that *Gastonia,* which is close to *Polacanthus,* appears more ankylosaurid than it is nodosaurid. The Polacanthinae is therefore considered to be a subfamily of the Ankylosauridae (Kirkland 1998; however, see Carpenter 2001).

The dermal armor of the polacanthine group is here discussed under the following regions: skull, cervicopectoral, presacral, sacral, and caudal.

Skull (Fig. 17.1)

Polacanthine skulls are rare; the only ones known come from *Pawpawsaurus campbelli*, *Gastonia burgei*, and *Gargoyleosaurus parkpinorum*. *Gastonia burgei* appears to have been common in the Cedar Mountain Formation fauna, with more than 10 partial skulls known (Kirkland, personal communication). In polacanthine skulls, the dorsal surface is covered in capitegulae arranged in varying degrees of bilateral symmetry that is either poorly developed (*Gastonia*) or developed to a moderate degree (*Gargoyleosaurus* and *Pawpawsaurus*). The dorsal and ventral coronuces are slightly developed in *Pawpawsaurus* (the "postorbital dermal plate" and "jugal dermal plate," respectively, of Lee 1996), and moderately well developed into small conical structures in *Gastonia* and *Gargoyleosaurus*.

Cervicopectoral Armor (Figs. 17.2, 17.3)

Polacanthines had at least two cervical rings (Kirkland 1998), but unlike the Edmontoniinae and the Ankylosauridae, in which the elements are linked into a continuous ring, the polacanthines had separate pieces, each with its own endodermal basal bone. *Gargoyleosaurus* cervical rings were composed of two separate quarter rings (Carpenter, personal communication).

The separate cervicopectoral spines of *Polacanthus* were first found by Blows in 1980 (Blows 1987; BMNH R9293) and are now also known in *Gastonia* (Kirkland 1998). In *Polacanthus*, the two known elements are large, laterally flat spines mounted on distinctively shaped bases (Figs. 17.2B, C, 17.2E, F), which differ from the bases of the presacral spines (Fig. 17.4C). The largest of the two spines is recurved posteriorly; the smaller has straight edges that are angled posteriorly. A third isolated specimen (GM 981.45) referred to *Polacanthus* is similar, but it differs by a groove extending down the posterior margin to accommodate the leading edge of the next spine along and a deep keellike endodermal base. *Gastonia* had smaller elements than *Polacanthus*, each of which forms a sharp point with a deep ventral keeled base (Fig. 17.3A–D).

Presacral Armor (Figs. 17.4A–C)

Since the first *Polacanthus foxii* was found by Fox in 1865 (Blows 1983), polacanthine discoveries have included spines varying in height from tall and conical to low and ridged. These have flat, solid bases, and the tallest spines are slightly laterally compressed and twisted 90° toward the summit. *Gastonia* has spines of very similar size and construction (Fig. 17.4A–C).

Sacral Armor (Figs. 17.5, 17.6)

Polacanthines are best known for their sacral shield, a very large sheet of fused tubercles and bosses covering both the ilia and sacrum, including the presacral rod (the "synsacrum"). In *Polacanthus foxii,* the shield is about 1 m² and is separated from the pelvis by about 1 cm of matrix (as preserved). The only elements of dermal armor found fused to the underlying endoskeleton in polacanthines were the caputegulae. In the holotype (BMNH R175) of *Polacanthus foxii,* ossified tendons are preserved in the space between the shield and the underlying sacrum.

In BMNH R175, a rough bilateral symmetry in the rows of bosses can be identified (Fig. 17.5), with the largest bosses placed in the posterolateral corners. Bilateral symmetry can also be traced in a partial shield of *Mymoorapelta maysi* (Fig. 17.6A). *Gastonia* also had a shield pattern as noted by Kirkland (1998), indicating that a definitive symmetry can be established in this taxon.

Caudal Armor (Figs. 17.7, 17.9)

The tail supports a bilateral series of laterally flattened, paired plates descending in size, from the largest behind the sacral shield to the smallest at the tail end. They have openings in the base that extend into the body of the plate and deep, rugose edges that indicate firm dermal attachment to overcome lateral forces. The sharp anterior and posterior edges of the dorsal keel unite at the peak, forming posteriorly directed tips. Viewed laterally, the bases of plates from BMNH R9293 have an asymmetrical curved edge at the point of dermal insertion, which probably accommodated a lateral or dorsolateral position on each side of a strongly curved tail surface. Variations in tail plate morphology in *Polacanthus foxii* exist between BMNH R175 and BMNH R9293, and this may be associated with sexual dimorphism (Blows 1996) or individual variation.

A caudal mass, part of BMNH R175, consists of vertebrae, ossified tendons, two flat, oval armor plates, and a central bony core (Fig. 17.9). The importance of ossified tendons in the tail of *Polacanthus* is that it is a feature previously known only in ankylosaurines as part of the tail modification associated with the caudal end club. The presence of ossified caudal tendons in a polacanthine may indicate modification of the tail for lateral swinging, as might be expected when carrying a tail-end weapon (currently unknown in this group) or if the caudal plates were projecting laterally, as postulated by Kirkland (1998) for *Gastonia.* Other ornithischians had similar tail-stiffening tendons, such as hadrosaurs, in which tail swinging may have been associated with mobility (e.g., swimming) or defense.

Function of Polacanthine Dermal Bone

The role of dermal bone can be placed into three categories; defense (including display), thermoregulation, and locomotion. Most dermal bones have traditionally been assigned to a defensive role in predator attacks, and for some armor elements, this is probably true. Spines and spikes, rising as they do to sharp points on a broad base, covered with thick keratin, seem to have been well able to keep predators away from the trunk or risk impalement, a form of passive defense for the dinosaur. Similar pectoral spines and specialized pelvic splates may have been important in protecting the animal at sites of enlargements of the spinal cord and emergence of limb plexi. Cranial coronuces and lateral caudal plates (or caudal caudorbitosa, if they existed in this group) could be regarded as active defense, used as offensive weaponry in those parts of the body mobile enough to launch an attack. If this assessment is accurate, the animal had active defense at both the front and rear ends, with passive defense in between.

Thermoregulation has been postulated as a function of dermal plates in *Stegosaurus* (Buffrénil et al. 1986; Farlow et al. 1976), and a similar function is possible for the caudal plates of the polacanthine dinosaurs. A surface pattern of fine grooves on these dermal bones is usually interpreted as vascular channels, but alternatively, these grooves could be present to improve the bone–keratin interface. Carpenter (personal communication) has proposed a blood-flushing mechanism for the plates of some ankylosaurs, causing a color change to occur as a signaling device or as a display for defensive or mating purposes. Blood circulating over the surface of the plates in large quantity would be subject to extensive losses if any trauma occurred to the keratin cover, and this would be a particular problem if the plates were mounted laterally on the tail in polacanthines. Hollow internal cavities suggest the opportunity for some pooling of blood that could then warm or cool in close contact with the environment before draining into the body for redistribution, as suggested for the dorsal scutes of *Alligator mississippiensis* (Seidel 1979). An alternative hypothesis is that the hollowed cavities of the base may have housed a tough tendon running from inside the plate for deep anchorage into the tail. This may have been important if the plates were laterally placed on the tail and used as a weapon.

Frey (1984) proposed a biomechanical role for the dorsal scutes of the crocodile on the basis of the ventral attachment of epaxial muscles to the scute edges. The medial pair of scutes that lies between the dorsal vertebral neural spines, the vertebrae themselves, and the underlying epaxial musculature form a functional unit (Frey's terminology). This unit serves a similar function to that of the tall neural spines found over load-bearing segments of the vertebral column in many mammalian quadrupeds (i.e., increased epaxial muscle attachment). Crocodiles and ankylosaurs have uniformly low neural spines throughout the vertebral column and transverse rows of dermal bones. It would appear that possession of transverse rows of dermal bones equates with uniformly low neural spines in large reptiles. Stegosaurs had tall neural spines and no transverse rows of dermal bones. Attachment of these muscles to the scutes in crocodiles may provide greater spinal strength by arching the spine during the high walk, allowing the limbs to be brought under the body for improved gait, increasing the stride and ground clearance of the trunk to improve speed. In polacanthine ankylosaurs, a similar musculature attachment to ossicles could serve to lift the tail, especially when the tail is required as an active weapon.

The polacanthine dinosaurs probably had a mixed-function armor, with specialized elements and other elements, possibly including the sacral shield, with dual roles. The shield appears to be a passive defense structure with bosses providing some sharper points, protecting the area of the sacral spinal enlargement and hindlimb plexus. It occupied that part of the trunk that because of the large horizontal ilia was already inflexible. If the shield was used as a thermoregulatory heat-collecting panel, its position on the animal's dorsum facing the sun would have been the optimum site for warming the blood passing through it. It would have provided a massive surface area for blood flow to trap solar heat.

Thyreophoran Dermal Bone Morphology

Taking the armored thyreophora as a whole, it is possible to trace a change of dermal bone morphology over the later half of the Mesozoic. As an Upper Jurassic stegosaur, *Stegosaurus* is atypical as far as dermal bone is concerned. The plates occupied over 90% of the total postcranial length, leaving the caudal end only with spines. The transition from anterior plates to posterior spines was sudden, with no splates involved. Most other stegosaurs—for example, *Kentrosaurus* and *Tuojiangosaurus*—had more equal postcranial lengths of plates anteriorly and spines posteriorly, with splates between creating a gradual transition.

In the Lower Cretaceous *Polacanthus foxii,* the armor morphologies are again close to equal but reversed, with 36% of the anterior postcranial length occupied by spines and 43% of the posterior postcranial length occupied by plates, with the sacral shield and its surrounding splates accounting for 21% of the postcranial body length, again acting as a transition in between. This potentially indicates a change in thyreophoran thermoregulatory or defensive behavior strategies across the Jurassic–Cretaceous transition. At the same time, the move from stegosaurs as the dominant thyreophoran within the Upper Jurassic to polacanthines in the Lower Cretaceous fauna is further suggestive of a significant environmental shift of unknown nature. A discussion about this Jurassic to Cretaceous transition phenomenon, including all the relevant dinosaur groups, can be found in Bakker (1998). By the Upper Cretaceous, the dominant thyreophorans were the ankylosaurids, with postcranial armor that appears to be mostly for defensive purposes. By Upper Cretaceous times, other unknown conditions may have changed that could have forced a need for ankylosaurs to modify their defensive strategy, along with the development of other dinosaur groups with extensive means of defense (e.g., the ceratopsia), which used horns for self-protection. Increased size of the Upper Cretaceous tyrannosaurid predators is probably an oversimplified answer to the massive defensive role of armor found in the ankylosaurids. Both large and small predators were also present in all the Middle to Late Mesozoic faunas, interacting with Jurassic stegosaurs and Lower Cretaceous polacanthines alike. The gradual increased emphasis placed on defense over the Cretaceous period, and therefore the corresponding need to use bone for this purpose, in all the quadrupedal ornithischian dinosaur groups is worthy of further study.

Acknowledgments. My gratitude goes to Dr. Angela Milner and Sandra Chapman of the Natural History Museum, London, and to Dr. J. Kirkland of Salt Lake City, Utah, for permission to study material in their care. I also thank Kenneth Carpenter of Denver Natural History Museum, Colorado, for the opportunity to submit this paper and for his comments on ankylosaurs.

References Cited

Bakker, R. 1998. Dinosaur mid-life crisis: The Jurassic–Cretaceous transition in Wyoming and Colorado. In Lucas S. G., Kirkland J. I., and Estep J. W. (eds.), *Lower and Middle Cretaceous Terrestrial Ecosystems. New Mexico Museum of Natural History and Science Bulletin* 14: 67–78.

Blows, W. T. 1983. William Fox, a neglected dinosaur hunter of the Isle of Wight. *Archives of Natural History* 11: 299–313.

———. 1987. The armored dinosaur *Polacanthus foxi* from the Lower Cretaceous of the Isle of Wight. *Palaeontology* 30: 557–580.

———. 1996. A new species of *Polacanthus* (Ornithischia; Ankylosauria) from the Lower Cretaceous of Sussex, England. *Geological Magazine* 133(6): 671–682.

———. 1998. A review of Lower and Middle Cretaceous dinosaurs of England. In S. G. Lucas, J. I. Kirkland, and J. W. Estep (eds.), *Lower and Middle Cretaceous Terrestrial Ecosystems. New Mexico Museum of Natural History and Science Bulletin* 14: 29–38.

Buffrénil, V. de, J. O. Farlow, and A. de Ricqles. 1986. Growth and function of *Stegosaurus* plates: Evidence from bone histology. *Paleobiology* 12: 459–473.

Carpenter, K. 1984. Skeletal reconstruction and life restoration of *Sauropelta* (Ankylosauria: Nodosauridae) from the Cretaceous of North America. *Canadian Journal of Earth Sciences* 21: 1491–1498.

———. 2001. Phylogenetic analysis of the Ankylosauria. In K. Carpenter (ed.), *The Armored Dinosaurs.* Bloomington: Indiana University Press [this volume, Chapter 21].

Carpenter, K., and J. I. Kirkland. 1998. Review of Lower and Middle Cretaceous Ankylosaurs from North America. In S. G. Lucas, J. I Kirkland, and J. W. Estep (eds.), *Lower and Middle Cretaceous Terrestrial Ecosystems. New Mexico Museum of Natural History and Science Bulletin* 14: 249–270.

Carpenter, K., C. Miles, and K. Cloward. 1998. First known skull of a Jurassic ankylosaur (Dinosauria). *Nature* 393(6687): 782–783.

Coombs, W. P., Jr. 1978. The families of the Ornithischian dinosaur order Ankylosauria. *Palaeontology* 21: 143–170.

Eaton, T. H. 1960. A new armored dinosaur from the Cretaceous of Kansas. *University of Kansas Paleontological Contributions, Vertebrata* 8: 1–24.

Farlow, J. O., C. V. Thompson, and D. E. Rosner. 1976. Plates of the dinosaur *Stegosaurus*: Forced convection heat loss fins? *Science* 192: 1123–1125.

Frey, E. 1984. Aspects of the biomechanics of crocodilian terrestrial locomotion. In W. Reif and F. Westphal (eds.), *Third Symposium on Mesozoic Terrestrial Ecosystems, Short Papers.* Tubingen: Verlag.

Gilmore, C. W. 1914. Osteology of the armored Dinosauria in the United States National Museum, with special reference to the genus *Stegosaurus*. *Bulletin of the United States National Museum* 89: 1–143.

Hulke, J. W. 1881. *Polacanthus foxii*, a large undescribed dinosaur from the Wealden formation in the Isle of Wight. *Philosophical Transactions of the Royal Society Series B, London* 178: 169–172.

Kirkland, J. I. 1998. A polacanthine ankylosaur (Ornithischia: Dinosauria) from the Early Cretaceous (Barremian) of Eastern Utah. In S. G. Lucas, J. I. Kirkland, and J. W. Estep (eds.), *Lower and Middle Cretaceous Terrestrial Ecosystems. New Mexico Museum of Natural History and Science Bulletin* 14: 271–281.

Kirkland, J. I., and K. Carpenter. 1994. North America's first pre-Cretaceous Ankylosaur (Dinosauria) from the Upper Jurassic Morrison formation of Western Colorado. *Brigham Young University Geology Studies* 40: 25–42.

Kirkland, J. I., K. Carpenter, A. P. Hunt, and R. D. Scheetz. 1998. Ankylosaur (Dinosauria) specimens from the Upper Jurassic Morrison Formation. *Modern Geology* 23: 145–177.

Lapparent, A. F., and R. Lavocat. 1955. Dinosauriens. In J. Piveteau (ed.), *Traite de Paleontologie, Amphibiens, Reptiles, Oiseaux*, Vol. 5, pp. 785–962. Paris: Masson et Cie.

Lee, J. E. 1843. Notice of saurian dermal plates from the Wealden of the Isle of Wight. *Annual and Magazine of Natural History* 2: 5–7.

Lee, Y.-N. 1996. A new nodosaurid ankylosaur (Dinosauria: Ornithischia) from the Paw Paw Formation (late Albian) of Texas. *Journal of Vertebrate Paleontology* 16: 232–245.

Lucas, F. A. 1901. A new dinosaur *Stegosaurus marshi* from the Lower Cretaceous of South Dakota. *Proceedings of the United States National Museum* 23 (1224): 591–592.

Lydekker, R. 1891. On part of the pelvis of *Polacanthus*. *Quarterly Journal of the Geological Society* 48: 148–149.

Mantell, G. A. 1833. *Geology of the South-East of England*. London: Longman.

———. 1841. Memoir on a portion of the lower jaw of the *Iguanodon*, and on the remains of the *Hylaeosaurus* and other saurians discovered in the strata of Tilgate Forest in Sussex. *Philosophical Transactions of the Royal Society, London* 131: 131–151.

Marsh, O. C. 1889. Notice of gigantic horned Dinosauria from the Cretaceous. *American Journal of Science* 5(138): 173–175.

———. 1890. Additional characters of the Ceratopsidae, with notice of new Cretaceous dinosaurs. *American Journal of Science* 3(391): 418–426.

Norman, D. B. 1985. *The Illustrated Encyclopedia of Dinosaurs*. London: Salamander Books.

Norman, D. B., and T. Faiers. 1996. On the first partial skull of an ankylosaurian dinosaur from the Lower Cretaceous of the Isle of Wight, southern England. *Geological Magazine* 133(3): 299–310.

Ostrom, J. H. 1970. Stratigraphy and palaeontology of the Cloverly Formation (Lower Cretaceous) of the Bighorn Basin area, Wyoming and Montana. *Peabody Museum of Natural History Bulletin* 5(35): 1–234.

Owen, R. 1863. Monograph of a fossil dinosaur (*Scelidosaurus harrisonii* Owen) of the Lower Lias. *Palaeontographical Society Monograph, British Fossil Reptilia of the Liassic Formation*, Part 2: 1–26.

Pereda-Suberbiola, J. 1994. *Polacanthus* (Ornithischia, Ankylosauria), a transatlantic armored dinosaur from the early Cretaceous of Europe and North America. *Palaeontographica Abteilung A* 232: 133–159.

Russell, L. S. 1940. *Edmontonia rugosidens* (Gilmore), an armored dinosaur from the Belly River Series of Alberta. *University of Toronto Studies (Geology)* 43: 1–28.

Seeley, H. G. 1881. The reptile fauna of the Gosau formation preserved in the geological museum of the University of Vienna. *Quarterly Journal of the Geological Society* 38(148): 620–702.

Seidel, M. R. 1979. The osteoderms of the American alligator and their functional significance. *Herpetologica* 35: 375–380.

Sternberg, C. M. 1928. A new armored dinosaur from the Edmonton formation of Alberta. *Transactions of the Royal Society of Canada* 4: 93–106.

———. 1929. A toothless armored dinosaur from the Upper Cretaceous of Alberta. *National Museum of Canada Bulletin* 54: 28–33.

Wieland, G. R. 1911. Notes on the armored Dinosauria. *American Journal of Science* 4(Art. 15): 112–124.

Williston, S. W. 1905. A new armored dinosaur from the Upper Cretaceous of Wyoming. *Science* 22(564): 503–504.

18. Mounted Skeleton of the Polacanthine Ankylosaur *Gastonia burgei*

ROBERT W. GASTON,
JENNIFER SCHELLENBACH, AND
JAMES I. KIRKLAND

Abstract

The mounted skeleton of the Early Cretaceous polacanthine ankylosaur *Gastonia burgei* combines bones from the Gaston Quarry and the Dalton Wells Quarry. By use of the holotype skull and the largest postcranial elements, an adult mount was assembled. The mount, with elaborate armor, includes both casts of original material as well as reconstructions based on similar specimens.

Introduction

Gastonia burgei (Kirkland 1998), as well as *Gargoyleosaurus parkpinorum* (Carpenter et al. 1998) at the Denver Museum of Natural History, Denver, Colorado, are the first complete mounted skeletons of polacanthid ankylosaurs. *Gastonia burgei* was found in the Yellow Cat Member of the Lower Cretaceous Cedar Mountain Formation (Kirkland 1998; Kirkland et al. 1997), whereas *Gargoyleosaurus* was found in the Morrison Formation (Carpenter et al. 1998).

Gastonia is the best known polacanthid, with nearly every skeletal element represented. A mounted skeleton was prepared that used casts of elements recovered from two quarries in Grand County, Utah: the Gaston Quarry, operated by the College of Eastern Utah Prehistoric Museum (Price, Utah), and the nearby Dalton Wells Quarry, operated by the Brigham Young University Earth Sciences Museum (Provo, Utah) and the Museum of Western Colorado. A minimum of five individuals were recovered, although the majority of the Dalton Wells material represents smaller individuals than those recovered from the Gaston Quarry.

The largest skull, the holotype, was combined with the largest postcranial skeletal elements of the paratype, resulting in a reconstruction of an adult 4.6 m in length (Kirkland 1998). Comparisons made with additional elements from Dalton Wells confirm the correct proportions of the skeletal elements used in the mount.

Institutional Abbreviations. BYU: Brigham Young University Earth Sciences Museum, Provo, Utah. CEUM: College of Eastern Utah Prehistoric Museum, Price, Utah. MWC: Museum of Western Colorado, Grand Junction, Colorado.

Techniques

The overall reconstruction of the internal skeleton is based principally on the work of Carpenter (1982, 1984, personal communication). Additional references include Coombs (1978, 1979), and Coombs and Maryańska (1990), as well as numerous illustrations by Paul (1995) and Ford in Kirkland (1998). In designing the mount, we referred to the mounted skeletons of *Talarurus plicatospineus* (Great Russian Dinosaurs Exhibit, as exhibited at the Mesa Southwest Museum, Mesa, Arizona; mount by the Russian Paleontological Institute) and *Euoplocephalus tutus* (Royal Tyrrell Museum, Drumheller, Alberta, Canada; mount by Research Casting International).

Although *Gastonia* is well represented by abundant material, much of it is crushed and distorted. Although every effort was made to use unmodified casts of original material, many of the bones proved to be unsuitable for mounting. Badly crushed material was molded, then hollow-cast with a thin layer of urethane resin. Just before setting up, when the resin was still soft and leathery, it was affixed to foam armatures that were sculpted to approximate the original size and shape of the bone. The end result is an "inflated" version of the original crushed bone, with all the surface details and features. Other missing or incomplete elements were sculpted. When mounting these casts, an internal armature was used wherever possible to allow the skeleton to be clearly viewed.

Measurements of the skeleton and individual elements are given in Table 18.1.

Cranial Skeleton

The holotype skull (CEUM 1307) exhibited minimal distortion, and with the exception of replacing all maxillary teeth, the cast was not modified for the mount (Kirkland 1998). An original tooth (CEUM 5145) was used to replace the missing teeth. Although several skulls have been excavated, to date, no lower jaws have been found from either quarry. The dentary and predentary have been reconstructed on the basis of *Gargoyleosaurus* and *Tarchia kielanae*. The skull is mount-

TABLE 18.1.
Measurements of the mounted skeleton.

	Width (cm)	Length (cm)	Height (cm)
Full skeleton without armor			
At hips:	115	459	112
Full skeleton with armor			
At shoulders:	60	459	138
Skeletal element:			
Skull	28	28	
Ilium, right	33	89	
Ischium, right	19	53	
Femur, right	20	59	
Tibia, right	14	30	
Scapulocorocoid, right	21	47	
Humerus, right	14	32	
Ulna, right	15	24	

ed in a fully raised position, but in life, it would have normally been held in a lowered, more vertical position when the animal was at rest or feeding (Kirkland 1998).

Axial Skeleton

Most of the paratype cervical vertebrae are severely crushed. However, many smaller specimens from the Dalton Wells Quarry that are less distorted provided an accurate basis for reconstruction of those elements. Some anterior and posterior cervical ribs have been found, and the missing ribs were sculpted by referring to the available specimens.

Numerous dorsal vertebrae were recovered, but most were compacted and unusable. Casts of a few anterior and middorsal vertebrae were used with minor reconstruction. Most of the posterior vertebrae were inflated, whereas others were sculpted, again by using better preserved but smaller vertebrae from Dalton Wells as reference.

Dorsal ribs are well represented. Although the original rib material displayed distortion both posteriorly and anteriorly, the integrity of the ventrolateral arcs is generally intact, which allowed us to use most of the original dorsal ribs with only slight modification. The anterior ribs, robust and L-shaped in cross section, were mounted so that each consecutive rib expands posterolaterally (Fig. 18.1), with the medial rib surface nearly aligning with the lateral surface of the preceding rib, as in the mount of *Talarurus*. More anterior ribs are T-shaped in cross section, as in other ankylosaurs.

A partial crushed synsacrum (CEUM 1163) with missing neural spines from a smaller individual was used as the basis for sculpting the synsacrum for the mount. The pelvises of *Polacanthus foxii* (Hulke 1888) and *Euoplocephalus* (Coombs 1979) were used as an additional

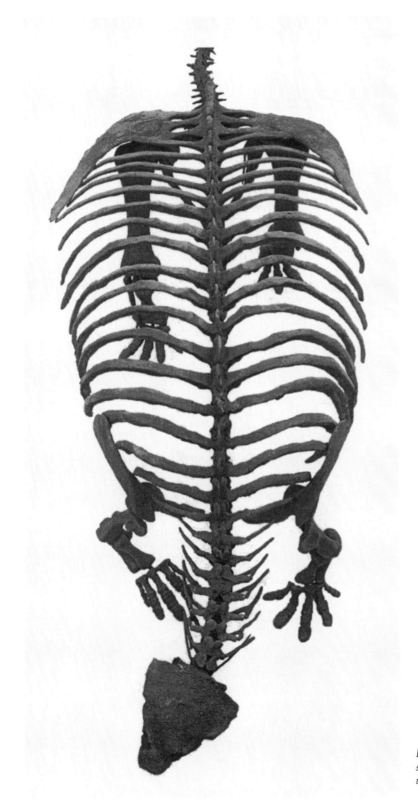

Figure 18.1. Gastonia burgei *skeletal mount in anterodorsal view.*

reference. Five fused dorsosacral vertebrae, three sacrals, and one cau-
dosacral (Kirkland 1998) were constructed for the mount.

Thirty-eight caudal vertebrae were used in the *Gastonia* tail, as in
the reconstruction of *Sauropelta edwardsi* (Carpenter 1984). A small
number of missing caudal elements were sculpted. The shape of the
proximal caudal vertebrae indicates that immediately behind the sacrum, the tail turned downward before extending out horizontally. The
tail is flexed laterally to reflect the primary lateral motion that it was
adapted for, as determined by the long caudal ribs and zygapophyses
(Kirkland et al. 1998).

Appendicular Skeleton

The left paratype scapula (CEUM 1354) and the right paratype
scapulocoracoid (CEUM 1238) were both found flattened; casts were
inflated to approximate the original shapes. Two undistorted scapulocoracoids were used as reference for restoring the proper curvature
between the scapula and the coracoid and to re-elevate the acromion process. The first was from a smaller individual *Gastonia* (BYU
VP2080) and the second (BYU 725–12963) from a Late Jurassic polacanthine, probably *Mymoorapelta* (Kirkland et al. 1998), from the
Dry Mesa Quarry in western Colorado. When the *Gastonia* mount is

Figure 18.2. Gastonia *mount in
lateral view (A) without armor
and (B) with armor.*

A

B

Figure 18.3. Gastonia *mount in anteroventral view.*

viewed laterally, the coracoids are not visible (Fig. 18.2A, B) contrary to the depiction in many two-dimensional skeletal reconstructions.

Although the coracoids are placed relatively close together on the mount, the precise distance between them, as well as the placement of the sternal plates (Fig. 18.3), is controversial (Carpenter, personal communication; Paul, personal communication).

The only recovered humerus (BYU VP136) was believed to be too small for use in the mount on the basis of Ankylosauria limb bone ratios (Pereda-Suberbiola 1993, 1994), and it was enlarged accordingly. The largest crushed ulna, from the right side (CEUM 3565), and a matching crushed radius (CEUM 3514) were inflated to proportions typical of other ankylosaurs. The articulation of the front leg bones is based on research by Carpenter (Carpenter et al. 1994; Carpenter, personal communication; Coombs 1978).

The front and rear limbs were positioned (Fig. 18.1) in accordance with the trackway *Tetrapodosaurus borealis* (Carpenter 1984), which shows approximately one footprint's width or less between the right and left tracks.

The right paratype ilium (CEUM 3880) was in excellent condition, and a cast of this was used with minor reconstruction. For the left side, a second incomplete ilium missing the preacetabular process was restored. Both right (CEUM 1371) and left ischia were used with minimal restoration. A right pubis (CEUM 5060) was recovered, and the miss-

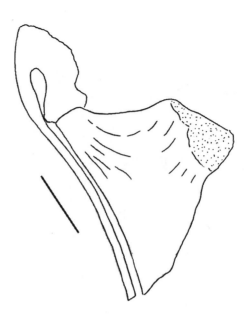

Figure 18.4. Cactus Park Mymoorapelta (MWC 2610) showing proximal end of left ischium (IS) with articulated pubis (P) in lateral view. Scale = 5 cm.

ing postpubic process was reconstructed by referring to the *Mymoor-apelta maysi* specimen from Cactus Park (Fig. 18.4). The positioning of the pubes was determined by Carpenter (personal communication); we also looked at the Cactus Park specimen.

The right paratype femur (CEUM 3646) was severely compacted. It was inflated with reference to the well-preserved femur of *Gargoyleosaurus* (Carpenter, personal communication). A single tibia (CEUM 9464) with severely crushed distal and proximal ends was restored. Reference to the tibiae of *Polacanthus* (Pereda-Suberbiola 1993) and *Euoplocephalus* (Coombs 1979), as well as advice from Carpenter (1998, personal communication), aided in the restoration. Two smaller tibiae recovered later seem to confirm that the sculpted element is of the correct size for the mount. The matching flattened fibula was also inflated.

The reconstruction of the animal's feet and hands used numerous miscellaneous unguals, phalanges, metacarpals, and metatarsals. Missing elements were sculpted on the basis of an articulated set of left metatarsals from the Cactus Park *Mymoorapelta* (MWC 2610) and the manus and pes of *Nodosaurus textilis* (Carpenter and Kirkland 1998) and *Sauropelta* (Carpenter 1984).

Armor

The major armor elements on this mount were placed according to comparisons with what is known for other polacanthids (Blows 1987; Carpenter et al. 1998; Coombs and Maryańska 1990; Kirkland 1998; Kirkland et al. 1998; Pereda-Suberbiola 1993). *Gastonia* has two major cervical armor elements. The anteriormost consists of a two-piece bony ring with fused scutes, and the posteriormost consists of two

Figure 18.5. Gastonia *mount showing dorsal view with armor. The sacral shield has been cut away to show the pelvis.*

Figure 18.6. Gastonia *mount in right oblique, anterior view.*

deeply rooted, laterally directed triangular spines and large, unfused scutes. Hollow-based lateral armor extends along the entire length of the body (Fig. 18.5), starting with laterally elongate, triangular spines with a posterior groove on the shoulders that decrease in size and both shorten and lengthen posteriorly to the hips (Fig. 18.6). The dorsal hip region of *Gastonia* is covered by a sacral shield of fused armor (Kirkland et al. 1991; Kirkland 1998), and small, anteroposteriorly elongate, hollow-based armor elements (not displayed on the mount) are thought to extend along the ventrolateral margin of the sacral shield.

As in *Polacanthus* (Blows 1987), *Gastonia* exhibits large, paired solid spines in the dorsal shoulder region. Placement of this armor in an erect position is supported by the radial ornament around its base, as in the larger elements on the fused sacral shield (Kirkland 1998). On the mount, the first three pairs of spines increase in height posteriorly, with the fourth and posteriormost pair being significantly smaller (Fig. 18.6). Like *Gargoyleosaurus* (Carpenter, personal communication), *Gastonia* has large, asymmetrical scutes forming bands laterally across the dorsal region (Fig. 18.7A, B). These scutes have low-profile keels extending along the medial edge. In *Gastonia*, however, these scutes are separated by single rows of smaller scutes (Fig. 18.7A). These small dermal ossicles, found by the hundreds in both quarries, probably covered all the spaces between major armor elements in the dorsal and dorsolateral areas of the body.

Large, triangular, hollow-based plates extend laterally along the

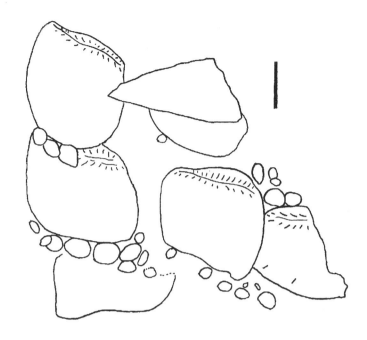

Figure 18.7. Gastonia.
(A) Articulated left dorsal armor adjacent to sacral shield as found (CEUM 1379, 1382, 1384, 1385, 1387, 1392), dorsal view. Scale = 5 cm. (B) Interpretation of armor on mount, intervening ossicles omitted to permit viewing of skeleton. Anterior to left.

tail (Kirkland 1998), each decreasing in size posteriorly. It is thought that large, rounded scutes, similar to those on the dorsal region, are paired along the dorsal surface of the tail (Kirkland 1998). However, additional preparation is needed before these can be included on the mount.

Because *Gastonia* has only been found in bone beds of multiple, mostly disarticulated individuals, additional reconstruction on the dorsal armor is difficult for two reasons: first, it is not known if differing sizes of similar armor elements represent graduated sizes belonging to one particular animal, or various growth stages on different individuals; and second, certain armor elements, in particular the large dorsal spines, may not be present in both sexes, especially if one purpose of the armor was for display and bluffing behavior (Kirkland 1998). The number and precise arrangement of the dorsal spines used for the mount is therefore speculative. Observations by one of us (J.I.K.) of the known *Polacanthus* material suggest that there are more large, erect spines in that taxon as compared with *Gastonia*.

Conclusion

Enough material has been recovered for the skeleton of *Gastonia* to be reconstructed with a reasonable degree of accuracy. This material includes a range of proportions reflecting differing growth stages from subadult to adult. In comparison with other polacanthine ankylosaurs, the general placement of body armor can be ascertained for *Gastonia*. However, because only small portions of this material were found articulated, the exact placement of many armor elements, particularly the larger dorsal armor, is uncertain. We hope that work underway at the Dalton Wells and Gaston Quarries will uncover additional material that will increase our understanding of the armor in *Gastonia*. Given a large representation of skeletal elements to refer to, laboratory techniques to inflate and restore crushed material can produce a skeletal reconstruction that closely approaches the accuracy of a skeleton mounted from uncrushed materials and contribute to the study of this animal.

Acknowledgments. Thanks are due to the staff of the College of Eastern Utah Prehistoric Museum, especially Don Burge, John Bird, and Duane Taylor, for the endless hours of excavation and preparation and without whom this mount would not have been possible. Many thanks also to Brooks Britt, Rod Scheetz, and Ken Stadtman for invaluable access to the Dalton Wells material and to the staff at the Museum of Western Colorado for the hours spent in the laboratory preparing the material. We are grateful to Kenneth Carpenter for taking so much time to consult with us on the reconstruction and mounting. We also thank Kent Hupps and Martin Lockley of the Dinosaur Trackers Research Group for access to, and assistance with, the Cactus Park material. Thanks are also extended to Dinamation International Society for assisting in work at the Gaston Quarry, as well as to Steven and Sylvia Czerkas and Greg Paul for valuable input on the mount, as well as to Ralph Molnar for sharing information on *Minmi*. Our gratitude also goes to Lee Welpley who first taught us molding techniques and provided critical assistance with mold-making for the project. Additional thanks to those who gave us much-needed help in reviewing this manuscript. Last, we thank Robert J. Gaston, Judi T. Gaston, David Gaston, and Susan Wilcox for providing the financial support necessary to undertake this project.

References Cited

Blows, W. T. 1987. The armored dinosaur *Polacanthus foxi* from the Lower Cretaceous of the Isle of Wight. *Palaeontology* 30: 557–580.

Carpenter, K. 1982. Skeletal and dermal armor reconstructions of *Euoplocephalus tutus* (Ornithischia, Ankylosauria) from the Late Cretaceous Oldman Formation. *Canadian Journal of Earth Sciences* 19: 689–697.

———. 1984. Skeletal reconstruction and life restoration of *Sauropelta* (Ankylosauria: Nodosauridae) from the Cretaceous of North America. *Canadian Journal of Earth Sciences* 21: 1491–1498.

Carpenter, K., and J. I. Kirkland. 1998. Review of Lower and Middle Cretaceous ankylosaurs from North America. In S. G. Lucas, J. I. Kirkland, and J. W. Estep (eds.), *Lower and Middle Cretaceous Terrestrial Ecosystems. New Mexico Museum of Natural History and Science Bulletin* 14: 249–270.

Carpenter, K., C. Miles, and K. Cloward. 1998. Skull of a Jurassic ankylosaur (Dinosauria). *Nature* 393: 782–783.

Carpenter, K., J. H. Madsen, and A. Lewis. 1994. Mounting of fossil vertebrate skeletons. In P. Leiggi and P. May (eds.), *Vertebrate Paleontology Techniques,* Vol. 1, pp. 285–322. Cambridge: Cambridge University Press.

Coombs, W. P., Jr. 1978. Fore limb muscles of the Ankylosauria (Reptilia, Ornithischia). *Journal of Paleontology* 52: 642–657.

———. 1979. Osteology and myology of the hind limb in the Ankylosauria (Reptilia, Ornithischia). *Journal of Paleontology* 53: 666–684.

Coombs, W. P., Jr., and T. Maryańska. 1990. Ankylosauria. In D. B. Weishampel, P. Dodson, and H. Osmólska (eds.), *The Dinosauria,* pp. 456–483. Berkeley: University of California Press.

Hulke, J. W. 1888. Supplemental note on *Polacanthus foxii,* describing the dorsal shield and some parts of the endoskeleton imperfectly known in 1881. *Philosophical Transactions of the Royal Society* 178: 169–172.

Kirkland, J. I. 1998. A Polacanthine ankylosaur (Ornithischia: Dinosauria) from the Early Cretaceous (Barremian) of eastern Utah. In S. G. Lucas, J. I. Kirkland, and J. W. Estep (eds.), *Lower and Middle Cretaceous Terrestrial Ecosystems. New Mexico Museum of Natural History and Science Bulletin* 14: 271–281.

Kirkland, J. I., D. I. Burge, and K. Carpenter. 1991. A nodosaur with a distinct sacral shield from the Lower Cretaceous of eastern Utah [abstract]. *Journal of Vertebrate Paleontology* 11(3, Suppl.): 40A.

Kirkland, J. I., K. Carpenter, A. P. Hunt, and R. D. Scheetz. 1998. Ankylosaur (Dinosauria) specimens from the Upper Jurassic Morrison Formation. *Modern Geology* 23: 145–177.

Kirkland, J. I., B. Britt, D. L. Burge, K. Carpenter, R. Cifelli, F. DeCourten, J. Eaton, S. Hasiotis, and T. Lawton. 1997. Lower to middle Cretaceous dinosaur faunas of the central Colorado Plateau: A key to understanding 35 million years of tectonics, sedimentology, evolution, and biogeography. *Brigham Young University Geology Studies* 42(2): 69–103.

Kirkland, J. I., R. Cifelli, B. Britt, D. L. Burge, F. DeCourten, J. Eaton, and J. M. Parrish. 1999. Distribution of vertebrate faunas in the Cedar Mountain Formation, east-central Utah. In D. Gillette (ed.), *Vertebrate Paleontology in Utah. Utah Geological Survey, Miscellaneous Publications* 99-1: 201–217.

Paul, G. S. 1995. Fat ankylosaurs—Really, really fat ankylosaurs. *Dinosaur Report* 6–7.

Pereda-Suberbiola, J. 1993. *Hylaeosaurus, Polacanthus* and the systematics and stratigraphy of Wealden armored dinosaurs. *Geological Magazine* 130: 767–781.

———. 1994. *Polacanthus* (Ornithischia, Ankylosauria), a transatlantic armored dinosaur from the Early Cretaceous of Europe and North America. *Palaeontographica, Abt. A* 232: 133–159.

19. An Ankylosaurian Cololite from the Lower Cretaceous of Queensland, Australia

RALPH E. MOLNAR AND
H. TREVOR CLIFFORD

Abstract

A cololite (preserved gut contents) was discovered in a specimen of *Minmi,* an Early Cretaceous armored dinosaur from north-central Queensland. This is the first report of a cololite in any thyreophoran dinosaur and the most reliable report to date of gut contents for any herbivorous dinosaur. The cololite consists of plant material representing fragments of vascular tissue, seed-bearing organs, seeds, and possible sporangia. The finding confirms that ankylosaurs were herbivorous and fed on soft vegetation. The small size of the segments of vascular tissue indicates considerable mastication of the material before swallowing, a finding that suggests the presence of cheeks.

Introduction

Preparation of an ankylosaurian skeleton (QM F18101) found in the Albian marine Allaru Formation of north-central Queensland (Molnar 1996) revealed the presence of gut contents. A preliminary report has already appeared (Molnar and Clifford 2000), and here, we present further details of the occurrence, compare the condition of the ingested material with that of gut contents of some living amniote tetrapods, and discuss the implications for the processes of feeding and digestion in what is presumably a typical advanced thyreophoran dinosaur.

Although there has been speculation regarding the diet of anky-losaurian dinosaurs (surveyed in Molnar and Clifford 2000), this is the first cololite found in association with any thyreophore. It thus represents the first opportunity to test hypotheses regarding the diet of ankylosaurs and also has implications concerning their digestion. Ankylosaurs have generally been considered herbivores on the basis of tooth form (see Barrett 2001; Rybczynski and Vickaryous 2001). Nopcsa's (1928) interpretation that they fed on large insects was not generally accepted. Maryańska (1977) considered it favorably, but Carpenter (1982), in a footnote, argued against it.

Cololites have been recorded in skeletons of small carnivorous dinosaurs, such as *Coelophysis* (Colbert 1989), *Compsognathus* (Ostrom 1978), and *Sinosauropteryx* (Ackerman 1998; Chen et al. 1998; Unwin 1998), showing that when well preserved, articulated specimens of theropods are found with gut contents often present. Cololites have also been reported for some herbivorous dinosaurs (Kräusel 1922; Stokes 1964; Weigelt 1927), but these occurrences have been questioned (Abel 1922; Chin 1997; Currie et al. 1995; Molnar and Clifford 2000), raising the possibility that the supposed gut contents were adventitiously introduced after the death of the animal. The cololite associated with *Minmi* is the best substantiated association of any with an herbivorous dinosaur.

Institutional Abbreviation. QM: Queensland Museum, Fortitude Valley, Australia.

Description

As described by Molnar and Clifford (2000), the gut contents were initially found as a dense scatter, 30 mm wide and 43 mm long, of plant material in the calcareous matrix on the left side of the vertebral column in the abdominal region (Fig. 19.1). Subsequent preparation revealed further material more sparsely distributed between the vertebral column and the preacetabular process of the left ilium. This material forms a mass at least 5 mm (possibly as much as 20 mm) thick and approximately 110 mm long (anteroposteriorly) by 170 mm wide. Altogether, the cololite probably covered an area of at least 200 by 240 mm, and thus occupied a volume of at least approximately 240 cm^3—perhaps as much as 960 cm^3. Apparently a substantial mass of plant material was present in the abdominal cavity at the time of the animal's death. This situation is to be expected for an herbivore and is consistent with the situation in large mammalian herbivores, such as ungulates, and at least some lizards (Farlow 1987). This mass consists largely of segments of vascular tissue of light color, with a preponderance of dark fragments medially.

The skeleton QM F18101 was discovered lying on its back with the right forelimb across the belly (Molnar 1996). Antebrachial elements lay across the vertebral column, and the manual phalanges rest against a layer of small ossicles, themselves resting between and on the ventral surfaces of the ribs. Most of the left ribs have been distorted so that,

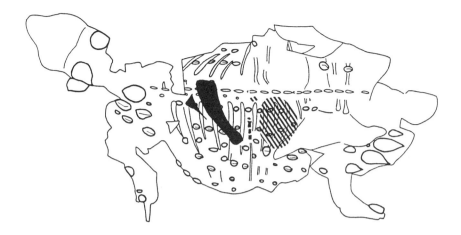

except proximally, they are now almost straight. The right ribs retained their original curvature but were rotated about 90° to lie flat (cf. Molnar 1996: fig. 1). The gut contents are preserved in the region of the abdominal cavity; the matrix of the anterior part of the rib cage lacks any sign of plant material.

The pieces of plants are embedded in the fine-grained (micritic) limestone matrix. Fresh broken surfaces of this matrix show no indication of bedding, and no traces of the soft tissues remain. This matrix is uniform over the entire neck and trunk regions and is overlain by sheets of broken inoceramid shell, about 10 mm above the plant debris. It is interesting in this context that the holotype specimen of *Minmi paravertebra* had large blobs of clay embedded in an otherwise uniform, unbedded, fine-grained limestone matrix dorsal to the ventral armor (Molnar 1980). These may be casts of digestive structures, similar to enterospirae, but the specimen is too incomplete to be certain.

The plant material comprises segments of fibrous or vascular tissue of light and dark color, fruiting bodies apparently preserved as yellowish-pink calcite, small spherical seeds, and dark masses of vesicular tissue that may represent fern sporangia. The light and dark fragments of fibrous tissue are by far the most abundant. These pieces are uniformly small, usually 0.6 to 2.7 mm long. Although the fibrous pieces resemble twigs or stems, in view of their size, we interpret them as the remnants of vascular bundles from leaves. Where they may plainly be seen, they terminate in clean breaks more or less perpendicular to the direction of the fibers. The light pieces appear to have deteriorated more than the dark, as they have a powdery appearance when examined under a microscope, whereas the dark pieces retain the fibrous structure. The seeds and fruiting bodies seem to have been swallowed whole, not surprising in view of their size (the seeds are about 0.3 mm across, and the fruiting body is about 4.5 mm across).

Figure 19.1. The position of the skeletal elements of Minmi *(QM F18101), from the Albian Allaru Mudstone, north-central Queensland, as seen on the dorsal surface. The position of the right antebrachial elements and one manual digit is given in black. The position of the cololite material is shown with diagonal hatching.*

Interpretation

This material is identified as gut contents for the following reasons. First, it is found within the abdominal region of the specimen. Second, the Allaru Formation, being marine, does not otherwise produce plant macrofossils, with the exception of a single specimen of *Nilssonia mucronatum* (Rozefelds 1986). Third, the fibrous pieces are reasonably uniform in size, which is not expected for plant material adventitiously introduced. And fourth, these pieces have all been cut, also unexpected for adventitious material. Optical microscopic examination of both the matrix with the gut contents and latex peels of the contents showed no trace of marine organisms. The possibility that a mass of uniformly cut pieces of plant material was accidentally introduced only into the intestinal region of a carcass in marine sediments is regarded as unlikely, so without doubt, we consider these to be gut contents. Furthermore, the single macroscopic plant specimen found in these sediments is preserved in a different fashion.

As mentioned previously, the plant pieces are scattered through a fine-grained, limey matrix indistinguishable from that exposed elsewhere in the specimen. If the soft tissues have entirely decayed without leaving any obvious trace, we may ask whether some of the gut contents have been lost. Certainly some kinds of plant material are more quickly digested than others, and we would expect this differential breakdown to continue during fermentation of the gut contents after the death of the animal. Here, only the vascular tissue shows deterioration, although the fruiting bodies are represented by pseudomorphs. Destruction of the plant material presumably may also continue during the diagenetic process in the early stages of fossilization. Thus, other plant material, not preserved, may also have been ingested. In view of the variety of the small amount of material remaining (Molnar and Clifford 2000) and the preservation of the seeds and fruiting bodies, it seems unlikely that large pieces would have been lost without trace. Thus, the preserved material is probably representative of the gut contents during life.

The position in the abdominal region and the elongate disposition of the plant material suggest that the plant material derived from the intestinal tract, consistent with the advanced state of deterioration of the vascular fragments. Pieces of vascular tissue interpreted as twigs have been recorded from the abdominal cavities of the large herbivorous marsupial *Diprotodon* found at Lake Callabonna, South Australia (Stirling 1900). In that occurrence also, leaves were absent and the vascular tissues showed signs of deterioration. The material was interpreted as intestinal contents.

Although the fragments of vascular tissue cannot be taxonomically identified, the fruiting bodies probably derive from an undescribed species of angiosperm. They are clearly unlike any known gymnosperm fruiting bodies, but they show a similar structural organization to the globular, fleshy synflorecences found among many of the Monimiacaeae. At the time the Allaru Formation was deposited, angiosperms were a well established component of the land surrounding the basin in which it was laid. They occur infrequently in the Styx Coal Measures

(Walkom 1919), which is regarded as Albian in age, but are a common component of the flora preserved in the Cenomanian-aged Winton Formation (McLoughlin et al. 1995). In both floras, the leaves are relatively small, attaining lengths of up to 125 mm, and their laminae vary in shape from linear to circular. From their shape, size, and venation patterns, these leaves are likely to have been soft in texture.

The structures representing possible sporangia, if correctly identified, would indicate that other plants, presumably ferns, were also eaten. These were identified from the abundant round vesicles of an appropriate size to represent fern sporangia. Spores indicate that ferns were present during the Albian in this region (Dettmann et al. 1992). No insect or other animal material, even such as might have been inadvertently ingested while taking foliage, has been found. Although gastroliths have been reported for some ankylosaurs (Carpenter 1997; Glut 1997), there is no indication of gastroliths with this specimen, which, given that gut contents were preserved, strongly suggests that this animal did not have gastroliths.

Paleobiologic Implications

This discovery of a cololite has both ecological and physiologic implications. For some time, there has been speculation that some groups of herbivorous dinosaurs evolved in concert with, or in response to, the evolution of angiosperms (e.g., Bakker 1978, 1986; Wing and Tiffney 1987), or at least that they ate angiosperms (e.g., Ryan and Vickaryous 1997). It was simply assumed, without independent supporting evidence, that given the variety of herbivorous dinosaurs, some must have eaten angiosperms. This cololite shows this assumption to be true. The presence of at least three fruiting bodies suggests that they were not consumed incidentally and that these kinds of dinosaurs may have played a role in plant dispersal, contrary to the opinion of Wing and Tiffney (1987). The possible presence of sporangia is consistent with the suggestion that *Minmi* may have consumed ferns (Dettmann et al. 1992), although because the sporangia have not been conclusively identified—much less identified taxonomically—the conclusion is tentative.

The nature of the breaks of the fibrous fragments indicates that they were cut, rather than torn by the teeth or beak, or by the grinding of gastroliths. Farlow (1987) suggested that the common conception that relatively simple teeth of many herbivorous dinosaurs necessitated further grinding by gastroliths or grit was overemphasized. The sizes of the plant pieces in the *Minmi* cololite in the absence of gastroliths or grit support his observation. The nature of the breaks indicates that *Minmi* was capable of using its teeth or beak to cut plants into small pieces and of retaining the pieces in the mouth. The pieces may have been cut by nibbling from the plant or by repeated chopping of material held in the mouth. In either case, this was probably accomplished by tooth-to-tooth contact. Ankylosaurian teeth have often been described as weak (e.g., Bakker 1986), and Haas (1969) regarded them as "reduced and feeble" and "not very effective." Although none of the (possibly replacement) teeth associated with QM F18101 show wear, this opinion is belied by European ankylosaurian teeth figured by Nopcsa (1929) that show extensive wear of a nature consistent with that expected from tooth-to-tooth contact. Wear has also recently been reported in other thyreophoran, including ankylosaurian, teeth (Barrett 1998, 2001; Rybczynski and Vickaryous 1998, 2001). Haas proposed that the cutting of food was done by the beak, which is certainly possible, but this does not account for the worn teeth. Unfortunately, both the tip of the snout and the symphyseal region of the jaws are not preserved in this specimen of *Minmi*, so no observations regarding the role of the presumptive beak are possible. The state of the plant fragments is consistent with the presence of a retractive jaw movement such as that postulated for *Euoplocephalus tutus* by Rybczynski and Vickaryous (1998, 2001). However, the jaw adductor musculature may also have been used for repetitive elevating of the jaw at small excursions for the nibbling of plant material.

Coombs and Maryańska (1990) proposed that the small size of the teeth (as opposed to their presumed weakness) implied that most of the maceration of the plant food occurred after swallowing. The cut ends of the plant fragments seen here imply that most of the comminution was done in the mouth and that small size does not imply weak or nonfunctional teeth. Barrett (1998) postulated that in some thyreophores, the teeth puncture-crushed the food. Two of the fruiting bodies seem to be neither cut nor crushed, indicating that these were swallowed intact, but the third was broken. This is consistent with the occurrence of some puncture-crushing.

Reduction of plant material to small pieces is also seen in what are probably coprolites from the Mygatt-Moore Quarry in the Morrison Formation of Colorado (Chin and Kirkland 1998). The similarity of the size and condition of the plant material seen there supports the observation of Chin and Kirkland (1998) that the coprolites possibly derived from a nodosaurid. The comminuted plant pieces associated with *Minmi* (as well as those from Colorado) are substantially smaller than those found in hadrosaur skeletons, which consist of twig segments 10 to 40 mm long (Currie et al. 1995). This observation suggests that either hadrosaurs were less effective at the comminution of plant material or that the material was indeed adventitiously introduced into the hadrosaur specimens. Chin and Gill (1996), however, observed plant fragments about 1 mm long in a coprolite attributed to the hadrosaur *Maiasaura*, supporting the latter alternative.

To provide some perspective on the efficacy of the reduction of the plant material by *Minmi*, the plant fragments in the cololite (Fig. 19.2) were compared to those in readily available gut contents or droppings of some living Australian tetrapods. It is assumed that the pieces remaining in the cololite are representative of all the pieces ingested in size. The degree of comminution of the plant material is substantially greater than that seen in modern herbivorous lizards, at least as represented by the agamid *Pogona barbata*. In this species, examination of stomach (Fig. 19.3) and intestinal contents (Fig. 19.4) shows that leaves, flowers, fruits, buds, and thin stems (to about 1 mm in diameter) are cropped into relatively large pieces about 5–18 mm long. Plant material taken by the emu (*Dromaius novaehollandiae*) also includes noticeably larger pieces (Fig. 19.5). Pieces found in scats are from stems or vascular bundles and range from 1.5 to at least 16 mm in length. The larger of these show similar abrupt perpendicular terminations where, presumably, they were cut by the beak. Insect remains may easily be seen among the plant material: emus eat angiosperm leaves, flowers, fruits and seeds, cycad seeds, podocarp fruit, and fern leaves, as well as insects (Barker and Vestjens 1979).

The degree of comminution seen in the *Minmi* cololite (as well as in the Mygatt-Moore Quarry and presumed *Maiasaura* material) is closer to that seen in geese. Crop contents of *Cereopsis novaehollandiae* (the Cape Barren goose) are similar in size, with pieces 2 to 4 mm long, although they include a small proportion of larger pieces, to 10 mm in length (Fig. 19.6). The pieces include those of leaves with abrupt, cut ends, basically similar to those seen in the *Minmi* cololite. There are also complete small leaves that were plucked and swallowed whole. *C. novaehollandiae* feeds on grasses, angiosperm leaves and seeds, and some algae (Barker and Vestjens 1979). In addition to plant material, the crop dissected also contained a substantial proportion of granules, mostly quartz, up to the size of large sand grains, in contrast to the cololite, which includes no such material.

The sizes of the pieces in both the goose and emu is markedly less uniform than in the Cretaceous cololite, giving the impression of a more sophisticated process of cutting the material for enzymatic digestion in *Minmi*. In the absence of gastroliths, the presence of small pieces in the crop contents of *Cereopsis* shows that it is possible to cut small pieces of plant material by nibbling with the beak. A similar process was seemingly used by *Minmi*, although as mentioned above, we think it

Figure 19.2. A fragment of the cololite associated with Minmi *(QM F18101), from the Albian Allaru Mudstone, north-central Queensland, Australia. The white and dark pieces are fragments of ingested plant material, interpreted as mostly representing vascular bundles. Scale in millimeters.*

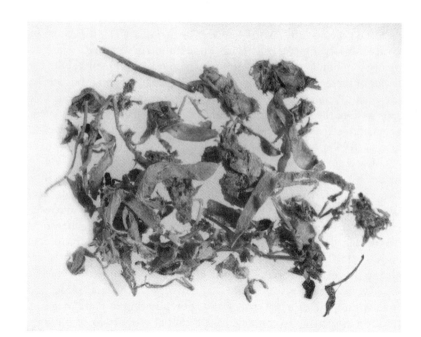

Figure 19.3. Stomach contents from a specimen of Pogona barbata, *from the Brisbane region, Australia. Scale in millimeters.*

Figure 19.4. Intestinal contents from the same specimen of Pogona barbata. *Scale in millimeters.*

likely that the teeth were used. The general condition of the *Minmi* cololite suggests that the preliminary processes of digestion were similar to those of modern herbivorous birds, specifically in reducing the sizes of food pieces, but that it was done more uniformly. The greater uniformity of the sizes of the pieces is more similar to what is seen in the droppings of *Trichosurus vulpecula*, the common brushtail possum (Fig. 19.7). In this case, almost all pieces are 1 to 2 mm in length.

Figure 19.5. Plant fragments from the scats of an Emu (Dromaius novaehollandiae), from the Brisbane region. Scale in millimeters.

Figure 19.6. Crop contents of a Cape Barren Goose (Cereopsis novaehollandiae), Bass Strait, Australia. Scale in millimeters.

Galton's (1973) hypothesis that advanced herbivorous ornithischians possessed cheeks has recently been questioned (e.g., Papp and Witmer 1998). This cololite presents novel evidence suggesting the possession of cheeks in at least some ankylosaurs. In the absence of gastroliths, the comminuted pieces would have been reduced in the oral region. The presence of similarly small pieces in the crop contents of

Figure 19.7. Plant fragments
*Figure 19.7. Plant fragments from the scats of a Common Brushtail Possum (*Trichosurus vulpecula*), from Brisbane. Scale in millimeters.*

geese shows that cheeks are not necessary if the plants are nibbled— that is, cut off as small pieces and immediately swallowed. Nibbling implies that the creature must remain at the food source for prolonged periods of time, a potentially risky procedure because sources of food may be in areas exposed to predation. Biting off large segments and reducing them by cutting in the mouth facilitates surveillance for predators because the head need not remain close to the food source throughout the period during which food is cut for swallowing. Cutting of food in the mouth suggests that cheeks would have been necessary to retain the resulting small pieces in the oral cavity before swallowing (cf. Carpenter 1997). The only alternative would have been to have possessed saliva sufficiently adhesive to have effectively glued the pieces to the tongue until they were swallowed, but we see no evidence to support this hypothesis. The scenario of biting and subsequent reduction in the mouth is plausible, but entirely hypothetical. However, the possession of dermal ossifications in the cheek region of the holotype skull of *Panoplosaurus mirus* (Lambe 1919) provides independent evidence that nodosaurids, at least, possessed some kind of cheek tissue in which this ossification took place. These cheeks may have been composed of connective tissue and may not have included the muscular tissue seen in mammalian cheeks.

Ankylosaurs (and other dinosaurs) are often said to have fed on soft vegetation, but "soft" is rarely clearly defined. In the sense that *Minmi* consumed leaves and fruiting bodies rather than twigs and other woody structures, this observation is verified here. Tiffney (1997) categorizes plant material as resistant to digestion (e.g., leaves and bark) and relatively easily digestible (e.g., fruits, seeds, and tubers). This cololite includes material of both categories, but by far the bulk is in the former category.

If this material was in the intestine, as its position suggests, then fermentation might have occurred in the hindgut. It has been thought by nutritional ecologists that animals that fermented food in the fore-stomach had the advantage over animals that fermented food in the hindgut in that the high-quality protein of the fermenting microorganisms could be extracted by digesting them in the gastric stomach (Foley and Cork 1992). Recent work (summarized in Foley and Cork 1992) indicates that for some herbivorous birds, this is not true and microbial protein was absorbed through the caecum. It is thus plausible to suggest that a similar digestive process occurred in ankylosaurs (Bakker 1986; Farlow 1987; Tiffney 1997), although we see no way to test this hypothesis.

We may, of course, question whether this material is representative of a typical diet, in view of the contention of Guthrie (1990) that some Pleistocene carcasses with abundant gut contents represent individuals that died of starvation. A similar point is made by Carpenter (1987). Starving herbivores do not always die from a lack of plants to eat but from a lack of plant material that is digestible. The animals tend to eat whatever is available and so fill their digestive tracts with indigestible material. This is possibly the case here: perhaps the *Minmi* consumed so much indigestible plant material because it was starving. However, the degree of deterioration of much of the plant material suggests that it was digestible, and because there is no other way to test this possibility—unlike at least some of the instances reported by Guthrie, for which there is independent evidence of death by starvation—we will consider the cololite to represent the remains of a normal diet.

Conclusions

The cololite may have occupied an area of 130 by 170 mm, about 6% of the estimated area of the flattened rib cage. It consists of plant material, including fragments of fibrous or vascular tissue, seemingly complete angiosperm fruiting bodies or their endocarps, small spherical seeds, and what may represent sporangia. The pieces of fibrous tissue are uniformly small and terminate in breaks approximately perpendicular to the direction of the fibers. The position of the plant material between the presacral rod and the left preacetabular process of the ilium suggests that it derived from the intestinal tract.

The mass of uniformly cut plant material, occurring only in the intestinal region of a skeleton found in marine sediments, indicates that these are gut contents. The presence of several fruiting bodies suggests that they were consumed deliberately and thus implies that *Minmi* may have played a role in dispersal of their parent plant. No gastroliths were found with the specimen, and the nature of the breaks on the plant fragments indicates that they were not torn by the grinding of gastroliths.

The breaks of the fibrous fragments show that they were cut from the plants. Thus, it is assumed that *Minmi* used its teeth (or beak) to cut the plants. Although ankylosaurian teeth have often been described as weak, tooth wear seen in some specimens suggests that tooth-to-tooth contact occurred during the cutting of the pieces. This in turn implies that the comminution took place in the mouth and that cheeks were present to retain the small pieces in the oral cavity. Dermal ossifications in the cheek region of the skull of the nodosaurid *Panoplosaurus mirus* also suggest some kind of cheek was present.

The comminution of the plant material is significantly greater in *Minmi* than in at least some modern herbivorous lizards, showing that ankylosaurs processed their food differently from these lizards, in a fashion more like that of some modern birds. The *Minmi* cololite shows that ankylosaurs effectively and uniformly reduced the plant material taken in to a degree that has not been surpassed among the modern birds here examined.

Acknowledgments. Angela M. Hatch discovered and recognized the gut contents. The assistance of Andrew Amey (Queensland Museum), Mary Dettmann (University of Queensland), Greg Erickson (Stanford University), Geoff Monteith (Queensland Museum), Steve Summers (Tasmanian Parks and Wildlife), and Joanne Wilkinson (Queensland Museum) in various aspects of this project is much appreciated. Kenneth Carpenter and Karin Chin provided useful and interesting comments on the manuscript.

References Cited

Abel, O. 1922. Diskussion zu den Vorträgen R. Kräusel und F. Versluys. *Paläontologische Zeitschrift* 4: 87.

Ackerman, J. 1998. Dinosaurs take wing. *National Geographic* 194: 74–86, 89–99.

Bakker, R. T. 1978. Dinosaur feeding behaviour and the origin of flowering plants. *Nature* 274: 661–663.

———. 1986. *The Dinosaur Heresies.* New York: William Morrow.

Barker, R. D., and W. J. M. Vestjens. 1979. *The Food of Australian Birds,* Vol. 1, *Non-Passerines.* Melbourne, Australia: Commonwealth Scientific and Industrial Research Organisation.

Barrett, P. M. 1998. Feeding in thyreophoran dinosaurs [abstract]. *Journal of Vertebrate Paleontology* 18(3, Suppl.): 26A.

———. 2001. Tooth wear and possible jaw action of *Scelidosaurus harrisonii* Owen and a review of feeding mechanisms in other thyreophoran dinosaurs. In K. Carpenter (ed.), *The Armored Dinosaurs.* Bloomington: Indiana University Press [this volume, Chapter 2].

Carpenter, K. 1982. Skeletal and dermal armor reconstruction of *Euoplocephalus tutus* (Ornithischia: Ankylosauridae) from the Late Cretaceous Oldman Formation of Alberta. *Canadian Journal of Earth Sciences* 19: 689–697.

———. 1987. Paleoecological significance of droughts during the Late Cretaceous of the Western Interior. In P. J. Currie and E. H. Koster (eds.), *Fourth Symposium on Mesozoic Terrestrial Ecosystems, Drumheller, August 10–14, 1987. Occasional Paper of the Tyrrell Museum of Palaeontology* 3: 42–47.

———. 1997. Ankylosaurs. In J. O. Farlow and M. K. Brett-Surman (eds.), *The Complete Dinosaur,* pp. 307–316. Bloomington: Indiana University Press.

Chen P., Dong Z., and Zhen S. 1998. An exceptionally well-preserved theropod dinosaur from the Yixian Formation of China. *Nature* 391: 147–152.

Chin, K. 1997. What did dinosaurs eat? Coprolites and other direct evidence of dinosaur diets. In J. O. Farlow and M. K. Brett-Surman (eds.), *The Complete Dinosaur,* pp. 371–382. Bloomington: Indiana University Press.

Chin, K., and B. D. Gill. 1996. Dinosaurs, dung beetles, and conifers: Participants in a Cretaceous food web. *Palaios* 11: 280–285.

Chin, K., and J. I. Kirkland. 1998. Probable herbivore coprolites from the Upper Jurassic Mygatt-Moore Quarry, western Colorado. *Modern Geology* 23: 249–275.

Colbert, E. H. 1989. The Triassic dinosaur *Coelophysis. Museum of Northern Arizona Bulletin* 57: 1–160.

Coombs, W. P., Jr., and T. Maryańska. 1990. Ankylosauria. In D. B. Weishampel, P. Dodson, and H. Osmólska (eds.), *The Dinosauria,* pp. 456–483. Berkeley: University of California Press.

Currie, P. J., E. B. Koppelhus, and A. F. Muhammad. 1995. "Stomach" contents of a hadrosaur from the Dinosaur Park Formation (Campanian, Upper Cretaceous) of Alberta, Canada. In A. Sun and Y. Wang (eds.), *Sixth Symposium on Mesozoic Terrestrial Ecosystems and Biota, Short Papers,* pp. 111–114. Beijing: China Ocean Press.

Dettmann, M. E., R. E. Molnar, J. G. Douglas, D. Burger, C. Fielding, H. T. Clifford, J. Francis, P. Jell, T. Rich, M. Wade, P. V. Rich, N. Pledge, A. Kemp, and A. Rozefelds. 1992. Australian Cretaceous terrestrial faunas and floras: Biostratigraphic and biogeographic implications. *Cretaceous Research* 13: 207–262.

Farlow, J. O. 1987. Speculations about the diet and digestive physiology of herbivorous dinosaurs. *Palaeobiology* 13: 60–72.

Foley, W. J., and S. J. Cork. 1992. Use of fibrous diets by small herbivores: how far can the rules be "bent"? *Trends in Ecology and Evolution* 7: 159–162.

Galton, P. M. 1973. The cheeks of ornithischian dinosaurs. *Lethaia* 6: 67–89.

Glut, D. F. 1997. *Dinosaurs, the Encyclopedia.* Jefferson, N.C.: McFarland.

Guthrie, R. D. 1990. *Frozen Fauna of the Mammoth Steppe.* Chicago: University of Chicago Press.

Haas, G. 1969. On the jaw muscles of ankylosaurs. *American Museum Novitates* 2399: 1–11.

Kräusel, R. 1922. Die Nahrung von *Trachodon. Paläontologische Zeitschrift* 4: 80.

Lambe, L. M. 1919. Description of a new genus and species (*Panoplosaurus mirus*) of armored dinosaur from the Belly River beds of Alberta. *Transactions of the Royal Society of Canada.* Series 3, 13: 39–50.

Maryańska, T. 1977. Ankylosauridae (Dinosauria) from Mongolia. *Palaeontologia Polonica* 37: 85–151.

McLoughlin, S., A. N. Drinnan, and A. C. Rozefelds. 1995. A Cenomanian flora from the Winton Formation, Eromanga Basin, Queensland, Australia. *Memoirs of the Queensland Museum* 38: 273–313.

Molnar, R. E. 1980. An ankylosaur (Ornithischia: Reptilia) from the Lower Cretaceous of southern Queensland. *Memoirs of the Queensland Museum* 20: 77–87.

———. 1996. Preliminary report on a new ankylosaur from the Early Cretaceous of Queensland, Australia. In F. A. Novas and R. E. Molnar (eds.), *Proceedings of the Gondwanan Dinosaur Symposium. Memoirs of the Queensland Museum* 39: 653–668.

Molnar, R. E., and H. T. Clifford. 2000. Gut contents of a small ankylosaur. *Journal of Vertebrate Paleontology* 20: 188–190.

Nopcsa F. 1928. *Scolosaurus cutleri*, a new dinosaur. *Geologica Hungarica, Series Paleontologica* 1: 54–74.

———. 1929. Dinosaurierreste aus Siebenbürgen V. *Geologica Hungarica, Series Paleontologica* 4: 1–76.

Ostrom, J. 1978. The osteology of *Compsognathus longipes* Wagner. *Zitteliana* 4: 73–118.

Papp, M. J., and L. Witmer. 1998. Cheeks, beaks, or freaks: A critical appraisal of buccal soft-tissue anatomy in ornithischian dinosaurs [abstract]. *Journal of Vertebrate Paleontology* 18(3, Suppl.): 69A.

Rozefelds, A. 1986. Type, figured and mentioned fossil plants in the Queensland Museum. *Memoirs of the Queensland Museum* 22: 141–153.

Ryan, M. J., and M. K. Vickaryous. 1997. Diet. In P. J. Currie and K. Padian (eds.), *Encyclopedia of Dinosaurs*, pp. 169–174. San Diego: Academic Press.

Rybczynski, N., and M. K. Vickaryous. 1998. Evidence of complex jaw movement in a Late Cretaceous ankylosaurid (Dinosauria: Thyreophora) [abstract]. *Journal of Vertebrate Paleontology* 18(3, Suppl.): 73A.

———. 2001. Evidence of complex jaw movement in the Late Cretaceous ankylosaurid *Euoplocephalus tutus* (Dinosauria: Thyreophora). In K. Carpenter (ed.), *The Armored Dinosaurs*. Bloomington: Indiana University Press [this volume, Chapter 14].

Stirling, E. C. 1900. Fossil remains of Lake Callabonna. Part II, 2. The physical features of Lake Callabonna. *Memoirs of the Royal Society of South Australia* 1: 1–15.

Stokes, W. L. 1964. Fossilized stomach contents of a sauropod dinosaur. *Science* 143: 576–577.

Tiffney, B. H. 1997. Land plants as food and habitat in the age of dinosaurs. In J. O. Farlow and M. K. Brett-Surman (eds.), *The Complete Dinosaur*, pp. 352–370. Bloomington: Indiana University Press

Unwin, D. M. 1998. Feathers, filaments and theropod dinosaurs. *Nature* 391: 119–120.

Walkom, A. B. 1919. Mesozoic floras of Queensland. Parts III and IV. The floras of the Burrum and Styx River Series. *Queensland Geological Survey Publication* 263: 1–77.

Weigelt, J. 1927. *Rezente Wirbeltierleichen und ihre paläobiologische Bedeutung.* Leipzig: Verlag von Max Weg.

Wing, S. L., and B. H. Tiffney. 1987. The reciprocal interaction of angiosperm evolution and tetrapod herbivory. *Review of Palaeobotany and Palynology* 50: 179–210.

20. Global Distribution of Purported Ankylosaur Track Occurrences

RICHARD T. MCCREA,
MARTIN G. LOCKLEY, AND
CHRISTIAN A. MEYER

Abstract

Until recently, reports of footprints attributable to ankylosaurs have been rare. Before the late 1990s, almost all discoveries were of small sites and isolated footprints with all but one site from rocks of Cretaceous age. One reason for the apparent rarity of ankylosaur track reports is that some of the footprints that we now regard as ankylosaurian were initially attributed to other quadrupedal dinosaurs such as ceratopsians or sauropods. Tracksites recently discovered near Grande Cache in western Canada and the Cal Orcko site in Bolivia significantly changed our previous perceptions of the abundance of ankylosaurid trackways—and even provided evidence of running behavior by one individual and skin impressions in others. These sites also give some insight into the paleoecology of ankylosaurs and indicate the variety of animals that left tracks that frequented the paleoenvironments in which their tracks are found. Probable ankylosaur tracks are now known from at least 14 localities in North America, South America, Europe, and Asia. Despite being poorly known, until very recently, possible ankylosaur tracks have been assigned to five ichnospecies, which, in order of historical naming, are: *Tetrapodosaurus borealis,* from the Lower Cretaceous of Canada; *Macropodosaurus gravis,* from the Lower Cretaceous of Central Asia; *Metatetrapous valdensis,* from the Lower Cretaceous of Germany; *Ligabuichnium bolivianum* from the Upper

Cretaceous of Bolivia; and *Deltapodus brodericki* from the Middle Jurassic of England, UK. These ichnotaxa and their probable affinities are reviewed, as well as the basis for assigning certain quadruped footprints to the ankylosaurs.

Introduction

Despite the relative abundance of skeletal remains of ankylosaurid dinosaurs, their footprints, such as those of other quadrupedal ornithischians (stegosaurids and ceratopsids) are relatively uncommon (Le Loeuff et al. 1998; Lockley 1991; Thulborn 1990). However, recent discoveries, especially in western Canada, where the first probable ankylosaurid footprints were recorded (Carpenter 1984; Sternberg 1932), and in Bolivia (McCrea et al. 1998), have revealed a number of new tracks that we describe in more detail than those from other sites, because of their much greater abundance.

On the basis of the known geographic and stratigraphic distribution of thyreophoran dinosaurs (nodosaurids and ankylosaurids), we predict widespread distribution of trackways in space and time (Carpenter 1997a, 1997b; Carpenter and Kirkland 1998; Coombs and Maryańska 1990; Lockley and Matsukawa 1998). To date, however, with the possible exception of a single Middle Jurassic site, all purported ankylosaur trackways are Cretaceous in age. The oldest purported thyreophoran prints have been found in the Lower Jurassic of central France (Le Loeuff et al. 1999). This evidence is supported by similar prints from the Liassic of Poland (Gierliński 1999; Le Loeuff et al. 1998; Lockley and Meyer 1999). We discuss these occurrences in relation to the sedimentary facies (where possible) in which the footprint assemblages occur.

Criteria for Identifying Ankylosaur Footprints

Given the value of the information that can be obtained from footprints, such as functional morphology, behavior, and paleoenvironment (Farlow and Chapman 1997; Lockley 1986), it is desirable to study footprints and attempt to identify the maker. Establishing the identity of any trackmaker can be a difficult undertaking (Farlow and Chapman 1997) because many different animals may be able to produce very similar looking traces. Identification may be further complicated when these groups of animals have overlapping geographic and stratigraphic ranges.

It is important to realize that the identification of footprints produced by ankylosaurs and the similar-size ceratopsians have the potential to be easily confused, particularly those footprints that possess a pentadactyl manus and tetradactyl pes, which are the most commonly found. Both of these groups are from the Cretaceous and overlap in time (to some extent) and geography. A two-pronged approach of morphologic and biostratigraphic analysis is therefore necessary to discriminate between the tracks produced by ankylosaurs and those produced by ceratopsians.

Morphologic Analysis

Skeletal Morphology. The morphology of the foot and hand skeletons should be considered, including the number and nature of the digits. A prediction of what types of footprints these animals might produce is possible by use of this type of evidence. Most taxa of ceratopsians (neoceratopsians) and ankylosaurs (ankylosaurids and nodosaurids) have five manual digits and four functional pedal digits. One exception (but not the only one) is *Euoplocephalus,* which has only three pes digits (Coombs and Maryańska 1990).

By using *Sauropelta* to represent nodosaurid ankylosaurs (Fig. 20.1a) and *Centrosaurus* to represent the neoceratopsians (Fig. 20.1b) as contrasting examples, some differences in the manual and pedal skeletons are evident. It is important to note that relative proportions are compared when referring to size, not absolute measurements.

The pes of *Sauropelta* has four digits, with digit I being significantly shorter than digits II–IV (Fig. 20.1a). The phalanges are relatively short, as are the metatarsals. The unguals of the pes digits are bluntly pointed. The digits are relatively wide and short. The foot skeleton is wider than it is long, but the animal could still produce a footprint that is longer than wide with the addition of a substantial metatarsal pad, which is quite likely to have existed, as it does in some modern graviportal mammals. From skeletal material, the natural position of the digits is difficult to know with precision. They would have been spread to different degrees in a living animal, depending on its weight and gait, as well as the composition and consistency of the substrate.

The pes skeleton of *Centrosaurus* also has four digits (Fig. 20.1b). Digit I is shorter than digits II–IV, but not as short in proportion to the other pes digits as digit I is in *Sauropelta* (Fig. 20.1a). The phalanges are

Figure 20.1. Comparison of skeletal and footprint morphology of ankylosaurs and ceratopsians. This illustration is to facilitate overall morphological comparisons between manus and pes skeletal morphology of ankylosaurs and ceratopsians (using *Sauropelta* and *Centrosaurus, respectively*) and how these skeletal differences may appear in footprints (using *Tetrapodosaurus* for ankylosaurs and *Ceratopsipes* for ceratopsians). To this end, the skeletal elements of *Sauropelta* and *Centrosaurus* have been adjusted in size for comparative purposes, although relative dimensions and size relationships between the manus and pes remain intact. The footprints have been similarly adjusted. (a) *Tetrapodosaurus* prints (Sternberg 1932) with skeletal elements of *Sauropelta* (modified from Carpenter 1984) superimposed on the left manus and pes. (b) *Ceratopsipes* prints (Lockley and Hunt 1995) with skeletal elements of *Centrosaurus* (manus and pes modified from Dodson and Currie 1990) superimposed on the left manus and pes.

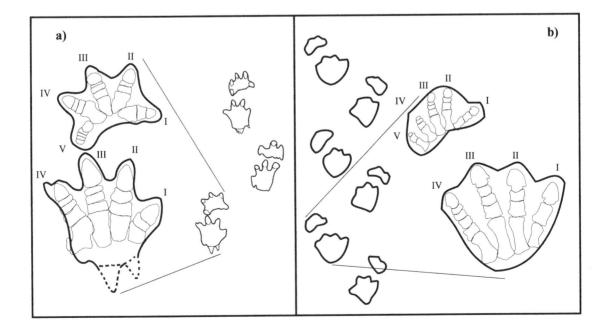

relatively longer than those in *Sauropelta,* as are the metatarsals. The unguals are rounded and hooflike. The foot appears to be nearly as long as it is wide (again, this depends on several factors). The digits are relatively more slender and lengthier when compared to those of *Sauropelta.*

Although there are some skeletal differences present, the pes footprint of these two genera would be very difficult to separate unless there were good indications of soft morphology differences (such as presence or absence of a fleshy pad that nearly envelops the digits) or significant difference in the positioning of the digits. These differences would be hard to recognize if the footprint was poorly preserved or was produced in any but ideal substrates.

The manus of *Sauropelta* has five digits, with digits II–IV the longest, digit I slightly shorter, and digit V the shortest (Fig. 20.1a). All manual digits are composed of relatively short phalanges and metacarpals. The manus of *Sauropelta* is about two-thirds the size of the pes. Manual prints of this animal would display short, stubby digits, with digit V the most reduced. The unguals are subangular to rounded in shape.

The manus of *Centrosaurus* also has five digits, with digit I of intermediate length, digits II and III the longest, and digits IV and V significantly shorter (Fig. 20.1b). The manual digits are composed of relatively short phalanges with a few being of medium length (digit II) and long length (digit I) as well. The metacarpals of digits I–III are long, but those of digits IV and V are short. The unguals of digits I–III are rounded and hooflike, but those of digits IV and V are smaller than the metacarpals with which they articulate. This size difference gives digits IV and V the appearance of tapering distally. The manus is approximately half the size of the pes. Manus prints of this animal would show relatively long and slender digits I–III and reduced digits IV and V.

The skeletal differences between the manus of *Sauropelta* and *Centrosaurus* are more likely to be of use in identifying prints than those of the pes. Digit V in *Sauropelta* is the only reduced manual digit, compared with the *Centrosaurus* manus, where both digits IV and V are reduced. Also, the manual digits of *Centrosaurus* are longer and more slender in proportion to those in *Sauropelta.*

Footprint Morphology. The pes prints of both ankylosaurs and ceratopsians are the most frequently preserved. Both ankylosaurs and ceratopsians (for the most part) have a tetradactyl pes, with the inner digit, digit I, the shortest. On the basis of a comparison of what we might consider the type material (*Tetrapodosaurus* and *Ceratopsipes,* respectively), it appears that *Tetrapodosaurus* prints display longer toe impressions with less enclosing flesh (Fig. 20.1a). This distinction may be due to ankylosaurs generally having shorter metatarsals relative to toe length, whereas the ceratopsians have longer metatarsals relative to toe length, as described above. These ratios appear to have a morphogenetic origin (Lockley 1999a, 1999b, in press). In addition, *Tetrapodosaurus* pes prints have a shorter inner digit than *Ceratopsipes,* which is also seen in the skeletal anatomy between *Sauropelta* and *Centrosaurus* (Fig. 20.1a, b). This may be why ceratopsian (*Ceratopsipes*) pes prints appear to be more symmetrical than those of ankylosaurs (Fig. 20.1a, b).

The orientation of the footprint may also be of some significance in making the distinction between ankylosaurs and ceratopsians. The middle digits, II and III of the pes, generally project forward, parallel to the parasagittal plane, in both *Tetrapodosaurus* and *Ceratopsipes*. However, the manus impression is located in front of the pes in *Tetrapodosaurus*; in *Ceratopsipes,* it is located slightly lateral.

Tetrapodosaurus manus prints are larger in proportion to the pes (about two-thirds the size of the pes), whereas *Ceratopsipes* manus prints are significantly smaller in proportion to the pes (about half the size of the pes). The size relationships between manus and pes prints are consistent with our comparison of *Sauropelta* and *Centrosaurus* manual and pedal skeletons. *Tetrapodosaurus* manus prints often have the outer digits (I and V) arranged so that they almost completely project backward in relation to the direction of travel (Figs. 20.1, 20.12, 20.24b). In some specimens, digit I of the *Tetrapodosaurus* manus seems much more prominently displayed than the other manual digits (Figs. 20.1a, 20.13). Manual digit I of *Ceratopsipes* does not seem quite as prominently displayed as that of *Tetrapodosaurus* (Fig. 20.1b). Well-defined manus and pes prints from the Gething, Gates, and Dunvegan Formations fit the ankylosaur pattern, as described above, very well.

To our knowledge, to date, only one purported ceratopsian ichnospecies has been named as *Ceratopsipes goldenensis* from the Maastrichtian of Colorado (Lockley and Hunt 1995). In comparison with the ankylosaur prints described herein, *Ceratopsipes* has very blunt toes and is more transverse (wider than long) than forms such as *Tetrapodosaurus*. The pes is also less symmetrical in ankylosaurian prints, with digit IV located much more anteriorly in relation to digit I than in *Ceratopsipes*, where the configuration of these digits is essentially symmetrical (Fig. 20.1b).

The approach of comparing these two genera has the advantage of having well-preserved foot skeletons. A similar approach was used by Thulborn (1990). However, one of the disadvantages is that we do not know how representative the foot skeletons are of the groups as a whole. Clearly, there is variation within both groups (ankylosaurs and ceratopsians), and we can predict that early, ancestral, or primitive forms were smaller, with more slender toes than later, larger, derived forms. To date, no purported prints of small ancestral forms have been reported, leaving us to compare footprints that belong to large species whose lower level taxonomic affinities are unknown.

A new and generalized approach has been suggested recently by Lockley (1999a, 1999b). First, it appears that the larger and more derived the trackmaker, the more flesh it has on its feet. Though this is only a general rule, it suggests that footprints may not always reflect the morphology of trackmaker's foot skeletons in detail, especially in the case of larger animals. Although this does not help identify the trackmaker, it does point to the need to describe footprint morphology in its own right regardless of the trackmaker's identity. This point is often overlooked by nonichnologists.

Pursuing this suggestion that footprint morphology must be examined in its own right and in relation to other footprints (not only in relation to possible trackmakers), some general rules for footprint morphology have been empirically formulated. If we place all major dinosaur groups in a standard phylogenetic or cladistic arrangement (cf. Lockley 1999a: figs. 4.5, 7.5), we see a systematic (and size-related) gradient in the relative width of the pes print in all clades, as well as a reiteration of this same gradient across the dinosaurian clade as a whole. In all cases, the ancestral forms have narrower feet and the derived forms have wider feet. In addition, there is a tendency for more primitive forms to have less fleshy feet than derived clades and for the step length to be shorter in derived clades.

Finally, it appears that flesh is first added to the rear of the foot in more primitive forms and progressively encloses more and more of the distal extremities of the toes. Thus, a sauropod has wider, fleshier feet with more enclosed toes and a shorter step than a theropod. The same holds for comparison between primitive and derived thyreophorans and cerapodans.

Any application of these principles to a comparison of ankylosaurian and cerapodan footprints must be considered tentative. However, one observation appears to be useful and capable of being tested as more data becomes available: ankylosaurian footprints of the *Tetrapodosaurus* type reveal more slender and less enclosed digit impressions than ceratopsian footprints of the *Ceratopsipes* type. However, we must remember here that we have only a single named example of purported ceratopsid footprints with which to compare purported ankylosaur footprints.

If this preliminary integrated model of footprint and trackway morphology is valid, it predicts that there will be convergence between forms such that the print of a large, highly derived ankylosaur would resemble a ceratopsian print more than a print of a smaller primitive species. Likewise, a small primitive ceratopsian might leave footprints that are convergent with those of ankylosaurs. Although this approach is generalized, it is empirically derived and is based on the principle of comparative anatomy of footprints. Preliminary analysis suggests that in trackways of the same size, ankylosaurs will have slightly longer steps and slightly narrower, less fleshy feet (i.e., with more clearly separated digits) than ceratopsians (Fig. 20.1a, b). We do not suggest that this is a model for an easy or foolproof differentiation of tracks of these two groups, only that it provides some generalized clues to comparative footprint morphology that may be useful when considered in conjunction with other morphologic and geologic evidence.

Biostratigraphic Range of Body Fossils of Potential Track Producers

Neoceratopsians and ankylosaurs (nodosaurids and ankylosaurids) are large enough to produce the types of footprints we are surveying here. Traditionally, it was assumed that ceratopsians, and thus their tracks, were confined to the upper part of the Late Cretaceous (Campanian and Maastrichtian), so that any large tetradactyl prints could not be considered ceratopsian in origin if they were pre-Campanian. This perspective has not significantly changed because the earliest record of large ceratopsians are from deposits only dating back to the Turonian (Wolfe and Kirkland 1998). Chinnery et al. (1998) reported the occurrence of neoceratopsian teeth from the Lower Cretaceous Cedar Mountain Formation (Albian–Cenomanian boundary) and the Arundel Formation (late Aptian). As there are not yet any complete skeletons from these sites with which to indicate size or anatomy, we still have to consider large neoceratopsians to be exclusively from the Upper Cretaceous; therefore, it is unlikely that pre-Turonian "ankylosaur" tracks are of ceratopsian origin. Furthermore, some of the most extensive and numerous tracksites are pre-Turonian—too early for the ceratopsians to be considered responsible for producing them. On the other hand, the abundance and diversity of pre-Turonian ankylosaurs shows that they were present in the same time periods and geographic locations as many of our tracksites, strongly implying that they were the trackmakers (Fig. 20.2). Future discoveries of more complete, pre-Turonian neoceratopsians may cast our assertion into doubt. However,

Figure 20.2. Cretaceous occurrences of ankylosaur (nodosaurid and ankylosaurid) and ceratopsian (excluding protoceratopsians and psittacosaurids) genera as well as purported ankylosaur and ceratopsian ichnogenera (Carpenter and Kirkland 1998; Coombs and Maryańska 1990; Dodson and Currie 1990; Haubold 1971; Kirkland 1998; Leonardi 1984; Lockley and Hunt 1995; Sternberg 1932; Wolfe and Kirkland 1998; Zakharov and Khakimov 1963).

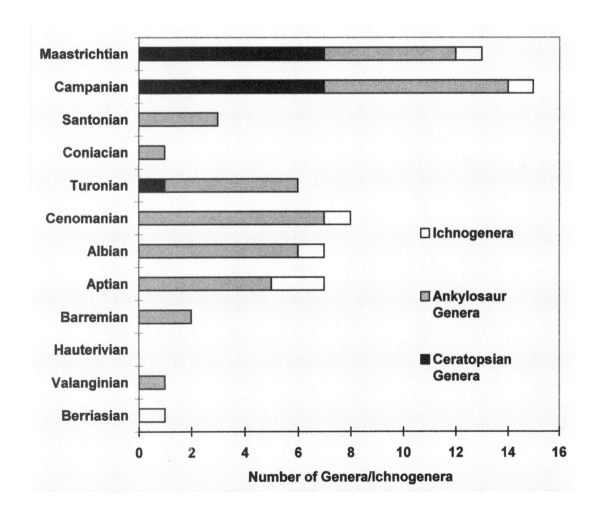

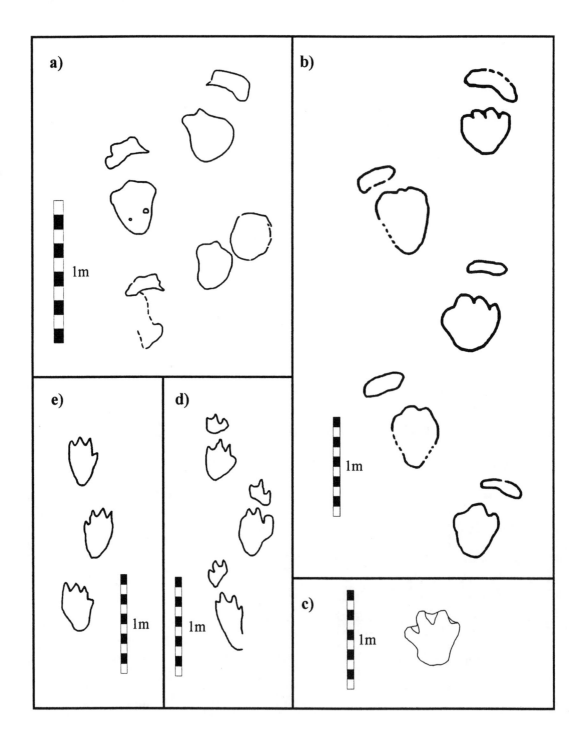

Figure 20.3. (a) Deltapodus brodericki (modified from Whyte and Romano 1994). (b) Purported ankylosaur track from the Purbeck Beds (after Ensom 1988 and Lockley 1991). (c) Isolated natural cast of possible ankylosaur pes from the Purbeck Beds (after Ensom 1987). (d) Metatetrapous valdensis (Haubold 1971). (e) Macropodosaurus gravis (Zakharov 1964).

on the basis of current evidence, there is a much stronger case for the ankylosaurs producing pre-Turonian footprints than there is for ceratopsians.

Institutional Abbreviations. CEUM: College of Eastern Utah Museum, Price, Utah. NMC: National Museums of Canada, Ottawa, Ontario, Canada; TMP: Royal Tyrrell Museum of Palaeontology, Drumheller, Alberta, Canada.

List of Footprint Sites in Ascending Chronostratigraphic Order

Saltwick Formation, England (Aalenian–Bajocian)

Prints from the Saltwick Formation of Yorkshire, England (Fig. 20.3a), were named *Deltapodus brodericki* by Whyte and Romano (1994), who thought they were sauropod prints. Lockley et al. (1994) expressed doubt, suggesting instead that they may be of thyreophoran origin because the pes prints appear too symmetrical to be of sauropodan origin and were therefore either ankylosaurian or stegosaurian. The lack of inward rotation of the pes and the bluntness of the digits is more reminiscent of ankylosaurs than other stegosaurlike thyreophorans (Lockley 1999a; Thulborn 1990), but preservation is not good enough to allow fine discrimination of morphologic detail. The conservative approach would be to label these as probable thyreophoran prints. Whyte and Romano (1994) interpret the depositional environment where the footprints were found in to be a marshy, fluvio-deltaic environment with abundant, lush vegetation.

Purbeck Beds, England (Berriasian)

Footprints from the Purbeck beds of England (Fig. 20.3b, c) were considered as possible sauropod prints, and the beds were originally assigned to the Upper Jurassic (Ensom 1987). Stratigraphic revisions now place these prints at the base of the Cretaceous (Berriasian), and most authors consider them to be ankylosaurian (Ensom 1987, 1988; Lockley 1991; Lockley and Meyer 1999; Wright 1996). Ensom (1987) referred skeletal elements to nodosaurid ankylosaurs; although this does not prove the identity of the trackmaker as ankylosaurian, it does suggest that the footprint assignments are reasonable and probably correct. The footprints are found in a freshwater depositional environment, which also contain the remains of freshwater fish, amphibians, turtles, and crocodiles (Ensom 1987, 1988).

Wealden Beds, Germany (Berriasian)

Metatetrapous valdensis Haubold (1971) from the Lower Cretaceous of Germany is interpreted as ankylosaurian (Fig. 20.3d). This ichnospecies is similar to *Tetrapodosaurus borealis* from Canada (discussed below) in having a tetradactyl pes with elongate heel and interior digit (digit I) shorter than the outer digit (digit IV). The manus appears to be tridactyl, but was probably pentadactyl with the inner and outer toes (digits I and V) not impressed.

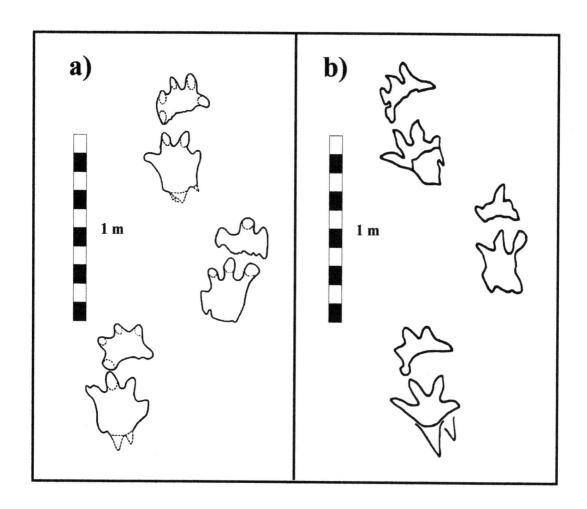

Figure 20.4. Two interpretative illustrations of the Tetrapodosaurus borealis holotype (NMC 8556) from the Peace River Canyon (Gething Formation). (a) Tetrapodosaurus borealis, NMC 8556 (Sternberg 1932). (b) Tetrapodosaurus borealis, NMC 8556 (Currie, unpublished data).

The trackway is from the Bueckeburg Formation, which is part of the Wealden Group and consists of a series of siliciclastic sandstones with minor shales. The sequence is rich in boreal plant fossils and also contains some freshwater bivalves and other invertebrate remains and traces. The depositional environment is considered to be a lowland fluvial coastal plain system. There is abundant evidence of shallow water in the form of both wave and current ripples (Mutterlose et al. 1997).

Gething Formation, British Columbia (Aptian–Albian)

Sternberg (1932) described quadrupedal prints from the Lower Cretaceous (Aptian) Gething Formation of the Peace River Valley, British Columbia, and named them *Tetrapodosaurus borealis* (Fig. 20.4a, b). They were originally interpreted as being ceratopsian, but they were later identified as ankylosaurian in origin (Carpenter 1984). We agree with this assignment on the basis of both age and morphology of the prints. The type specimen was described from a series of 14 footprints (seven manus and seven pes). A portion of this trackway (Fig. 20.4a, b) is preserved as a plaster cast (NMC 8556). In the 1970s, expeditions to

422 • Richard T. McCrea, Martin G. Lockley, and Christian A. Meyer

the Peace River Canyon by the Provincial Museum of Alberta were unable to relocate the *Tetrapodosaurus* holotype. This single ankylosaur trackway was found among an ichnofauna dominated by *Amblydactylus* footprints, which have been attributed to hadrosaurs (Currie 1983, 1995; Currie and Sarjeant 1979). The Gething Formation was part of a major deltaic complex and contains major coal deposits (Gibson 1985; Stott 1972).

Gates Formation, Grande Cache, Alberta (Lower Albian)

In recent years, several tracksites have been excavated at the Smoky River Coal Mine, near Grande Cache, Alberta (McCrea and Currie 1998). Currently, 16 ankylosaur tracksites are known from this coal mine, most of which occur on steeply dipping footwall slopes (McCrea and Currie 1998). Most are similar to *Tetrapodosaurus borealis* (Sternberg 1932), which at present is a monospecific ichnogenus. The 16 sites are spread out over approximately 25 km² and are treated as separate localities. The sites are labeled with local mine designations: E-2 Pit, W1a, W1b, W1c, W2, W3 Main, W3 Extension, W3 Corner, 9 Mine, 9 Mine West Extension (A and B), 9 Mine West Extension Fold Axis, Mine Dump, Center Limb Pit, 8 Mine and 12 Mine South A-Pit, E-2 Pit, W1a, W1b, W1c, W2, W3 Main, W3 Extension, and W3 Corner are associated with a continuous outcrop along the limb of an anticline.

E-2 Pit. This tracksite was figured in Psihoyos and Knoebber (1994: 189) and shows several footprint sequences, some crossing over others (Fig. 20.5). A few beds above this footprint layer is another footprint layer that shows several trackways navigating through a stand of fossil tree stumps with radiating roots (Fig. 20.6). This site entered paleontological folklore when a large part of the cliff collapsed (McCrea and Currie 1998; Psihoyos and Knoebber 1994). Psihoyos and Knoebber (1994: 189) show before and after photographs revealing that most of the footprints at this spectacular site were destroyed. The remaining footprints were later destroyed as a result of continued slope failure at this site.

W1a. A newly discovered site containing a dinoturbated area with a profusion of *Tetrapodosaurus* footprints (Fig. 20.7).

W1b. A new site with several short ankylosaur trackways that show good impressions of the feet and digits, preserved on friable substrate.

W1c. A small site with a few *Tetrapodosaurus* footprints in one trackway.

W2. Another newly discovered, dinoturbated area with abundant *Tetrapodosaurus* footprints (Fig. 20.8). There is evidence of gregarious behavior of two groups of animals displayed at this site. Each group traveled in a different direction, with the individuals (three per group) within the respective groups walking in the same direction. The spacing between individuals is consistent and the tracks do not cross, indicating that these animals were traveling together at the time the footprints were made. All of these preliminary observations were made from the ground. Further research is needed at this site.

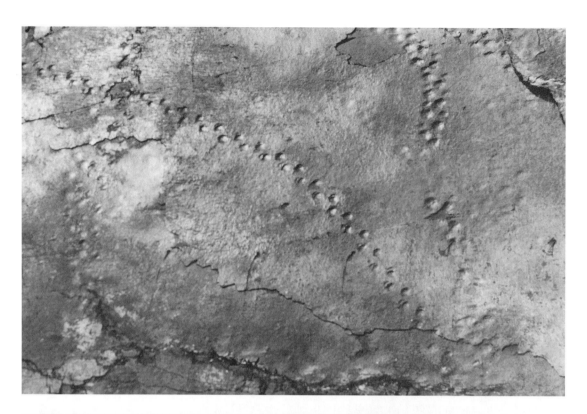

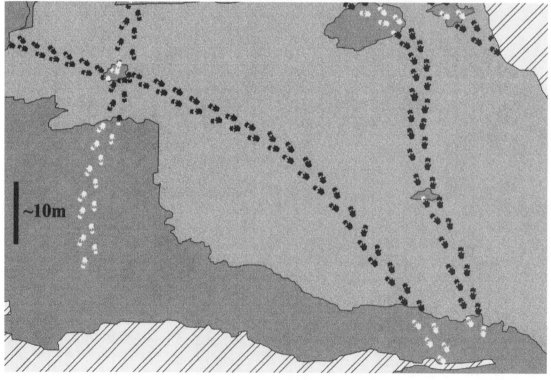

Figure 20.5. (opposite page) Tetrapodosaurus *trackways (photograph and illustration with stylized footprints) from E-2 Pit. White prints indicate underprints. Photograph provided by Dr. Philip J. Currie.*

Figure 20.6. E-2 Pit site showing faint Tetrapodosaurus *footprints associated with in situ tree stumps with radiating roots. Photograph provided by Dr. Philip J. Currie.*

W3 Main. The W3 Main footwall contains more than 6000 footprints, with over 20 *Tetrapodosaurus* trackways, some with more than 50 consecutive manus and pes prints (Fig. 20.9, 20.10). Some footprints preserved as natural molds reveal skin impressions, making this a particularly significant site. Skin impressions were found on the posterior margin of a manus print that had been cast. The tubercles are elliptical and measure 7 mm by 2 mm. These skin impressions are on the edge of a mud bulge at the heel of the manus print, and it appears that the tubercles were compressed anterior to posterior, but they were probably originally rounded, as are those of other prints observed at this site and within the same footprint sequence.

10m

Figure 20.7. W1a site showing intense dinoturbation of Tetrapodosaurus *footprints.*

There are at least five layers of strata that contain variably preserved footprints at this site. Some footprints were made in substrate that had a relatively high water content, evidenced by the large mud bulges around the footprints (Fig. 20.11). However, on the same bedding plane, just above the central prints shown in Figure 20.11, are footprints that appear to have been made after the substrate had dried out somewhat. These prints are faint, but the digits are well defined and do not display mud bulges. These prints are the same size as the other *Tetrapodosaurus* prints on this footwall. The pes prints show four elongate and slender digits, with digits I and II being the most deeply impressed and with the tips of digits II–IV being very deeply impressed. Manus digits III–V are well impressed, and digits I and II are present, though less deeply impressed (Fig. 20.12).

There is a pair of parallel trackways that trend in the same direction that could be interpreted as additional evidence of gregarious behavior of (adult) ankylosaurs on the basis of footprints. The W3 Main ankylosaur footprints are associated with several other dinosaur (theropod) and bird ichnotaxa, making this among the most diverse ichnofaunas in the world (McCrea and Sarjeant 1999).

W3 Extension. There are several *Tetrapodosaurus* trackways found high up on the footwall. A natural cast of a left manus print (Fig. 20.13) that was recently collected from this site (TMP 99.49.2) shows well-defined digits (digits I–V) and fits our criteria for ankylosaurian prints.

W3 Corner. A few *Tetrapodosaurus* trackways are found very high up on the footwall.

12 Mine South, A-Pit. Located on the anticline limb opposite the E-2 Pit-W3 Corner tracksites, this site had the longest *Tetrapodosaurus*

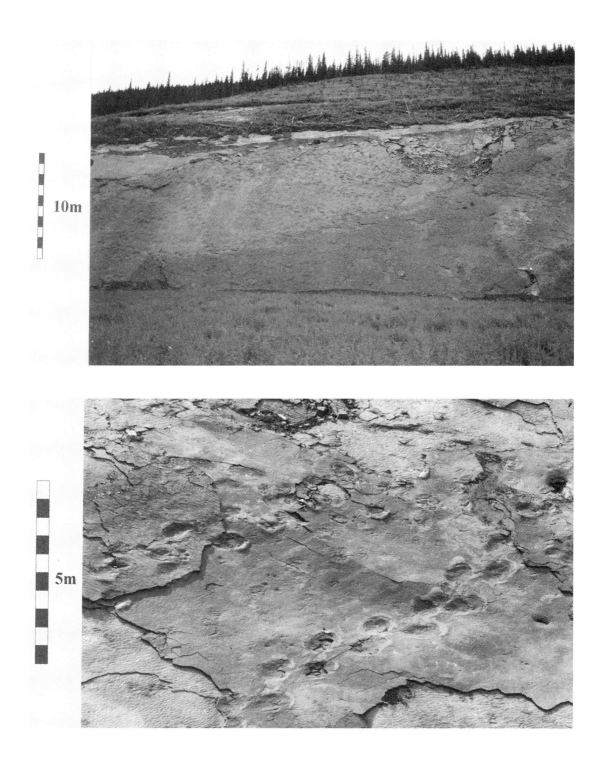

10m

5m

Figure 20.8. (top) W2 *site showing dinoturbation of* Tetrapodosaurus *footprints.*

Figure 20.9. (bottom) Tetrapodosaurus *trackways at the W3 site associated with faint tridactyl (*Ornithomimipus*) footprints.*

Global Distribution of Purported Ankylosaur Tracks • 427

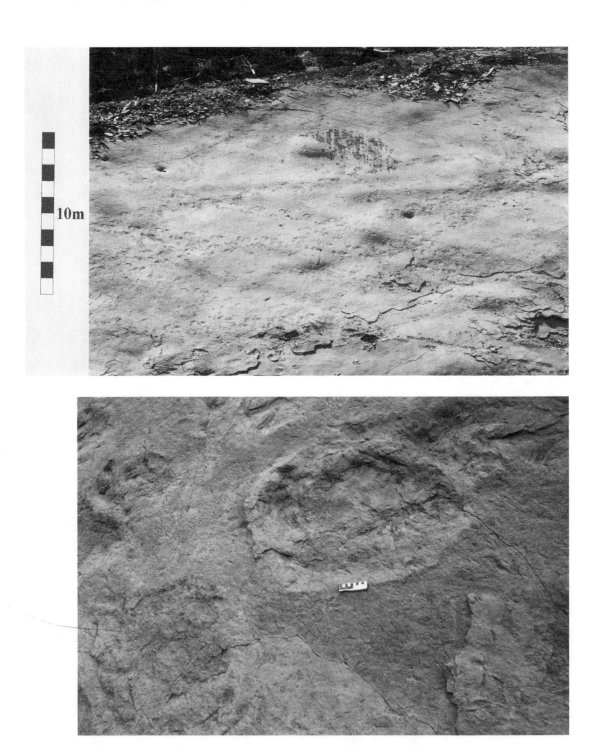

10m

Figure 20.10. (top) Lengthy Tetrapodosaurus *trackways at the W3 site associated with equally lengthy* Irenesauripus *trackways.*

Figure 20.11. (bottom) Tetrapodosaurus *manus and pes with mud bulges at the W3 site. Scale = 10 cm.*

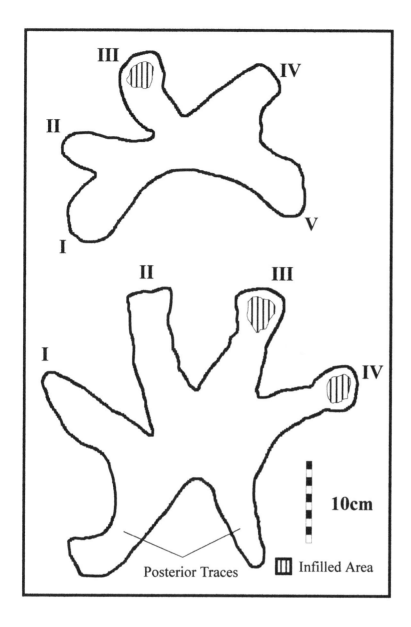

II III

I

II III

I IV

10cm

Posterior Traces   ▦ Infilled Area

trackways yet reported before its destruction. The trackway consisted of over 120 consecutive manus and pes prints (McCrea and Currie 1998) and was adjacent to another *Tetrapodosaurus* trackway of nearly the same length, but which trended in the opposite direction (Fig. 20.14). One possible interpretation is the coming and going of one animal along a preferred route (McCrea and Currie 1998). There were several other *Tetrapodosaurus* trackways of lesser length at this site, but all were lost when the site collapsed over the course of the summer of 1998.

9 Mine. A solitary ankylosaur trackway (Fig. 20.15) was illustrated by Grady (1993: 221) and Psihoyos and Knoebber (1994: 186). A block containing a manus and pes was cut out of this track with

Figure 20.12. Outline of Tetrapodosaurus *right manus and pes from the W3 Footwall, grid B17/18 (McCrea and Sarjeant 1999). A portion of these illustrated prints are seen in Figure 20.11 in the top right corner. Drawn from TMP 98.89.4 (replica cast).*

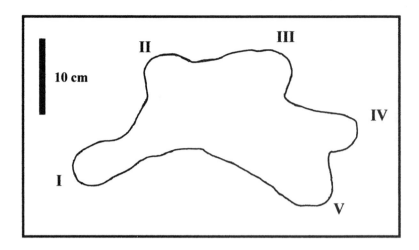

Figure 20.13. Illustration of a natural cast of a right manus print from the W3 Extension site (TMP 99.49.2).

rocksaws and is now in the collection of the Royal Tyrrell Museum of Palaeontology (TMP 92-107-1). There are a few additional *Tetrapodosaurus* trackways at this site, but they are very faint and h ave not yet been mapped or illustrated.

Mine Dump and 9 Mine West Extension (A and B) and 9 Mine West Extension Fold Axis are found exposed along the same limb of an anticline.

Mine Dump. Several natural casts of *Tetrapodosaurus* footprints (manus and pes) are eroding from the top of a high slope (Fig. 20.16). One pes print (specimen on the far right of Fig. 20.16) was collected from this site (TMP 97.5.20).

9 Mine West Extension A. Discovered in the summer of 1998, there are at least two *Tetrapodosaurus* trackways present, but so far, these have only been observed from a distance.

9 Mine West Extension B. There is a solitary sequence of *Tetrapodosaurus* footprints that have only been observed from a distance. Some of the footprints appear to be infilled with a light orange mineral, possibly limonite, which has also been observed in some of the footprints from the 12 Mine South A-Pit site (McCrea and Currie 1998).

9 Mine West Extension Fold Axis. A few *Tetrapodosaurus* footprints were found at this newly discovered tracksite, which is the only site to date that has been found on level ground. The track-bearing surface is lightly covered with mud and pebbles that have filled in most low-lying areas at this site, including the footprints.

Center Limb Pit. Another recently discovered site that displays several *Tetrapodosaurus* trackways on one limb of an anticline.

8 Mine. A few probable *Tetrapodosaurus* footprints are found on the wall of a steep, water-filled pit (McCrea and Currie 1998).

Current research on Gates Formation footprints is focused on the ichnologically diverse W3 footwall, where *Tetrapodosaurus* prints measure 40 cm long and 38 cm wide to 52 cm long and 46 cm wide. The average pace is close to 100 cm, and the average stride nears 160 cm. Step angle is generally under 120°, with the average being 110°. Preliminary speed calculations based on Alexander's (1976) formula indi-

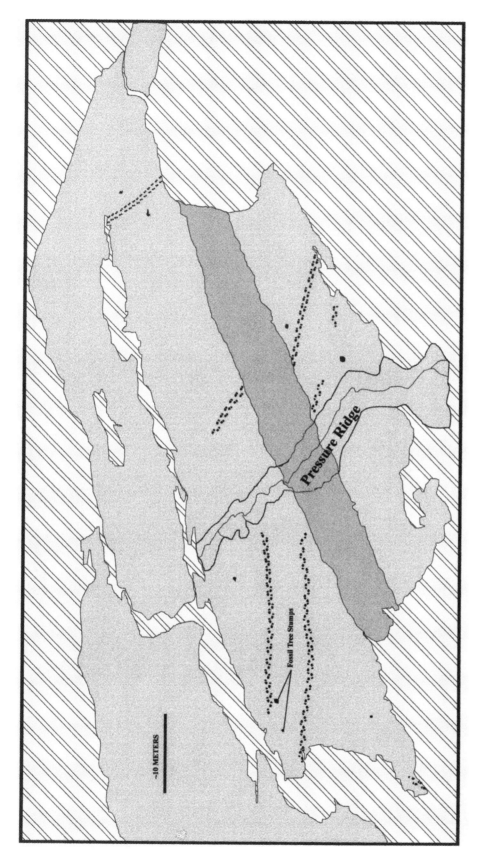

Figure 20.14. Illustration of the 12 Mine South, A-Pit, Tetrapodosaurus trackside. The formation of the pressure ridge figured herein split the two longest trackways almost in half.

Within the figure:
Pressure Ridge
Fossil Tree Stumps
–10 METERS

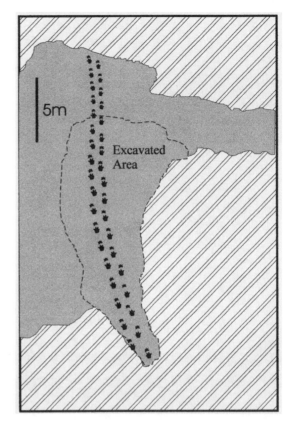

Figure 20.15. (above)
Tetrapodosaurus *trackway*
(photograph and illustration with
stylized footprints) from the 9
Mine site. Photograph provided
by Dr. Philip J. Currie.

Figure 20.16. (right) Natural
casts of Tetrapodosaurus *prints*
from the Mine Dump site. TMP
97.5.20 (pes) is at the far right.

cates the animals proceeded at a relatively unhurried walk (approximately 3 km/h).

The Gates deposits were formed in a coastal plain or deltaic environment (Langenberg et al. 1987) with abundant plant remains including ginkgoes, cycads, ferns, conifers, and at least two species of angiosperms (Wan 1996). *Tetrapodosaurus* prints dominate the Gates ichnofauna exposed in this region, with footprints numbering in the thousands.

Shirabad Suite, Tadjikistan (Albian)

Footprints from Albian carbonates and evaporites of the Shirbad suite, discovered by F. H. Khakimov in 1963 (Zakharov and Khakimov 1963) include the very distinctive track *Macropodosaurus gravis* (Zakharov 1964), which we consider similar to *Metatetrapous valdensis* (Haubold 1971).

The tracksite is situated in Shirkent National Park in the Hissar Range of central Tadzhikistan. The footprints occur in Albian coastal carbonates and evaporites, and they comprise two trackways, including that of *Macropodosaurus gravis* (Zakharov 1964). One trackway consists of alternating series of at least seven large tetradactyl prints. The prints are 50 cm long and 29 cm wide, with a pace of 75 cm and a stride of 146 cm (see Fig. 20.3e). These were previously attributed, incorrectly in our opinion, to a theropod (Zakharov 1964).

The tracks of *Macropodosaurus gravis* are similar to the pes prints of *Metatetrapous valdensis* (Haubold 1971) from the Lower Cretaceous of Germany. Similarities include pes size (length 50 cm and 44 cm, respectively) and pes shape (pes longer than wide, with a shorter digit impression on the inside). The pace angulation (about 145° and 135°, respectively) is also very similar. The main difference between these two ichnotaxa is that *M. valdensis* is clearly the trackway of a quadruped, with clear manus impressions as well as pes. The lack of manus prints in *M. gravis*, however, could be due to overprinting, where the pes print overlaps the manus print partially or completely, as is common in large quadrupeds.

M. valdensis has been attributed to an ankylosaur (Haubold 1971; Lockley and Meyer 1999; Thulborn 1990), and it is possible that *M. gravis* is probably also attributable to this group. Although this assignment is not proven, the Tadjikistan tracks are evidently not attributable to sauropods; thus, it is reasonable to assign them to a large quadrupedal ornithischian. If *M. gravis* is indeed the same as or similar to *M. valdensis* at least at the ichnogenus level, the question arises as to whether the latter European material might not be considered a junior synonym of the material from Central Asia. Both ichnotaxa are in need of further detailed study if the type material can be relocated.

Cedar Mountain Formation, Utah (Albian–Cenomanian)

Footprints from the Mussentuchit Member of the Cedar Mountain Formation (Kirkland et al. 1997) of east central Utah include two tetradactyl footprints of probable ankylosaurian affinity (Fig. 20.17). These prints were discovered by Frank DeCourten (1991) and were first

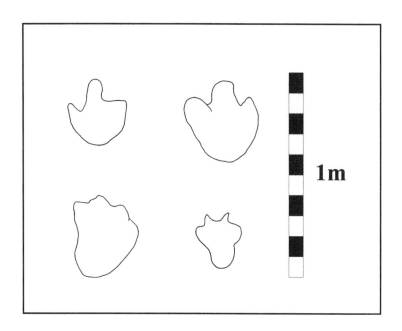

Figure 20.17. (top) Cedar Mountain Formation prints, Utah (Albian–Cenomanian), after Lockley et al. (1999).

Figure 20.18. (bottom) Probable ankylosaur pes print from the Dakota Group of Colorado (Baca County).

described by Lockley et al. (1999). Their age is close to the Albian–Cenomanian boundary, and so they are similar in age to footprints from the Dunvegan Formation. According to Kirkland et al. (1997), the Mussentuchit Member consists predominantly of drab gray, highly smectitic mudstone; it probably represents sedimentation that is almost continuous with the overlying, more carbonaceous Dakota Group. The absence of calcareous nodules representing paleosols is taken to indicate that this member represents a much wetter depositional environment than the underlying members of the Cedar Mountain Formation, in part due to the transgression of the Mowry Sea (Kirkland et al.

434 • Richard T. McCrea, Martin G. Lockley, and Christian A. Meyer

1997). The fauna of this member include small nodosaurid (*Animantarx ramaljonesi*) and neoceratopsian teeth (Carpenter et al. 1999; Chinnery et al. 1998). These teeth are among the oldest neoceratopsian remains known, and they imply that it may be possible to find ceratopsian tracks as early as the mid-Cretaceous (Albian–Cenomanian Boundary). However the origin of ceratopsians is still poorly understood (Chinnery et al. 1998), and it would be premature to speculate on the size or distribution of the trackmakers. Below the Mussentuchit Member are the Ruby Ranch Member, the Poison Strip Sandstone, and the Yellow Cat Member, which contain the skeletal remains of ankylosaurs, including ?*Sauropelta* and *Gastonia*, respectively (Carpenter et al. 1999). Notwithstanding the significant new discoveries of ceratopsians, it is still fair to state that ankylosaurs appear to have been more abundant and diverse at this time.

Dakota Group (Albian–Cenomanian)

Dinosaur tracks occur abundantly in the Dakota Group of Colorado and northeastern New Mexico. Until recently, the only known types were attributed to ornithopods and gracile theropods (Lockley et al. 1992). Recently, however, we have recorded a single four-toed print from Baca County, Colorado. The specimen is about 27 cm long by 25 cm wide, with one lateral digit (the one on the right—presumably digit I) shorter and more pointed than the other three (Fig. 20.18). This print could be attributed to an ankylosaur, although given the recent discoveries cited above, a neoceratopsian origin cannot be ruled out.

Probable ankylosaur tracks have recently been reported by William Kurtz from the Plainview Formation (Dakota Group) near Cañon City, Colorado. These tracks have been observed by one of us (M.G.L.) and are now the subject of further investigation.

Chandler Formation, Northslope, Alaska (Albian–Cenomanian)

Natural casts of tridactyl dinosaur footprints from the Chandler Formation near the Colville River in Alaska have recently been reported and were tentatively identified as belonging to *Amblydactylus* (Gangloff 1998). However, close examination of one of the footprints revealed the presence of an additional reduced digit (digit I), making this a tetradactyl footprint (Fig. 20.19). Gangloff (personal communication) reported additional collections in 1998 that have further substantiated the presence of tetradactyl footprints: at least one natural cast of a partial manus print (KCM 98-5) with skin impressions (Fig. 20.20). The tubercles on the manus print are round, measuring 4–6 mm diameter. These prints were very likely produced by ankylosaurs. The form of the natural cast figured in Gangloff (1998: 217; see also Fig. 20.19) is reminiscent of the natural casts of *Tetrapodosaurus* pes prints (Fig. 20.16) of the Mine Dump near Grande Cache, Alberta (McCrea and Currie 1998). Footprints are nearly the only record of terrestrial vertebrates in the Chandler Formation and are found among coal-bearing deltaic sediments (Gangloff 1998).

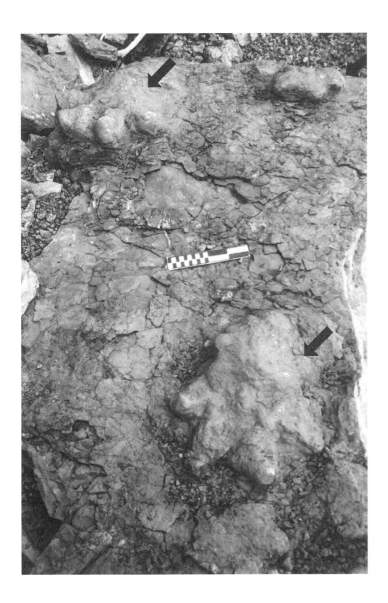

Figure 20.19. (opposite page top) Probable ankylosaur pes print (right) from the Chandler Formation (Albian–Cenomanian) of Alaska. Pen in photograph is approximately 15 cm in length. Photograph provided by Dr. Roland Gangloff.

Figure 20.20. (opposite page bottom) Probable ankylosaur manus print (natural cast, KCM 98-5) with skin impressions from the Chandler Formation (Albian–Cenomanian) of Alaska. Photograph provided by Dr. Roland Gangloff.

Figure 20.21. (above) Natural casts of probable ankylosaur prints, Dunvegan Formation, northeastern British Columbia (Cenomanian). The right pes is at the bottom right, the partial right manus is at the top right, and the left pes is at the top left. Scale is in centimeters. Photograph provided by Dr. A. Guy Plint.

Dunvegan Formation, Alberta and Northeast British Columbia (Cenomanian)

Currie (1989) reported the collection of a left manus print of *Tetrapodosaurus* (TMP 81.32.1) from the Murray River in northeastern British Columbia. Although Currie expressed uncertainty about the stratigraphic position from which these prints originated, Dr. A. Guy Plint (personal communication) has confirmed that similar prints are found throughout the Dunvegan Formation of northeastern British Columbia (Fig. 20.21). This natural cast trackslab, found along the banks of the Pine River, includes a right and left tetradactyl pes and a right (incomplete) manus. The right pes is 37 cm wide and 40 cm long, and the left pes is 37 cm wide and 39 cm long. The eroded remains of two manus prints are present in front of both pes prints, but the quality is too poor to get measurements from. This particular trackslab has

Figure 20.22. Probable print of an ankylosaur with skin impressions, Dunvegan Formation, northeastern British Columbia (Cenomanian). Photograph provided by Dr. A. Guy Plint.

Figure 20.23. Illustration of a natural cast of a pes print (TMP 99.59.2) from the Dunvegan Formation, northeastern British Columbia (Cenomanian).

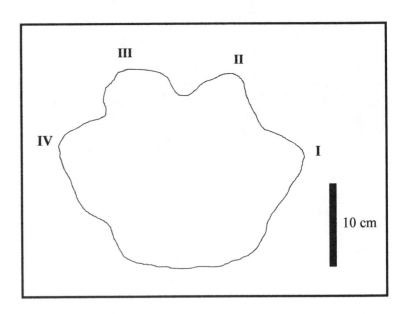

Figure 20.24. (a) Natural cast of right manus and pes prints with skin impressions (TMP 94.183.1). Dunvegan Formation, western Alberta (Cenomanian). Scale is in centimeters. (b) Close-up of right manus from Figure 20.24a, showing slide marks produced by tubercles on the posterior margin of the print. (c) Close-up of right pes from Figure 20.24a, showing faint digit I with patch of skin impressions.

since been collected by a resident with the use of rocksaws. Additional ankylosaur prints are known from this area, some of which have skin impressions. One partial print, probably a pes, has skin impressions with round tubercles measuring 3–4 mm in diameter (Fig. 20.22). A small pes print recently collected from a nearby creek bed (TMP 99.59.2) is significantly smaller than many other specimens and may be of a juvenile ankylosaur (Fig. 20.23).

A large sandstone slab with natural casts of a right manus and pes and a partial left pes from the Dunvegan Formation near Pouce Coupe, Alberta (TMP 94.183.1), is in collections at the Royal Tyrrell Museum of Palaeontology (Fig. 20.24a). This specimen has the first recorded skin impressions from footprints of ankylosaurs (McCrea et al. 1998). The tubercles are round, measuring 2–4 mm diameter on the pes and 1–2 mm on the manus. The right pes print appears tridactyl, but under oblique lighting, a short but well-defined digit I is visible with a small patch of skin impressions on it (Fig. 20.24c). Digit I is also preserved on the partial left pes, but digits II–IV are absent, having been eroded away. The manus print (Fig. 20.24b) has skin impressions as well, but it also displays slide marks produced by tubercles on the posterior margin of the manus.

We feel that it is improbable that any of the Dunvegan prints could have been produced by a ceratopsian because of inferred skeletal differences (particularly in the manus) and the fact that ceratopsians large enough to have made these prints are not yet known from this time.

Plint and Lumsden (personal communication) observed that footprints found in the Dunvegan Formation are often preserved as natural casts on the base of tabular crevasse-splay sandstones, overlying dark gray mudstone of backswamp or lacustrine origin (see also Plint 1996; McCarthy and Plint 1998).

Blackhawk Formation, Utah (Campanian)

Several natural casts of tetradactyl prints previously identified as ceratopsian in the College of Eastern Utah (CEUM 746, CEUM F-16, and CEUM 1834) are probably the pes prints of ankylosaurs (Fig. 20.25). This identification is based on similarities to *Tetrapodosaurus* prints, asymmetry of the pes, and reduction of the presumed digit I. Other footprint specimens in the museum collections have been identified as ceratopsian on the basis of the symmetry of the pes and their resemblance to *Ceratopsipes* prints. Carpenter (1992: fig. 5) illustrated a tracing of a footprint from Parker and Rowley (1989: fig. 40.23) with a *Styracosaurus* foot skeleton within it demonstrating the possibility that some of these prints were produced by ceratopsians.

The footprints are from coal mines near Price, Utah, where many footprints have been found in the past (Parker and Balsley 1989; Parker and Rowley 1989). Large tetradactyl prints of this age, such as those from other upper Cretaceous sites described below, could have been made by ceratopsians, which were extant at the time. The sediments in which the footprints are found were formed in brackish and freshwater swamps as well as along lake margins (Parker and Balsley 1989).

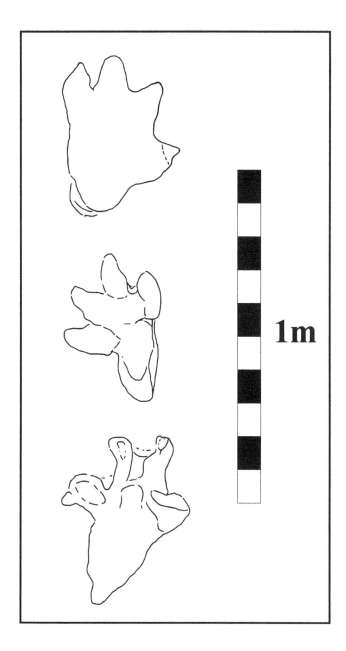

1m

Djadokhta Formation, Mongolia (Campanian)

Ishigaki (1999) recently reported the presence of massive quadrupedal footprints (digits 30–90 cm in length) from the Gobi Desert that could possibly be ankylosaurian. Ishigaki (1999) noted that the footprints are found in a region known for its ankylosaur remains (Jerzykiewicz 1997), but where large ceratopsians are absent. Ishigaki considers the pace angulation (80–90°) to be that of wide-bodied trackmakers, such as ankylosaurids. The footprints are found in fluvial-derived sediments along with the fossilized remains of freshwater animals including molluscs, crocodiles, turtles, and fish (Ishigaki 1999).

Figure 20.25. Tetradactyl footprints from the College of Eastern Utah Museum collection including footprints of both ankylosaurid and probable ceratopsid origin. Top, CEUM 746; middle, CEUM F-16; bottom, CEUM 1834.

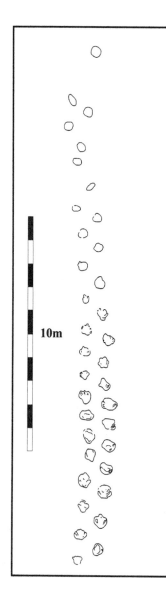

10m

Toro Toro Formation, Bolivia (Campanian)

Footprints described by Leonardi (1984) from the Upper Cretaceous (Campanian) Toro Toro Formation of the Potosi Group, Bolivia,
are possibly of ankylosaur origin. They were originally attributed to
sauropods (Campbell 1983) but subsequently were described in detail
by Leonardi (1984) and named *Ligabuichnium bolivianum* (Fig. 20.26).
Leonardi (1984) indicated that they were not well enough preserved to
determine if they were ankylosaurid or ceratopsid. Subsequently, however, Leonardi (1994: 39) restated his position, inferring that the prints
"may be attributed to an unusually large Ankylosauria," but went on
to suggest that "if the footprint is interpreted as tetradactyl instead of
pentadactyl, the trackway might have been made by a ceratopsian."
The reasons for this statement are not altogether clear because in most
cases, both ankylosaurs and ceratopsians have a tetradactyl pes and a
pentadactyl manus.

Cal Orcko Site, El Molino Formation, Bolivia (?Maastrichtian)

Approximately a half-dozen trackways of probable ankylosaurid
origin have been found in the El Molino Formation at the Cal Orcko
Limestone Quarry, near Sucre, in southern Bolivia (Figs. 20.27, 20.28).
It is outside the scope of this review to describe these trackways in
detail; they are currently under investigation as part of a detailed study
of this site. However, the lack of convincing evidence for ceratopsians
in South America (Huene 1929; Weishampel 1990) and the resemblance of the Bolivian footprints to other ankylosaurian prints described herein suggest that they may be ankylosaurian. Further work is
necessary to determine whether the prints from Cal Orcko should be
attributed to the ichnogenus *Ligabuichnium, Tetrapodosaurus,* or some
other ichnotaxon.

As shown in Figure 20.27, there are several long trackway segments known from the Cal Orko site. One (T/3/4/2) appears at first
sight to be that of a bipedal animal (Figs. 20.27, 20.28). The trackway
reveals a long step and stride (1.5–1.65 m and 3.0–3.3 m, respectively),
which evidently indicates an individual running or moving at a fast trot
of 11–12 km/h, based on the formula of Alexander (1976). Tracings
were made of all the footprints to produce the composite shown in
Figure 20.27 (T/2/3/8), which shows an ankylosaurian footprint morphology comparable to many of the examples cited above.

Trackway T/3/5/2 is characterized by footprints in which the rear
margin is angled in an anteromedial to posterolateral direction, especially on the right side of the trackway. Toe impressions, however, point
anteriorly. Footprints on the left side tend to have inwardly rotated toe
impressions and less transverse posterior margins.

Trackways T/4/5/1 and T/4/5/4, which occur in the same layer as T/
3/5/2, are remarkably wide, with transverse posterior margins. This
mode of preservation seems to suggest that the prints were perhaps
made on an overlying layer of sediment and represent the penetration
of the distal portion of the toes into an underlayer. The short stride

Figure 20.26. Ligabuichnium
bolivianum *trackway (Leonardi
1984) from the Toro Toro
Formation, Toro Toro site,
Bolivia.*

*Figure 20.27. (opposite page)
Ankylosaur trackways from the
El Molino Formation, Cal Orko
Site, Bolivia.*

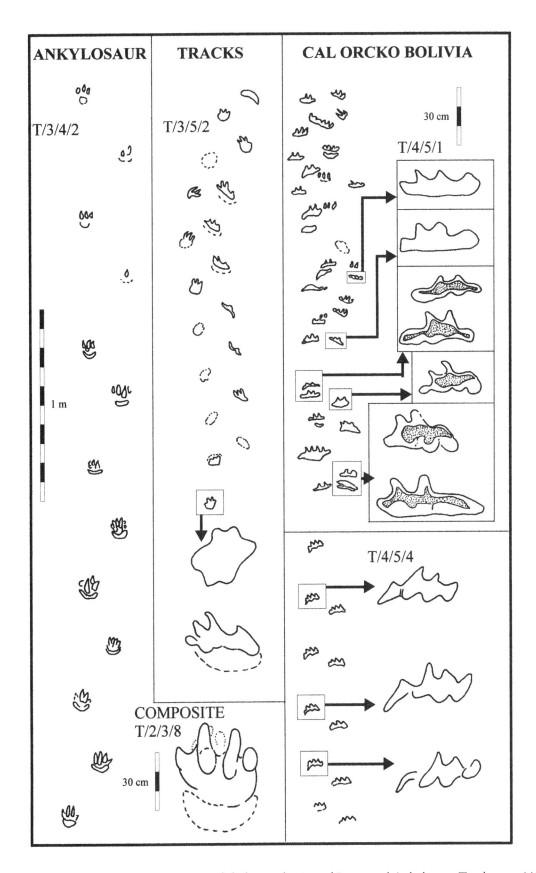

ANKYLOSAUR   **TRACKS**   **CAL ORCKO BOLIVIA**

T/3/4/2

T/3/5/2

30 cm

T/4/5/1

1 m

T/4/5/4

COMPOSITE
T/2/3/8

30 cm

Figure 20.28. Photographs of an ankylosaur trackway from the El Molino Formation, Cal Orko Site, Bolivia (T/3/4/2). The horizontal trackway across the top of the left photograph has been attributed to a sauropod. Dr. Giuseppe Leonardi is standing next to the ankylosaur trackway (T/3/4/2) in the right photograph.

444

indicates that these animals were not running. Many manus impressions are overprinted, but where pes sets are visible, both show that the central toes (digits II and III of the pes and also ?II and ?III of the manus) are the longest and most prominent.

These footprints are found around the margins of lakes that are believed to be ephemeral and perennial on the basis of episodic pedogenesis and lacustrine stromatolites. The ankylosaur tracks occur in two levels that also yield remains of freshwater organisms, such as catfish, turtles, and crocodiles, as well as snails and charophytes (Meyer et al. 1999, in press).

Discussion

To date, there are 14 probable ankylosaur tracksites known worldwide (Fig. 20.29). Although few sites have been studied in detail, there are five ichnospecies names assigned, all in different ichnogenera. It is outside the scope of this article to discuss the systematics in detail. However, the first step, to record occurrence and briefly describe and illustrate the material, is accomplished herein. It remains to be seen to what extent tracks from all 14 sites are comparable in detail; whether the morphologic features warrant interpretations of all tracks as ankylosaurian; and whether the morphologic features of various ichnites are sufficiently distinctive to warrant assignment of different ichnogenera.

At present, there is no obvious pattern discernible in the geographic distribution of ankylosaur tracksites, although there is a temporal clustering of sites in the Albian–Cenomanian. It is also apparent that almost all known sites are Cretaceous in age. As indicated above, it is not certain if the Jurassic tracks from Yorkshire, England, are ankylosaurian in origin (Lockley et al. 1994). The Gething, Gates, and Dunvegan Formations are within two of the three Cretaceous cyclic, clastic sequences found in western Alberta and northwestern British Columbia (Stott 1975, 1982, 1984). This sequence shows evidence of the continuous presence of ankylosaurs, as evidenced by footprints, during the early part of the Cretaceous in this part of the world. The high concentration of Aptian–Cenomanian sites (50%) in western North America could be considered a real paleogeographic phenomenon indicating a higher concentration of trackmakers in this region, but it might also be an artifact of preservation.

Current biostratigraphic constraints would seem to suggest that it is unlikely to find ceratopsians in pre-Cenomanian or pre–late Albian deposits. Such inferences seem to rule out the possibility that any of the older ichnofaunas (1–10; Fig. 20.29) could perhaps be ceratopsian rather than ankylosaurian. The large size of many of the later Albian and Cenomanian tracks described herein, and the fact that ankylosaurs were particularly abundant at this time, also makes it unlikely that any of the footprints might be those of early ceratopsians that had already attained large body sizes. The possibility that some South American footprints might be ceratopsian in origin has already been discussed and is largely ruled out by paleobiogeographic considerations. In fact, to date, it is only in the Blackhawk Formation assemblage that we find

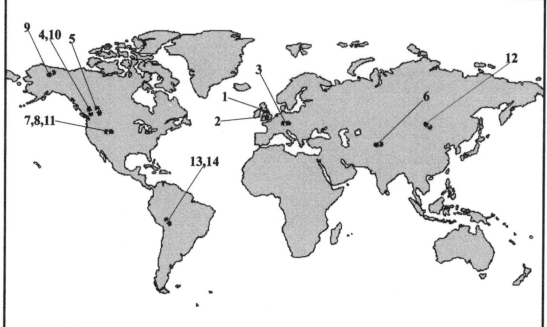

14. El Molino Formation, Cal Orcko site, Bolivia (?Maastrichtian)
13. Toro Toro Formation, Bolivia (Campanian): type locality for *Ligabuichnium bolivianum* (Leonardi, 1984)
12. Djadokhta Formation, Mongolia (Campanian).
11. Blackhawk Formation, Utah (Campanian).
10. Dunvegan Formation, Alberta and N.E. British Columbia (Cenomanian).
9. Chandler Formation, Northslope, Alaska (Albian - Cenomanian).
8. Dakota Group, Colorado and New Mexico (Albian-Cenomanian)
7. Cedar Mountain Formation, Utah (Albian - Cenomanian).
6. Shirabad Suite, Tadjikistan (Albian): type locality for *Macropodosaurus gravis* (Zakharov, 1964).
5. Gates Formation, Grande Cache, Alberta (Lower Albian).
4. Gething Formation, British Columbia (Aptian-Albian): type locality for *Tetrapodosaurus borealis* (Sternberg, 1932).
3. Wealden Beds, Germany (Berriasian): type locality for *Metatetrapous valdensis* (Haubold, 1971).
2. Purbeck Beds, England (Berriasian).
1. Saltwick Formation, England (Aalenian - Bajocian): type locality for *Deltapodus brodericki* (Whyte and Romano, 1994).

Figure 20.29. Location and list of purported ankylosaur tracksites.

a collection of isolated and somewhat out-of-context footprints that might be of both ankylosaurian and ceratopsian origin.

Another valuable by-product of the increase in discovery of ankylosaur footprints has been the recognition of prints with skin impressions, found in the Albian–Cenomanian (Gates, Chandler and Dunvegan Formations). It is outside the scope of this paper to fully describe this material, but the tubercles all appear to be round, though of varying diameter. This moderately well preserved material will allow us to compare skin surface textures of ankylosaurs with those of contemporary or near-contemporary ornithopods (Currie et al. 1991), as well as noncontemporary large herbivorous dinosaurs, such as sauropods (Czerkas 1992, 1994). It is also theoretically possible that the study of these tracks will ultimately help distinguish them from ceratopsian footprints, although to date, no ceratopsian footprint skin impressions have been described.

Ankylosaur Paleoenvironments and Paleoecology

There has been some discussion about the habitat preference of ankylosaurs. Carpenter (1997b) observes that ankylosaur distribution was dependent on the distribution of the plants that made up their diet. He also noted that North American ankylosaurs were found in moist, coastal areas, whereas Asian ankylosaurs tend to be found in arid to semiarid environments. The track record of ankylosaurs clearly has a significant bearing on this discussion and shows that they did not shy away from moist environments. In fact, signs of ankylosaurs, with footprints numbering in the thousands, dominate the Smoky River ichnofauna of the Gates Formation (McCrea and Currie 1998), which is a coastal plain depositional environment that had once supported large coal swamps (Langenberg et al. 1987).

Ankylosaur footprints are concentrated in coal-bearing and floodplain facies (Blackhawk, Cedar Mountain, ?Chandler, Dunvegan, Gates, Gething, ?Saltwick, and Wealden Formations) that seem to suggest a strong facies preference for well-vegetated, well-watered lowlands. The Purbeck, Toro Toro, and Cal Orcko settings were not necessarily very different in terms of food and water supplies available to large dinosaurs. The El Molino Formation in Bolivia provides evidence that ankylosaurs lived around freshwater depositional settings.

It is evident that ankylosaur tracks are no longer nearly as rare as once thought (Lockley 1991; Schult and Farlow 1992). It has been noted that tracks of large ornithischians (stegosaurs, ankylosaurs, and ceratopsians) are rarer than might be predicted on the basis of relatively abundant skeletal remains. It has been suggested that perhaps this was due to the preference of these animals for dry (e.g., upland or inland) habitats, where tracks would not be preserved, rather than humid (or lowland or coastal) settings (e.g., Lockley and Conrad 1987). Although this may be true in the case of stegosaurs (Buffrénil et al. 1986; Dodson et al. 1980; Galton 1990), it appears that this generalization does not extend to ankylosaurs or ceratopsians. For example, ceratopsian footprints are best known from coastal plain facies, where they were made on very wet substrates (Lockley and Hunt 1995). Retallack (1997: 355) goes so far as to refer to "soggy ceratopsians" and indicates that "some ceratopsians such as *Styracosaurus* lived in swamps." It now appears that ankylosaur tracks are associated with low-lying floodplain and coastal-plain facies assemblages and so cannot be considered solely dry land or upland animals on the basis of ichnologic evidence.

The paleoenvironmental setting of a few ankylosaur skeletal sites generally support the paleoenvironmental interpretation derived from tracksites (i.e., preference for low-lying wetlands). The depositional environment of the Arundel Formation has been compared with the Mississippi delta, with oxbow swamps (Kranz 1998; Lipka 1998), and contains the remains of nodosaurid ankylosaurs. The Cloverly Formation, where the skeletal remains of *Sauropelta* are found, also contains the remains of aquatic animals, such as fish (*Amia* and lungfish), as well as crocodiles and turtles (Ostrom 1970), indicating that *Sauropelta* frequented the margins of freshwater environments. The Judith River

Formation of Alberta contains the fossil remains of ankylosaurs (Brinkman 1990), as well as a variety of terrestrial and aquatic taxa (Brinkman 1990: table 1). Eberth (1990: 4) describes the Judith River Formation as a "channel-facies dominated, coastal plain sequence."

Significant numbers of ankylosaurs have also been found in marine deposits, suggesting that they were washed out to sea (Eaton 1960; Horner 1979; Lee 1996). Such taphonomic contexts suggest that the animals may have been transported by rivers or floods into the marine environment and that they had presumably lived in coastal areas. We note, however, that the ankylosaur (*Pinacosaurus*) from the Campanian age Djadokhta Formation of Mongolia is found "in semi-arid, alluvial-to-eolian settings," but some dune beds may have been "organically rich and seasonally moist" (Jerzykiewicz 1997: 188). The recent report of probable ankylosaur footprints from fluvial-derived substrates in Mongolia (Ishigaki 1999) suggests that some Mongolian ankylosaurs lived on the margins of freshwater environments.

Given the increase in ankylosaur tracksite reports from 2 in 1991 (Lockley 1991) to 14 (above), it is clear that previous perceptions of scarcity were an artifact of limited data rather than scarcity in the track record. With the abundance of footprints reported at certain sites, such as the Smoky River ichnofauna (Grande Cache), we can ask if there is a possibility that trackways might indicate gregarious behavior among ankylosaurs. Trackways and inferred ankylosaurian dinoturbation are sufficiently abundant at Grand Cache sites W3, 12 Mine South, A-Pit, W1a, and W2 to raise this question. *Pinacosaurus* skeletons from Dja-dokhta Formation (Jerzykiewicz 1997) also indicate a social group or clutch of juveniles that died together and were buried by a sandstorm. Like other large herbivorous dinosaurs, some ankylosaur taxa were probably gregarious. Further study is required to determine the extent to which this can be demonstrated by use of ichnologic data.

It is outside the scope of this summary to provide detailed discussion of the ichnotaxa associated with the ankylosaur track assemblages. However some general observations are pertinent. It appears that only in the examples from the Purbeck, Wealden, Toro Toro, and Cal Orko sites is there evidence of sauropods in association with ankylosaur footprints. This may reflect the sauropod preference for carbonate substrates rather than an association between these two groups (Lockley et al. 1994). Theropod footprints also occur at these localities, and at the two European sites we also find ornithopod (iguanodontid) prints. At the North American sites, quite rich, though as yet little described, ichnofaunas are present, including ornithopod prints, theropod prints, and, at all the Canadian sites, bird footprints. At most sites, sauropod footprints are conspicuous by their absence, owing in large part to the late Albian–Maastrichtian sauropod hiatus in North America (Lucas and Hunt 1989).

Acknowledgments. We thank Dr. Philip Currie and Darren Tanke of the Royal Tyrrell Museum of Palaeontology for help with study of the Grande Cache specimens. Funding for studying the Grande Cache tracksites was provided by the Royal Tyrrell Museum of Palaeontology, the Heaton Student Support Grant, Smoky River Coal Limited, and the Jurassic Foundation. Smoky River Coal Limited and its employees have been of considerable assistance in facilitating access to the Gates Formation tracksites. Study of the Cal Orko site, Bolivia, was made possible by the support of the Swiss National Science Foundation (grant 21-52649.97) and the cooperation of the Fancesa cement quarry operators. Dr. Giuseppe Leonardi and the Cal Orko research team also helped considerably with the documentation of ankylosaur tracks at this site. We thank Dr. A. Guy Plint and Matthew Lumsden for bringing some specimens from the Dunvegan Formation to our attention. We also thank Dr. Roland Gangloff for information on the Alaskan footprints and William Kurtz, Jr. (University of Colorado at Denver), for sharing his newly discovered tracksite with us. We thank Rob Gaston for helping us with some references. We are grateful to the staff of the Price Museum in Utah for allowing access to their impressive footprint collections. R.T.M. acknowledges the assistance and advice of Dr. William A. S. Sarjeant during the course of his research.

References Cited

Alexander, R. McN. 1976. Estimates of speeds of dinosaurs. *Nature* 261: 129–130.

Brinkman, D. B. 1990. Paleoecology of the Judith River Formation (Campanian) of Dinosaur Provincial Park, Alberta, Canada: Evidence from vertebrate microfossil localities. *Palaeogeography, Palaeoclimatology, Palaeoecology* 78: 37–54.

Buffrénil, V. de, J. O. Farlow, and A. de Ricqles. 1986. Growth and function of *Stegosaurus* plate: Evidence from bone histology. *Paleobiology* 12: 459–473.

Campbell, K. 1983. Trackways, clues to passing dinosaurs. *Terra, Museum of Natural History of Los Angeles County* 6: 12–13.

Carpenter, K. 1984. Skeletal reconstruction and life restoration of *Sauropelta* (Ankylosauria–Nodosaurida) from the Cretaceous of North America. *Canadian Journal of Earth Sciences* 21: 1491–1498.

———. 1992. Behavior of hadrosaurs as interpreted from footprints in the "Mesaverde" Group (Campanian) of Colorado, Utah, and Wyoming. *Contributions to Geology, University of Wyoming* 29: 81–96.

———. 1997a. Ankylosauria. In P. J. Currie, and K. Padian (eds.), *Encyclopedia of Dinosaurs*, pp. 16–20. New York: Academic Press.

———. 1997b. Ankylosaurs. In J. O. Farlow and M. K. Brett-Surman (eds.), *The Complete Dinosaur*, pp. 307–316. Bloomington: Indiana University Press.

Carpenter, K., and J. I. Kirkland. 1998. Review of lower and middle Cretaceous ankylosaurs from North America. In S. G. Lucas, J. I. Kirkland, and J. W. Estep (eds.), *Lower and Middle Cretaceous Terrestrial Ecosystems. New Mexico Museum of Natural History and Science Bulletin* 14: 249–270.

Carpenter, K., J. I. Kirkland, D. Burge, and J. Bird. 1999. Ankylosaurs (Dinosauria: Ornithischia) of the Cedar Mountain Formation, Utah, and their stratigraphic distribution. In D. D. Gillette (ed.), *Vertebrate Paleontology in Utah. Utah Geological Survey, Miscellaneous Publication* 99-1: 243–251.

Chinnery, B. J., T. R. Lipka, J. I. Kirkland, J. M. Parrish, and M. K. Brett-Surman. 1998. Neoceratopsian teeth from the Lower to Middle Cretaceous of North America. In S. G. Lucas, J. I. Kirkland, and J. W. Estep (eds.), *Lower and Middle Cretaceous Terrestrial Ecosystems. New Mexico Museum of Natural History and Science Bulletin* 14: 297–302.

Coombs, W. P., Jr., and T. Maryańska. 1990. Ankylosauria. In D. B. Weishampel, P. Dodson, and H. Osmólska (eds.), *The Dinosauria,* pp. 456–483. Berkeley: University of California Press.

Currie, P. J. 1983. Hadrosaur trackways from the lower Cretaceous of Canada. In Second Symposium on Mesozoic Terrestrial Ecosystems, Jadwisin, 1981. *Acta Palaeontologica Polonica* 28: 63–73.

———. 1989. Dinosaur footprints of western Canada. In D. D. Gillette and M. G. Lockley (eds.), *Dinosaur Tracks and Traces,* pp. 293–300. Cambridge: Cambridge University Press.

———. 1995. Ornithopod trackways from the lower Cretaceous of Canada. In W. A. S. Sarjeant (ed.), *Vertebrate Fossils and the Evolution of Scientific Concepts,* pp. 431–443. Singapore: Gordon and Breach.

Currie, P. J., G. C. Nadon, and M. G. Lockley. 1991. Dinosaur footprints with skin impressions from the Cretaceous of Alberta and Colorado. *Canadian Journal of Earth Sciences* 28: 102–115.

Currie, P. J., and W. A. S. Sarjeant. 1979. Lower Cretaceous dinosaur footprints from the Peace River canyon, British Columbia, Canada. *Palaeogeography, Palaeoclimatology, Palaeoecology* 28: 103–115.

Czerkas, S. 1992. Discovery of dermal spines reveals a new look for sauropod dinosaurs. *Geology* 20: 1068–1070.

———. 1994. The history and interpretation of sauropod skin impressions. In M. G. Lockley, V. F. Santos, C. Meyer, and A. P. Hunt (eds.), *Aspects of Sauropod Paleobiology. Gaia: Revista de Geociencias, Museu Nacional de Historia Natural* (Lisbon, Portugal) 10: 173–182.

DeCourten, F. 1991. New data on early Cretaceous dinosaurs from the Long Walk Quarry and tracksite, Emery County, Utah. *Utah Geologists Association Publication* 19: 311–324.

Dodson, P., R. T. Behrensmeyer, R. T. Bakker, and J. S. McIntosh. 1980. Taphonomy and paleoecology of the dinosaur beds of the Jurassic Morrison Formation. *Paleobiology* 6: 208–232.

Dodson, P., and P. Currie. 1990. Neoceratopsia. In D. B. Weishampel, P. Dodson and H. Osmólska (eds.), *The Dinosauria,* pp. 593–618. Berkeley: University of California Press.

Eaton, T. H., Jr. 1960. A new armored dinosaur from the Cretaceous of Kansas. *University of Kansas Paleontological Contributions* 25: 1–24.

Eberth, D. A. 1990. Stratigraphy and sedimentology of vertebrate microfossil sites in the uppermost Judith River Formation (Campanian), Dinosaur Provincial Park, Alberta, Canada. *Palaeogeography, Palaeoclimatology, Palaeoecology* 78: 1–36.

Ensom, P. 1987. Dinosaur tracks in Dorset. *Geology Today* 31(November–December): 182–183.

———. 1988. Excavations at Sunnydown Farm, Langton Matravers, Dorset: Amphibians discovered in the Purbeck Limestone. *Proceedings of the Dorset Natural History and Archaeological Society* 109: 148–150.

Farlow, J. O., and R. E. Chapman. 1997. The scientific study of dinosaur footprints. In J. O. Farlow and M. K. Brett-Surman (eds.), *The Complete Dinosaur,* pp. 519–553. Bloomington: Indiana University Press.

Galton, P. M. 1990. Stegosauria. In D. B. Weishampel, P. Dodson, and H. Osmólska (eds.), *The Dinosauria,* pp. 435–455. Berkeley: University of California Press.

Gangloff, R. A. 1998. Arctic dinosaurs with emphasis on the Cretaceous record of Alaska and the Eurasian–North American connection. In S. G. Lucas, J. I. Kirkland, and J. W. Estep (eds.), *Lower and Middle Cretaceous Terrestrial Ecosystems. New Mexico Museum of Natural History and Science Bulletin* 14: 211–220.

Gibson, D. W. 1985. Stratigraphy and sedimentology of the Lower Cretaceous Gething Formation, Carbon Creek Coal Basin, northeastern British Columbia. *Geological Survey of Canada* Paper 80-12: 1–29.

Gierliński, G. 1999. Tracks of a large thyreophoran dinosaur from the Early Jurassic of Poland. *Acta Palaeontologica Polonica* 44: 231–234.

Grady, W. 1993. *The Dinosaur Project: The Story of the Greatest Expedition Ever Mounted*. Toronto: Macfarlane, Walter Ross.

Haubold, H. 1971. *Ichnia Amphibiorum et Riptiliorum Fossilium*. In O. Kuhn (ed.), *Handbuch der Palaoherpetolgie*. Vol. 18. Stuttgart: Gustav Fischer Verlag.

Horner, J. R. 1979. Upper Cretaceous dinosaurs from the Bearpaw Shale (Marine) of south central Montana, with a checklist of Upper Cretaceous dinosaur remains from marine sediments in North America. *Journal of Paleontology* 53: 566–577.

Huene, F. von. 1929. Los saurisquios y ornithisquios de Cretaceo Argentino. *Anales de Museo De la Plata*, Series 2, 3: 1–196.

Ishigaki, S. 1999. Abundant dinosaur footprints from Upper Cretaceous of Gobi Desert, Mongolia [abstract]. *Journal of Vertebrate Paleontology* 19(3, Suppl.): 54A.

Jerzykiewicz, T. 1997. Djadokhta Formation. In P. J. Currie and K. Padian (eds.), *Encyclopedia of Dinosaurs*, pp. 188–191. New York: Academic Press.

Kirkland, J. I. 1998. A polacanthine ankylosaur (Ornithischia: Dinosauria) from the Early Cretaceous (Barremian) of Eastern Utah. In S. G. Lucas, J. I. Kirkland, and J. W. Estep (eds.), *Lower and Middle Cretaceous Terrestrial Ecosystems. New Mexico Museum of Natural History and Science Bulletin* 14: 271–281.

Kirkland, J. I., B. B. Britt, D. L. Burge, K. Carpenter, R. Cifelli, F. De-Courten, J. Eaton, S. Hasiotis, M. Kirshbaum, and T. Lawton. 1997. Lower to Middle Cretaceous dinosaur faunas of the Central Colorado Plateau: A key to understanding 35 million years of tectonics, sedimentology, evolution and biogeography. *Brigham Young University Geology Studies* 42: 69–103.

Kranz, P. M. 1998. Mostly dinosaurs: A review of the vertebrates of the Potomac Group (Aptian Arundel Formation), USA. In S. G. Lucas, J. I. Kirkland, and J. W. Estep (eds.), *Lower and Middle Cretaceous Terrestrial Ecosystems. New Mexico Museum of Natural History and Science Bulletin* 14: 235–238.

Langenberg, C. W., W. Kalkreuth, and C. B. Wrightson. 1987. Deformed Lower Cretaceous coal-bearing strata of the Grande Cache area, Alberta. *Geological Survey Department, Alberta Research Council Bulletin* 56: 1–56.

Lee, Y.-N. 1996. A new nodosaurid ankylosaur (Dinosauria: Ornithischia) from the Paw Paw Formation (late Albian) of Texas. *Journal of Vertebrate Paleontology* 16: 232–245.

Le Loeuff, J., M. G. Lockley, C. A. Meyer, and J.-P. Petit. 1998. Earliest tracks of Liassic Basal Thyreophorans [abstract]. *Journal of Vertebrate Paleontology* 18(3, Suppl.): 58A–59A.

———. 1999. Discovery of a thyreophoran trackway in the Hettangian of central France. *Compte Rendus de l'Academie des Sciences, Paris, Earth and Planetary Science* 328: 215–219.

Leonardi, G. 1984. Le impromte fossili di dinosauri. In J. F. Bonaparte, E. H. Colbert, P. J. Currie, A. de Ricqles, Z. Kielan-Jaworowska, G. Leonardi, N. Morello, and P. Taquet (eds.), *Sulle Orme dei Dinosauri*, pp. 165–186. Venice: Errizo Editrice.

———. 1994. Annotated atlas of South America tetrapod footprints (Devonian to Holocene). *Companhia de Pesquisa de Recursos Minerais, Brasilia* 248: 1–35.

Lipka, T. R. 1998. The affinities of the enigmatic theropods of the Arundel Clay Facies (Aptian) Potomac Formation, Atlantic Coastal Plain of Maryland. In S. G. Lucas, J. I. Kirkland, and J. W. Estep (eds.), *Lower and Middle Cretaceous Terrestrial Ecosystems. New Mexico Museum of Natural History and Science Bulletin* 14: 229–234.

Lockley, M. G. 1986. The paleobiological and palaeoenvironmental importance of dinosaur footprints. *Palaios* 1: 37–47.

———. 1991. *Tracking Dinosaurs—A New Look at an Ancient World*. New York: Cambridge University Press.

———. 1999a. *The Eternal Trail: A Tracker Looks at Evolution*. Reading: Perseus Books.

———. 1999b. Insights into the biology of form as expressed in dinosaurs and their trackways. *Tycho de Brahe—Jahrbuch für Goetheanismus*, pp. 135–166.

———. In press. Dinosaur trackways. In D. E. G. Briggs and P. Crowther (eds.) *Paleobiology: A Synthesis*.

Lockley, M. G., and K. Conrad. 1987. Mesozoic tetrapod tracksites and their application in paleoecological census studies. In P. J. Currie and E. H. Koster (eds.), *Mesozoic Terrestrial Ecosystems, 4th Symposium (Short Papers). Tyrrell Museum of Paleontology Occasional Paper* 3: 144–149.

Lockley, M. G., J. Holbrook, A. P. Hunt, M. Matsukawa, and C. Meyer. 1992. The dinosaur freeway: A preliminary report on the Cretaceous megatracksite, Dakota Group, Rocky Mountain front range and highplains; Colorado, Oklahoma and New Mexico. In R. Flores (ed.), *Mesozoic of the Western Interior, SEPM Midyear Meeting Fieldtrip Guidebook*, pp. 39–54. Fort Collins, Colo.: SEPM.

Lockley, M. G., and A. P. Hunt. 1995. Ceratopsid tracks and associated ichnofauna from the Laramie Formation (Upper Cretaceous: Maastrichtian) of Colorado. *Journal of Vertebrate Paleontology* 15: 592–614.

Lockley, M. G., J. I. Kirkland, F. DeCourten, B. B. Britt, and S. Hasiotis. 1999. Dinosaur tracks from the Cedar Mountain Formation of Eastern Utah: A preliminary report. In D. D. Gillette (ed.), *Vertebrate Paleontology in Utah. Utah Geological Survey, Miscellaneous Publication* 99-1: 253–257.

Lockley, M. G., and M. Matsukawa. 1998. Lower Cretaceous vertebrate tracksites of East Asia. In S. G. Lucas, J. I. Kirkland, and J. W. Estep (eds.), *Lower and Middle Cretaceous Terrestrial Ecosystems. New Mexico Museum of Natural History and Science Bulletin* 14: 135–142.

Lockley, M. G., and C. Meyer. 1999. *Dinosaur Tracks and Other Vertebrate Footprints of Europe*. New York: Columbia Press.

Lockley, M. G., C. Meyer, A. P. Hunt, and S. G. Lucas. 1994. The distribution of sauropod tracks and trackmakers. *Gaia: Revista de Geociencias, Museu Nacional de Historia Natural* 10: 233–248.

Lucas, S. G., and A. P. Hunt. 1989. *Alamosaurus* and the sauropod hiatus in the Cretaceous of the North American Western Interior. In J. O. Farlow (ed.), *Paleobiology of the Dinosaurs. Geological Society of America Special Paper* 238: 75–85.

McCarthy, P. J., and A. G. Plint. 1998. Recognition of interfluve sequence boundaries: Integrating paleopedology and sequence stratigraphy. *Geology* 26: 387–390.

McCrea, R. T., and P. J. Currie. 1998. A preliminary report on dinosaur tracksites in the lower Cretaceous (Albian) Gates Formation near Grande Cache, Alberta. In S. G. Lucas, J. I. Kirkland, and J. W. Estep (eds.), *Lower and Middle Cretaceous Terrestrial Ecosystems. New Mexico Museum of Natural History and Science Bulletin* 14: 155–162.

McCrea, R. T., M. G. Lockley, and A. G. Plint. 1998. A summary of purported ankylosaur track occurrences [abstract]. *Journal of Vertebrate Paleontology* 18(3, Suppl.): 62A.

McCrea, R. T., and W. A. S. Sarjeant. 1999. A diverse vertebrate ichnofauna from the Lower Cretaceous (Albian) Gates Formation near Grande Cache, Alberta [abstract]. *Journal of Vertebrate Paleontology* 19(3, Suppl.): 62A.

Meyer, C., M. G. Lockley, G. Leonardi, and F. Anaya. 1999. Late Cretaceous vertebrate ichnofacies of Bolivia—Facts and implications [abstract]. *Journal of Vertebrate Paleontology* 19(3, Suppl.): 63A.

Meyer, C. A., D. Hippler, and M. G. Lockley. In press. Late Cretaceous vertebrate ichnofacies of Bolivia—Facts and implications. *Proceedings of the VIIth Symposium on Mesozoic Terrestrial Ecosystems.* Buenos Aires.

Mutterlose, J., M. G. E. Wippich, and M. Geisen (eds.). 1997. Cretaceous depositional environments of northwest Germany. *Buchumer Geologishe und Geotechnische Arbeiten* 46: 1–134.

Ostrom, J. H. 1970. Stratigraphy and paleontology of the Cloverly Formation (Lower Cretaceous) of the Bighorn Basin Area, Wyoming and Montana. *Peabody Museum of Natural History, Yale University, Bulletin* 35: 1–234.

Parker, L. R., and J. K. Balsley. 1989. Coal mines as localities for studying dinosaur trace fossils. In D. D. Gillette and M. G. Lockley (eds.), *Dinosaur Tracks and Traces,* pp. 353–360. Cambridge: Cambridge University Press.

Parker, L. R., and R. L. Rowley, Jr. 1989. Dinosaur footprints from a coal mine in east-central Utah. In D. D. Gillette and M. G. Lockley (eds.), *Dinosaur Tracks and Traces,* pp. 361–366. Cambridge: Cambridge University Press.

Plint, A. G. 1996. Marine and nonmarine systems tracts in forth order sequences in the early-middle-Cenomanian, Dunvegan alloformation, northeastern British Columbia, Canada. In J. Howell, and J. D. Aitken (eds.), *High Resolution Sequence Stratigraphy: Innovations and Applications. Geological Society of London, Special Paper* 104: 159–191.

Psihoyos, L., and J. Knoebber. 1994. *Hunting Dinosaurs.* New York: Random House.

Retallack, G. 1997. Dinosaurs and dirt. In D. Wolberg (ed.), *Dinosfest International,* pp. 345–359. Philadelphia: Academy of Natural Sciences.

Schult, M. F., and J. O. Farlow. 1992. Vertebrate trace fossils. In C. G. Maples and R. R. West (eds.), *Trace Fossils. Paleontological Society, Short Course* 5: 34–63.

Sternberg, C. M. 1932. Dinosaur tracks from Peace River, British Columbia. *National Museum of Canada, Annual Report* 1930: 59–85.

Stott, D. F. 1972. The Cretaceous Gething Delta, north-eastern British Columbia. In Proceedings of the Geological Conference on Western Canadian Coal. *Research Council of Alberta Information Series* 60: 151–163.

———. 1975. The Cretaceous system in north-eastern British Columbia. *Geological Association of Canada, Special Paper* 13: 441–467.

———. 1982. Lower Cretaceous Fort St. John Group and Upper Cretaceous Dunvegan Formation of the Foothills and Plains of Alberta, British Columbia, District of Mackenzie and Yukon Territory. *Geological Survey of Canada Bulletin* 328: 1–124.

———. 1984. Cretaceous sequences of the foothills of the Canadian Rocky Mountains. In D. F. Stott and D. J. Glass (eds.), *The Mesozoic of Middle North America. Canadian Society of Petroleum Geologists, Memoir* 9: 85–107.

Thulborn, R. A. 1990. *Dinosaur Tracks.* London: Chapman and Hall.

Wan, Z. 1996. *The Lower Cretaceous Flora of the Gates Formation from Western Canada.* Ph.D. thesis. University of Saskatchewan, Saskatoon.

Weishampel, D. B. 1990. Dinosaur distribution. In D. B. Weishampel, P. Dodson, and H. Osmólska (eds.), *The Dinosauria,* pp. 63–139. Berkeley: University of California Press.

Whyte, M. A., and M. Romano. 1994. Probable sauropod footprints from the Middle Jurassic of Yorkshire, England. *Gaia: Revista de Geociencias, Museu Nacional de Historia Natural* 10: 15–26.

Wolfe, D. G., and J. I. Kirkland. 1998. *Zuniceratops christopheri* n. gen. and n. sp., a ceratopsian dinosaur from the Moreno Hill Formation (Cretaceous, Turonian) of west-central New Mexico. In S. G. Lucas, J. I. Kirkland, and J. W. Estep (eds.), *Lower and Middle Cretaceous Terrestrial Ecosystems. New Mexico Museum of Natural History and Science Bulletin* 14: 303–317.

Wright, J. L. 1996. *Fossil Terrestrial Trackways: Preservation, Taphonomy and Palaeoecological Significance.* Ph.D. thesis. University of Bristol, England.

Zakharov, S. A. 1964. On the Cenomanian dinosaur, the tracks of which were found in the Shirkent River Valley [in Russian]. In V. M. Reiman (ed.), *Paleontology of Tadzhikistan,* pp. 31–35. Dushanbe: Ajadennuta Bayj Tadzhikskoi SSR.

Zakharov, S. A., and Khakimov, F. H. 1963. About Cenomanian dinosaur footprints in Western Tadzhikistan [in Russian]. *Doklady Academy of Sciences of Tadzhikistan S.S.R.* 6(9): 25–27.

21. Phylogenetic Analysis of the Ankylosauria

Kenneth Carpenter

Abstract

A hypothesis regarding the phylogeny of the Ankylosauria is presented. The phylogenetic analysis uses node-based definitions for higher categories. By definition, these nodes represent monophyletic assemblages of taxa, and some nodes are equivalent to the family in Linnaean parlance. On the basis of several synapomorphies, *Scelidosaurus* is united with more traditional ankylosaurs (*Minmi* + Polacanthidae + Nodosauridae + Ankylosauridae) at the node Ankylosauromorpha. This basal node is introduced to include all thyreophorans closer to *Scelidosaurus* than to *Stegosaurus*. The phylogenetic analysis was conducted by the software PAUP, with the thyreophoran *Huayangosaurus* and the primitive ornithischian *Lesothosaurus* as successive outgroups. A modified compartmentalization technique is used to minimize the effects of distant taxa on one another. The analysis indicates that *Minmi* is the closest sister taxon to the Nodosauridae but cannot be assigned to any known family either because of its primitiveness or the loss of typical ankylosaur synapomorphies. The Polacanthidae is a valid family and is the closest sister group to the Ankylosauridae. Two data sets for the ankylosaur genera of each major node (Nodosauridae, Polacanthidae, and Ankylosauridae) were analyzed: a smaller set containing taxa for which 50% or more of the characters were known, and a larger set containing all taxa of each node. Although differences between the resultant phyletic trees were apparent, some similarities were also noted, and these similarities suggest that parts of the phyletic trees are robust.

Introduction

Ankylosaurs are a group of armor-plated, quadrupedal ornithischians. They were poorly known throughout most of the 19th century, despite their being one of the first dinosaurs named (*Hylaeosaurus armatus* Mantell, 1833). In his review of dinosaurs, O. C. Marsh (1896) placed the known taxa of ankylosaurs within the better-known Stegosauria on the basis of the presence of body armor. However, in the early 20th century, numerous ankylosaur specimens (including articulated skeletons with armor preserved in situ; see Carpenter 1982, 1990) led Osborn (1923) to elevate ankylosaurs to their own subordinal status, the Ankylosauria. A formal diagnosis for the order was given by Romer (1927) primarily on the basis of the structure of the pelvis, which included such characters as horizontal ilium, reduced pubis, and closed acetabulum. Nopcsa (1929) proposed Thyreophora to encompass both stegosaurs and ankylosaurs, whereas Lapparent and Lavocat (1955) called this superfamily the Stegosauroidea. The order Ankylosauria was widely accepted after Romer (1956) presented a revised and longer diagnosis. Recently, however, Cooper (1985) and Sereno (1986) have argued for uniting the Ankylosauria and Stegosauria again, plus *Scelidosaurus* and *Scutellosaurus,* in the Thyreophora, a position now widely accepted (e.g., Coombs and Maryańska 1990; Sereno 1999b).

The familial-level taxonomy of ankylosaurs, however, remained in disarray until Coombs (1978) placed all known ankylosaurs into one of two families, Nodosauridae or Ankylosauridae. This bipartite division has been universally accepted (e.g., Carpenter 1982, 1984; Maryańska 1977; Tumanova 1987). Recent discoveries, however, now cast doubt on this dichotomy (Carpenter et al. 1998; Carpenter and Kirkland 1998; Kirkland 1998). These and other ankylosaurs share features that have been considered as either nodosaurid or ankylosaurid as diagnosed by Coombs (1978). Kirkland (1998) has placed these new ankylosaurs into the subfamily Polacanthinae, and on the basis of cranial characters, he assigned this subfamily to the Ankylosauridae as the closest sister group to Tumanova's (1983) subfamily Shamosaurinae. However, new analyses indicate that the polacanthines should be elevated to family status (see below; cf. Table 21.1 with Table 21.2). The family Polacanthidae was named by Wieland (1911: 118), but it was not defined. Most later paleontologists have ignored Wieland, although Gilmore (1930: 30) suspected that the coossified sacral armor of *Polacanthus* would prove diagnostic of a distinct subfamily. This suggestion was formalized by Lapparent and Lavocat (1955: 872) as the major character of the subfamily Polacanthinae. As will be shown below, the coossified sacral armor is indeed unique to this group of ankylosaurs, but numerous other characters warrant elevating the subfamily to family level, as originally suggested by Wieland (1911).

Until now, in the field of dinosaur studies, phyletic analyses have been performed as part of analyses of new taxa (Carpenter et al. 1998; Kirkland 1998; Lee 1996; Sullivan 1999). These analyses are all pre-

TABLE 21.1.
Ankylosaur taxa accepted as valid.
Compare with Table 21.2, nonvalid taxa.

Thyreophora Nopcsa, 1915
    Eurypoda Sereno, 1986
    Ankylosauromorpha
        Scelidosaurus harrisonii Owen 1860
        Ankylosauria Osborn, 1923
        Polacanthidae Wieland 1911
            Gargoyleosaurus parkpinorum Carpenter et al., 1998
                (amended from *parkpini, International Code of Zoo-logical Nomenclature*, 1999: art. 31.1.2.A).
            Gastonia burgei Kirkland, 1998
            Hoplitosaurus marshi (Lucas, 1901)
            Hylaeosaurus armatus Mantell, 1833
            Mymoorapelta maysi Kirkland et Carpenter, 1994
            Polacanthus foxii Owen, 1865
            Polacanthus rudgwickensis Blows, 1996
        Ankylosauridae Brown, 1908
            Aletopelta coombsi Ford and Kirkland, 2001
            Ankylosaurus magniventris Brown, 1908
            Cedarpelta bilbeyhallorum Carpenter et al., 2001
            Euoplocephalus tutus (Lambe, 1902)
            Nodocephalosaurus kirtlandensis Sullivan, 1999
            Pinacosaurus grangeri Gilmore, 1933
            Pinacosaurus mephistocephalus Godefroit, Pereda Su-berbiola, and Dong, 1999.
            Saichania chulsanensis Maryańska, 1977
            Shamosaurus scutatus Tumanova, 1983
            Talarurus plicatospineus Maleev, 1952
            Tarchia gigantea (Maleev, 1956)
            Tsagantegia longicranialis Tumanova, 1993
        Nodosauridae Marsh, 1890a
            Animantarx ramaljonesi Carpenter et al., 1999
            Anoplosaurus curtonotus Seeley, 1879
            Edmontonia longiceps Sternberg, 1928
            Edmontonia rugosidens (Gilmore, 1930)
            Niobrarasaurus coleii Carpenter et al., 1995
            Nodosaurus textilis Marsh, 1889
            Panoplosaurus mirus Lambe, 1919
            Pawpawsaurus campbelli Lee, 1996
            Priconodon crassus Marsh, 1888
            Sauropelta edwardsorum Ostrom, 1970a
            Silvisaurus condrayi Eaton, 1960
            Stegopelta landerensis Williston, 1905
            Struthiosaurus austriacus Bunzel, 1871
            Struthiosaurus transylvanicus Nopcsa, 1915
            Texasetes pleurohalio Coombs, 1995
        Family *incertae sedis*
            Minmi paravertebra Molnar, 1980

TABLE 21.2.
Ankylosaur taxa not accepted as valid because the material does not allow the taxon to be diagnosed separately from other taxa (references in Carpenter et al., in press).

Nomen dubia	Material
Acanthopholis eucercus Seeley, 1869	Caudal centra
Acanthopholis horrida Huxley, 1867	Braincase, isolated postcranial elements
Acanthopholis macrocercus Seeley, 1869	Scutes
Acanthopholis platypus Seeley, 1869 in part	Phalanx, caudal centra
Acanthopholis stereocercus Seeley, 1869 in part	Caudal vertebrae, spine
Amtosaurus magnus Kuzanov et Tumanova, 1978	Partial braincase
Anodontosaurus lambei Sternberg, 1929	Skull, mandible, scutes (= *Euoplocephalus tutus*)
Anoplosaurus major Seeley, 1879 in part	Cervical vertebra
Brachypodosaurus gravis Chakravarti, 1934	Humerus
Chassternbergia rugosidens Bakker, 1988	Skull
Crataeomus lepidophorus Seeley, 1881	Vertebrae (= *Struthiosaurus austriacus*)
Crataeomus pawlowitschii Seeley, 1881	Vertebrae (= *Struthiosaurus austriacus*)
Cryptodraco eumerus (Seeley, 1869)	Femur
Danubiosaurus anceps Bunzel, 1871	Scapula, ilium (= *Struthiosaurus austriacus*)
Denversaurus schlessmani Bakker, 1988	Skull (= *Edmontonia* sp.)
Dracopelta zbyszewskii Galton, 1980	Anterior portion of skeleton
Dyoplosaurus acutosquameus Parks, 1924	Tail with armor (= *Euoplocephalus tutus*)
Dyoplosaurus giganteus Maleev, 1956	Fragmentary tail (*Tarchia gigantea*)
Edmontonia australis Ford, 2000	Cervical scutes
Glyptodontopelta mimus Ford, 2000	Pelvic osteoderms
Hieshansaurus pachycephalus Bohlin, 1953	Fragmentary skeleton
Hierosaurus coleii Mehl, 1936	Partial skeleton (= *Niobrarasaurus coleii*)
Hierosaurus sternbergi Wieland, 1909	Numerous scutes
Hoplosaurus ischyrus Seeley, 1881	Fragmentary postcrania, scutes (= *Struthiosaurus austriacus*)
Loricosaurus scutatus Huene, 1929	Scutes (titanosaurid)
Lametasaurus indicus Matleyi, 1923	Various postcrania
Leipsanosaurus noricus Nopcsa, 1918	Tooth (= *Struthiosaurus austriacus*)
Maleevus disparoserratus Tumanova, 1987	Two maxillae
Nodosaurus ischyrus (Seeley, 1881)	Fragmentary postcrania, scutes
Onychosaurus hungaricus Nopcsa, 1902	Vertebrae, scutes
Palaeoscincus africanus Broom, 1910	Snout (stegosaur)
Palaeoscincus asper Lambe, 1902	Tooth
Palaeoscincus costatus Leidy, 1856	Tooth
Palaeoscincus latus Marsh, 1892	Tooth (pachycephalosaur)
Polacanthus marshi (Lucas, 1901)	Fragmentary skeleton (= *Hoplitosaurus marshi*)
Polacanthoides ponderosus Nopsca, 1929	Scapula
Peishansaurus philemys Bohlin, 1953	Jaw fragment (pachycephalosaur)
Pinacosaurus ninghsiensis Young, 1935	Fragmentary skeleton
Pleuropeltus suessii Seeley, 1881 in part	Ilium fragment (= *Struthiosaurus austriacus*)

TABLE 21.2. *(cont.)*

Nomen dubia	Material
Priodontognathus phillipsi (Seeley, 1969)	Maxilla
Regnosaurus northamptoni Mantell, 1848	Jaw fragment (stegosaur)
Rhadinosaurus alcinus Seeley, 1881	Fragmentary bones (= *Struthiosaurus austriacus*)
Rhodanosaurus ludgunensis Nopcsa, 1929	Scutes
Sangonghesaurus Chao, 1983	Not given
Sarcolestes leedsi Lydekker, 1893	Mandible
Scolosaurus cutleri Nopcsa, 1928	Partial skeleton with armor in situ (= *Euoplocephalus tutus*)
Shanxia tianzhenensis Barrett et al., 1998	Fragmentary skull and skeleton
Stegosaurides excavatus Bohlin, 1953	Fragmentary bones
Struthiosaurus ludgunensis (Nopcsa, 1929)	Scutes
Syngonosaurus macrocercus Seeley, 1879	Scutes
Syrmosaurus viminocaudus Maleev, 1952	Partial skeleton with armor in situ
Talarurus disparoserratus Maleev, 1952	Two maxillae
Tarchia kielanae Maryańska, 1977	Partial cranium (= *Tarchia gigantea*)
Tianchisaurus nedegoapeferima Dong, 1993	Partial skeleton
Tianzhenosaurus youngi Pang and Cheng, 1998	Skull, partial skeleton (= *Saichania chulsanensis*)

liminary and do not have the kind of detail that the study of ankylosaurs warrants. To better understand the interrelationship of the ankylosaurs, an in-depth analysis has been undertaken. The results provide a hypothetical framework against which future discoveries can be assessed.

Institutional Abbreviations. AMNH: American Museum of Natural History, New York. BMNH: Natural History Museum (formerly British Museum [Natural History], London. DMNH: Denver Museum of Natural History, Denver, Colorado.

Methods

In the real world, there is only one true phyletic tree, and goal of any phyletic analysis is to closely approximate that true tree (or part of the tree). Unfortunately, the vagaries of fossilization (including taphonomic processes) result in very incomplete data matrices. The traditional methods of phyletic analysis, in which all the taxa and all the data are treated equally, typically result in hundreds of parsimonious trees, as researchers have frequently lamented. Part of the problem lies in the traditional method itself. As Vermeij (1999: 431) has noted,

> the phylogenetic method derives phylogeny in one part of the tree by relying on character taxon relationships in another part of the tree, in direct violation of the principle that clades, once they have diverged, are independent of one another. The lack of independence among clades inferred in a cladistic analysis becomes evident whenever the addition of a taxon to, or the removal of a taxon from, the matrix changes the topology of the tree.

Although techniques have been developed to reduce the number of trees (e.g., DeBry and Olmstead 2000), the results have not been very satisfactory. Vermeij (1999: 431) notes that

> Some workers might mistakenly believe that if only we had more data—more outgroup taxa, more ingroup taxa, and more characters—the trees would stabilize around the "true" phylogeny. This kind of robustness, in which the addition or subtraction of taxa no longer has a destabilizing effect on the topology of the tree, means only that the interdependencies established in the matrix have become so numerous and so widespread that inference becomes insensitive to additional empirical data.

A possible solution to the problem of divergent clades affecting each other "unnaturally" in a phylogenetic analysis was called by Mishler (1994) compartmentalization. This technique reduces large data sets into ones of smaller, more manageable sizes to eliminate the spurious homoplasy problem articulated by Vermeij (1999).

Although Mishler did not state it, a compartment may be defined by a node—for example, "*Ankylosaurus* and all ankylosaurs closer to it than to *Edmontonia.*" *Ankylosaurus* and *Edmontonia* serve as reference taxa to determine which taxa are closer to *Ankylosaurus* than to *Edmontonia* (Sereno 1999a). What those "ankylosaurs" are can be determined by selecting some apomorphies of *Ankylosaurus* not found in *Edmontonia* but shared by the ingroup; these apomorphies are the synapomorphies for the node. The node is essentially equivalent to the family in Linnaean taxonomy, and the family name (Ankylosauridae) can be used as the name of the node (e.g., de Queiroz and Gauthier 1992; Sereno 1999a). For convenience of this example, only the tail club of *Ankylosaurus* is used as the apomorphy to identify *Edmontonia* as the outgroup because it does not have one (in actuality, multiple characters would be used to define the node). Sorting through all the taxa of ankylosaurs, *Euoplocephalus, Pinacosaurus, Saichania,* and *Tarchia* are known to have a tail club and form the ingroup to *Ankylosaurus*. For this node, the Ankylosauridae, the synapomorphy shared by all of them, is the tail club (the tail club is not yet known for several poorly known primitive ankylosaurids, e.g., *Shamosaurus* and *Cedarpelta*).

The process is repeated until all of the nodes (e.g., Ankylosauridae and Nodosauridae) of the better known taxa have been identified. These nodes are fairly stable, although slight refinement will occur as new specimens and taxa are discovered. These nodes are relatively stable because the inclusive taxa are limited to those that are well represented by fossil material and because the synapomorphies are not liable to change in any major way (e.g., the tail club with always be a synapomorphy for the Ankylosauridae). Because they are stable, the nodes also serve as reference nodes for other, poorly known taxa. Such taxa will be difficult to assign to a node when many of the synapomorphies, which might place it within one or another family, are missing. The chance of finding a shared synapomorphy with the node is increased by use of multiple synapomorphies. Even so, these less complete taxa will still add many unknowns (i.e., question marks) to the data matrix. Because increasing the number of taxa with missing data will increase the number of parsimonious trees, as pointed out by Vermeij (1999), the data poor taxon is removed before the next is added so that the missing data in the matrix do not influence each other. Once all of the taxa have been processed, those that can be assigned to a node are added to that node's taxa list. Remnant taxa that cannot be assigned to any node with confidence are also those that potentially cause the greatest number of phylogenetic trees (Vermeij 1999); these are excluded from further analysis because they provide little phyletic information.

The addition of new taxa to the original set of reference taxa requires modification of the original synapomorphies. The criteria used to separate autapomorphies of the genera from the synapomorphies of the node are their occurrence in multiple taxa. The amended synapomorphies of the nodes (e.g., the tail club) are analyzed by use of an outgroup and exclude the autapomorphies of the genera. Because the data matrix will be smaller than it would be if all taxa were treated as equal (i.e., apomorphic analysis of all taxa), it will be easier to find the most parsimonious tree for the nodes.

Each node is also composed of different taxa, and their phyletic relationship to one another can be determined. Separate data matrices are constructed of the characters for the taxa in each node, exclusive of the synapomorphies of the node's diagnosis. For example, the synapomorphy for the node Ankylosauridae is the tail club. But for each genus in this node, the tail club is symplesiomorphic and therefore would not be used in the data matrix of the genera. In traditional apomorphic analyses, these dual synapomorphic–plesiomorphic characters would be used in a data matrix, and this redundancy may be partially responsible for the creation of multiple parsimonious phylogenetic trees. The data matrix of the genera is then analyzed and a mini-tree is generated. It is easier to identify the most parsimonious tree from the smaller data set Ankylosauridae than from the larger data set Ankylosauria. Mini-trees are created for nodes containing two or more genera (because a tree cannot be made from a single genus). Adding these mini-trees to the larger tree of the nodes results in an overall phyletic tree for the ankylosaurs (e.g., Figs. 21.6–21.10).

The advantages of the compartmental method are several. First, the discovery of a new ankylosaurid species may change the relationship of the other ankylosaurids in the mini-tree, but its discovery should have no bearing on, or be affected by, the characters of the nodosaurid *Edmontonia* or the node Nodosauridae. Second, compartmentalization avoids comparing apples and oranges, which can happen by comparing synapomorphies of a node with the apomorphies or autapomorphies of the genera in a multitaxon node. Third, added uncertainties due to missing data will be compounded if the phyletic placement of, for example, the ankylosaurid *Talarurus* is influenced or affected by the nodosaurids *Edmontonia* and *Silvisaurus* in a traditional data matrix. Why? because *Edmontonia* and *Silvisaurus* are only distantly related to the ankylosaurids, so why assume otherwise by branch swapping, etc.

TABLE 21.3.
Data matrices for the ankylosaurs (see Tables 21.1 and 21.2).
Characters are given in Appendix 21.1. Taxa indicated with asterisks
have more than 50% missing characters.

I. Ankylosauromorpha

Lesothosaurus
00000 00000 00000 00000 00000 00000 00000 0000

Scutellosaurus
00000 00000 00000 00000 00000 00000 00010 0000

*Emausaurus**
10000 00000 00000 ?00?? ????? ????? ????2? ?0??

Huayangosaurus
00010 00000 00000 00000 00000 01000 00020 0000

Scelidosaurus
00010 00000 00011 00010 00000 01110 00011 0100

Minmi
111?? 111?? 1?011 012?1 ??2?0 01110 11?21 000?

Polacanthidae
111?1 11012 12012 11211 01101 01111 12011 1110

Nodosauridae
100?0 11001 11112 22111 01210 01111 12022 2200

Ankylosauridae
11110 11112 22012 11211 12001 11111 11111 3201

II. Polacanthidae

Scelidosaurus	00000	00000
Gargoyleosaurus	01001	?0??0
Gastonia	12110	21000
*Hoplitosaurus**	?????	10??1
*Hyaleosaurus**	???01	3???0
*Mymoorapelta**	????1	??110
Polacanthus	???0?	20001

III. Ankylosauridae

Scelidosaurus	00000	00000	00000	000
Ankylosaurus	11010	10000	10100	?10
Cedarpelta	00001	00000	00???	?0?
Euoplocephalus	11211	00100	11111	111
Nodocephalosaurus	?11?1	10?10	100??	???
Pinacosaurus	11111	11100	10110	111
Saichania	11210	01001	12111	210
Shamosaurus	01000	01001	10000	210
*Talarurus**	???10	?0?00	10???	?21
Tarchia	11110	00010	12110	2??
Tsagantegia	11111	01001	11???	???

TABLE 21.3. *(cont.)*

IV. Nodosauridae

Scelidosaurus	00000	00000	00000	00
Animantarx	0010?	?1101	??111	1?
*Anoplosaurus**	?????	?0??0	1?1?0	??
Edmontonia	00001	11101	11001	12
*Niobrarasaurus**	????1	1???1	?1??1	??
*Nodosaurus**	?????	?????	?????	??
Panoplosaurus	01101	11010	0011?	00
*Pawpawsaurus**	11100	0????	?????	??
Sauropelta	1100?	10010	01101	11
Silvisaurus	11100	01011	01???	?1
*Stegopelta**	?????	????0	????0	12
Struthiosaurus	0011?	?0000	0???1	01
*Texasetes**	?????	????1	0?1?1	??

(which could be viewed as claiming shared genes)? Finally, a low consistency index for the mini-tree of a particular node will not drag down the consistency index for the mini-trees of other nodes. The advantage for this, of course, is that weakly supported mini-trees can more readily be identified. For example, if the consistency index for the mini-trees at the node Nodosauridae is significantly lower than for the mini-trees at Ankylosauridae and Polacanthidae, then a more concerted effort can be directed at improving the data for the Nodosauridae, even if it means something as minor as removing one taxon from the analysis.

Analysis of the Ankylosauromorpha

Traditionally, *Scelidosaurus* has been treated as the outgroup for the Stegosauria and Ankylosauria (e.g., Norman 1984; Sereno 1986, 1997, 1998, 1999b). Such practice has been based in part on historical treatment of *Scelidosaurus* as a primitive ornithischian (e.g., Thulborn 1977) and in part on its antiquity and its Early Jurassic age (the confusing history of the type of *Scelidosaurus harrisonii* is given by Newman 1968). But the type of *Scelidosaurus* has been acid prepared so that much of the skeleton can be studied in detail (Norman 1996). In addition, new specimens of *Scelidosaurus* have been found in recent years that supplement the type, especially in regard to the skull. Examination of the type, as well as published photographs of the new skull, show that *Scelidosaurus* has apomorphies that have traditionally been considered diagnostic of the Ankylosauria (Table 21.3; Appendix 21.1).

One of these synapomorphies of *Scelidosaurus* occurs on the skull. The skull bones of most nonthyreophoran ornithischians, as well as stegosaurian thyreophorans, are smooth, although older individuals of *Stegosaurus* may show faint rugosities on the parietals and pits on the supraorbitals (Carpenter, unpublished data). The skulls of ankylosaurs, however, have been characterized by an extensive covering of dermal armor, except in *Minmi* and juvenile *Pinacosaurus*. The origin for the armor has been controversial. Coombs (1971) questioned the traditional account of coossification of armor to the surface of the skull, suggesting instead that the "armor" was a modification of the skull bones' surface. Coombs (1978) later abandoned this hypothesis, although not completely (Coombs and Maryańska 1990). There is now supporting evidence for this hypothesis for at least two ankylosaurs, *Cedarpelta* (see Carpenter et al. 2001; but see also Vickaryous et al. 1998, 2001) and *Pinacosaurus*. The paratype skull of *Cedarpelta* is mostly disarticulated, yet the surface of many elements has the rugose texture that has been identified as coossified dermal armor in other ankylosaurs. This cranial ornamentation in *Cedarpelta* is the result of extensive remodeling of the cortical surface by the overlying periosteum. A similar condition exists in *Pinacosaurus* on the basis of comparing the juvenile skull described by Maryańska (1971, 1977) and a slightly older juvenile specimen described by Godefroit et al. (1999). The older specimen, with open cranial sutures, shows extensive remodeling of the individual cranial bones. Godefroit et al. (1999) assumed the cranial sculpturing was coossification of thin dermal scutes, but considering how well developed this sculpting is and how well coossified it appears to be, it now seems more likely that it is remodeling of the skull bone surfaces.

Cortical remodeling resulting in ornamentation is known on the jugal, maxilla, postorbital, and posterolateral corner of the mandible of *Scelidosaurus* (Fig. 21.1; see Barrett 2001; Lambert 1993: 112; Norman 1988: 77). The ornamented jugal and postorbital regions correspond to areas with well-developed periosteal derived armor in *Pinacosaurus* and *Minmi* (Vickaryous et al. 2001). Thus, whereas some of the cranial armor in ankylosaurs may be dermal in origin (Vickaryous et al. 2001), remodeling of individual cranial bones to produce surface ornamentation is a synapomorphy of the ankylosaurs.

Other ankylosaur synapomorphies in *Scelidosaurus* occur in the pelvis (Fig. 21.2). First is the horizontal expansion of the entire ilium. Although other ornithischians have independently acquired wholly or partially horizontal ilium (e.g., Neoceratopsia, Stegosauria, Pachycephalosauria, and the ornithopod *Planicoxa*; DiCroce and Carpenter 2001), the details differ. The precursor for the horizontal ilium may be the development of the brevis shelf along the medial side of the posterior iliac blade (Fig. 21.2A). The shelf is probably for the M. caudi–femoralis brevis (Galton 1969), although there is reason to believe that it was for the two heads of the M. ilio-fibularis in *Planicoxa* (DiCroce and Carpenter 2001). In stegosaurs (including *Huayangosaurus*), the supra-acetabular and postacetabular portions of the ilium are expanded horizontally, whereas the preacetabular portion remains almost vertical (Fig. 21.2B). Furthermore, an antitrochanter is developed along the lateral edge of the postacetabular portion. In older individuals, the dorsal surface of the sacral ribs expand posteriorly as a sheet of bone that coalesces over the intersacral rib fenestra (Carpenter, unpublished data). In ankylosaurs, the entire length of the ilium is expanded horizontally, with the preacetabular portion divergent. Furthermore, the postacetabular portion is greatly reduced. The intersacral rib fenestra

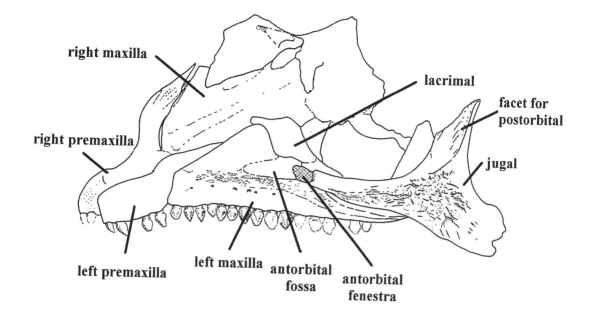

right maxilla

lacrimal

facet for postorbital

right premaxilla

jugal

left premaxilla   left maxilla   antorbital fossa   antorbital fenestra

retain the plesiomorphic open condition. The ilium of *Scelidosaurus* in lateral view resembles the primitive ornithischian condition (Fig. 21.2C) and is a view frequently emphasized to place *Scelidosaurus* as the sister group to Stegosauria + Ankylosauria (e.g., Norman 1984; Sereno 1986, 1997, 1998; Thulborn 1977). However, a dorsal view (Fig. 21.2C) shows that the ilium is in fact horizontally expanded and slightly divergent. The ilium, then, shows a mixture of primitive and derived characters, as would be expected for a basal ankylosaur.

The second ankylosaur synapomorphy in the pelvis is the massive body of the pubis (Fig. 21.3). The primitive condition, seen in *Lesothosaurus*, is for the body of the pubis to be narrow, with a long, slender postpubic process and a short ischiadic peduncle extending posteriorly to partially enclose an obturator foramen; the prepubic process is absent (Fig. 21.3A). Although both *Scelidosaurus* and ankylosaurs retain the overall primitive pubis having a long postpubic rod and no prepubic process, the body of the pubis and ischiadic peduncle are expanded into a large mass, or block, that is rotated dorsolaterally so as to partially close the acetabulum (Fig. 21.3B–D). In rotating, the pubis partially or completely hides the obturator foramen in lateral view. In stegosaurs, the pubic body remains narrow, although a thin sheet projects dorsoposteriorly to partially close the acetabulum (Fig. 21.3E). Because the pubis does not rotate, the obturator foramen retains its primitive position.

The body armor of *Scelidosaurus* is another synapomorphy shared with ankylosaurs. In stegosaurs, the armor consists mostly of erect plates and spines arranged in two parasagittal, medially placed rows. In addition, posterolaterally projecting spikes occur at the end of the tail; parascapular spikes occur on *Huayangosaurus, Tuojiangosaurus,* and

Figure 21.1. Skull of Scelidosaurus *showing the remodeled cortical surface of the jugal and maxilla. Similar remodeling also occurs on other cranial bones and is also seen on several ankylosaurid specimens (see text). Redrawn from photograph in Lambert (1993).*

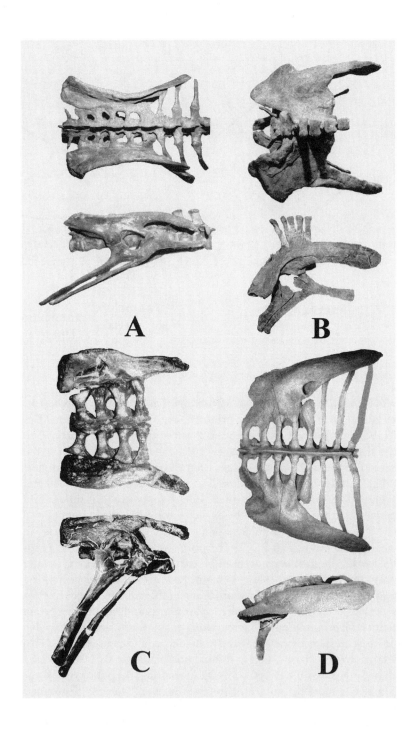

Figure 21.2. Dorsal and lateral comparative views of the pelvises. (A) General primitive pattern as reflected by Thescelosaurus *(AMNH 117). (B)* Stegosaurus *(DMNH 1483). (C)* Scelidosaurus *(BMNH R1111). (D)* Euoplocephalus *(AMNH 5409). Although* Thescelosaurus *shows the overall primitive ornithischian pelvic plan, it does show widening of the pelvis by development of the brevis shelf. This widening of the postacetabular process of the ilium occurs independently several times in ornithischian evolution—for example, stegosaurs (B) and ankylosaurs (C, D).*

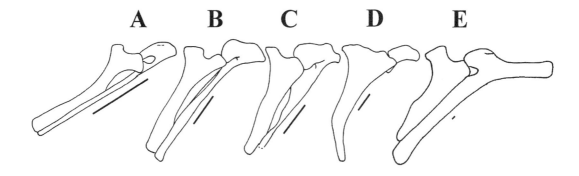

A   B   C   D   E

Kentrosaurus (Sereno and Dong 1992), and a parasagittal row of lateral body scutes occurs in *Huayangosaurus*. These plates and spikes, as well as the scutes of *Huayangosaurus*, are solid, as are the plates and scutes of the primitive thyreophoran *Emausaurus* (Haubold 1990). The body armor of ankylosaurs and *Scelidosaurus*, in contrast, are composed of multiple transverse or parasagittal rows of keeled scutes and plates. In nodosaurids, these scutes are mostly solid and may be supplemented with solid spikes projecting from the neck and shoulder regions (Carpenter 1984, 1990). Ankylosaurids, on the other hand, have thin-walled, conical scutes (Carpenter 1982; Coombs 1978). *Scelidosaurus* also has a large, dorsoventrally compressed, asymmetrically hollow-based, triangular lateral body and caudal plates that are indistinguishable from those of the polacanthids *Gargoyleosaurus* and *Gastonia* (Fig. 21.4), as well as from *Mymoorapelta* and *Polacanthus*. These scutes in no way resemble those of any stegosaur, including *Huayangosaurus*, or the primitive thyreophoran *Emausaurus*.

Finally, *Scelidosaurus* has at least two rows of cervical armor (Fig. 21.5). The first row consists of a cluster of three spiked scutes (Fig. 21.5A, B). The second row consists of a band of two spiked scutes per side (Fig. 21.5C), with a possible laterally projecting scute below this band (Fig. 21.5D). The presence of cervical armor arranged in bands is a synapomorphy of ankylosaurs (Sereno 1997, 1999b).

On the basis of the above synapomorphies, *Scelidosaurus* is elevated from being the sister taxon of Stegosauria + Ankylosauria to being the closest sister taxon of the Ankylosauria (i.e., *Scelidosaurus* + Ankylosauria). Nevertheless, *Scelidosaurus* lacks many of the other synapomorphies of the Ankylosauria, such as a skull wider than it is tall (see below). To accommodate *Scelidosaurus*, a new taxon, Ankylosauromorpha, is proposed. This node is defined in the caption of Figure 21.6, along with those of other nodes.

Before an analysis of the ankylosaurs was possible, a review of the ankylosaurs was conducted to identify valid taxa (Tables 21.1, 21.2). At least part of this review resulted in similarities with Coombs and Maryańska (1990), although notable exceptions occurred (e.g., Polacanthidae). Characters were identified in part from the literature, but mostly from specimens (including casts). These characters are identified

Figure 21.3. Lateral view of the pubis and ischium: the primitive pattern reflected by Lesothosaurus *(A), where the pubic body is a narrow, vertical blade. The primitive ankylosaur pattern is seen in* Scelidosaurus *(B) and* Minmi *(C), where the main pubic body is an expanded mass, laterally rotated to form a part of the medial wall of the acetabulum. The primitive condition also occurs in polacanthids (see text). The advanced ankylosaur pubis is seen in* Edmontonia *(D), where the pubis is a narrow, vertical blade and the postpubic process a short nub. In* Stegosaurus *(E), the pubis is a narrow, vertical blade pierced by the obturator foramen. Ankylosaur evolution is marked by a reduction of the pubis and the expansion of the ischium; the reverse occurs in stegosaurs. See text for discussion.*

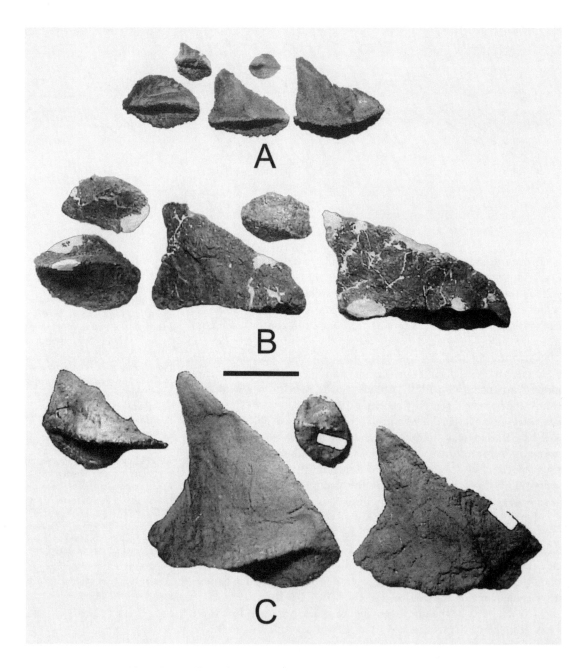

Figure 21.4. Oval keeled scutes from the body and large, hollow-based, dorsoventrally compressed plates from the lateral sides of the body in (A) Scelidosaurus, (B) Gargoyleosaurus, and (C) Gastonia. Bar = approximately 10 cm.

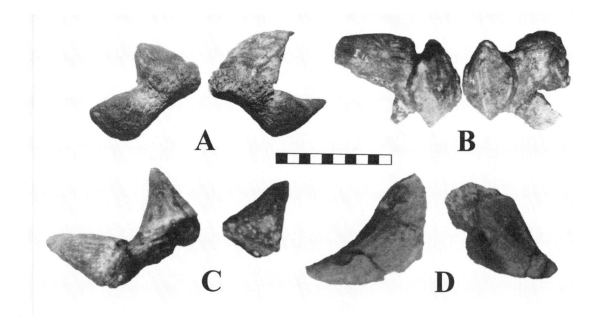

in Appendix 21.1 and the data matrices presented in Table 21.3. Analysis was conducted with PAUP version 4 beta (Swofford 2000) under accelerated transformation (ACCTRAN), with the characters unordered.

Results

For analysis of the Ankylosauromorpha, *Lesothosaurus* was used as the outgroup. Two data sets were analyzed, one with the enigmatic ankylosaur *Minmi* and the other without (Fig. 21.6). The results showed minor differences in the branches to *Scutellosaurus* and *Emausaurus*. Without *Minmi*, there are 11 parsimony uninformative characters, and the consistence index (CI) = 0.89, retentive index (RI) = 0.89, and homoplasy index (HI) = 0.11. With *Minmi*, there are nine parsimony uninformative characters, with CI = 0.83, RI = 0.85, and HI = 0.17. Although the scoring is slightly reduced, it is nevertheless high. Regardless, *Scelidosaurus* is the sister taxon to the Ankylosauria (*Minmi* + Nodosauridae + Polacanthidae + Ankylosauridae) at the node Ankylosauromorpha. The polacanthids are the closest sister taxon to the ankylosaurids, as previously implied by Carpenter et al. (1998) and as stated by Kirkland (1998); these results are contrary to those of Wilkinson et al. (1998).

Analysis of the taxa comprising the Ankylosauria used *Scelidosaurus* as the outgroup. Two data sets of the Nodosauridae were analyzed, the first of taxa having at least 50% of the characters (*n* = 6), and the other of all taxa (*n* = 12). The Bootstrap command with the Branch and Bound option resulted in a polytomous tree for the first set and was therefore discarded; the second set was terminated after six hours. The

Figure 21.5. Cervical armor of Scelidosaurus. The first cervical ring is composed of three coossified scutes in (A) posterior and (B) dorsal views. (C) Second armor ring in anterior view. (D) Lateral cervical armor in dorsal view. Scale in centimeters.

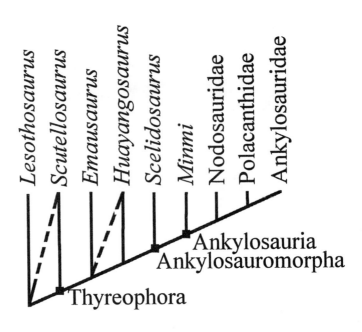

Lesothosaurus
Scutellosaurus
Emausaurus
Huayangosaurus
Scelidosaurus
Minmi
Nodosauridae
Polacanthidae
Ankylosauridae

Ankylosauria
Ankylosauromorpha
Thyreophora

data were therefore analyzed with the heuristic command (Hsearch), which resulted in six trees for the first set. A consensus of these trees produced the same tree, as did the command Hsearch with the nchuck = 1 chuckscore = 1 options (Fig. 21.7A) (technique recommended by PAUP support). Excluding three parsimony uninformative characters, CI = 0.57, RI = 0.45 and HI = 0.42. These results are not very good, but they can serve as a starting point for future analysis. The second data set resulted in 250 heuristic trees, but this was reduced to 1 by the nchuck–chuckscore option as before (Fig. 21.7B). When excluding one parsimony uninformative character, CI = 0.56, RI = 0.58, and HI = 0.43.

Again, these results are not very good. With the smaller data set, *Struthiosaurus* is the most primitive nodosaurid, but in the larger data set, it is much more derived (cf. Fig. 21.7A, B); the reverse occurs to *Panoplosaurus*. In both analyses, *Sauropelta* and *Silvisaurus* are the most derived, and *Animantarx* is always the closest sister taxon to *Edmontonia*; both occur in about the same relative position within the phyletic trees (in two of the six trees generated by the heuristic command of the first data set, *Animantarx* and *Edmontonia* occupy the terminal position).

The smaller data set was also analyzed with *Minmi* because the initial analysis identified it as the closest sister group to the Nodosauridae, implying the possibility that *Minmi* was in fact a nodosaurid. However, analysis of data by a heuristic search resulted in *Minmi* falling outside the nodosaurid genera, rather than among them. For that reason, the possibility that *Minmi* is a nodosaurid is rejected.

Analysis of the Polacanthidae (*n* = 6) used the Bootstrap with the

Figure 21.6. (opposite page) Phyletic tree for the Thyreophora and the ankylosauromorphs, using Lesothosaurus *as the outgroup. The nodes are defined as follows.*

(1) Ankylosauromorpha *are thyreophorans that are closer to* Scelidosaurus, Minmi, Polacanthidae, Nodosauridae, *and* Ankylosauridae, *than to* Stegosaurus. *(1a) Diagnosis: cranial ornamentation surface sculpturing of some or most external cranial bones; horizontally developed ilium (preacetabular and postacetabular portions); expanded pubic body that is rotated into the acetabulum, hiding the obturator foramen from lateral view; extensive neck, body, and tail scutes, not erect plates; at least two cervical rings; large, dorsoventrally compressed, asymmetrically hollow-based, triangular lateral body plates.*

(1b) Reference taxa: Scelidosaurus, Minmi, Polacanthidae, Nodosauridae, Ankylosauridae.

(2) Ankylosauria *are ankylosauromorphs that are closer to* Minmi, Gastonia, Edmontonia, *and* Euoplocephalus *than to* Scelidosaurus. *(2a) Diagnosis: skull width greater than height; antorbital and supratemporal fenestra closed; maxillary tooth row lingually inset forming maxillary shelf; postocular wall developed; interpterygoid fenestra closed; nasal septum divides rostrum parasagittally; mandible with ventrolateral ornamentation; coronoid process of mandible proportionally lower than in many ornithopods; synsacrum formed of caudals, dorsals and sacrals; posterior chevrons anteroposteriorly elongated and contacting one another; scapula dorsoventrally deep, with acromion; acetabulum closed; pubis almost completely excluded from acetabulum. (2b) Reference taxa:* Polacanthidae, Nodosauridae, Ankylosauridae.

(3) Polacanthidae *are ankylosaurs that are closer to* Gastonia *than to* Edmontonia *and* Euoplocephalus. *(3a) Diagnosis: skull almost as wide as long; postorbital and jugal horns prominent; inverted U-shaped notch at front of premaxillaries; maxillary shelf extending onto jugal; well-developed postocular wall formed by postorbital, jugal, and possibly the frontal; quadrate straight and steeply sloped anteroventrally; pterygoids separated posteriorly; occipital condyles crescent-shaped and set on very short neck; coronoid process very low and rounded; scapula with well-developed acromion flange originating from the dorsal margin; distal end of scapula flared; deltopectoral crest extending to midlength of humerus; ischium bent anteriorly at midshaft; sacral armor coossified; lateral shoulder spines triangular and grooved along posterior margin. (3b) Reference taxa:* Gargoyleosaurus, Gastonia.

(4) Nodosauridae *are ankylosaurs that are more closely related to* Edmontonia *than to* Gastonia *or* Euoplocephalus. *(4a) Diagnosis: hourglass-shaped palate; cutting edge of premaxillary connected to maxillary tooth row; maxillary shelf restricted to maxilla; postocular shelf narrower than in Ankylosauridae; basipterygoid process rounded and rugose; occipital condyle hemispherical, formed only by basioccipital and set on short neck, angled about 50° posteroventrally from plane of maxillary tooth row; jugal and postorbital horns small if present, usually rounded; single ornamentation between external nares; cranial ornamentation usually well defined, including large frontoparietal ornamentation and anteroposteriorly narrow nuchal ornamentation; coronoid process on mandible relatively tall; acromion formed by a ridge ending in a knob dorsal to the glenoid; deltopectoral crest extend to less than midlength of humerus; coracoid large, anteroposteriorly long relative to dorsoventral width; ischium bent anteriorly near midlength; cervical and shoulder armor consisting of projecting spines, three rows of cervical armor; body scutes thick walled or solid. (4b) Reference taxa:* Edmontonia, Panoplosaurus, Sauropelta.

(5) Ankylosauridae *are ankylosaurs that are more closely related to* Euoplocephalus *than to* Gastonia *or* Edmontonia. *(5a) Diagnosis: maximum width of skull equal to or greater than length; snout arches above level of skull roof; maxillary shelf extending onto jugal; well-developed postocular wall composed of postorbital, jugal, and frontal; lateral temporal and paroccipital process not visible in lateral view; usually prominent jugal and squamosal "horn"; supraorbitals slightly domed; coronoid process on mandible low and rounded; interlocking distal caudal prezygapophyses and postzygapophyses; posterior chevrons elongate and overlapping; terminal tail club; deltopectoral crest extend to midlength of humerus; deltopectoral crest and transverse axis through distal epicondyles in same plane; postacetabular process of ilium short; ischium straight and near vertical below acetabulum; fourth trochanter on distal half of femur; two rows of cervical armor; loss of large, dorsoventrally compressed, asymmetrically hollow-based, triangular lateral plates.*

(5b) Reference taxa: Euoplocephalus, Saichania, Pinacosaurus.

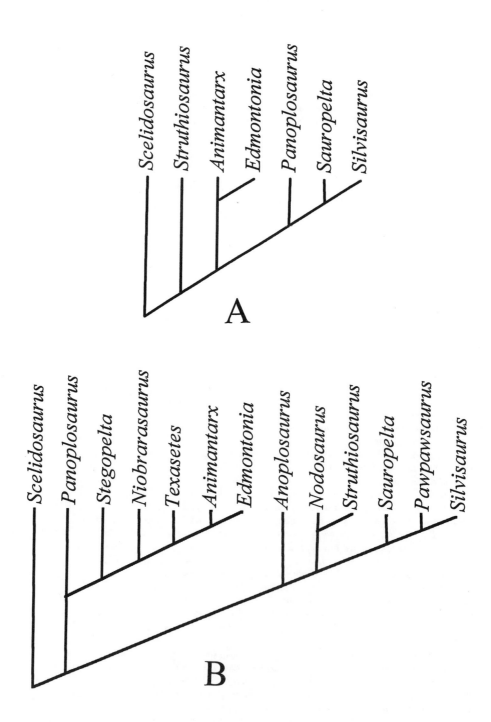

Figure 21.7. Phyletic trees for the Nodosauridae using Scelidosaurus *as the outgroup. (A) Taxa having 50% or more characters. (B) All taxa, including those with less than 50% characters known.* Animantarx *and* Edmontonia *remain closely associated in all analyses.*

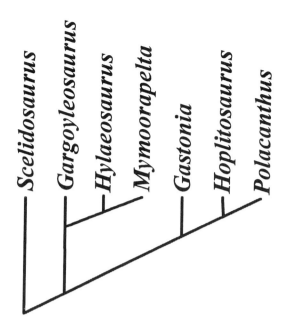

Branch and Bound option, but this resulted in a polytomous tree (50% majority), which was rejected. The Hsearch command with the nchuck–chuckscore option gave a single tree (Fig. 21.8). The same tree was created with the heuristic search command followed by the neighbor-joining method. Deletion of *Polacanthus* (<50% characters known) resulted in the same tree, excluding *Polacanthus*. Scoring of the tree, excluding eight parsimony uninformative characters, is CI = 1.0, RI = 1.0, and HI = 0. Although the results appear very good, only two characters were parsimony informative. The family Polacanthidae is clearly valid, so I conclude that the low number of parsimony informative characters indicates that the taxa are conservative.

Finally, analysis of the Ankylosauridae also used two data sets: one in which taxa with 50% unknown characters were excluded (*n* = 7) and the other of all taxa (*n* = 10). The heuristic search command resulted in a single tree for each data set (Fig. 21.9A, B). For the smaller data set, only one character was parsimony uninformative, resulting in CI = 0.71, RI = 0.65, and HI = 0.29. For the all inclusive data set, all characters were informative, with CI = 0.66, RI = 0.67, and RI = 0.67. The addition of *Nodocephalosaurus* to the smaller data set caused all ankylosaurs from *Ankylosaurus* to *Pinacosaurus* to collapse into an unresolved polytomy. The score for this tree was CI = 0.54, RI = 0.35, and HI = 0.46. Therefore, the clade of *Saichania–Tarchia–Nodocephalosaurus*, as advocated by Sullivan (1999), is not supported. The trees produced by the two data sets share a considerable amount of similarities: note relative positions of *Cedarpelta, Shamosaurus, Ankylosaurus, Euoplocephalus* and *Pinacosaurus*; *Saichania* and *Tarchia* al-

Figure 21.8. Phyletic tree for the Polacanthidae. The clade formed by Gargoyleosaurus, Hylaeosaurus, *and* Mymoorapelta *is weakly supported and is not named. Additional analysis may collapse this clade.*

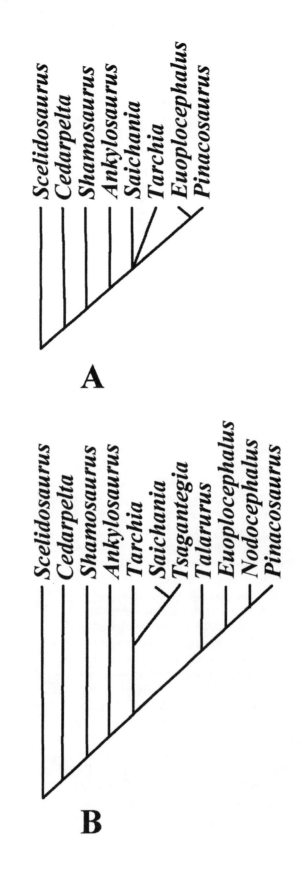

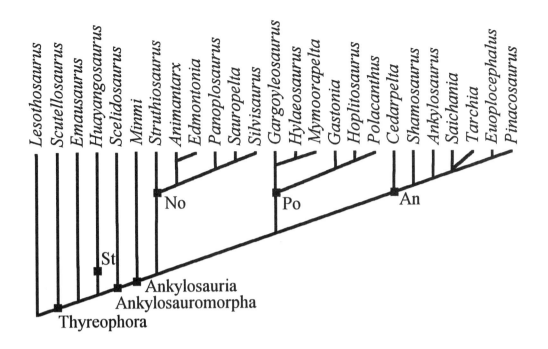

Lesothosaurus
Scutellosaurus
Emausaurus
Huayangosaurus
Scelidosaurus
Minmi
Struthiosaurus
Animantarx
Edmontonia
Panoplosaurus
Sauropelta
Silvisaurus
Gargoyleosaurus
Hylaeosaurus
Mymoorapelta
Gastonia
Hoplitosaurus
Polacanthus
Cedarpelta
Shamosaurus
Ankylosaurus
Saichania
Tarchia
Euoplocephalus
Pinacosaurus

No

Po

An

St

Ankylosauria
Ankylosauromorpha
Thyreophora

so remain close phyletically. The subfamily Shamosaurinae of Tumanova (1983) is not supported at this time.

Discussion

The results of the phyletic analyses can be summarized in one tree (Fig. 21.10), where the individual mini-trees (Figs. 21.6–21.9) are combined for those taxa for which more than 50% of the characters are known (assumed to be the best trees). This summary tree presents a hypothesis about the phyletic relationships of the ankylosauromorphs. The data can then be used to show the temporal ranges of the Thyreophora and ankylosauromorphs (Fig. 21.11).

Scelidosaurus is the closest sister taxon to the Ankylosauria on the basis of numerous shared synapomorphies (Table 21.3; Appendix 21.1). The implication for this placement is that the thin-walled scutes of the Polacanthidae and Ankylosauridae are a plesiomorphic character for the Ankylosauria and the solid scutes and spines of the Nodosauridae an autapomorphy. Furthermore, the loss of the premaxillary teeth occurred independently in each family.

Minmi is an enigmatic ankylosaur that is more derived than *Scelidosaurus* (see caption Fig. 21.6) and is the closest sister group to the Nodosauridae. It represents a primitive ankylosaur or an ankylosaur that has secondarily lost numerous more derived ankylosaur characters. At the present, it is not possible to ascertain the more likely of the two.

Figure 21.9. (opposite page) Phyletic trees for the Ankylosauridae. (A) Tree of taxa with less than 50% characters removed. (B) Tree of all taxa. Note the high similarity in the relative position of taxa of A in B. These similarities suggest phyletically stable portions.

Figure 21.10. (above) Composite tree formed from Figures 21.6–21.9 for taxa with 50% or more characters. Nodes:
St = Stegosauria;
No = Nodosauridae;
Po = Polacanthidae;
An = Ankylosauria.

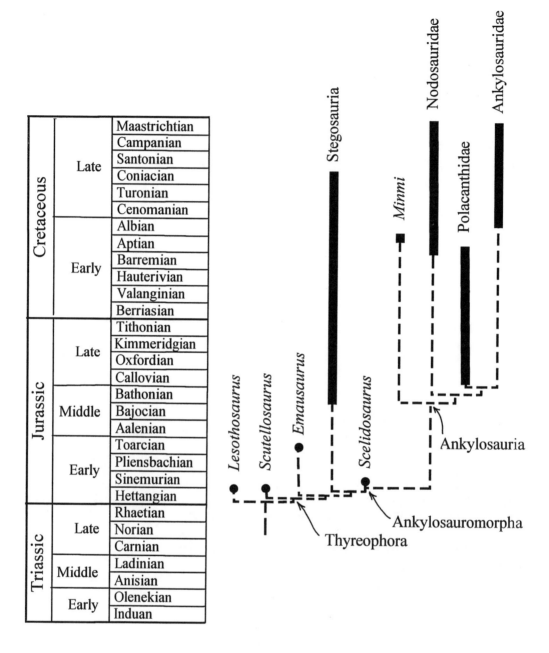

Figure 21.11. Temporally calibrated phylogeny of the Thyreophora, with emphasis on the Ankylosauromorpha. Known temporal durations shown by solid bar. No attempt has been made to determine missing ranges (ghost lineages); however, it is clear that the Stegosauria had a much earlier origin than previously thought (see Sereno 1999b: fig. 1).

The Nodosauridae comprise taxa known from the Early and Late Cretaceous, but its phyletic position indicates an origin during the Jurassic, along with the Polacanthidae and Ankylosauridae.

The Polacanthidae appear to be restricted to the Late Jurassic and Early Cretaceous and comprise the oldest, but not most primitive, ankylosaurs (*sensu stricto*) to date. They apparently became extinct at about the time that the land bridge opened between Asia and North America (see Kirkland 1996).

The Ankylosauridae range from the latest Albian to the end of the Maastrichtian, although the phyletic analysis (Fig. 21.11) suggests an even longer record. They comprise the most derived ankylosaurs, although they retain the plesiomorphic thin-walled scutes found in *Scelidosaurus* and polacanthids.

Summary

The compartmentalization technique used in the phyletic analysis of the ankylosaurs consists of four steps. First, reference nodes are identified by use of those taxa that are best represented by fossil material (reference taxa). The diagnoses for the nodes are the synapomorphies that unite the taxa, although nodes may also contain a single taxon. Second, the synapomorphies for each node form a reference data matrix, which, when analyzed, generates a reference mini–phyletic tree. Third, the remaining taxa with a significantly higher number of missing characters (about >50%) are added to the data matrix and analyzed. The resultant mini-tree is not as reliable because of the problems associated with missing data. Nevertheless, similarities between the trees may indicate more phyletically robust portions of the trees. Fourth, the mini–phyletic trees replace their respective nodes to make an overall phyletic tree.

The compartmentalization technique makes assumptions about monophyly by the definition of each node. In addition, at least part of the overall phyletic pattern is already established by these nodes (e.g., taxa of the Nodosauridae are separate from those of the Ankylosauridae and will cluster together in the phyletic mini-tree).

The results of the study demonstrate that *Scelidosaurus* shares several key synapomorphies with "traditional" ankylosaurs + *Minmi*, requiring that it be united with them at the node Ankylosauromorpha. *Minmi* is an ankylosaur united with traditional ankylosaurs at the node Ankylosauria, but it is not a member of a known family. Three families are recognized on the basis of numerous autapomorphies: Nodosauridae, Polacanthidae, and Ankylosauridae.

Acknowledgments. I extend thanks to the many people involved in the Vertpaleo shared list Internet discussion about Vermeij's (1999) provocative article and the resultant discussions about phyletic compartmentalization. The exchanges were thought provoking, and although not everyone may agree with the methodology, I offer the results as both a hypothesis for the method and for the resultant phylogeny of the ankylosaurs. I also thank Jim Kirkland for sharing the preliminary results of his ongoing analyses, but especially for sharing his frustrations in trying to do the analyses the traditional way. Critical comments by Jim Kirkland on an earlier draft are appreciated.

References Cited

Carpenter, K. 1982. Skeletal and dermal armor reconstruction of *Euoplocephalus tutus* (Ornithischia: Ankylosauridae) from the Late Cretaceous Oldman Formation of Alberta. *Canadian Journal of Earth Sciences* 19: 689–697.

———. 1984. Skeletal reconstruction and life restoration of *Sauropelta* (Ankylosauria: Nodosauridae) from the Cretaceous of North America. *Canadian Journal of Earth Sciences* 21: 1491–1498.

———. 1990. Ankylosaur systematics: Example using *Panoplosaurus* and *Edmontonia* (Ankylosauria, Nodosauridae). In K. Carpenter and P. Currie (eds.), *Dinosaur Systematics: Perspectives and Approaches,* pp. 282–298. Cambridge: Cambridge University Press.

Carpenter, K., and J. I. Kirkland. 1998. Review of Lower and Middle Cretaceous ankylosaurs from North America. In S. G. Lucas, J. I. Kirkland, and J. W. Estep (eds.), *Lower and Middle Cretaceous Terrestrial Ecosystems. New Mexico Museum of Natural History and Science Bulletin* 14: 249–270.

Carpenter, K., J. I. Kirkland, D. Burge, and J. Bird. 2001. Disarticulated skull of a new primitive ankylosaurid from the Lower Cretaceous of eastern Utah. In K. Carpenter (ed.), *The Armored Dinosaurs.* Bloomington: Indiana University Press [this volume, Chapter 11].

Carpenter, K., C. Miles, and K. Cloward. 1998. Skull of a Jurassic ankylosaur (Dinosauria). *Nature* 393: 782–783.

Carpenter, K., T. Maryańska, and D. B. Weishampel. In press. The Ankylosauria. In D. B. Weishampel, P. Dodson, and H. Osmólska (eds.), *The Dinosauria.* Berkeley: University of California Press.

Coombs, W. P., Jr. 1971. The Ankylosauria. Ph.D. diss. Columbia University, New York.

———. 1978. The families of the ornithischian dinosaur order Ankylosauria. *Journal of Paleontology* 21: 143–170.

Coombs, W. P., Jr., and T. Maryańska. 1990. Ankylosauria. In D. B. Weishampel, P. Dodson, and H. Osmólska (eds.), *The Dinosauria,* pp. 456–483. Berkeley: University of California Press.

Cooper, M. R. 1985. A revision of the ornithischian dinosaur *Kangasaurus coetzeei* Haughton, with a classification of the Ornithischia. *Annals of the South African Museum* 95: 281–317.

DeBry, R. W., and R. C. Olmstead. 2000. A simulation study of reduced tree-search effort in bootstrap resampling analysis. *Systematic Biology* 49: 171–179.

de Queiroz, K., and J. Gauthier. 1992. Phylogenetic taxonomy. *Annual Review of Ecology and Systematics* 23: 449–480.

DiCroce, T., and K. Carpenter. 2001. New ornithopod from the Cedar Mountain Formation (Lower Cretaceous) of Eastern Utah. In D. H. Tanke and K. Carpenter (eds.), *Mesozoic Vertebrate Life.* Bloomington: Indiana University Press.

Ford, T. L., and J. I. Kirkland. 2001. Carlsbad ankylosaur (Ornithischia, Ankylosauria): An ankylosaurid and not a nodosaurid. In K. Carpenter (ed.), *The Armored Dinosaurs.* Bloomington: Indiana University Press [this volume, Chapter 12].

Galton, P. M. 1969. The pelvic musculature of the dinosaur *Hypsilophodon* (Reptilia: Ornithischia). *Postilla, Peabody Museum* 131: 1–64.

Gilmore, C. W. 1930. On dinosaurian reptiles from the Two Medicine Formation of Montana. *Proceedings of the United States National Museum* 77: 1–39.

Godefroit, P., X. Pereda Suberbiola, H. Li, and Z. Dong. 1999. A new species of the ankylosaurid dinosaur *Pinacosaurus* from the Late Cretaceous of Inner Mongolia (P.R. China). *Bulletin de l'Institut des Sciences Naturelles de Belgique, Sciences de La Terra* 69(Suppl. B): 17–36.

Haubold, H. 1990. Ein neuer Dinosaurier (Ornithischia, Thyreophora) aus dem unteren Juras des Nördlichen Mitteleuropa. *Revue do Paléobiologie* 9: 149–177.

International Code of Zoological Nomenclature. 1999. London: Natural History Museum.

Kirkland, J. I. 1996. Biogeography of North America's mid-Cretaceous faunas—Loosing European ties and the first Asian–North American interchange [abstract]. *Journal of Vertebrate Paleontology* 16(3, Suppl.): 45A.

———. 1998. A polacanthine ankylosaur (Ornithischia: Dinosauria) from the Early Cretaceous (Barremian) of Eastern Utah. In S. G. Lucas, J. I. Kirkland, and J. W. Estep (eds.), *Lower and Middle Cretaceous Terrestrial Ecosystems. New Mexico Museum of Natural History and Science Bulletin* 14: 271–281.

Lambert, D. 1993. *The Ultimate Dinosaur Book.* London: Dorling Kindersley.

Lapparent, A. F. de, and R. Lavocat. 1955. Dinosauriens. In J. Piveteau (ed.), *Traité de Paléontologie* 5: 785–962. Paris: Masson et Cie.

Lee, Y.-N. 1996. A new nodosaurid ankylosaur (Dinosauria: Ornithischia) from the Paw Paw Formation (late Albian) of Texas. *Journal of Vertebrate Paleontology* 16: 232–345.

Marsh, O. C. 1896. The dinosaurs of North America. *U.S. Geological Survey, 1894–1895 Annual Report* 16: 133–244.

Maryańska, T. 1971. New data on the skull of *Pinacosaurus grangeri* (Ankylosauria). *Palaeontologica Polonica* 25: 45–53.

———. 1977. Ankylosauridae (Dinosauria) from Mongolia. *Palaeontologia Polonica* 37: 85–151.

Mishler, B. D. 1994. Cladistic analysis of molecular and morphological data. *American Journal of Physical Anthropology* 94: 143–156.

Newman, B. H. 1968. The Jurassic dinosaur *Scelidosaurus harrisonii*. *Palaeontology* 11: 40–43.

Nopcsa, F. 1929. Dinosaurierreste aus Siebenbügen. *Geologica Hungarica, Series Palaeontologica* 4: 1–76.

Norman, D. B. 1984. A systematic reappraisal of the reptile order Ornithischia. In W. E. Reif and F. Westphal (eds.), *Third Symposium on Mesozoic Terrestrial Ecosystems*, pp. 157–162. Tubingen: Atempto.

———. 1988. *The Prehistoric World of the Dinosaur.* New York: Gallery Books.

———. 1996. The cranial morphology of *Scelidosaurus harrisonii* [abstract]. *Journal of Vertebrate Paleontology* 16(3, Suppl.): 56A.

Osborn, H. F. 1923. Two Lower Cretaceous dinosaurs from Mongolia. *American Museum of Natural History Bulletin* 95: 1 –10.

Romer, A. S. 1927. The pelvic musculature of ornithischian dinosaurs. *Acta Zoologica* 8: 225–275.

————. 1956. *Osteology of the Reptilia.* Chicago: University of Chicago Press.

Sereno, P. C. 1986. Phylogeny of the bird-hipped dinosaurs (order Ornithischia). *National Geographic Research* 2: 234–256.

————. 1997. The origin and evolution of dinosaurs. *Annual Review of Earth and Planetary Sciences* 25: 435–489.

————. 1998. A rationale for phylogenetic definitions, with applications to the higher-level taxonomy of Dinosauria. *Neues Jahrbuch für Geologie und Paläontologie, Abhandlungen* 210: 41–83.

————. 1999a. Definitions in phylogenetic taxonomy: Critique and rationale. *Systematic Biology* 48: 329–351.

————. 1999b. The evolution of dinosaurs. *Science* 284: 2137–2147. Supplementary material at: http://www.sciencemag.org/feature/data/ 1041760.shl; accessed December 22, 2000.

Sereno, P. C., and Z. Dong. 1992. The skull of the basal stegosaur *Huayangosaurus taibaii* and a cladistic diagnosis of Stegosauria. *Journal of Vertebrate Paleontology* 12: 318–343.

Sullivan, R. M. 1999. *Nodocephalosaurus kirtlandensis,* gen. et sp. nov., a new ankylosaurid dinosaur (Ornithischia: Ankylosauria) from the Upper Cretaceous Kirtland Formation (upper Campanian), San Juan Basin, New Mexico. *Journal of Vertebrate Paleontology* 19: 126–139.

Swofford, D. L. 2000. *PAUP: Phylogenetic analysis using parsimony.* Version 4. Sunderland, Mass.: Sinauer Associates.

Thulborn, R. A. 1977. Relationships of the Lower Jurassic dinosaur *Scelidosaurus harrisonii. Journal of Paleontology* 51: 725–739.

Tumanova, T. 1983. The first ankylosaurs from the Lower Cretaceous of Mongolia [in Russian]. *Transactions of the Joint Soviet–Mongolia Paleontological Expedition* 24: 110–120.

————. 1987. Armored dinosaurs of Mongolia [in Russian]. *Transactions of the Joint Soviet–Mongolian Paleontological Expedition* 32: 1–77.

Vermeij, G. J. 1999. A serious matter with character-taxon matrices. *Paleobiology* 25: 431–433.

Vickaryous, M. K., A. P. Russell, P. J. Currie, K. Carpenter, and J. I. Kirkland. 1998. The cranial sculpturing of ankylosaurs (Dinosauria: Ornithischia): Reappraisal of developmental hypotheses [abstract]. *Journal of Vertebrate Paleontology* 18(3, Suppl.): 83A–84A.

Vickaryous, M. K., A. P. Russell, and P. J. Currie. 2001. Cranial ornamentation of ankylosaurs (Ornithischia: Thyreophora): Reappraisal of developmental hypotheses. In K. Carpenter (ed.), *The Armored Dinosaurs.* Bloomington: Indiana University Press [this volume, Chapter 15].

Wieland, G. 1911. Notes on the armored Dinosauria. *American Journal of Science* 4: 112–124.

Wilkinson, M., P. Upchurch, P. M. Barrett, D. J. Gower, and M. J. Benton. 1998. Robust dinosaur phylogeny? *Nature* 396: 423–424.

APPENDIX 21.1.
Characters used to create the data matrix shown in Table 21.3.

I. Ankylosauromorpha

Cranial character

1. Skull width greater than height
2. Skull width and length: width = length—1; width > length—2
3. Snout arches above level of skull roof
4. 2+ supraorbitals (where visible)
5. Inverted U-shaped notch at front of premaxillaries
6. Antorbital fenestra closed
7. Supratemporal fenestra closed
8. Lateral temporal and paroccipital process not visible in lateral view
9. Cutting edge of premaxillary connected to maxillary tooth row
10. Maxillary tooth row lingually inset forming maxillary shelf: restricted to maxilla—1; extend onto jugal—2
11. Nasal septum divides rostrum parasagittally: partially—1; extends to roof of snout roof—2
12. Postocular wall: partial—1; well-developed, formed by postorbital, jugal, and possibly the frontal—2
13. Hourglass-shaped palate
14. Cheek teeth with cingulum
15. Cranial ornamentation surface sculpturing: jugal, maxilla, postorbital—1; most external cranial bones—2
16. Cranial ornamentation well defined: multiple small to medium—1; large frontoparietal ornamentation and anteroposteriorly narrow nuchal ornamentation—2
17. Ornamentation between external nares: multiple, small—1; single large—2
18. Postorbital/squamosal and jugal horns: small and rounded—1; prominent—2
19. Mandible with ventrolateral ornamentation

Postcranial character

20. Synsacrum formed of caudals, dorsals and sacrals
21. Interlocking distal caudal prezygapophyses and postzygapophyses
22. Posterior chevrons anteroposteriorly elongated and contact one another—1; overlapping = 2
23. Scapula with acromion: flange—1; knob—2; swelling on dorsal edge—3
24. Coracoid large, anteroposteriorly long relative to dorsoventral width
25. Deltopectoral crest extending to midlength of humerus
26. Deltopectoral crest and transverse axis through distal epicondyles in same plane
27. Postacetabular portions ilium horizontally developed
28. Preacetabular portions ilium horizontally developed
29. Body of pubis massive and rotated dorsolaterally
30. Acetabulum closed
31. Pubis almost completely excluded from acetabulum
32. Ischium straight and near vertical below acetabulum—1; bent anteriorly at midshaft—2
33. Fourth trochanter on distal half of femur

APPENDIX 21.1. *(cont.)*

Osteodermal armor character

34 Body scutes: thin-walled, cones—1; thick-walled or solid—2
35 Cervical rings: two—1; three—2
36 Lateral cervical-shoulder spines: triangular and grooved along posterior margin—1; solid, oval in cross section—2; secondary loss—3
37 Large, dorsoventrally compressed, asymmetrically hollow-based, triangular lateral body plates—1; secondary loss—2
38 Sacral armor coossified
39 Terminal tail club

II. Polacanthidae

40 Loss of premaxillary teeth
41 Premaxillary scoop: narrow—1; wider than long—2
42 Premaxillary scoop wider than long
43 Elongate basipterygoid
44 Prominent jugal horn
45 Acromion flange: short—1; to middle of blade—2; down to glenoid—3
46 Anterior trochanter fused to greater trochanter
47 Preacetabular blade folded laterally
48 Spined plate

III. Ankylosauridae

49 Premaxillary beak wider than long
50 Loss of premaxillary teeth
51 External nares face: anterolaterally—1; completely forward—2
52 Complex sinus chamber
53 Skull roof domed
54 Wide internarial bar
55 Paroccipital process fused to quadrate
56 Secondary palate extends to second maxillary tooth
57 Large nodular cranial ornamentation
58 Prominent raised scale encircling orbit
59 Prominent cranial posterior cranial horns
60 Nuchal ridge along posterior rim of skull roof: low—1; vertically tall—2
61 Postsymphseal edentulous section of dentary long
62 Dorsal and ventral margins of dentary sinuous and parallel
63 Alveolar margin rises to height of coronoid process
64 Large, pocketlike internal mandibular fenestra: opened posteriorly—1; closed posteriorly—2
65 Anteriormost caudal ribs angled >10°: ventrolaterally—1; strongly anteroventrally—2
66 Scapula dorsal and ventral margins parallel

IV. Nodosauridae

67 Prominent emargination for lateral temporal fenestra on skull roof
68 Nuchal ornamentation along posterior rim of skull roof
69 Quadrate strongly concave posteriorly (arced)
70 Long basisphenoid process
71 Secondary palate, more than half tooth row
72 Nasal chamber split completely by vomer–nasal plate

73 Postsymphseal edentulous section of dentary long
74 Tooth row highest at midpoint, lower at each end
75 Dorsal and ventral margins of dentary sinuous and parallel
76 Cervical centra shorter than wide
77 Anterior caudal neural spines laterally expanded (swollen)
78 Anterior caudal articular face heart-shaped
79 Coracoid longer than deep
80 Coracoid half or more length of scapula
81 Anterior trochanter fused
82 Distal end of scapula tapered to cervicodorsal spine: dorsoposterior—1; anterior—2
83 Neck–shoulder armor spine: projecting dorsoposteriorly—1; projecting anteriorly—2

Index

Page numbers in boldface type refer to illustrations.

292, 293; description of, 270; as
holotype specimen of *Scolosaurus
cutleri*, 263, 270, 293; material
and horizon of, 263
BMNH R9293, 379, 380; caudal
plates from, **373**; dermal armor
from, **365**; splate from, **372**
BMNH R9533: description of, 131
BMNH R13002, **367**
BMNH R15950, **133**; description of,
131–132
BMNH R15951, 131, **134**; description
of, 132–133
BMNH R.16010, 41
Bock, Walter, xi
Bolivia: ankylosaur footprints from,
413, 414, **442–443**, 442–445,
444
Bone Cabin Quarry: *Stegosaurus*
specimens in, 90
Bone fractures: healing of, 142–143,
153–154; osteomyelitis with,
142–144
Bones: postfracture healing of, 142–
143, 153–154
"Bone wars": between Cope and
Marsh, 4
Bosses: defined, 370–371; of *Polacanthus foxii*, 368, 371. *See also*
Dermal armor
Brachiosauridae: from Cedar
Mountain Formation, 212
Brachypodosaurus gravis: material and
validity of, 458
Braincases: anatomical abbreviations
for, 105–106; of *Cedarpelta
bilbeyhallorum* gen. et sp. nov.,
228–229, 229–230; of *Hesperosaurus mjosi* gen. et sp. nov., **60**,
62; of *Hylaeosaurus armatus*,
170, 171; in Peabody *Stegosaurus*
mount, 80; of *Stegosaurus*, 97,
103–127, **107, 108, 111, 112,
118, 120–121, 124–125**; of
Struthiosaurus, 199; of *Struthiosaurus austriacus*, 184–185; of
Struthiosaurus transylvanicus,
185; of USNM 4936, 110–111,
118, 120–121, 124–125. *See also*
Basioccipitals
Brain endocasts: anatomical abbreviations for, 105–106; of *Euoplocephalus*, 185; of *Gargoyleosaurus*, 185; of *Kentrosaurus*,
104, 105, **109**; of *Polacanthus*,
185; of *Stegosaurus*, 97, 103–
127, **108, 111, 113, 114, 118,
124–125**; of *Struthiosaurus
austriacus*, 185; of *Struthiosaurus
transylvanicus*, 185

Brains: of stegosaurs, 113–115
Brigham Young University Earth
Sciences Museum: Dalton Wells
Quarry of, 386–387. *See also*
BYU 725–12963; BYU VP2080
Bristol City Museum and Art Gallery.
See BRSMG Ce12785
British Association for the Advance-
ment of Science: "Hippocampus
debate" within, 18; Richard
Owen's 1840 report to, 20;
Richard Owen's 1842 report to,
4–5, 20; Richard Owen as
president of, 18
British Columbia: ankylosaur
footprints from, 437–440, **437,
438**
British Empire: dinosaurs and, 5–6, 22
British fossil reptiles: Richard Owen's
reviews of, 4–5, 20
British Museum (Natural History):
construction of, 4; Richard Owen
at, 18–19; as Richard Owen's
monument, 22; *Struthiosaurus
austriacus* specimens at, 180. *See
also* BMNH 2422; BMNH 3789;
BMNH 36515–36517; BMNH
37713; BMNH 37714; BMNH
39496; BMNH 39533; BMNH
40458; BMNH 42068; BMNH
46320; BMNH 46322; BMNH
47338; BMNH R133; BMNH
R175; BMNH R202; BMNH
R643; BMNH R.1111; BMNH
R1875; BMNH R.2682; BMNH
R3775; BMNH R3848; BMNH
R4947; BMNH R4966; BMNH
R.4992; BMNH R5161; BMNH
R9293; BMNH R9533; BMNH
R13002; BMNH R15950;
BMNH R15951; BMNH
R.16010; BMNH miscellaneous
specimens
Brontosaurus amplus: from Quarry
11, 85, **85**
Brooklyn College: Walter Coombs at,
xii
Brown, Barnum, xii, 261
Brown, Frederick, 79; at Quarry 11,
84; at Quarry 12, 87, 88
BRSMG Ce12785, 27; jaw joint of,
33–34, 35; premaxillary teeth of,
32. *See also* Juvenile scelidosaur
Brushtail possum: gut contents of,
406, **408**
Buccal mandibular emargination: in
ankylosaurs, 41, 43; cheeks and,
45–46; in ornithischians, 45–46;
in stegosaurs, 37
Buccal maxillary emargination: in

ankylosaurs, 41; cheeks and, 45–46; in ornithischians, 45–46; in stegosaurs, 36
Buckland, William, 20
Buffalo, Wyoming: *Hesperosaurus* gen. nov. from, 56
Bungil Formation: *Minmi* from, 342
Bunzel, Emmanuel: Gosau material described by (tables), 175–177; *Struthiosaurus* described by, 174
Burge, Don, 211
BYU 725–12963, 390
BYU VP2080, 390

Cactus Park: *Mymoorapelta* from, 392, **392**
Calcaneum: of juvenile scelidosaur, **10**
California, xiii; *Aletopelta* gen. nov. from, 239–257
Callus development: after bone fractures, 142–143, 150
Cal Orcko tracksite: ankylosaur footprints from, 413, **442–443**, 442–445, **444**, 447
Camarasaurus: from Quarry 12, 91
Campanian: *Aletopelta* gen. nov. from, 239–257; ankylosaur footprints from, 437–442; ceratopsian versus ankylosaur footprints from, **419**; ceratopsian versus ankylosaur footprints in, 418–421; multivariate analysis of ankylosaurids from, 261; North American ankylosaurids from, 261–263; South American ankylosaur material from, **160**, 160–166, **162, 163, 164, 165**; *Struthiosaurus austriacus* from, 173–206
Camptosaurus: *Diracodon laticeps* versus, 98; from Quarry 13, 93
Canada: ankylosaur footprints from, 413, 414, **422**
Canadian Museum of Nature. *See* National Museum of Canada
Cañon City, Colorado: ankylosaur footprints from, 435; *Stegosaurus stenops* from, 94, 106. *See also* Felch Quarry 1
Caputegulum (caputegulae): defined, 371. *See also* Dermal armor
Carlsbad, California: *Aletopelta* gen. nov. from, 239–257
Carnegie Museum of Natural History: *Struthiosaurus austriacus* specimens at, 180. *See also* CM 106; CMNH 41681
Carpenter, Kenneth, 55, 76, 141, 169, 211, 455
Carpi: in Peabody *Stegosaurus* mount,

80; of *Stegosaurus* and *Diracodon laticeps*, 99
Caudal armor: of polacanthines, 380
Caudal plates: of *Gastonia*, 380; of *Polacanthus*, 380; of *Polacanthus foxii*, **373**. *See also* Dermal armor
Caudal vertebrae: of juvenile scelidosaur, **10**; of *Scelidosaurus harrisonii*, **12–13**. *See also* Vertebrae
Caudorbitos (caudorbitosa): defined, 371; of polacanthines, 375; of *Polacanthus foxii*, **375**. *See also* Dermal armor
Cedar Mountain Formation: ankylosaur footprints from, 433–435, **434**, 447; ankylosaurs from, 212; *Cedarpelta* gen. nov. from, 211–235; ceratopsian versus ankylosaur footprints from, 419; *Gastonia burgei* from, 379, 386–396; location of, 446; polacanthines from, **364–365, 366–367**; stratigraphy of, **214–215**
Cedarpelta bilbeyhallorum gen. et sp. nov., 211–235; cranial material of, 211–235, **218–219, 220–221, 222, 224–225, 226, 227, 228–229, 231, 235**; diagnosis of, 216; holotype specimen of, 216, **218–219, 220–221, 225, 228–229**; horizon and age of, 213; paratype material for, 216–217; quarry map of, **215**; *Shamosaurus scutatus* versus, 212–215, 234; size of, 235; stratigraphy of, **214–215**; as valid taxon, 457
Cedarpelta gen. nov., 211–235; in ankylosaur phylogenetic analysis, 473; data matrix for, 462; dermal armor of, 211; *Euoplocephalus* versus, 230; *Minmi* versus, 220; phylogeny of, **474–475**; *Pinacosaurus* versus, 217, 220, 223, 227, 233; *Polacanthus* versus, 230; as reference taxon, 460; *Scelidosaurus* versus, 464; *Shamosaurus* versus, 212, 227, 233, 234; *Stegosaurus* versus, 234; stratigraphy of, **214–215**; *Struthiosaurus austriacus* versus, 233; systematic paleontology of, 215–232; *Talarurus* versus, 223
CEM site: quarry map of, **215**
Cenomanian: ankylosaur footprints from, 433–440, 445–446; ankylosaurs from (table), 213; ceratopsian versus ankylosaur footprints from, 419, **419**; North American ankylosaurs from, 212

China: stegosaur dermal armor from, 137–138
Chronic osteomyelitis: features of, 143; in *Stegosaurus* tail spikes, 141–154
Chungkingosaurus: dermal armor of, 137; *Hesperosaurus mjosi* gen. et sp. nov. versus, 58, 68, 69; ilium of, **69**; ischium of, **71**; pelvis of, **69**; in stegosaur phylogeny, **73**; vertebrae of, **65, 66**, 68, 69
Chungkingosaurus jiangbeiensis: character matrix for, 73
Clifford, H. Trevor, 399
Cloward, Karen, 55
CM 106, **107**, 110; braincase and endocast from, 104, 105, **108, 114**, 121–126; cranial nerves of, 115–116; description of, 106; inner ear of, 116
CMNH 41681: skull and mandibles of, 37, 38
Cnemidophorus soki, 339
Coelophysis: cololites from, 400
Coelurus: from Quarry 12, 91
Colbert, Edwin H., xi, xii, 4
College of Eastern Utah (CEU) Prehistoric Museum: dinosaur prospecting by, 212; Gaston Quarry of, 386. *See also* CEM site; CEUM 746; CEUM 1163; CEUM 1238; CEUM 1307; CEUM 1354; CEUM 1371; CEUM 1379; CEUM 1382; CEUM 1384; CEUM 1385; CEUM 1387; CEUM 1392; CEUM 1834; CEUM 3514; CEUM 3565; CEUM 3646; CEUM 3880; CEUM 5060; CEUM 5145; CEUM 9464; CEUM 10259; CEUM 10266; CEUM 10267; CEUM 10270; CEUM 10345; CEUM 10352; CEUM 10405; CEUM 10421; CEUM 10560; CEUM 10561; CEUM 12360; CEUM F-16; CEUM specimens
Cololites: from *Coelophysis*, 400; from *Compsognathus*, 400; from *Minmi*, 399–410, **405**; from *Sinosauropteryx*, 400; thyreophoran cheeks and, 407–408
Colorado, 261; ankylosaur footprints from, **434**, 435; ceratopsian footprints from, 417; stegosaurs from, 55
Columbia University: Walter Coombs at, xi–xii
Colville River, Alaska: ankylosaur footprints from, 435

Combat: dermal armor in ankylosaur, 290; dermal armor in stegosaur, 153, 154
Como anticline: Quarry 12 and, 85–86
Como Bluff: Marsh's quarries at, 79–81, 81–84, 84–85, 85–91, 91–94; pathological stegosaur tail spikes from, 144–145; stegosaurs from, 104, 111. *See also* Quarry 8; Quarry 9; Quarry 11; Quarry 12; Quarry 13
Compartmentalization: in phylogenetic analysis, 460–463, 477
Compsognathus: cololites from, 400
Computed tomography (CT) scans: of pathological tail spikes, 145, 148, **148**, 149, 150, **152**, 153
Coniacian: ceratopsian versus ankylosaur footprints from, **419**
Conybeare, William, 20
Coombs, Alexandra, xii
Coombs, Margery Chalifoux, xi, xii
Coombs, Matthew, xii
Coombs, Walter P., Jr., 241; on ankylosaur cranial ornamentation development, 327–329, 330–331; ankylosaur systematics of, 262, 456; bibliography of, xiv–xv; career of, xi–xiv
Cooper, David, 367
Coossification: defined, 322; in development of cranial ornamentation, 324–330, 334–335, 335–336. *See also* Dermal armor; Dermal ossifications
Cope, Edward Drinker: in "bone wars," 4; discovery of *Laelaps* by, 8; stegosaurs described by, 55–56
Coprolites: of *Maiasaura*, 404; from Mygatt-Moore Quarry, 404
Coracoids: of *Crataeomus lepidophorus*, 198; of *Gastonia burgei*, 388, 390–391; of *Mymoorapelta*, 390; in Peabody *Stegosaurus* mount, **78–79**, 80; of *Struthiosaurus austriacus*, 180, 189–192, **190–191**, 192; of *Struthiosaurus transylvanicus*, **192**. *See also* Forelimb bones; Pectoral girdle
Coria, Rodolfo A., 159
Cornish, L., 32
Coronux (coronuces): as defense, 381; defined, 371; of *Gargoyleosaurus parkpinorum*, 376; of *Hylaeosaurus armatus*, 377; of nodosaurids, 375; of *Pawpawsaurus campbelli*, 377; of polacanthines, 374–375. *See also* Dermal armor
Corytophanes cristatus, 339

Foulke, William Parker: *Hadrosaurus foulkii* and, 6
Fox, William, 131, 132
France: *Struthiosaurus* from, 202; thyreophoran footprints from, 414
Frankenhof: *Leipsanosaurus noricus* from, 178, 183
Freshfields Quarry: *Polacanthus foxii* from, 367
Frontals: of *Cedarpelta bilbeyhallorum* gen. et sp. nov., **218–219**, 222, **224–225**; of *Hesperosaurus mjosi* gen. et sp. nov., **60, 61, 62**; of *Scelidosaurus harrisonii*, 29
Frontoparietals: of *Hesperosaurus mjosi* gen. et sp. nov., 62
Fullerian Professor: Richard Owen as, 19
Fungi: in osteomyelitis, 142

Gaffney, Eugene, xi
Galápagos Islands, xiii
Galton, Peter M., 45–46, 76, 103, 104, 173, 407
Garden Park, Colorado: dinosaur quarrying at, 81–82; Late Jurassic climate of, 152; stegosaurs from, 104, 106, **118, 120–121, 122, 123, 124–125**, 144. *See also* Felch Quarry 1
Gargoyleosaurus: in ankylosaur phylogenetic analysis, 467, 471; brain endocast of, 185; cervicopectoral armor of, 379; data matrix for, 462; dermal armor of, 287, **468**; femur of, 392; *Gastonia* versus, 387; *Hylaeosaurus armatus* versus, 169, 170, 171; *Minmi* versus, 361; phylogeny of, **473, 475**; premaxillary teeth of, 41; size of, 235; *Struthiosaurus austriacus* versus, 198; *Struthiosaurus* versus, 185, 201; teeth of, 185
Gargoyleosaurus parkpinorum: as ankylosaurid, 376; description of, 376–377; as polacanthine, 376–377; skeleton of, 386; skull of, 379; as valid taxon, 457
Gaston, Robert W., 386
Gastonia, 435; *Aletopelta* gen. nov. versus, 248, 250; in ankylosaur phylogenetic analysis, 467, 471; caudal plates of, 380; data matrix for, 462; dermal armor of, **468**; *Hylaeosaurus armatus* versus, 171; *Minmi* versus, 360, 361; mounting skeleton of, 387; phylogeny of, **473, 475**;

Polacanthus versus, 378, 379, 396; *Struthiosaurus austriacus* versus, 193, 198; *Struthiosaurus* versus, 199, 200
Gastonia burgei: Aletopelta coombsi gen. et sp. nov. versus, 248; coracoids of, 390–391; dermal armor of, **366–367**, 390, 392–396, **393, 394, 395**; femur of, 247–248; *Gargoyleosaurus* versus, 387; hindlimb proportions of, 249; horizon and age of, 213; mandibles of, 387; measurements of, 388; *Nodosaurus textilis* and, 392; pelvis of, 388–390; as polacanthine, 376; ribs of, 388, **389**; *Sauropelta* and, 392; skeleton of, 386–396, **389, 390, 391, 393, 394, 395**; skull of, **364–365**, 379, 387, 388; *Tarchia kielanae* versus, 387; teeth of, 387; *Tetrapodosaurus borealis* and, 391; as valid taxon, 457; vertebrae of, 390
Gaston Quarry, 396; *Gastonia burgei* from, 386
Gastroliths: with *Minmi*, 403, 409
Gates Formation: ankylosaur footprints from, 417, 423–433, 445, 446, 447; footprints from, **424, 425, 426, 427, 428, 429, 430, 431, 432**; location of, **446**
Gecko gekko, 339
Geese: crop contents of, 405, **407**
Gekkota, 339; in amniote phylogeny, **321**; development of cranial ornamentation in, 324, 326; ornamentation in, 322
Geological Sciences Museum. *See* GSM 10956; GSM 109660
Germany: ankylosaur footprints from, 421–422
Gerrhonotus multicarinatus, 339
Gerrhosaurus: development of cranial ornamentation in, 324
Gerrhosaurus major, 339; ornamentation of, 322
Gerrhosaurus nigrolineatus, 339
Gething Formation: ankylosaur footprints from, 417, **422**, 422–423, 445, 447; location of, **446**
Gilmore, Charles Whitney, 92, 104; *Dyoplosaurus* skull described by, 262; *Euoplocephalus tutus* skull described by, 262; Quarry 13 map by, 91, **93**; on stegosaur dermal armor, 154; *Stegosaurus* specimens discovered by, 106
Glenoid fossae (mandibular): of ankylosaurs, 41, 45; of *Euoplo-*

cephalus, 45; of *Euoplocephalus tutus,* 305; of *Scelidosaurus harrisonii,* 34, 45; of stegosaurs, 37–38, 45

Glossopharyngeal nerve (cranial nerve IX): of stegosaurs, 116. *See also* Cranial nerves

Glyptodontopelta mimus: material and validity of, 458

Glyptops: from Quarry 12, 91

Gobi Desert: ankylosaur footprints from, 441

Golden, Colorado: Arthur Lakes at, 86

Goniopholis: from Quarry 12, 91

Gosau Beds: fauna from, 173; history of, 174–178; *Struthiosaurus austriacus* from, 173–206; *Struthiosaurus austriacus* specimens from, **178–179,** 178–183, 184–198, **188–189, 190–191, 192, 194–195, 196, 197**

Gosau Formation. *See* Gosau Beds

Grande Cache, Alberta: ankylosaur footprints from, 413, 423–433

"Gravid Scelidosaur," 15

Great Britain. *See* British Museum (Natural History); England

Great Exhibition Hall, 5; *Hylaeosaurus armatus* model at, 9. *See also* Crystal Palace

Gregory, Joseph T.: Walter Coombs and, xi

Gruber, Jacob, 18

Grünbach Basin, Austria: *Struthiosaurus austriacus* from, 180

GSM 10956, **10**

GSM 109660, **10**

Gymnosperms: angiosperms versus, 402

Haas, Georg: on ankylosaurid feeding mechanisms, 300–301

Haberlein, Karl, 19

Haddonfield, New Jersey: *Hadrosaurus foulkii* discovered at, 6

Hadrosauria: beaks of, 45; commonness of, 368; osteomyelitis in, 143

Hadrosauridae: South American, 159, 160, 165, 166

Hadrosaurus: Joseph Leidy and, 4, 6–8, 20, 21; in marine deposits, 15

Hadrosaurus foulkii: discovery of, 4, 6; early skeletal restoration of, 7

Hall, Evan, 212, 216

Hallett, Richard: at Quarry 12, 86–87

Harrison, James: *Scelidosaurus* discovered by, 8–9

Hastings, England: stegosaur dermal armor from, 131, **132**

Hateg Basin, Romania: *Struthiosaurus transylvanicus* from, 178, **192**

Hatteria: Stegosaurus versus, 97

Hauterivian: ceratopsian versus ankylosaur footprints from, **419;** stegosaur dermal armor from, 131, **132**

Haversian systems: cranial ornamentation and, 329; osteomyelitis and, 143, 150–152, **151**

Hawkins, Benjamin Waterhouse: Central Park dinosaur exhibit by, 6; Great Exhibition Hall dinosaur models by, 5, **9**

Hayashibara Museum of Natural History. *See* HMNH 001

Heishansaurus pachycephalus: material and validity of, 458

Hell Creek Formation: ankylosaurids from, 261

Heloderma, 339; development of cranial ornamentation in, 324, 325

Heloderma horridum: development of cranial ornamentation in, 324–325

Heloderma suspectum, 339; development of cranial ornamentation in, **325;** ornamentation of, 322

Hemal arches. *See* Vertebrae

Hematogenous osteomyelitis, 142

Herbivorous dinosaurs: ankylosaurs as, 400; feeding mechanisms of, 25–47

Herbivory: of *Minmi,* 399–410; paleobiology of, 402–409

Hesperornis: brain endocast of, 117

Hesperosaurus gen. nov., 55–74; description of, 58–70; diagnosis of, 56; holotype specimen of, **61, 62, 63, 64, 65, 66, 68, 69, 71, 72;** ilium of, **69;** pelvis of, **69;** quarry map for, **57;** in stegosaur phylogeny, 55; systematic paleontology of, 56–58

Hesperosaurus mjosi gen. et sp. nov., 55–74; character matrix for, 73; cranial material of, **60, 61;** dermal armor of, 70, **72;** diagnosis of, 58; holotype specimen of, 56–70; ilia of, 70; ilium of, **68;** ischium of, **68, 71;** measurements for, 59; ossified tendons in, 68; pelvis of, **68, 69;** pubis of, 59, **68,** 70, **71;** quarry map for, **57;** scapula of, 70; skull of, **62;** in stegosaur phylogeny, **73;** stratigraphy of, 70; taxonomy of, 72; teeth of, **63;** type locality of, 56; vertebrae of, **63, 64, 65, 66,** 67–68, **68**

Leonardi, Giuseppe, **444**

Lesothosaurus, 300; in ankylosaur phylogenetic analysis, 465, 469, 471; braincase of, 110; data matrix for, 462; occlusion in, 44; in phylogenetic analysis, 455; phylogeny of, **470, 475;** predentary-dentary joint of, 314; predentary of, 37; pubis of, **467;** *Scutellosaurus* versus, 36; temporal phylogeny of, **476;** tooth wear in, 31, 35, 304

Lexovisaurus: dermal armor of, 137; *Hesperosaurus mjosi* gen. et sp. nov. versus, 58, 67, 69; ilium of, **69;** ischium of, **71;** pelvis of, **69;** pubis of, **71;** in stegosaur phylogeny, **73;** vertebrae of, **64, 66, 67, 69**

Lexovisaurus durobrivensis: character matrix for, 73

Lexovisaurus vetustus: Hesperosaurus mjosi gen. et sp. nov. versus, 67–68; vertebrae of, 67–68

Lialis burtonis, 339

Liassic: thyreophoran footprints from, 414

Liassic cliffs: discovery of *Scelidosaurus* in, 3

Liassic theropod: in *Scelidosaurus harrisonii* material, 9; with *Scelidosaurus harrisonii* material, **10**

Ligabuichnium, 442

Ligabuichnium bolivianum, 413–414; footprints of, **442**

Limb bones: of *Struthiosaurus austriacus*, 183, 189–195, **190–191, 192, 194–195, 196.** *See also* Forelimb bones; Hindlimb bones

Lockley, Martin G., 413

Locomotion: dermal armor for, 380, 381; of dinosaurs, 21–22; ornamentation versus, 319; of *Scelidosaurus harrisonii*, 15, 16–17; of *Stegosaurus*, 99–100. *See also* Bipeds; Quadrupeds

London: British Museum in, 4; *Hylaeosaurus armatus* model in, 9

Loricosaurus scutatus: material and validity of, 458; *Saltasaurus loricatus* versus, 159

Los Angeles County Museum of Natural History. *See* LACM 123344

Lower Cretaceous. *See* Early Cretaceous

Lower Horseshoe Canyon Formation:

ankylosaurid specimens from (table), 263

Lower Jurassic. *See* Early Jurassic

Lull, Richard Swann, 76, 84; on Peabody *Stegosaurus ungulatus*, 77, **78–79**

Lydekker, Richard, 131

Lyell, Charles, 18

Lyme Regis: juvenile scelidosaur material from, **10**

Maastrichtian: ankylosaur footprints from, 442–445; ceratopsian footprints from, 417; ceratopsian versus ankylosaur footprints from, 419; ceratopsian versus ankylosaur footprints in, 418–421; North American ankylosaurids from, 261–263; South American ankylosaur material from, **160**, 160–166, **162, 163, 164, 165;** *Struthiosaurus* from, 202, 205–206; *Struthiosaurus transylvanicus* from, 178

MacLeod, Roy, 18

Macropodosaurus gravis, 413; footprints of, **420,** 433; *Metatetrapous valdensis* versus, 433

Magyarosaurus dacus: dwarfism in, 204

Maiasaura, 405; coprolites of, 404

Maleevus disparoserratus: material and validity of, 458

Mammalia, 340; in amniote phylogeny, **321;** cranial ornamentation in, 323; Quaternary, xi; Tertiary, xi, xii

Mandibles: of ankylosaurs, 41; of *Cedarpelta bilbeyhallorum* gen. et sp. nov., 230–231, **231;** of *Crataeomus*, 185; of *Diracodon laticeps*, **98,** 98–99, **99;** of *Euoplocephalus tutus*, **40, 306,** 306–309, **307, 313;** of *Gastonia burgei*, 387; of *Hesperosaurus mjosi* gen. et sp. nov., 59, **60, 62;** of *Huayangosaurus*, 37, 62; of *Huayangosaurus taibaii*, **62;** of *Hylaeosaurus armatus*, **170,** 171; of *Lesothosaurus*, 37; of *Regnosaurus*, 37, 130; of *Scelidosaurus*, 186; of *Scelidosaurus harrisonii*, 27–29, **28;** of *Stegosaurus*, 37; of *Stegosaurus stenops*, 37, 62, **62;** of *Struthiosaurus*, 200–201; of *Struthiosaurus austriacus*, **178–179,** 180, 181, 185–186, 203. *See also* Angulars; Articulars;

phylogeny of, **470**, 475, **475**, 477; *Polacanthus* versus, 361; pubis of, **467**; *Scelidosaurus* versus, 360, 463, 464; size of, 204; skull of, **346–347**; skull of juvenile, 212; *Struthiosaurus* versus, 204, 205; taphonomy of, 356; temporal phylogeny of, **476**; types of dermal armor on, 342–343, 343–344, **344**

Minmi paravertebra: dermal armor of, **345**; holotype specimen of, 342, 401; horizon and age of, 213; as valid taxon, 457

MIWG 1191a, 131, 138; description of, 136

MIWG 5307, **135**, **136**, 138; description of, 133–134, 136

Mjos, Ronald G., 56

Mollusca: in Gosau Beds, 174

Molnar, Ralph E., 341, 399; Gosau material described by (tables), 176–177

Moloch horridus, 339

Mongolia. *See* Djadokhta Formation

Monimiacaeae, 402

Monkonosaurus: Hesperosaurus mjosi gen. et sp. nov. versus, 70; ilium of, **69**, 70; pelvis of, **69**; taxonomy of, 72

Montana, xiii; ankylosaurids from, 261–263; ankylosaurids from (table), 263; multivariate analysis of ankylosaurids from, 261

MOR 433, 264, **282–283**, 298; dermal armor from, 284, **285**, **286**, 287; description of, 270–271, 281–283, 293; material and horizon of, 263; measurements of, 274; in multivariate analysis, 275, 276, **276**, 277, **277**, 278; skull from, **267**, 268, 289, 291

Morphologic analysis: of footprints, 415–418

Morphometric analysis: of dinosaurs, 271–273; of *Euoplocephalus tutus,* 272–273, 273–278

Morrison, Colorado: stegosaurs from, 55; *Stegosaurus armatus* from, 81–84

Morrison Formation, 261; coprolites from, 404; *Hesperosaurus* gen. nov. from, 55–74; pathological stegosaurs from, 144–145; stegosaurs from, 55–56, 70, 72–74, 104; *Stegosaurus armatus* from, 81–84; *Stegosaurus duplex* from, 84–85; *Stegosaurus stenops* from, 91–94; *Stegosaurus sulca-*

tus from, 91–94; *Stegosaurus ungulatus* from, 85–91, 91–94; type section of, 81

Morrison Museum of Natural History: *Stegosaurus armatus* preparation at, 81, 83

MPCA-Pv specimens, **160**, 160–166, **162**, **163**, **164**, **165**

MPCA-SM 1, 160, **164**, 166

MU 650 VP, **242–243**, 247, 248

Mudge, Benjamin: at Felch Quarry 1, 94; at Garden Park, Colorado, 81

Multivariate analysis: of *Euoplocephalus tutus,* 261

Museo Provincial Carlos Ameghino, Cipolleti, Argentina: ankylosaur material in, 159–166, **160**, **162**, **163**, **164**, **165**

Museum of Isle of Wight Geology. *See* Isle of Wight; MIWG 1191a; MIWG 5307

Museum of the Rockies. *See* MOR 433

Museum of Western Colorado, Grand Junction: Dalton Wells Quarry of, 386–387. *See* MWC 1825; MWC 2610; MWC 2843

Mussentuchit Member: ankylosaur footprints from, 433–435; *Tenontosaurus* from, 212

Muthmannsdorf, Austria, **174**; *Struthiosaurus austriacus* from, 173–206; *Struthiosaurus austriacus* specimens from, **178–179**, 178–183, 184–198, **188–189**, **190–191**, **192**, **194–195**, **196**, **197**

MWC 1825, 373, 376

MWC 2610, **392**

MWC 2843, 376

Myelencephalon: of stegosaurs, 115

Mygatt-Moore Quarry, 405; coprolites from, 404

Mymoorapelta: in ankylosaur phylogenetic analysis, 467; coracoids of, 390; data matrix for, 462; metatarsals of, 256; pelvic elements of, **392**; phylogeny of, **473**, **475**; *Struthiosaurus austriacus* versus, 193, 198; *Struthiosaurus* versus, 200, 201, 204; tricorns of, 373; ulna of, **246**

Mymoorapelta maysi, 201; pelvic buckler of, **369**; pelvis of, 392; as polacanthine, 376; *Scelidosaurus harrisonii* versus, 376; tricorns of, 376; as valid taxon, 457

Nasals: of *Cedarpelta bilbeyhallorum* gen. et sp. nov., 220; of

416; of *Ceratopsipes,* **415,** 416–
417; of *Dyoplosaurus,* **249;** of
Euoplocephalus, 415; of
Euoplocephalus tutus, **282–283,**
283; of *Hypsilophodon,* 110; of
juvenile scelidosaur, **10;** of
nodosaurids, 415; osteomyelitis
in *Deinonychus,* 144; of
Sauropelta, **249,** 415–416; of
Scelidosaurus harrisonii, **12,** 17;
of *Stegosaurus duplex,* 110; of
Stegosaurus stenops, **122;** of
Tetrapodosaurus, **415,** 416–417.
See also Metatarsals; Phalanges
Peishansaurus philemys: material and
validity of, 458
Pelves: of *Aletopelta coombsi* gen. et
sp. nov., 245–247; of *Chungkingosaurus,* **69;** of *Dacentrurus,*
69; of *Dyoplosaurus acutosquameus,* **279,** 279–280; of
Euoplocephalus, 388–390; of
Euoplocephalus tutus, **279,** 279–
280; of *Gastonia burgei,* 388–
390; of *Hesperosaurus* gen. nov.,
69; of *Hesperosaurus mjosi* gen.
et sp. nov., **68;** of *Huayangosaurus,* **69;** of *Kentrosaurus,* **69;**
of *Lexovisaurus,* **69;** of *Monkonosaurus,* **69;** of *Mymoorapelta,*
392; of ornithischians, **466;** of
Polacanthus foxii, 388–390; of
Scolosaurus cutleri, **279,** 279–
280; of stegosaurs, **69;** of
Stegosaurus, **69;** of *Wuerhosaurus,* **69.** *See also* Hindlimb
bones; Ilia; Ischia; Pubes
Pelvic bucklers: absent from *Minmi,*
356; of *Gargoyleosaurus,* 361;
of *Gastonia,* 361; of *Polacanthus,*
356, 361. *See also* Dermal armor
Peninsular Ranges Terrane: *Aletopelta*
gen. nov. from, 240
Penkalski, Paul, 261
Pereda Suberbiola, Xabier, 173
Peterson, O. A.: *Stegosaurus* specimens
discovered by, 106
Phalanges: of *Crataeomus lepidophorus,* 195; of juvenile scelidosaur, **10;** in Peabody *Stegosaurus*
mount, **78–79,** 80; of stegosaurs,
110; of *Struthiosaurus austriacus,*
180, 182, 190, **196**
Philadelphia: *Hadrosaurus foulkii* in,
6, 7
Philpott Museum, Lyme Regis: juvenile
scelidosaur in, **10**
Phrynosoma: ornamentation of, 322,
331

Phrynosoma cornutum, 339; ornamen-
tation of, 322
Phrynosoma hernandesi, 339; orna-
mentation of, 322
Phrynosoma mcalli, 339
Phrynosoma modestum, 339;
ornamentation of, 322, 331, **332**
Phrynosomatidae, 339; in amniote
phylogeny, **321**
Phylogenetic analysis: methods of,
459–463
Phylogeny: of Ankylosauria, 455–477.
See also Crown group phylogenetic bracketing
Pickersgill, H. W.: portrait of Richard
Owen by, 5
Piesting: *Leipsanosaurus noricus* from,
178
Pinacosaurus, 448; *Aletopelta* gen.
nov. versus, 251, 257; in ankylosaur phylogenetic analysis, 471,
473; *Cedarpelta* gen. nov. versus,
217, 220, 223, 227, 233; cranial
dermal armor of, 233; data
matrix for, 462; dermal armor of,
284; humerus of, **245,** 255;
metatarsals of, 256; ontogeny of,
289; phylogeny of, **474–475;**
quadrates of, 40–41; as reference
taxon, 460; *Scelidosaurus* versus,
463, 464; *Struthiosaurus
austriacus* versus, 204; tooth
wear in, 43; ulna of, **246**
Pinacosaurus grangeri, 323; cranial
ornamentation of, 318–319, **323,**
323–324; cranial sutures of, 327;
ontogeny of cranial ornamentation in, 327–328, 330, 332; skull
of juvenile, 212; as valid taxon,
457
Pinacosaurus mephistocephalus: as
valid taxon, 457
Pinacosaurus ninghsiensis: material
and validity of, 458
Piscivory: of *Echinodon,* 16
Pituitary fossae: of stegosaurs, 103,
105, 119–120, 124–125, 126,
127
PIUW 2349/6: as *Struthiosaurus
austriacus* holotype, **178–179,**
180
PIUW specimens. *See* Paläontologisches Institut der Universität
Vienna
Plainview Formation: ankylosaur
footprints from, 435
Planicoxa, 464
Plateosaurus: morphometric analysis
of, 271

305–309, **320**; fusion in *Cedarpelta* gen. nov., 233–234; of *Gargoyleosaurus parkpinorum,* 379; of *Gastonia burgei,* **364–365,** 379, 387, 388; of *Hesperosaurus mjosi* gen. et sp. nov., 58–62, **60**; of *Huayangosaurus,* 36, 58–62; of *Huayangosaurus taibaii,* **62**; of *Hylaeosaurus armatus,* 169–171, **170**; of *Minmi,* 212, 345–347, **346–347**; ornamentation on ankylosaur, 318–336, **320**; of *Pawpawsaurus campbelli,* **364–365,** 379; in Peabody *Stegosaurus* mount, 78–79, **80**; of *Pinacosaurus grangeri,* 212; of polacanthines, **364–365,** 374–375, 379; of *Saichania chulsanensis,* **40**; of *Scelidosaurus,* 8–9, **11**, **465**; of *Scelidosaurus harrisonii,* 15–16, 27–34, **28**; of stegosaurs, 36–40, **37,** 104–105; of *Stegosaurus,* 36, 99, 104–105, 106–111; of *Stegosaurus stenops,* **37,** 58–62, **62,** 97, **124–125**; of *Struthiosaurus,* 198–199, **200**; of *Struthiosaurus austriacus,* **178–179,** 180, 181, 184–186, 203; of *Struthiosaurus transylvanicus,* 203; of *Talarurus,* 40–41; teeth in ankylosaur, 41; of *Tuojiangosaurus,* 36. *See also* Cranial material
Small, Bryan, 144
SMC B.28814: as *Craterosaurus pottonensis* holotype, 130–131
SMC B53408, **42, 44**
Smithsonian Institution, xii; Quarry 13 *Stegosaurus* specimens at, 91–94; *Stegosaurus stenops* at, 94. *See also* USNM 2274; USNM 4752; USNM 4934; USNM 4936; USNM 4937; USNM 6645; USNM 6646; USNM 7414; USNM 7943; USNM 8610; USNM 11892; USNM 419656; USNM miscellaneous specimens
Smoky River Coal Mine: ankylosaur footprints from, 423–433
Snout width: thyreophorans' diets and, 45
Soft vegetation: as ankylosaur food, 408–409
South America: ankylosaur footprints from, 413–414; ankylosaur material from, 159–166, **160, 162, 163, 164, 165**; first report of ankylosaurs from, 159–160; North American dinosaur inter-

change with, 159, 166; stegosaur material from, 166
South Dakota: polacanthines from, **372**
South Kensington, London: British Museum in, 4, 22
Spaculae. *See* Pectoral girdle
Spain: Wealden stegosaurs from, 130, 138. *See also* Iberian Peninsula
Sphenodon: Stegosaurus versus, 97; tooth wear in, 301
Spicules: in pathological bone, 149, **149**
Spikes: as defense, 380–381; defined, 372. *See also* Spines; Tail spikes
Spines: as defense, 380–381; defined, 372; of polacanthines, 376, 379. *See also* Parascapular spines; Shoulder spines; Spikes; Tail spikes
Splates: as defense, 381; defined, 372–373; of *Hoplitosaurus marshi,* **372,** 372–373; of *Kentrosaurus,* 372; of polacanthines, 376; of *Polacanthus foxii,* **372,** 372–373. *See also* Dermal armor; Plates; Spines
Sporangia: in *Minmi* cololite, 403
Squamata: in amniote phylogeny, **321;** ornamentation in, 322
Squamosals: of *Cedarpelta bilbeyhallorum* gen. et sp. nov., **227,** 227–229; of *Euoplocephalus tutus,* 305, 332–333, **334;** of *Minmi,* 345, **346–347;** of *Scelidosaurus harrisonii,* 29; of *Stegosaurus,* 97
Stagonolepis: Scelidosaurus versus, 17
Staphylococcus aureus: in osteomyelitis, 142
Stegoceras: morphometric analysis of, 271
Stegopelta: Aletopelta gen. nov. versus, 239, 240, 257; data matrix for, 462; dermal armor of, 256; phylogeny of, **472;** *Struthiosaurus* versus, 200
Stegopelta landerensis: Aletopelta gen. nov. versus, 256; description of, 377; excluded from Polacanthinae, 377; horizon and age of, 213; pelvic buckler of, **369;** as valid taxon, 457
Stegosauria, 104; ankylosaur dermal armor versus armor of, 136–139; Ankylosauria versus, 130, 382, 456; in ankylosaur phylogenetic analysis, 465–467; beaks of, 46; brain anatomy of, 113–115;

braincases of, 104–105; brain endocasts of, 104–105, 112–113; character matrix for, 73; characters of British Wealden armor of, 138–139; cheeks of, 46–47; classification of, 137–139; cranial blood vessels of, 116–117; cranial nerves of, 115–116, 119–120; dermal armor of, 363; dermal armor of Wealden, 130–139, **132, 133, 134, 135, 136;** diets of, 45; feeding mechanisms of, 36–40, 45, 47, 299; ilia of, **69;** inner ears of, 116; locomotor functions of plates in, 382; in Morrison Formation, 72–74; ontogeny in, 289; ossified tendons in, 68; pelves of, **69;** phylogeny of, 55, **73, 475;** pituitary fossae of, 103, 105, 119–120; *Scelidosaurus* in, 26; sexual dimorphism in, 289; skulls of, 36–40, **37,** 104–105; South American, 166; synapomorphies of, 464; tail spike pathology in, **146, 147, 148, 149, 151, 152;** tail spikes of, 138–139; teeth of, 41; temporal phylogeny of, **476;** in thyreophoran phylogeny, **27,** 47; vertebrae of, 67–70, **69**

Stegosauria monograph: by O. C. Marsh (unpublished), 91–92

Stegosauridae, 104, 374; Ankylosauridae versus, 382; dermal armor of, 138; diversity of North American, 55–56; *Hesperosaurus mjosi* gen. et sp. nov. in, 72; in stegosaur phylogeny, **73;** undescribed Morrison, 56

Stegosaurides excavatus: material and validity of, 459

Stegosauroidea: creation of, 456

Stegosaurus, 261; in ankylosaur phylogenetic analysis, 463, 471; beak of, 46; behavior of, 99–100; bones in Peabody mount, 80; braincase of, 62; braincases of, 103–127, **107, 108, 111, 112, 118, 120–121, 124–125;** brain endocasts of, 97, 103–127, **108, 111, 113, 114, 118, 124–125;** broken tail spikes in, 144, 153–154; buccal mandibular emargination in, 37; buccal maxillary emargination in, 36; *Cedarpelta* gen. nov. versus, 234; cranial nerves of, 115–116; dermal armor of, 70, 137, 138, 383; *Diracodon laticeps* referable to, 99; *Diracodon laticeps* versus,

98; *Hesperosaurus* gen. nov. versus, 58–62; *Hesperosaurus mjosi* gen. et sp. nov. versus, 67, 68, 69, 70; *Hylaeosaurus armatus* versus, 170; ilium of, **69,** 70; ischium of, **71;** Late Jurassic climate and, 152–153; mandibles of, 37; ontogeny of, 289; pelvis of, **69, 466;** phylogenetic analysis of, 455; plate thermoregulation in, 381; pubis of, 70, **71, 467;** sexual dimorphism in, 110; skeletal reconstructions of, 76–100, **78–79;** tail spike pathology in, 141–154; tail spikes of, 76–100; teeth of, 38, 39–40; uses of dermal armor in, 153; vertebrae of, **64, 65,** 67, 68, 69

Stegosaurus armatus: braincase and endocast of, 104–105, 117; definition of, 77, 83–84; diagnosis of, 83–84; *Diracodon laticeps* versus, 98–99; *Hesperosaurus mjosi* gen. et sp. nov. versus, 58; holotype specimen of, 81–84; in Marsh's views of *Stegosaurus,* 94–95, 96; from North America, 55; pituitary fossa of, 105; postcranial material of, 104; quarrying of, 81–84; *Stegosaurus stenops* versus, 84; *Stegosaurus ungulatus* versus, 84; stratigraphy of, 70; tail spikes of, 77, 96; teeth of, 97–98; USNM 4936 referred to, 106

Stegosaurus duplex: bones in Peabody *Stegosaurus* mount from, 80; holotype specimen of, 84–85; quarrying of, 84–85, **85;** *Stegosaurus ungulatus* versus, 84, 110; tail spikes of, 97

Stegosaurus longispinus: stratigraphy of, 70

Stegosaurus stenops: braincase and endocast of, 103, 104–105, 117, **118, 120–121,** 120–126, **124–125,** 126–127; character matrix for, 73; definition of, 77; dermal armor of, 97, **122, 123;** *Diracodon laticeps* versus, 99; from Felch Quarry 1, 94, **95;** femur of, **122;** fibula of, **123;** *Hesperosaurus* gen. nov. versus, 55; *Hesperosaurus mjosi* gen. et sp. nov. versus, 58, 60, 67; holotype specimen of, 94, **95,** 97, 99, 106; *Huayangosaurus* versus, 104; humerus of, **123;** ilium of, **122;**

ischium of, **123**; Late Jurassic climate and, 152–153; in Marsh's views of *Stegosaurus*, 95, 96–97; metatarsals of, **122**; from North America, 55; osteomyelitis in, 144; pes of, **122**; from Quarry 13, 92; quarrying of, 94, **95**; ribs of, **123**; sacrum of, **122**, **123**; skull of, 37, 58–62, **62**, 104; in stegosaur phylogeny, **73**; *Stegosaurus armatus* versus, 84; *Stegosaurus ungulatus* versus, 110–111; tail spike orientation in, 145; tail spike pathology in, 145–150, **146**, **147**, **148**, **149**, 150–152, **151**, 152–153; tail spikes of, 77, 96–97; teeth of, **39**; tibia of, **123**; USNM 4936 referred to, 106; vertebrae of, **66**, 67, **122**. *See also* USNM 4934; USNM 4936

Stegosaurus sulcatus: bones in Peabody *Stegosaurus* mount from, 80; dermal armor of, 138–139; holotype specimen of, 97; from Quarry 13, 92; tail spikes of, 96–97

Stegosaurus ungulatus, 100; bones in Peabody *Stegosaurus* mount from, 80; braincase and endocast of, 103, 104, **108**, **111**, 117, 120–126; braincase of, 62; character matrix for, 73; CM 106 referred to, 106, **107**; cranial nerves of, **108**, **111**, 115–116; definition of, 77; dermal armor of, 97; *Hesperosaurus mjosi* gen. et sp. nov. versus, 58, 67; holotype specimen of, 85–91, 103, 111, **111**, **112**, **113**, **114**; inner ear of, **108**, **114**, 116; *Kentrosaurus* versus, 103; in Marsh's views of *Stegosaurus*, 95–97; from North America, 55; quarrying of, 85–91, **86**, **88**, **89**, **90**, 91–94; skeletal reconstructions of, 77–79, **78–79**; skull of, 104; in stegosaur phylogeny, **73**; *Stegosaurus armatus* versus, 84; *Stegosaurus* behavior and, 99; *Stegosaurus duplex* versus, 84, 110; *Stegosaurus stenops* versus, 110–111; stratigraphy of, 70; syntypes of, 87, 90; tail spike orientation in, 145; tail spike pathology in, 144–145, 146, **146**, **152**; tail spikes of, 76, 96–97; vertebrae of, **66**, 67. *See also* CM 106; YPM 1853

Stereocephalus tutus, 261

Stirton, R. A.: Walter Coombs and, xi

Stoliczka, F., 174

Stonesfield, England: *Megalosaurus* from, 20

Streptococcus: in osteomyelitis, 142

Streptospondylus: *Scelidosaurus* versus, 16

Streptostyly: in ankylosaurs, 41; in *Scelidosaurus harrisonii*, 29; in stegosaurs, 36

Struthiosaurus: *Acanthopholis* versus, 201; acromion process of, 203; ankylosaurids versus, 201; in ankylosaur phylogenetic analysis, 470; *Ankylosaurus* versus, 200; *Anoplosaurus* versus, 192, 200, 201; braincase of, 199; *Cryptodraco* versus, 201; data matrix for, 462; diagnosis of, 199–202; *Dracopelta* versus, 204; dwarfism in, 174, 204–205, 206; *Edmontonia* versus, 199–200, 201, 204, 205; *Euoplocephalus* versus, 185, 199, 204; evolution of, 204–205; femur of, 201; *Gargoyleosaurus* versus, 185, 201; *Gastonia* versus, 199, 200; geographical distribution of, 202, 205–206; *Hoplitosaurus* versus, 201; *Hylaeosaurus* versus, 199–200, 201; from Iberian Peninsula, 198; mandibles of, 200–201; *Minmi* versus, 204, 205; *Mymoorapelta* versus, 200, 201, 204; *Niobrarasaurus* versus, 199–200; as nodosaurid, 173–174, 199, 205; ontogeny of, 202–203, 203–204; as originally described, 174; *Panoplosaurus mirus* versus, 199; *Panoplosaurus* versus, 199, 200, 201, 204, 205; *Pawpawsaurus* versus, 200; phylogeny of, 198–202, **472**, **475**; polacanthines versus, 201–202; *Polacanthus* versus, 185, 199–200, 201; primitive characters of, 200–201; quadrate of, 199; *Saichania* versus, 199–200; *Sarcolestes* versus, 185, 201; *Sauropelta* versus, 192, 199–200, 201; scapula of, 200; *Scelidosaurus* versus, 204; *Silvisaurus* versus, 199–200, 201; size of, 173–174, 204; species in, 202–203; *Stegopelta* versus, 200; synapomorphies of, 198–199; systematic paleontology of, 178–183; taxonomy of, 198–202, 202–203; teeth of, 200–201; *Texasetes* versus, 200; *Tianchisaurus* versus, 204; vertebrae of, 199–200, 201

Struthiosaurus austriacus: acromion process of, 200, 203; *Anoplosaurus curtonotus* versus, 203, 204; *Anoplosaurus* versus, 204; brain endocast of, 185; *Cedarpelta* gen. nov. versus, 233; cranial dermal armor of, 184, 233; cranial material of, 184–186; *Cryptodraco* versus, 193, 204; dermal armor of, 185–186, 195–198, **197**; diagnosis of, 183; *Edmontonia* versus, 184, 186, 193; femur of, 203–204; forelimb bones of, 189–193, **190–191, 192**; *Gargoyleosaurus* versus, 198; *Gastonia* versus, 193, 198; Gosau specimens referred to, **178–179**, 178–183, 184–198, **188–189, 190–191, 192, 194–195, 196, 197**; Gosau specimens referred to (tables), 175–177; hindlimb bones of, 193–195, **194–195, 196**; holotype of, **178–179**, 180; *Hoplitosaurus* versus, 193; mandibles of, 185–186, 203; *Mymoorapelta* versus, 193, 198; *Nodosaurus* versus, 189; as originally described, 174; *Panoplosaurus* versus, 184; *Pawpawsaurus* versus, 184; *Pinacosaurus* versus, 204; *Polacanthus* versus, 193, 198, 204; redescription of, 173–206; ribs of, 187–189, **188–189**; *Sauropelta* versus, 187, 189, 193; scapulae of, 189–192, **190–191, 192**; skull of, 203; *Struthiosaurus transylvanicus* versus, 184–185, 187, 189, 190; synonyms of, 458, 459; systematic paleontology of, 179–183; taxonomy of, 205; teeth of, 185–186; as valid taxon, 457; vertebrae of, 186–189, **188–189**, 203

Struthiosaurus ludgunensis: material and validity of, 459

Struthiosaurus transylvanicus: acromion process of, 200; brain endocast of, 185; coracoid of, 192, **192**; cranial material of, 185; *Edmontonia* versus, 185; as originally described, 178; *Panoplosaurus* versus, 185; partial skeleton of, 192; *Pawpawsaurus* versus, 185; *Sauropelta* versus, 185; scapula of, 190, 192, **192**; *Silvisaurus*

versus, 185; skull of, 203; *Struthiosaurus austriacus* versus, 184–185, 187, 189, 190, 202–203; as valid taxon, 457; vertebrae of, 187, 189

Styracosaurus, 447; footprints of, 440

Styx Coal Measures, 402–403

Suess: Gosau Beds prospected by, 174

Sundance Formation, 56

Supraoccipitals: of *Cedarpelta bilbeyhallorum* gen. et sp. nov., 229

Supraorbitals: of *Cedarpelta bilbeyhallorum* gen. et sp. nov., 218–**219**, 220; of *Euoplocephalus tutus*, 332–333, **333**; of *Hesperosaurus mjosi* gen. et sp. nov., 59, 60, **60, 61**; of *Struthiosaurus austriacus*, 178–179

Supratemporal fenestrae: secondary closure of, 319

Surangulars: of *Cedarpelta bilbeyhallorum* gen. et sp. nov., 218–**219**, 230–231, **231**; of *Hesperosaurus mjosi* gen. et sp. nov., 60, **61**; of *Scelidosaurus harrisonii*, 34

Sussex, England: *Polacanthus foxii* from, 367; *Regnosaurus northamptoni* from, 130; stegosaur dermal armor from, 131, **132**

Swimming: by dinosaurs, 20; by *Scelidosaurus harrisonii*, 15, 16

Sydenham, England: Great Exhibition Hall at, 5

Synapomorphies: of ankylosaurs, 460–461

Syngonosaurus macrocercus: material and validity of, 459

Syrmosaurus viminocaudus: material and validity of, 459

Tadjikistan: ankylosaur footprints from, 433

Tail clubs: of ankylosaurids, 280–281; caudorbitosa of, 371; of *Euoplocephalus tutus*, 280–281; of polacanthines, 375; of *Polacanthus foxii*, 375

Tail plates: of *Minmi*, 354, 354–355, **355**, 360–361, **361**. *See also* Dermal armor

Tail spikes: anatomy of, 145; broken *Stegosaurus*, 144, 153–154; of *Dacentrurus*, 137; discovery of first *Stegosaurus*, 86, **89**; of *Hesperosaurus mjosi* gen. et sp. nov., 72; horny sheaths of, 154; pathology in *Stegosaurus*, 141–

154, 146, 147, 148, 149, 151,
152; in Peabody *Stegosaurus*
mount, 78–79, 80; of Spanish
Wealden stegosaur, 131; of
stegosaurs, 138–139; of *Stegosaurus,* 76–100; of *Stegosaurus
armatus,* 77, 83, 96; of *Stegosaurus duplex,* 84, 97; *Stegosaurus*
sexual dimorphism and, 96–97;
of *Stegosaurus stenops,* 77, 96–
97; of *Stegosaurus sulcatus,* 96–
97; of *Stegosaurus ungulatus,* 77–
79, **78–79,** 86, 89–90, **90, 92,**
96–97
Talarurus: Cedarpelta gen. nov. versus,
223; data matrix for, 462; dermal
armor of, 289; *Euoplocephalus
tutus* versus, 264; humerus of,
245, 255; metatarsals of, 256;
phylogeny of, **474–475;** as
reference taxon, 461; ribs of,
388; skull of, 40–41; ulna of, **246**
Talarurus disparoserratus: material
and validity of, 459
Talarurus plicatospineus: skeleton of,
387; as valid taxon, 457
Tanzania: *Kentrosaurus* from, 103,
104, **109**
Taphonomy: of DMNH 2818, 152–
153; of *Minmi* specimens, 356;
osteomyelitis and, 143, 154; of
Scelidosaurus harrisonii, 17
Tarchia: in ankylosaur phylogenetic
analysis, 473–475; beak of, 309;
data matrix for, 462; phylogeny
of, **474–475;** as reference taxon,
460; tooth wear in, 43
Tarchia gigantea: synonyms of, 458,
459; as valid taxon, 457
Tarchia kielanae: Gastonia versus,
387; material and validity of, 459
Tarentola: development of cranial
ornamentation in, 324, 325, 326
Tarentola annularis, 339
Tarentola mauritanica, 339; ornamentation of, 322
Tarsi: dermal armor on *Minmi,* 354
Teeth: of *Aletopelta coombsi* gen. et
sp. nov., 241–243, **242;** of
Aletopelta gen. nov., 255; of
ankylosaurs, 41–44, **42,** 300–
301, 403–404; of *Ankylosaurus,*
255; of *Cedarpelta bilbeyhallorum* gen. et sp. nov., 217–
219, **218–219;** of *Cedarpelta* gen.
nov., 211; of *Crataeomus,* 186; of
Diplodocus lacustris, 97; of
Diracodon laticeps, 98, 98–99,
99; of *Dyoplosaurus acutosquameus,* 268; of *Edmontonia*

longiceps, **242–243;** of *Edmontonia rugosidens,* 42, 242–243,
255; of *Emausaurus,* 36; of
Euoplocephalus tutus, 255, 304–
305, **305, 306, 307, 308,** 309–
312, **310, 311, 312;** of *Gargoyleosaurus,* 185; of *Gastonia burgei,*
387; of *Hesperosaurus* gen. nov.,
58; of *Hesperosaurus mjosi* gen.
et sp. nov., 62, **63;** of *Huayangosaurus,* 37, 38, 40; of *Huayangosaurus taibaii,* **39;** of *Kentrosaurus,* 37; of *Leipsanosaurus
noricus,* 186; of *Niobrarasaurus
coleii,* 255; of ornithischians, 38,
309; of *Panoplosaurus mirus,* 42;
of *Priodontognathus phillipsi,* 42;
of *Regnosaurus northamptoni,*
130; of *Sarcolestes,* 185; of
Sauropelta, 185; of *Scelidosaurus,*
465; of *Scelidosaurus harrisonii,*
15–16, **28,** 29–31, **30,** 31–32,
32–33; of *Scutellosaurus,* 36; of
South American ankylosaur, 160,
160, 161, 165; of stegosaurs, 36–
40, **39;** of *Stegosaurus,* 37, 97–
99, **98, 99;** of *Stegosaurus
stenops,* **39,** 62; of *Struthiosaurus,* 200–201; of *Struthiosaurus austriacus,* **178–179,** 185–
186; of thyreophorans, 25–47.
See also Tooth wear
Telencephalon: of stegosaurs, 113
Telmatosaurus: dwarfism in, 205
Telmatosaurus transsylvanicus:
dwarfism in, 204
Temporal phylogeny: of ankylosaurs,
476
Tenontosaurus: from Cedar Mountain
Formation, 212
Terrestrial ecosystems: herbivorous
dinosaurs in Mesozoic, 25–26
Tertiary mammals, xi, xii
Tetradactyl footprints: ceratopsian
versus ankylosaur, 418–421
Tetrapodosaurus, 440, 442; footprints
of, **415,** 416–418, 423–433, **424–
425, 426, 427, 428, 429, 430,
431, 432**
Tetrapodosaurus borealis, 413; foot-
prints of, 421, **422,** 422–423;
Gastonia burgei and, 391
Texas: cranial ornamentation in
nodosaurids from, 333
Texasetes: data matrix for, 462;
excluded from Polacanthinae,
378; phylogeny of, **472;**
Struthiosaurus versus, 200
Texasetes pleurohalio: horizon and age
of, 213; as valid taxon, 457

Thackray, John, 18

Thermoregulation: dermal armor for, 380, 381; sacral shields in, 383

Theropoda: with *Scelidosaurus* material, 9–15, **10**

Thescelosaurus: pelvis of, **466**

Thyreophora: beaks of, 45–46; cheeks of, 45–47, 407–408; cranial material of, 26; creation of, 456; dermal armor changes among, 363; dermal armor morphology in, 383–384; diets of, 45, 400, 409; feeding mechanisms of, 25–47; jaw mechanisms of, 25–47, 299; ontogeny in, 289; phylogenetic analysis of, 455; phylogeny of, **27, 470, 475**; sexual dimorphism in, 289; South American, 166; *Struthiosaurus austriacus* in, 178–180; temporal phylogeny of, **476**; tooth wear in, 25–47; as valid taxon, 457. *See also* Ankylosauria; Basal Thyreophora; Stegosauria

Tianchisaurus: dermal armor of, 284; *Struthiosaurus* versus, 204

Tianchisaurus nedegoapeferima: material and validity of, 459

Tianzhenosaurus youngi: material and validity of, 459

Tibiae: of *Aletopelta coombsi* gen. et sp. nov., **247**, 248; of *Allosaurus*, 83, **83**; of *Brontosaurus amplus*, **85**; of *Crataeomus lepidophorus*, 192–193; of *Crataeomus pawlowitschii*, 193, 195; dermal armor on *Minmi*, 353, **353**; of *Euoplocephalus*, 392; of *Euoplocephalus tutus*, **249**; of *Gastonia burgei*, **249**, 388; of juvenile scelidosaur, **10**; of *Niobrarasaurus coleii*, **249**; of *Nodosaurus textilis*, **249**; osteomyelitis in hypsilophodont, 144; in Peabody *Stegosaurus* mount, **78–79**, 80; of *Polacanthus*, 392; of *Polacanthus foxii*, **249**; of *Sauropelta edwardsorum*, **249**; of *Stegosaurus duplex*, **85**; of *Stegosaurus stenops*, **123**; of *Struthiosaurus austriacus*, 180, 182, 190, **196**

Tilgate Forest: *Hylaeosaurus* from, 8; *Regnosaurus* from, 130

Tiliqua scincoides, 339

Titanosauria: dermal armor of, 159; South American ankylosaurs found with, 160, 165

TMP 67.19.4, 333

TMP 67.20.20, 329, 333

TMP 79.14.164, 333, **335**

TMP 82.1.1, 304

TMP 82.9.3, 280, 287

TMP 88.106.5, 332–333, **333**

TMP 89.36.183, 328, **328**

TMP 90.301.1, 332

TMP 91.127.1, 298, 305; description of, 266; material and horizon of, 263; measurements of, 274; in multivariate analysis, **276, 277**, 277–278; skull from, **267, 289–290, 293**

TMP 92.36.1226, **308**

TMP 93.36.79, 333, **334**

TMP 94.183.1, 440

TMP 96.75.1, 305, 311; teeth from, **311**

TMP 97.12.112, 250–251

TMP 97.132.1, 305

TMP 98.115.2, 329, **329**

TMP 98.89.4, 429

TMP 99.49.2, 430

TMP 99.59.2, 438, 440

Tooth replacement: in *Scelidosaurus harrisonii*, 31

Tooth wear: in ankylosaurs, 41–44, **42**, 44–47, 300–301, 403–404; in *Edmontonia rugosidens*, **42**, 309; in *Euoplocephalus*, 44–45; in *Euoplocephalus tutus*, 299–300, 301, 304–305, 309–312, **310, 311, 312**, 313–314; in *Huayangosaurus taibaii*, 39; in *Panoplosaurus mirus*, **42**, 44–45; in *Pinacosaurus*, 43; in *Saichania*, 43; in *Scelidosaurus harrisonii*, 26, 31–32, 32–33, **33**, 34–35, 43, 44–47; in stegosaurs, 38–40; in *Stegosaurus stenops*, 39; in *Tarchia*, 43; in thyreophorans, 25–47, **33**, 39, **42**, 313–314

Torosaurus: forelimb of, 256

Toro Toro Formation, 442; ankylosaur footprints from, **442**, 447; location of, **446**

Torrens, Hugh, 22

Transylvania: *Struthiosaurus transylvanicus* from, 178

Trauma-related injuries: in *Stegosaurus* tail spikes, 141–154

Triceratops: morphometric analysis of, 271

Trichosurus vulpecula: gut contents of, 406, **408**

Tricorns: defined, 373; of *Mymoorapelta maysi*, 376; of *Scelidosaurus harrisonii*, 376. *See also* Dermal armor

Trigeminal nerve (cranial nerve V): of stegosaurs, 115. *See also* Cranial nerves

Tripodal stance: of *Stegosaurus,* 100
Trochlear nerve (cranial nerve IV): of
    stegosaurs, 115. *See also* Cranial
    nerves
True phylogeny, 459–460
Tsagantegia: data matrix for, 462;
    phylogeny of, **474–475**
Tsagantegia longicranialis: as valid
    taxon, 457
Tubercles: defined, 374; of *Pola-
    canthus foxii,* 374. *See also*
    Dermal armor
Tuojiangosaurus: in ankylosaur
    phylogenetic analysis, 465–467;
    braincase of, 111; dermal armor
    of, 137, 383; *Hesperosaurus
    mjosi* gen. et sp. nov. versus, 58,
    68; in stegosaur phylogeny, 73;
    vertebrae of, **64, 65,** 68
Tuojiangosaurus multispinus: charac-
    ter matrix for, 73
Turonian: ceratopsian versus ankylo-
    saur footprints from, 419, **419**
Turtles: beaks of, 45, 46; from Gosau
    beds, 178; from Gosau Beds
    (tables), 175–177; resemblance of
    Stegosaurus armatus to, 94–95
12 Mine South A-Pit tracksite, 423,
    426–429; footprints from, **431**
Two Medicine Formation: ankylo-
    saurid specimens from (table),
    263

UA 31, 293, 298; dermal armor from,
    287; description of, 266; material
    and horizon of, 263; measure-
    ments of, 274; in multivariate
    analysis, **276, 277, 277,** 278;
    skull from, 267, **267, 294**
Ulnae: of *Aletopelta coombsi* gen. et
    sp. nov., 243, **244,** 244–245, **246;**
    dermal armor on *Minmi,* 352,
    353; of *Euoplocephalus,* **246;** of
    Gastonia burgei, 388, 391; of
    Mymoorapelta, **246;** pathology in
    Aletopelta coombsi gen. et sp.
    nov., 244, 245, **246,** 256; in
    Peabody *Stegosaurus* mount, 78–
    79, 80; of *Pinacosaurus,* **246;** of
    Saichania, **246;** of *Scelidosaurus
    harrisonii,* **12;** of *Scolosaurus,*
    246; of *Struthiosaurus austriacus,*
    180, 182, 189, **190–191,** 192–
    193; of *Talarurus,* **246**
Ungulates: thyreophorans versus, 45
United States: *Pawpawsaurus
    campbelli* from, 377; pola-
    canthines from, **364–365, 366–
    367;** stegosaurs from, 104

University of Alberta. *See* UA 31
University of California, Berkeley:
    Walter Coombs at, xi
University of Massachusetts: Margery
    Coombs at, xii; Walter Coombs
    and, xiii
Upper Horseshoe Canyon Formation:
    ankylosaurid specimens from
    (table), 263
Upper Jurassic. *See* Late Jurassic
Upper Two Medicine Formation:
    ankylosaurid specimens from,
    293; ankylosaurid specimens
    from (table), 263
Uromastyx: tooth wear in, 301
Uromastyx hardwickii, 339
USNM 2274, 110
USNM 4752, 247, 376; splate from,
    372
USNM 4934, 104, 105; brain endocast
    from, 120–126; description of,
    106; in Peabody *Stegosaurus*
    mount, 80; skull and mandibles
    of, 36, 37, **37, 124–125;** as
    Stegosaurus stenops holotype, 94,
    95, 97, 99, 110; teeth of, 38, **39**
USNM 4936, 104–105, 126–127;
    cranial material of, **118, 120–
    121, 124–125;** description of,
    106–111, 120–126; postcranial
    material of, **122, 123**
USNM 4937, 138; as *Stegosaurus
    sulcatus* holotype, 97
USNM 6645, 110
USNM 6646: tail spike pathology in,
    144–145, **146,** 150, **152, 153**
USNM 7414: from Quarry 12, 87; as
    Stegosaurus ungulatus holotype,
    90
USNM 7943: dermal armor from,
    284, **285,** 291; material and
    horizon of, 263
USNM 8610, **369**
USNM 11892, 262, 264, 290–291,
    298; description of, 267–268;
    material and horizon of, 263;
    measurements of, 274; in
    multivariate analysis, **276, 277,
    277,** 278; skull from, **267**
USNM 419656, 38
USNM miscellaneous specimens, 41–
    43
Utah: ankylosaur footprints from,
    433–435, **434;** ankylosaurs from,
    212; *Cedarpelta* gen. nov. from,
    211–235; Dalton Wells Quarry
    in, 386–387; Gaston Quarry in,
    386; *Mymoorapelta maysi* from,
    369

QE
862
.065
A76

2001